Physiology of Marine Mammals

Suppose you were designing a marine mammal. What would they need to live in the ocean? How would you keep them warm? What design features would allow them to dive for very long periods to extreme depths? Do they need water to drink? How would they minimize the cost of swimming, and how would they find their prey in the deep and dark?

These questions and more are examined in detail throughout *Marine Mammal Physiology*, which explores how marine mammals live in the sea from a physiological point of view. This undergraduate textbook considers the essential aspects of what makes a marine mammal different from terrestrial mammals, beyond just their environment. It focuses on the physiological and biochemical traits that have allowed this group of mammals to effectively exploit the marine environment that is so hostile to humans.

The content of this book is organized around common student questions, taking the undergraduate's point of view as the starting point. Each chapter provides a set of PowerPoint slides for instructors to use in teaching and students to use as study guides. New "Study Questions" and "Critical Thinking Points" conclude each chapter, which are each motivated by a "Driving Question" such as "How do mammals stay warm in a cold ocean?" or "How do mammals survive the crushing pressures of the deep sea?" Full-color images and comprehensive, accessible content make this the definitive textbook for marine mammal physiology.

CRC MARINE BIOLOGY SERIES

The late Peter L. Lutz, Founding Editor
Julie Mondon and Stephen Bortone, Series Editors

For more information about this series, please visit: https://www.crcpress.com/CRC-Marine-Biology-Series/book-series/CRCMARINEBIO?page=&order=pubdate&size=12&view=list&status=published,forthcoming

Physiology of Marine Mammals

Adaptations to the Ocean

Edited by
MICHAEL A. CASTELLINI
JO-ANN MELLISH

CRC Press
Taylor & Francis Group
Boca Raton London New York

CRC Press is an imprint of the
Taylor & Francis Group, an **informa** business

First edition published 2024
by CRC Press

6000 Broken Sound Parkway NW, Suite 300, Boca Raton, FL 33487-2742
and by CRC Press

4 Park Square, Milton Park, Abingdon, Oxon, OX14 4RN

CRC Press is an imprint of Taylor & Francis Group, LLC

Library of Congress Cataloging-in-Publication Data

Names: Castellini, Michael Angelo, editor. | Mellish, Jo-Ann, editor.
Title: Physiology of marine mammals : adaptations to the ocean / edited by
Michael Castellini, Jo-Ann Mellish.
Description: First edition. | Boca Raton : CRC Press, 2023. | Series: First edition. | Includes
bibliographical references and index. | Summary: "This unique undergraduate textbook considers
what makes a marine mammal different from terrestrial mammals, focusing on the physiological
and biochemical characteristics that have allowed this group of mammals to effectively exploit the
hostile marine environment. The contents are organised around common student questions such as
"How do mammals stay warm in a cold ocean?" while each chapter provides a set of PowerPoint slides
for instructors to use in teaching. Study Questions and Future Implications points conclude each
chapter, while full-colour images and comprehensive, accessible content make this the definitive
textbook for marine mammal physiology"-- Provided by publisher.
Identifiers: LCCN 2023006691 (print) | LCCN 2023006692 (ebook) |
ISBN 9781032285603 (paperback) | ISBN 9781032285702 (hardback) |
ISBN 9781003297468 (ebook)
Subjects: LCSH: Marine animals--Physiology. | Marine animals--Adaptation.
Classification: LCC QL713.2 .P49 2023 (print) | LCC QL713.2 (ebook) |
DDC 591.77--dc23/eng/20230405
LC record available at https://lccn.loc.gov/2023006691
LC ebook record available at https://lccn.loc.gov/2023006692

ISBN: 978-1-032-28570-2 (hbk)
ISBN: 978-1-032-28560-3 (pbk)
ISBN: 978-1-003-29746-8 (ebk)

DOI: 10.1201/9781003297468

Typeset in Janson Text LT Std
by KnowledgeWorks Global Ltd.

Printed in the UK by Severn, Gloucester on responsibly sourced paper

Access the eResources: https://www.routledge.com/9781032285603

Dedication

This book is designed for students and teachers, so it is only fitting that we dedicate the end product to our own teachers and students.

Our mentors were giants in their respective fields, and they paved the way for us to succeed in our own careers. Gerry Kooyman and George Somero introduced Mike to the world of diving physiology and biochemistry, and he has never looked back. Sara Iverson was and still is a trailblazer in marine mammal physiology, mentoring Jo-Ann in hands-on field, lab, and career skills. She also has a career's-worth of gratitude for Alan Pinder. A true physiologist with an approach to science and life in general that was infused with equal parts enthusiasm and devil's advocate that still sticks with her today.

Between us, we have had hundreds of students. The nature of our specialty means that at times we have relied on and learned from them as much as they relied on and learned from us. We have been the fortunate ones in that our class experience was not limited to a room in a building, but expanded to ice floes, sand dunes, boats, and helicopters.

Our families have been with us through the thick and thin of having a field biologist around, including the difficulties of long absences, the torture of repeatedly hearing the same stories of our exploits during those long absences (near or actual accidents and animal shenanigans) or ranting about the paperwork required before and after those long absences. They have been through it all with us, even when we have been physically apart. It is fundamentally their support that allows us to keep asking more questions and we would not be where we are without them.

CONTENTS

Mike and Jo-Ann started their academic careers a generation apart, but their paths have been down the same road. Their adventures studying the physiological ecology of marine mammals have taken them from Arctic to the Antarctic and back again. All the heavy seas, unstable ice floes and sideways snowstorms were a small price to pay for the privilege of learning more about these fantastic animals. They have shared the excitement, fun and adventure of this field with their own graduate students and hundreds of school children over the years. With this book, they aim to bring that feeling to a larger population of university level students as they launch their own careers in marine mammal biology.

PREFACE

When we published *Marine Mammal Physiology: Requisites for Ocean Living* in 2015, we asked what makes a marine mammal different from terrestrial mammals? We worked with an international team of physiologists and biochemists to provide insights into those differences. Our focus was to provide a consolidated set of deep literature reviews for graduate students, postdocs, and colleagues. With this new version, *Physiology of Marine Mammals: Adaptations to the Ocean*, we have shifted our target to serve the undergraduate classroom with more accessible content and organization. We asked students what they considered to be important and arranged the book chapters around their ideas.

In this book, we provide study questions, critical thinking concepts, fields for future studies, and electronic downloadable files that can be used by teachers and students alike. We introduce several new topics, provide expanded glossaries, and include further reading sections for those that wish a deeper dive.

We hope that the information in the *Physiology of Marine Mammals: Adaptations to the Ocean* helps to inspire a new generation of marine mammal biologists to find even more answers.

INTRODUCTION

How do marine mammals differ from terrestrial mammals? This book tackles a **Driving Question** in each chapter, ranging from basic survival in an aquatic environment to finer scale assessments of health or the impact of sharing a planet with humans.

We begin with marine mammal hydrodynamics by Jeremy Goldbogen, Frank Fish, and Jean Potvin (**Chapter 1: Hydrodynamics; Driving Question: How does hydrodynamics influence the swimming ability and means of procuring food in a highly dense and viscous fluid medium?**). They explore many of the adaptations for moving a body through water including drag, propulsion, swimming at various depths, and unique aspects such as what happens when a whale swimming at high speed opens its mouth to catch krill on its baleen.

Shawn Noren, Jennifer Maresh, and Terrie Williams explore how marine mammals find the energy necessary for underwater exercise (**Chapter 2: Energy for Exercise; Driving Question: How do marine mammals exercise while holding their breath?**). The authors explore the metabolic costs of resting, exercise, and diving in marine mammals. They discuss how to measure the metabolic rate (oxygen consumption) of marine mammals under various conditions and consider the apparent conflict of trying to save oxygen while diving and exercising at the same time.

Authors Cassondra Williams and Paul Ponganis expand on how marine mammals manage their oxygen while diving (**Chapter 3: Oxygen Stores and Diving; Driving question: How do marine mammals stay underwater so long?**) A marine mammal must be able to hold its breath for a very long time without the benefit of gills or other specialized organs to extract oxygen from seawater. Where do they carry the oxygen they need? How do they use that oxygen? and what happens if they stay underwater for longer than the amount of oxygen they have available?

Having discussed how a marine mammal moves through the water, the next question comes how do they withstand the pressures of deep diving **(Chapter 4: Under Pressure; Driving Question: How do marine mammals cope with huge pressures at depth?)**. How do they avoid "the bends"? What are the physiological and biochemical adaptations that allow their tissues and cells to survive such an insult? Sascha Hooker, J. Chris McKnight, and Andreas Fahlman explore those adaptations that allow marine mammals to dive far deeper than freely diving humans have ever even approached.

How do marine mammals keep warm in the water? Co-editors Michael Castellini and Jo-Ann Mellish **(Chapter 5: Thermoregulation; Driving question: How do marine mammals stay warm in the cold ocean?)** look at the normal body temperature of a marine mammal and how their fur or blubber helps to maintain that temperature. But how do they keep from overheating when on land? What about species that live on the frozen sea ice and give birth to pups at extremely cold temperature? Will some of the predicted changes in water temperature through climate change impact marine mammals?

Dorian Houser and Jason Mulsow focus on the acoustic aspects of the physiology and anatomy of hearing underwater **(Chapter 6: Acoustics; Driving Question: How do marine mammals hear and produce sound underwater?)**. Many marine mammals rely on underwater hearing and sound production for purposes of breeding, socializing, finding food, and navigating. What is the nature of their sound production? What are the differences in the acoustic biology of whales and sea otters? How does increasing background noise influence marine mammals?

Marine mammals must be able to see, both in air at the surface and deep underwater. The darkness of the ocean at several hundred meters and variable light under the sea ice are not conditions that we as humans are built to contend with. Kenneth Sørensen and Frederike Hanke **(Chapter 7: Vision and Touch; Driving Question: How can marine mammals find their food and navigate the oceans even in dark and murky waters?)** consider the difference in the neural structures of the vision system across marine mammals. They look at how they might sense particle flow fields from plankton and other small debris floating in the water. They also examine the extremely sensitive whiskers (vibrissae) of seals, and how they might be used to sense the disturbance in the water after a fish has swum nearby.

It is well known that humans cannot survive on the ocean without access to freshwater. Miwa Suzuki and Rudy Ortiz explore the fascinating question of water balance in marine species **(Chapter 8: Hydration; Driving Question: How do marine mammals hydrate their body in seawater?)**. You will find out that marine mammal kidneys look very different from a terrestrial mammal yet have many physiological and biochemical similarities. What hormones control the balance of water in their bodies, and if available, will a marine mammal drink freshwater, or ignore it?

In **Chapter 9: Nutrition (Driving Question: What kind of food do marine mammals need to stay healthy?)**, David Rosen and Lisa Hoopes explore the nutritional requirements of marine mammals. They discuss how different types of nutrients (fats, carbohydrate, and protein) are important to marine mammals and how that balance may vary seasonally. The authors compare the dietary requirements of animals in nature and those kept at aquaria. Also, they consider how fasting from food at various times of year is an important part of the natural history for many marine mammals.

Christopher Marshall and Jeremy Goldbogen expand on nutrition by examining the feeding mechanisms that have been perfected by various species **(Chapter 10: Feeding Mechanisms; Driving Question: How do marine mammals catch their prey?)**. What is the difference between how baleen and toothed whales feed? How do dugongs and manatees obtain food for their vegetarian diets? Are the teeth of marine mammals different than terrestrial species? How has evolution driven the many different types of feeding mechanisms we see in marine mammals?

Many marine mammals fast for very long periods either during migration or when on land for breeding purposes. Allyson Hindle and David Rosen explore these abilities in **Chapter 11: Fasting; Driving question: how do marine mammals survive long periods without feeding?** They consider the body stores of energy, how they are used, and the hormonal control of fasting. They make the critical distinction between fasting and starvation.

The feeding and fasting biology of newborn animals is covered by Daniel Crocker and Brigitte McDonald who consider how nursing mothers support rapidly growing newborns, yet do not eat or drink for many weeks (**Chapter 12: Post-delivery; Driving Question: How do marine mammals provision their young to succeed in the challenging marine environment?**). They explore the complex biochemical pathways that allow milk production by fasting females and explain their work on elephant seal rookeries.

Claire Simeone and Shawn Johnson examine disease and mortality events (**Chapter 13: Health and Disease; Driving Question: Why do some animals get sick while others stay healthy?**). They look at health challenges for all the major groups of marine mammals and ask if the respiratory system, which is highly adapted for exchanging large amounts of air very quickly, could be a weakness by trapping foreign particles. Can the fat diet of polar bears cause issues with contaminants? They explore possible diseases that could impact seals and sea lions when they haul out on beaches.

John Harley, Judith Castellini, and Todd O'Hara focus on potential toxic elements in the environment and how they relate to the health of marine mammals (**Chapter 14: Ecotoxicology; Driving Question: How are marine mammals exposed to and affected by toxicants?**). From organic pollutants to metals, marine mammals must deal with a wide range of potential contaminants in the ocean and in their food. They write how lipid-soluble contaminants can be transported to the calf though the high-fat milk from the mother. There are also a considerable number of natural toxins in the marine environment, including harmful algal blooms.

In **Chapter 15: Sharing Earth's Oceans (Driving Question: Do human activities influence marine mammal physiology?)**, co-editor Michael Castellini explores if interacting with humans can influence the physiology of marine mammals? From ship strikes to pollution, plastics, competition for food, noise, and climate change, he looks at how our human use of the ocean may alter the biology of marine mammals. He discusses actions that we can take to reduce some of those interactions and how everyone, not just scientists, have roles to play in those actions.

In the final concluding **Chapter 16: More Questions and Mysteries (Driving Question: What are new and exciting fields of study?)**, Michael Castellini writes about some of the many other fields of physiology and biochemistry that could not be covered in detail in this book. He looks at developing methods of using DNA in ocean water to track marine mammals, asks how marine mammals sleep, discusses new directions in diving biochemistry, assessing population health and the remarkable old age of some marine mammals.

In your reading of this book, we expect you to have as many questions as you find answers, which is why most of these authors have made a career out of studying this fascinating group of animals. It will take many more decades of work for us to even scratch the surface of their complex physiology. One of our favorite questions to ask students is "How would you design a marine mammal?" because the reality is sometimes even more fantastical than our imagination.

We hope you use this book to drive your imagination to find out more about the animals that have intrigued us for a lifetime.

Michael A. Castellini, Fairbanks, Alaska
Jo-Ann Mellish, Anchorage, Alaska

Frank E. Fish
Department of Biology, West Chester University, West Chester, PA

Jean Potvin
Department of Physics, Saint Louis University, St. Louis, MO

Jeremy A. Goldbogen
Oceans Department, Hopkins Marine Station, Stanford University, Pacific Grove, CA

Frank E. Fish

Frank received his Ph.D. in Zoology from Michigan State University, where he studied the energetics, biomechanics, and hydrodynamics of aquatic mammals. By examining the propulsive systems of aquatic mammals, he focused his work on the evolution and mechanics of propulsive systems. His work has been performed in both the field and laboratory. He has performed research on swimming performance from small beetles to the enormous blue whale. He has also been involved with the field of biomimetics, which attempts to take the lessons of animals and incorporate them into engineered technologies.

physics of the inflation process, be it in parachutes or animals. He obtained his Ph.D. in Physics in 1985 at the University of Colorado in Boulder. In addition to participating in whale bio-logging tag deployment expeditions, he has been involved in several US government-sponsored wind tunnel and flight test studies of parachutes. Dr. Potvin is an AIAA Associate Fellow and a member of the AIAA Aerodynamic Decelerator Systems Technical Committee. Since 2002, he has been directing the H.G. Heinrich Parachute Systems Short Course, a biennial week-long professional development course on parachute systems design. He is a parachutist with over 2600 jumps to his credit, and an FAA-rated senior parachute rigger.

Jean Potvin

Jean has been interested in whale bio-hydrodynamics since 2009, as a spin-off of his parachute research which began in 1994. As a physics professor at Saint Louis University, his main interest has focused on the

Jeremy Goldbogen

Jeremy Goldbogen is a comparative physiologist who studies the integrative biology of marine organisms. He started his research career studying the biomechanics of locomotion in hummingbirds and Antarctic sea butterflies (pteropods) as an undergraduate

DOI: 10.1201/9781003297468-1

student at the University of Texas at Austin. Jeremy then completed his M.Sc. in marine biology from the Scripps Institution of Oceanography at the University of California – San Diego. He later moved on to earn his Ph.D. from the University of British Columbia in Vancouver, Canada, with a thesis titled "Mechanics and energetics of rorqual lunge feeding". He returned to Scripps as a postdoctoral researcher followed by another postdoctoral fellowship at the Cascadia Research Collective in Olympia, Washington. He is now an associate professor of Oceans at Stanford University, located at the Hopkins Marine Station in Pacific Grove, California.

Driving Question: How does hydrodynamics influence the swimming ability and means of procuring food in a highly dense and viscous fluid medium?

BACKGROUND

The interaction of marine mammals with their natural fluid environment incurs physical forces associated with the flow of water (Goldbogen, Fish and Povin, 2016). The pattern of flow has molded the body morphology, determined the efficiency of propulsive movements, and influenced the effectiveness of feeding mechanisms for life in water. As opposed to the terrestrial environment occupied by the ancestors of the highly evolved marine mammals where gravity is the dominating force of locomotion, the aquatic realm presents an almost weightless environment regulated by the buoyancy of the animal. Unbound from gravitational considerations, various marine mammals (e.g., whales, elephant seals, sirenians) have experimented with gigantism (Goldbogen and Madsen 2018). Locomotion in water is dependent on the transfer of momentum from propulsive surfaces to a fluid to generate thrust to oppose shear and pressure forces from the motion of the fluid (Webb 1975). Furthermore, terrestrial mammals feed using physical contact with prey,

although some marine mammals still employ this technique, hydrodynamic differences between the environmental and oral pressures allow for capture of elusive and small prey (Goldbogen and Madsen 2018).

In this chapter, we explore the hydrodynamic mechanisms that marine mammals use to achieve different levels of swimming performance and facilitate increased effectiveness in foraging. It is closely aligned with several other chapters (**Chapters 2 and 10**) each of which can be referred to for more discussions on swimming and feeding metabolism in marine mammals. Previous hydrodynamic works have reviewed the mechanics of aquatic locomotion and feeding, but new techniques have been applied to show a more complete picture of the fundamental behaviors of locomotion and feeding for survival in the open ocean. These techniques include the use of drones, archival tags, hydrodynamic modeling, and digital **particle image velocimetry**. We will provide a basic description of relevant **hydrodynamics**, and specifically review the mechanics of locomotion among the various groups of marine mammals and fluid mechanics of feeding.

Hydrodynamic forces at play during locomotion: lift, drag, and thrust

Flow structures around a body moving underwater

As seen from a distance, the flow of water moving around a marine mammal while coasting or gliding appears streamlined and orderly (**Figure 1.1a**). The incoming flow – or **freestream** – parts ways laterally and dorsoventrally (top to bottom) at the nose (or rostrum), to accelerate toward the body's widest section. The parted flows begin to decelerate past this point, to rejoin at the speed of the freestream past the body. At smaller scales near the animal's body (millimeters to decimeters, **Figure 1.1b and c**), there is a thin layer of fluid "tucked" under the fluid streamlines nearest to

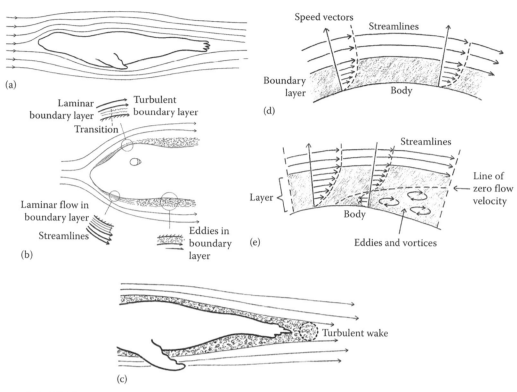

Figure 1.1 Fluid motions past a swimmer (relative to the body): (a) flow streamlines showing the paths of fluid particles, (b) laminar and turbulent boundary layers along the body surface, (c) boundary layer aft of the body, terminating into the turbulent wake, (d) fluid particles mean speeds within the boundary layer, at several stations above the body surface, and (e) boundary layer structure at the threshold of flow separation genesis.

the body known as the **boundary layer**. Typically moving at significantly slower speeds than the fluid around the animal, such layer may also appear streamlined, or **laminar**, but in the right conditions, it will also appear **turbulent**, i.e., as unsteady flow populated with **eddies** spinning in all directions and at varying rates. The thickness of the boundary layer generally increases toward the rear of the animal. Along the 30 m body of a blue whale, for example, it is estimated to be only a few millimeters thick at the rostrum and about 0.2 m at the end of the tail. Whether laminar or turbulent, the boundary layer at the tail separates from the body, becoming a **turbulent** wake past the tail. This wake is "contained" within the streamlines of fluid that are also re-joining behind the body (**Figure 1.1c**). Again,

these conditions describe the water flow around a gliding animal.

In contrast, the flow structure about actively swimming mammals is more complicated (Fish et al. 2014). Here an animal's appendages accelerate portions of the surrounding flows, to add large-scale tornado-like **vortices** that may persist long after the animal has passed by, similar to the vortices seen on the surface of the water when paddling in a kayak or canoe. More specifically, the **hydrofoil** shape of the flukes, flippers, and fins produces vortices from the appendages' oscillations in the fluid. Created at commensurate sizes, such vortical flows move in all directions, thereby creating significant (and additional) **drag** and, most importantly, the **lift** that in the horizontal translates into propulsion.

Drag generation in marine mammals is first related to how boundary-layer flows interact with large-scale outer flows. Such a link is crucial. The boundary layer arises from two basic forces: (1) shear forces in between adjacent layers of fluid caused by **viscosity**, which tend to resist the layers' relative motion as well as induce a rotation of both, and (2) the molecular forces at the body's surface causing the fluid layer closest to the body to literally stick to its surface – a phenomenon called the **no-slip condition** (**Figure 1.1d**). In other words, the body's surface atoms attract and meet the nearest atoms of the water, thereby slowing down the latter to a zero velocity with respect to the body. With increasing distance from the body surface, the attraction weakens and the flow velocity increases until it matches the velocity in the outer flow around the animal, i.e., at the top of the boundary layer.

The boundary layer is more dynamic and has more structure than the streamlined flows above it. It begins as laminar, with its thickness increasing posteriorly, thereby making it more unstable when encountering pronounced surface curvature and/or roughness, along with random perturbations from the flows above it. The result is often a laminar boundary layer changing into a turbulent boundary layer (**Figure 1.1b**). On a well-streamlined, **fusiform** animal, and depending on swim speed, this **laminar-to-turbulent transition** occurs approximately halfway along the body (Fish and Rohr 1999). This transition is dynamic and difficult to predict because it depends on the factors mentioned above, as well as on the body (and limb) postures adopted at any instant. This relationship is key because the structure of the boundary layer, i.e., how much of it is laminar and how much is turbulent, determines how much drag is produced.

Another important feature of the boundary layer is **flow separation**, which occurs when the flows within it are decelerated to a near zero-speed when approaching an area of greater pressure, as usually occurring along the posterior half of a tapered body. If these **adverse pressure gradients** are strong enough, the boundary layer flows may reverse direction relative to the streamlined flows above the layer-forming eddies. The interaction between the two flows creates a large zone of turbulent flow that interacts with the boundary layer and keeps the streamline flow away from the surface – effectively "separating" the boundary layer away from the body (**Figure 1.1e**). The creation of both the turbulent wake and turbulent zone underneath the separated flow generally leads to more drag but also less, as further explained next. Note that turbulent boundary layers are less prone to adverse pressure gradients than laminar layers and thus less conducive to flow separation. "Tripping" or inducing the boundary layer to transition from laminar to turbulent, for example, by dimples on a golf ball or sharp protrusions (by barnacles), will help reduce drag. Also note that in some forms of oscillatory swimming where more adverse pressure gradients appear along the body, boundary layer control becomes crucial for efficient propulsion.

Many types of drag

To move through the aquatic medium, marine mammals must apply a force to parts of the fluid around their body. In *reaction* to this force, and because of Newton's third law of motion, i.e., action-reaction, the fluid applies a force onto the animal, i.e., drag in the horizontal (or lift in the vertical). From the point of view of the swimmer, drag is an energy-dissipating force because it transfers the swimmer's kinetic or motion energy into water molecules kinetic energy, which in most cases is never retransferred back to the animal, instead ending up heating the fluid.

There are different types of drag depending on the mechanisms generating the resistive force. First, there is **friction drag** due to the shear force (or sliding "friction") among the sub-layers of the boundary layer, and **pressure drag** due to pressure differences along the body and the existence of the turbulent near wake at the tail end of the body. Generally, the smaller the width ("girth") of the turbulent wake (**Figure 1.1c**), the lesser is the pressure drag. With non-streamlined or *bluff* bodies, such as paddling paws, pressure drag dominates over friction drag. The opposite occurs with the streamlined bodies of marine mammal and their appendages (flukes and fins).

Another source of drag appears when a body is accelerated or decelerated. This is when **added mass drag** (or **acceleration reaction**, Vogel 1994) enters the picture, out of the necessity to accelerate a fluid mass roughly equivalent to that of the fluid displaced by the body along with the animal's body mass. Added mass is important, for example, with sea lions because of their accelerating-decelerating swimming style (Feldkamp 1987a, 1987b). It is also important during rorqual whales lunge feeding where large masses of engulfed seawater and prey are pushed forward by the whales' ventral pouch walls to be accelerated up to the speed of the whale (Potvin, Goldboegn and Shadwick,. 2009). Most added mass drag effects can be reduced by body streamlining, as achieved with the body shapes of many marine mammals which typically accelerate added mass in amount of 5–20% that of the displaced water. Another form of added mass appears when flukes, flippers, and tail are oscillated to produce propulsion. Here added mass drag appears mainly during the motion reversal of the appendages and limbs (i.e., a combined accelerative and declarative motion of the tail, flukes, flippers, etc.), resulting into added mass in the form of the vortices that are shed into the wake. In this case the drag is significant, going as much as three times the drag generated by the same body and appendages in a rigid state (Fish and Rohr 1999; Fish et al. 2014). This is another energy-dissipation mechanism, with the vortices' (rotational) kinetic energy originating from the swimmer's propulsive motions.

The motion of wing-like flippers and flukes in thrust production introduces yet a fourth source of drag, namely, **induced drag**. Induced drag appears because when canted at an angle to the flow, wing-like structures produce a large-scale vortex at the very tip of the wing, which extends far behind. (This is an effect originating from the pressure differential at the very tip.) Like the other vortices and eddies described previously, tip vortices are tornado-like structure of spinning fluid whose rotation feeds off the wing's own kinetic energy. Generally, induced drag *is* minimal when the wing is tapered, as is the case with most flukes and flippers. Induced drag is a major energy sink, for example, with aircraft, in comparison to the friction and pressure drag generated by the fuselage. This is because of the requirement for using large wings to generate enough lift to support an aircraft's heavy weight. Because marine mammals are close to neutral buoyancy, they do not require as much lift to "fly" in the ocean. Thus, marine mammal lifting surfaces are much smaller in relation to the rest of the body, but they are still important for the forces that effect rolls, turns, or propulsive forces (Segre et al. 2016).

Marine mammals that swim at the surface incur two extra sources of drag, namely, ventilation and **wave drag** (**Figure 1.2**). The former happens whenever portions of the body break the water surface, such as with a seal's head or a cetacean's blowhole. **Ventilation drag** arises because of the mass of water that is elevated on the leading side of the partially submerged body feature and also because of the depressed wake that follows behind (**Figure 1.2a**). Drag appears because the body has to apply a force to push water upward above the (mean) surface level, as well as downward below that surface. Ventilation drag can be minimized through streamlining and reduction of surface-breaking body features.

Wave drag

Wave drag is also generated from the lifting of water above the mean water level, to create a visible wave on the leading edge of the body when traveling at the water surface (**Figure 1.2b**), as commonly seen in ships and paddling water birds. This is another energy sink because portions of the body immediately under or breaking

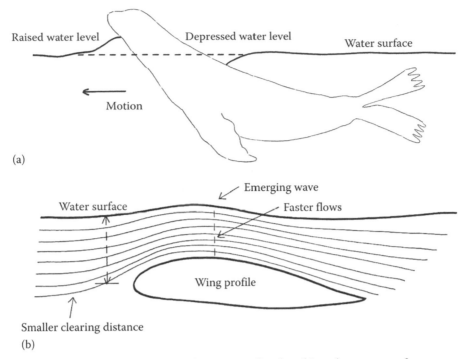

Figure 1.2 Hydrodynamics at the sea surface: (a) sea lion breaking the water surface, and (b) a streamlined object (appendage or body) moving just below it.

the surface (head, blowhole, dorsal fin, etc.) are effectively lifting water against gravity as well as moving it sideways. Thus, the higher and longer the waves, the more drag is generated and more of the swimmer's energy ends up lost. As a general rule, the faster the swim, the larger the wave drag, and the longer the body, the lower the wave drag (Fish 1996; Vennell, Pease, and Wilson, 2006). In addition, swimming depth has a significant effect so that at greater depths, less water is lifted (Hertel 1966; Vennell, Pease, and Wilson, 2006) (**Figure 1.2b and c**). An animal submerged below three body diameters experiences no wave drag.

Note that added mass, ventilation, induced, and wave drag all can exist even if no viscosity is present. These drags are generated due to the body forces and pressure gradients, which deflect fluid via direct contact. The drags are generated over scales that are commensurate with body size, in contrast to the viscous or cohesive forces affecting fluid motion at smaller scales (at 10^{-5} m or larger).

Body shape to limit drag

Generally, and during non-foraging travel, metabolic energy needed for transport is mainly invested in the generation of oscillatory limbs/appendage motions necessary for thrust production to oppose drag. Limiting propulsion expenditures and increasing propulsive efficiency thus involve reducing drag production wherever achievable. In this context, body **fineness ratio** (body length/body diameter) is a crucial determinant of drag (**Figure 1.3**). With the so-called **bluff body** where (lateral) body girth far exceeds body length in the direction of flow (i.e., very low fineness ratio), pressure drag is far more important than friction drag. Here the oncoming flow at maximum girth is tripped and separated into a sizable turbulent wake behind the body (**Figure 1.3a**). Pressure drag

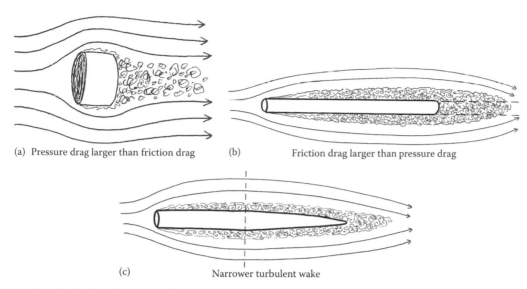

Figure 1.3 Body slenderness and tapering effects on flow streamlines, turbulent boundary layers and near wakes: (a) bluff body with its characteristic wide wake, (b) slender body with narrower wake, and (c) slender and aft-tapered body with narrower and shorter wake.

represents a tremendous energetic cost and can be tolerated (and nutritionally compensated) only for short-distance travel.

Animals that exhibit long distance migration or move at a high speed are never bluff but rather streamlined into a torpedo-like **fusiform** shape. Recall that the size of the turbulent wake depends on the width of the body where separation occurs; therefore, a more tapered posterior region yields a smaller turbulent wake and thus smaller pressure drag (**Figure 1.3b**). Additionally, to reduce friction drag, one also needs high taper over the anterior portion of the body to get the longest extension of the laminar boundary layer. Although, marine mammals ought to be shaped like javelins with fineness ratios exceeding 20, most species have a fineness ratio less than 8 (Fish and Rohr 1999; Ahlborn, Blake and Chan, 2009). Noting that friction drag increases with the body length and high-boundary layer thickness, high-fineness ratios would entail significant friction drag. Moreover, pressure drag is not eliminated by an extreme fineness ratio, because the boundary layer and hence the turbulent wake would also increase. In order to achieve minimum total body drag,

rather than just pressure or friction drag on a submerged body, the fineness ratio is limited to a range of 5.5–7.2, depending on the body mass (in the 10^3–10^5 kg range) (Ahlborn et al. 2009). Another benefit of high-aspect fusiform shapes is the low-associated added mass drag arising from the small amount of surrounding fluid mass that needs to be accelerated.

Lift and propulsion

Thrust for swimming is generated from the interaction of the water with the movement of a propulsive surface (e.g., paddles, flippers, flukes). Propulsors, therefore, are large in span and area to increase the volume of water accelerated by the excursion of the propulsor (Fish 1996). Oscillatory propulsion uses motions of the paired appendages (e.g., feet) or a highly modified lunate tail (e.g., flukes).

Lift-based oscillation is a high-performance swimming mode (Lighthill 1975; Webb 1975; Fish 1996). Several marine mammal taxa exhibit lift-based swimming modes including cetaceans and sirenians (caudal flukes), otariid seals (pectoral flippers), and phocid seals and walrus (pelvic flippers) (Fish 1998). This swimming mode, also described as **thunniform** or carangiform with lunate tail (Lighthill 1975; Webb 1975), is similar to propulsion by certain fast-swimming fish. Lift-based swimming produces force up to five times greater than drag-based propulsion (Goldbogen, Fish and Potvin, 2016).

As further explored in the next section, paddling involves using pressure drag to generate propulsion. This is a low-speed approach to swimming. For fast swimmers, thrust is generated more efficiently using flippers and flukes that act as lift-producing **hydrofoils**. These hydrodynamic-lifting surfaces generate propulsion when oscillated or raised and lowered in rhythmic fashion orthogonal to direction of transport.

Propulsive flippers and flukes have a planar geometry with a high-**aspect ratio**, where aspect ratio (= square of the span over the planar area of the hydrofoil) (Webb 1975; Feldkamp 1987a; Fish 1998). High-aspect ratio reduces (induced) drag while maximizing thrust. Aspect ratio works in concert with the sweep of the hydrofoil geometry. Sweep is the rearward inclination of the leading edge. A combination of low sweep with high-aspect ratio allows for high-efficiency swimming, whereas high sweep may compensate for the reduced lift production of low-aspect ratio hydrofoils (Fish and Rohr 1999).

Thrust is produced via the combination of the lift and drag forces that the hydrofoils produce. While drag always points in a direction opposite to the body or appendage in motion, lift is directed perpendicularly to that motion. Using a simplistic description, the lift generated by a wing or hydrofoil arises from the combination of the low-pressure region existing over the upper surface of the wing, in comparison to the higher pressure under the wing. This is the so-called Bernoulli lift, here created when the flow over the wing is faster than the flow underneath. The relatively higher speed of fluid over the upper-wing surface induces a reduction in pressure compared to the lower-wing surface, thereby giving rise to an upper-lift force. Lift is also the result of the downward defection of

the air (or water) imparted by both the reduced pressure over the upper surface and the deflecting action of the solid bottom surface.

With aircraft, wings typically have upper surfaces that are cambered compared to the lower surface that enhance the fluid velocity over the upper surface to generate the lift to compensate for the aircraft's weight. Flippers and flukes, on the other hand, are more symmetrical (i.e., have near-equal camber on both surfaces) because they are used to produce forward-directed propulsion, which involves the use of both downward- and upward-directed lift (Fish and Rohr 1999).

The basic design of a lift-producing wing-like surface hinges on minimizing drag and maximizing lift. In general, both lift and drag increase with increasing hydrofoil **angle of attack** (AOA). AOA characterizes the angle between the direction of the flow relative to the surface of the foil, and orientation of its mean chord line. An oscillating hydrofoil (i.e., flukes, flippers) generates lift at a controlled AOA (Lighthill 1975; Feldkamp 1987b; Fish 1998). The AOA should be small to avoid separation of the flow from the hydrofoil surface, which reduces lift and increases hydrodynamic drag.

To produce the most lift at the lowest-energetic (drag) cost, marine mammals must optimize the lift-over-drag ratio. At low AOA, drag is the result of friction, pressure, and induced drag. In this regime, the AOA for optimal lift-over-drag is at about 15°, depending on the morphological design of the hydrofoil. This is why dolphins, which use flukes as hydrofoils to produce lift, appear to swim almost effortlessly by barely moving their flukes at low AOA.

Many marine mammals that have flippers used them primarily for maneuvering, i.e., to effect turns and rolls. Moreover, many marine mammals that have flukes use them principally for propulsion. Thrust along the line of body motion arises from combining both drag and lift forces generated by the hydrofoil, that is, whenever the lift is angled forward enough, and with enough magnitude, to cancel the rearward action of the drag. Note that the AOA also changes with the sweeping motion of the tail and flukes. It is here that flukes' flexibility becomes advantageous, as it reduces the AOA closer to the optimal AOA.

Drag and propulsion

Drag-based propulsion is used by semi-aquatic and aquatic mammals (e.g., muskrat, polar bear, otter) for swimming and maneuvering (e.g., manatee, humpback whale) (Fish 1996). Propulsion by drag-based oscillation is produced by the motion of various combinations of the paired appendages (quadrupedal, pectoral, pelvic) either alternately or simultaneously and oriented in either the parasagittal or horizontal planes (Fish 1996). The stroke cycle includes power and recovery phases (Fish 1984). In the power phase, the posterior sweep of the limb generates a large pressure drag, which provides an anterior thrust.

Maximum thrust is generated with a broad paddle area that is configured as a circle or triangle with a constriction at the attachment point with the body (Fish 1996). The constriction minimizes interference drag with the body and provides a long lever arm to increase the velocity of the paddle during the power stroke. Paddle area is increased by abduction of the digits and by interdigital webbing or fringe hairs (Fish 1996). The increased paddle area allows for the production of a high-pressure drag on the paddle as it is swept posteriorly. The reaction force to the pressure drag is the thrust that moves the paddling animal forward. The size of the paddle accelerates a large mass of fluid to a low velocity, which is more efficient than accelerating a small mass of fluid to a high velocity (Fish 2004). The recovery phase repositions the limb, incurring a non-thrust-generating drag. To limit the reduction in thrust during the recovery phase, drag on the appendage is reduced by adducting the digits or rotating the appendage to reduce paddle area and by changing the timing of movement to reduce the relative velocity (Fish 1996).

Drag-based oscillation has a low-propulsive efficiency (thrust power/total mechanical power output) of ≤ 0.33 (Fish 1996). This low efficiency occurs because thrust is generated through only half of the stroke cycle (Fish 1996). Energy is lost to increased resistive drag as the foot is repositioned during the recovery phase. In addition, the total energy expended through the stroke is lost in acceleration of the mass of the limb and the water entrained to the foot (Fish 1996). Propulsive efficiency for the drag-based oscillation is highest at low speeds (Vogel 1994).

Locomotion of marine mammals

Thrust for swimming is generated from the interaction of the water with the movement of a propulsive surface (e.g., paddles, flippers, flukes). Propulsors, therefore, are large in span and area to increase the volume of water accelerated by the movements of the propulsor (Fish 1993). The propulsive movements can be classified broadly as drag-based oscillatory, lift-based oscillatory, and undulatory (Fish 1996). Oscillatory propulsion uses motions of the paired appendages (e.g., feet) or a highly modified wing-like tail (e.g., flukes), whereas undulatory propulsion uses movements of the body and tail.

Cetaceans

The first hydrodynamic calculation on the swimming energetics was performed for dolphins and porpoises (Gray 1936; Webb 1975). The model of Gray (1936) used estimates of drag based on rigid bodies. The model assumed that the thrust generated by swimming dolphins was equal to estimates of drag on a flat plate of equal surface area. In what became known as "Gray's paradox", calculations indicated that a dolphin swimming at 10 m/s would be impossible from high drag with a turbulent boundary layer with insufficient muscle mass to power the animal. Gray's (1936) resolution to the paradox was to consider that the dolphin had a lower drag with a laminar boundary layer but no viable mechanism was found. Despite more recent information on mass power output and acknowledgment of errors in Gray's analysis, research was stimulated for examination of potential drag-reduction mechanisms (Fish 2006).

Various mechanisms for drag reduction were proposed for cetaceans (Fish and Rohr 1999). These have included microvibrations from the dolphin's skin, active deformation of the skin, mobile skin folds, presence of cutaneous ridges in the skin, boundary layer heating, sloughing of epidermal cells containing lipid droplets, secretions from the eyes, boundary layer acceleration, and viscous dampening due to skin compliance. These mechanisms, although showing some success in mechanical applications, have not been shown to be effective for swimming cetaceans. The major mechanisms for reducing drag in cetaceans are due to the fusiform shape (i.e., elongate teardrop) of cetaceans and smoothness of the integument.

Drag is minimized primarily by streamlining the shape of the body and appendages (i.e., flukes, flippers, dorsal fin) (Fish 1993). The streamlined shape of most cetaceans falls within a fineness ratio (i.e., body length/maximum body diameter) between 4.5 and 7. It is within this optimal range where drag is at a minimum (Webb 1975). Furthermore, the position of the maximum thickness from the rostrum will maintain laminar flow over a large portion of the body, to minimize drag (Webb 1975; Fish and Rohr 1999). The position of the dolphin's maximum thickness is at 34–45% of body length (Fish and Hui 1991). The addition of appendages increases the drag. The body of the harbor porpoise (*Phocoena phocoena*) generates 64.3% of the drag with the remaining 35.7% of drag coming from the dorsal fin (4.3%), pectoral flippers (18.0%), and flukes (13.4%). The drag added by the appendages of *Tursiops* was estimated as 28% of the total drag (Lang and Pryor 1966). The added increase in drag due to the appendages is caused by interference drag as flow over the body is distorted by the appendages and induced drag from lift generation by the appendages (Vogel 1994; Fish 1998). The induced drag component is the energy lost due to tip vortices that are generated by pressure

differences between the two sides of the flukes during the oscillating propulsive strokes. In addition, induced drag can be produced by the flippers and dorsal fin when canted at an angle to the oncoming flow during maneuvers (Webb 1975; Fish 1993).

The naked skin of odontocetes was shown to have low roughness compared to other swimming animals (e.g., fish, sea snake, manatee) (Wainwright et al. 2019) and the dermal ridges were not great enough to influence the smoothness of the skin. A smooth skin surface will aid in reducing the frictional drag component for cetaceans.

Swimming motions, however, will increase drag on the body. Skin friction could increase up to a factor of five (Lighthill 1975). An additional increase in the pressure drag will occur from a deviation from a rigid streamlined body (Fish and Rohr 1999).

The swimming motions of cetaceans are characterized as the thunniform mode (Lighthill 1975; Webb 1975). In this mode, the posterior one-third of the body is oscillated dorsoventrally (heaving) in the sagittal plane with the lateral flukes pitched (i.e., rotation within the vertical plane) at their base (**Figure 1.4**; Fish and Hui 1991). These heaving and pitching motions are responsible for changes of angle of attack (i.e., incident angle to oncoming flow) of the flukes (Fish 1993). In this manner, the flukes act as oscillating wings that generate thrust as a component from a forward-directed lift.

The cross-sectional geometry of the flukes is similar to symmetrical engineered foil sections with an elongate teardrop design. This design has a rounded leading edge that increases to a maximum thickness at approximately 24–36% of the chord (i.e., linear distance from leading to trailing edges) (Fish 2004). From the maximum thickness, the hydrofoil slowly tapers to the trailing edge. This configuration prevents stalling due to boundary layer separation as the hydrofoil is oscillated.

The typical planform of cetacean flukes has a sweptback, wing-like shape with tapering tips. Aspect ratio varies from 2.0 for the Amazon

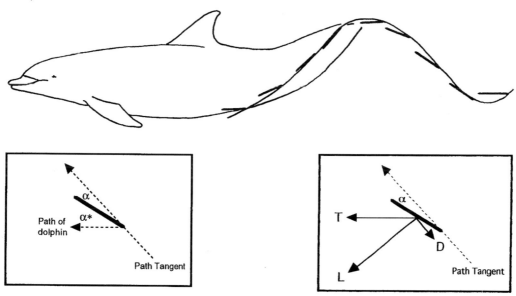

Figure 1.4 Path of oscillating dolphin flukes through a stroke cycle. The tips of the flukes move along a sinusoidal path. Sequential fluke positions along the path are illustrated as straight lines. The box on the left shows the relationship between the tangent to the path of the flukes with the angle of attack, α, and the pitch angle, α^*. Angle of attack is the angle between the tangent of the fluke's path and the axis of the fluke's chord; pitch angle is the angle between the fluke axis and the translation movement of the animal. The box on the right shows the relationship between the major forces produced by the motion of the fluke. D is the drag, L is the lift, and T is the thrust resolved from L (Fish 1993).

river dolphin (*Inia geoffrensis*) to 6.1 and 6.2 for fin whale (*Balaenoptera physalus*) and false killer whale (*Pseudorca crassidens*), respectively (Fish and Rohr 1999). The flukes of fast-swimming cetaceans have higher-aspect ratios than slow swimmers. Sweepback in flukes ranges from 4.4° for the killer whale (*Orcinus orca*) to 47.4° for the white-sided dolphin (*Lagenorhynchus acutus*) (Fish and Rohr 1999). The flukes of mature male narwhals (*Monodon monoceros*) have a slightly concave leading edge without sweepback.

Lift is produced from differential flow between the upper and lower surfaces of the flukes (i.e., Bernoulli effect). The flukes move along a sinusoidal path through the water due to a combination of the forward movement of the dolphin and the heaving motion of the tail. The pitch of the flukes relative to the pathway cants them at an angle of attack to the incident water flow. Lift is directed perpendicular to the

pathway traversed by the flukes and is resolved into an anteriorly directed thrust vector (Lighthill 1975). The action of the flukes in generating lift induces a rotation of fluid in the wake known as vorticity. The increase in vorticity transfers the kinetic energy of the dolphin to the water, which is eventually lost in the formation of eddies and frictional forces in the water. Vorticity shed into the wake reflects the transfer of momentum from the flukes to the fluid. Thrust is generated almost continuously throughout a stroke cycle. Although the hydrofoil produces some drag, it is small compared to the lift.

Computational fluid dynamic (CFD) approaches exploit the capacity of modern computer processing power to calculate the motion fluid particles down to scales of about one-hundredth millimeters, over body lengths spanning those up to the largest whales. This computational

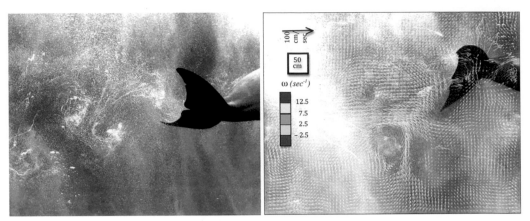

Figure 1.5 **Vortices in the wake of a swimming dolphin from Fish et al. (2014). The image on the left shows vortices generated from a microbubble curtain that are generated by oscillations of the flukes. The image on the right shows the velocity vectors from the displacement of the microbubbles. The concentration of red vectors indicates a vortex with a counterclockwise rotation.**

technique enables the visualization of fluid structures that cannot be resolved in the laboratory due to their extreme scale or ephemeral nature. An important example is the boundary layer found very close to the body, or the trail of vortices produced by the sweeping motions of a fluke. One can also simulate, although imperfectly, the turbulence trailing the body. These data allow the estimation of the force applied on and resulting motions by a swimmer. Tag design has also involved the use of CFD, to identify the flow characteristics moving past the sensors and the added drag associated with its attachment (Shorter et al. 2014). Such data can help inform tag placement-dependent hydrodynamic effects on the energetic cost of swimming (van der Hoop et al. 2014).

A more direct means of visualizing the fluid structure is by particle image velocimetry (PIV). Small neutrally buoyant particles are added to the water to be illuminated by a laser sheet to measure velocity of the flow with its corresponding pressure. The production of vortices from the movements of an oscillating propulsor can be observed and used to determine the thrust production. This technique was used to determine the thrust production by swimming dolphins using microbubbles (<1 mm) as the

particles, which were illuminated by natural sunlight (Fish et al. 2014). A pattern of strong vortices was generated in the wake of the oscillating flukes (**Figure 1.5**). Thrust production was measured from the strength of the vortices. At about 3.5 m/s, the dolphins produced 700 N during steady swimming with a small fluke oscillatory amplitude and up to 1458 N with a large fluke amplitude. See **Chapter 7** for examples of how natural particle imaging in the water can be used by marine mammals for navigation.

The lift-based oscillation of cetaceans is characterized by high-propulsive efficiencies. Propulsive efficiencies for cetaceans range from 0.75 to 0.9 (Fish and Rohr 1999), where maximum efficiencies are achieved within the range of normal cruising speeds (0.8–1.5 body lengths/s). Such a relationship is indicated by the thunniform Strouhal number (*St*). *St* is related to how fast vortices are shed from the action of the propulsive appendages (Rohr and Fish 2004). *St* is calculated as $St = Af/U$, where *A* is the width of the wake, taken as the peak-to-peak maximum excursion of the flukes, *f* is the frequency of oscillation, and *U* is the mean forward velocity of the animal. The maximum propulsive efficiency occurs within the range of *St* of 0.2–0.4 (**Figure 1.6**).

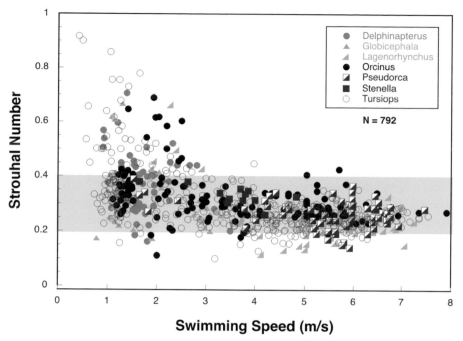

Figure 1.6 Strouhal number (*St*) as a function of length-specific swimming speed for various species of cetaceans. The shaded area indicates the region of *St* = 0.2–0.4, where propulsive efficiency is at maximal.

As the energy cost of swimming can be large due to the density and viscosity of water, marine mammals have adopted a variety of behavioral, morphological, and physiological mechanisms to swim economically. Such mechanisms rely upon energy management by capturing energy from external and internal sources. External sources of energy can be utilized from the prevailing physics of the environment (i.e., gravity, hydrodynamics, waves). In contrast, kinetic energy from muscular contraction that would typically be lost can be recycled internally from the elastic properties of the connective tissues of the body (Fish and Rohr 1999). In both cases, the available energy to perform the work of swimming is augmented or conserved to lower metabolic power consumption, increase dive time, and increase speed.

The capture of external forms of energy to add to the total energy budget for movement by marine mammals is known as **free-riding**. The simplest type of free-riding behavior is gliding while diving. The density of the body increases with depth due to increased hydrostatic pressure and compression of the lungs. When diving deeply, lung collapse reduces the net buoyant force causing the animal to sink (Moore et al. 2011). The animal can glide deeper as gravity now supplies the motile force. The gliding configuration of the body minimizes drag. Whales and dolphins intermittently switch between active swimming and gliding depending on dive depth and the net buoyancy of the body (Williams et al. 2000; Williams 2001). During deep dives, dolphins can reduce energy costs by approximately 20% when transiting to the bottom by using intermittent swimming behaviors (Williams 2001). During ascent, the reverse occurs and the animal accelerates by actively swimming until its lungs re-inflate sufficiently to provide positive buoyancy (Williams et al. 2000).

The occurrence of highly organized formations by cetaceans has been suggested as an adaptation for energy economy (Fish and Rohr 1999).

Formation swimmers are able to capture energy from the vortex patterns in the wake of conspecifics and decrease drag with a concomitant decrease in overall energy cost of locomotion. In addition, when two bodies are in close proximity, the water flow between them is accelerated resulting in an attractive force due to the Bernoulli effect (Fish et al. 2013). Pods of whales and dolphins swim in side-by-side and echelon formations to save energy (Weihs 2004; Fish et al. 2013). Small dolphins often position themselves beside and slightly behind the maximum diameter of a larger animal (Noren 2008; See also **Chapter 2**). While the larger dolphin will experience increased drag, the smaller gains an energetic benefit (Weihs 2004; Noren 2008). This effect is beneficial particularly for young whales in order to maintain speed with their mothers. A neonatal dolphin could use this mechanism to gain up to 90% of the thrust needed to move alongside its mother.

Dolphins are able to reduce energy costs by riding the bow waves generated by large whales and boats (Fish and Hui 1991; Williams et al. 1992). Williams et al. (1992) found that wave-riding dolphins could swim at a higher speed while reducing or maintaining metabolic rate, heart rate, lactate production and respiratory rate. Dolphins can either ride bow waves like a surfer, or they make use of the pressure front created by the boat (Fish and Hui 1991). This behavior is complex with any energy savings to the dolphin related to bow design, swimming depth, body orientation, and distance from the ship (Fish and Hui 1991). See **Chapter 2** for more background information on wave-riding by marine mammals.

Wind-wave riding and surf-wave riding can use gravity to reduce the energy cost of swimming (Fish and Hui 1991). These wave-riding behaviors differ from bow wave riding because they use the interaction of the dolphin's weight and slope of the wave front to produce movement analogous to human surfers (Fish and Hui 1991). Dolphins and sea lions have been observed to surf on in-shore waves (Norris and

Prescott 1961; Riedman 1990). Dolphins ride waves with a forward slope of 10–18° at velocities of 5–6 m/s (Fish and Rohr 1999). In the open sea, the flukes of large whales could absorb energy from ocean waves (Bose and Lien 1990). Whales absorb 25% of their propulsive power from head seas and 33% from following seas, by synchronizing the motion of the wave with the motion of the flukes.

Pinnipeds

Regarding the hydrodynamics of propulsion, pinnipeds, like cetaceans, use lift-based oscillations for swimming. However, as opposed to oscillations of flukes on the tail, the pinnipeds oscillate paired limbs. Members of the Phocidae and Odobenidae swim with the hindflippers, whereas the Otariidae swim with the foreflippers. The foreflippers of *Zalophus* have a high-aspect ratio of 7.9 (Feldkamp 1987a). Aspect ratio for the hindflippers of phocids is 3.4–4.0 (Fish, Innes and Ronald 1988).

The hindflippers of phocids and walrus are oscillated laterally. The oscillations are confined to flexion of the posterior half of the body. The two hindflippers are alternated so that the digits of the trailing flipper are fully abducted (spread) to generate thrust as the digits of the leading flipper adducted (folded) (Fish, Innes and Ronald, 1988). The flexion at the base of the propulsive flippers is controlled at the tibiofemoral and ankle joints. The trailing flipper is canted at an angle of attack to the incident flow as the flipper traces a sinusoidal path (**Figure 1.7**). The range of maximum angles of attack (17.8–33.2°) for harp (*Pagophilius groenlandicus*) and ringed seals (*Pusa hispida*) was found to be greater than non-oscillating hydrofoils. These high angles are consistent with other oscillating biological systems and allow the flippers to operate more efficiently.

The propulsive efficiency for harp and ringed seals was reported to be high at approximately at 0.85 (Fish, Innes and Ronald,1988). Alternating the hindflippers allowed the seals to generate thrust throughout the stroke cycle and maintain

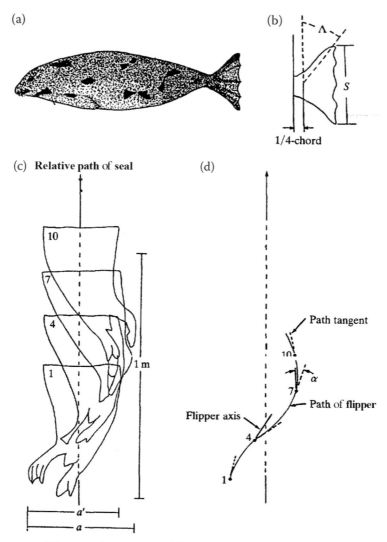

Figure 1.7 Diagram of the morphometric and kinematic measurements of a swimming phocid seal. (a) Lateral view of a swimming seal showing the profile of the body and hindflipper. (b) Planiform of hindflipper during power stroke showing span (S) and sweepback angle (L). (c) Sequential tracings from films of the dorsal view of the posterior half of a ringed seal (swimming at 1.42 m/s during half a complete stroke cycle. Numbers indicate the frame of the film and the one-half (a') and full cycle (a) amplitudes are labeled. (d) Orientation and movement of the hindflipper with respect to swimming direction of the seal used for estimation of the angle of attack (α) (Fish et al. 1988).

high efficiency. The drag coefficient for the actively swimming phocids was large ranging from 0.012 to 0.028. The drag coefficient for the seals was 2.2 – five to six times greater than a theoretical rigid body and greater than values measured for towed or gliding phocid seals (Innes 1984; Williams and Kooyman 1985).

The greater drag was due to the increased drag from the body and hindflipper movements.

Like whales and dolphins, seals can use a gliding body form and lung compression with depth to reduce drag and lower the metabolic cost of swimming. Seals exhale before diving. This exhalation has been considered as a mechanism to prevent decompression sickness; moreever, this behavior may effectively reduce buoyancy to decrease the energy cost of swimming during the initial descent (Kooyman 1973).

The otariid pinnipeds, including the sea lions and fur seals, swim by use of the enlarged wing-like foreflippers. The foreflippers are flapped in a manner reminiscent of underwater flight (Feldkamp 1987b). The foreflippers generate thrust throughout most of the stroke cycle. During steady swimming, the stroke cycle is divided into a forceful downstroke of the flippers during a power phase that terminates with a paddling phase before the flippers are lifted in a recovery phase (Feldkamp 1987b). For unsteady burst and glide swimming, sea lions use a propulsive clap to accelerate the body (Friedman and Leftwich 2014). The clap is characterized with the fore-flippers raised above the animal's midline and rapidly drawn ventrally to be pressed against the abdomen. The sea lion then glides while maintaining a streamlined configuration.

The propulsive foreflippers have a planar geometry with a high-aspect ratio. Aspect ratio is equal to span²/planar area (Feldkamp 1987b; Fish 1998). The foreflippers of *Zalophus* have a high-aspect ratio of 7.9. High-aspect ratio reduces drag while maximizing thrust. Oscillation of the foreflippers by *Zalophus* provide maximum efficiencies of 0.8 (Feldkamp 1987b).

While not used for aquatic propulsion, the hindflippers of otariids are used as control surfaces to aid in maneuverability. Porpoising sea lions have a maximum angle of attack for the hindflippers of 18.3° (Leahy et al. 2021). The hindflippers can have a maximum angle of attack of 50.8° during lateral turns. These high angles of attack indicate that the hindflippers could act as lifting surfaces to affect a turning moment (Leahy et al. 2021). The triangular configuration of the hindflippers alludes to a delta wing design. This design allows for the generation of lift while minimizing stall (i.e., dramatic loss of lift), which would be beneficial in executing rapid sharp turns.

Sirenians

Manatees (*Trichechus* sp.) possess a broad paddle-like tail that is used for swimming. The tail is undulated in the vertical plane (Kojeszewski and Fish 2007). Undulatory swimming entails throwing the body and tail into a traveling wave. The wave is propagated in the direction opposite of forward motion of the animal and at a velocity greater than the forward velocity of the animal. As the traveling wave moves along the body, its amplitude increases to a maximum at the tip of the tail. As each section on the body accelerates vertically, the wave faces caudally at an angle to the mean motion of the body. Fluid adjacent to the accelerated section produces a reaction force with a component in the direction of thrust (Lighthill 1975).

Although showing some overlap with the propulsive efficiency of lift-based swimmers (e.g., cetaceans, pinnipeds), the undulatory mode of manatees has a lower-propulsive efficiency. The propulsive efficiency of the manatee (*Trichechus manatus latirostris*) ranges between 0.67 and 0.81 (Kojeszewski and Fish 2007).

Sea otter and polar bear

Drag-based propulsion by paddling is used by the sea otter (*Enhydra lutris*) and polar bear (*Ursus maritimus*) for swimming (Fish 1996). Propulsion by drag-based paddling in the sea otter and polar bear is produced by the motion of hind feet and fore feet, respectively. The stroke cycle is divided into distinct power and recovery phases (Fish 1996). In the power phase, the posterior sweep of the limb generates an anterior thrust. The recovery phase generates no thrust and repositions the limbs to initiate another power stroke.

The sea otter has broad-webbed hind feet that maximize thrust production. The otter paddles while swimming underwater or on its back, while on the surface of the water (Tarasoff et al. 1972). Swimming at the water surface is energetically demanding due to wave drag (Hertel 1966). Kinetic energy from the animal's motion is transferred into potential energy in the formation of surface wave by the upward displacement of water. This energy loss can be up to a maximum of five times the frictional drag (Hertel 1966). The cost of transport for surface paddling sea otters was 69% greater than with submerged swimming (Williams 1989). In submerged swimming, the wave drag is negated permitting increased swimming speed. **Chapters 2** and **5** describe more aspects of the high energy and metabolic demands on otters.

Polar bears are generally considered strong swimmers that can swim distances up to 687 km (Pagano et al. 2019). The bears swim by alternate paddling of the forelimbs. The polar bears demonstrated high-swimming costs because of their paddle propulsion (Pagano et al. 2019).

Hydrodynamics of feeding
Feeding hydrodynamics by mysticete whales – a case study

The search for, and assimilation of food is key to survival and reproduction for all living organisms. In marine mammals, and along with morphological and behavioral adaptations, hydrodynamics often plays an outsize role in constraining **pelagic prey** acquisition strategy, kinematics and energetics. Nowhere are hydrodynamical constraints more important than with large cetaceans such as the **baleen** whales (Mysticeti) (~6–30 m in adult body sizes), for which the energy required for survival and reproduction has increased in the face of larger body mass (Potvin et al. 2021), altered maneuverability (Segre et al. 2022), and ability to capture energy-rich agile prey (Goldbogen et al. 2017). Spectacular examples include the rorqual (Balaenopteridae) and balaenid whales (Balaenidae), two sub-groups of cetaceans that use baleen instead of teeth to catch and retain prey (Werth et al. 2018). All mysticetes forage on large aggregations of small and low-mobility prey, typically patches of plankton (copepods and krill) or schools of forage fish (anchovies,

capelin and the like). Moreover, most distinguish themselves by using feeding and breeding grounds separated by hundreds or thousands of miles (Abrahms et al. 2019). As a result, these animals have evolved strategies and morphologies that not only enable efficient energy collection from ephemeral food resources but also promote accumulation of significant fat reserves. Even more remarkable has been their ability to choose strategies aimed at reducing drag and locomotor expenditures, and in ways that favored evolution to body sizes larger than typical Odontoceti (Goldbogen et al. 2019). **Chapter 10** considers the evolutionary aspects of feeding structures and functions for whales and other marine mammals.

Feeding strategies

Both balaenopterid and balaenid sub-groups use baleen, which are triangular thin plates made of keratinous tissue suspended from the palatal gingiva, with base width and height approximating 0.10 m and 0.77 m, respectively. Numbering at 200–400 on each side of the palate, these are longitudinally fibrous in structure, running and separating into mats of bristles (or "fringes") on the lingual side because of through-filter flow erosion (Werth et al. 2018). Most interestingly, and despite the similarities in their baleen morphology, each family has evolved distinct oropharyngeal morphologies and prey capture strategies that nevertheless allowed the evolution into commensurate levels of gigantism (Goldbogen et al. 2019). For example, balaenids swim through large patches of copepods with their mouth opened, while enabling the (temporally) continuous collection and separation of the prey from the seawater via through-baleen filtration (Goldbogen et al. 2017). Balaenopterids, on the other hand, are intermittent filter-feeders which accelerate towards a patch of prey, to then engulf large sections of the patch and the large volume of water in which the prey is embedded. The behavior resembles the raptorial "attack-and-capture" approach of the toothed whales but differs post mouth-closure with the additional stage of water expulsion from the (then) filled ventral pouch and attendant through-baleen filtration and prey retention (Werth and Ito 2017). The latter is enabled by the partial opening of the oral margins, along with the contraction of muscle embedded within the ventral groove blubber (VGB), which builds up pressure inside the filled pouch to push water out through baleen. Most importantly, engulfment of very large quantities of prey and water by the rorquals is enabled by the initially pleated VGB, which accordion-folds into a low-area surface on the ventral side of the body during non-feeding transport, but expand up to many times its original volume during engulfment (Goldbogen 2010; Goldbogen, Potvin and Shadwick 2010).

From the point of view of drag management in rorquals, a crucial distinction is to be made between lunges in which the prey-water engulfment stage is carried out without the heavy fluking usually found during prey approach, i.e., during the so-called "coasting engulfment" mode, as occurring in all lunges directed at krill (Potvin et al. 2021). (This is also called engulfment "on momentum" per Gough et al. 2022). This contrasts to the heavy fluking occurring during both prey approach and well into engulfment, i.e., "powered engulfment", as observed in fish lunges for which escape is to be tactically neutralized by matching swimming performance or bubble netting (Simon, Johnson and Madsen, 2012; Cade et al. 2020).

Managing hydrodynamics to enhance feeding efficiency

Considering locomotor expenditures as a major (metabolic) energy sink in mysticete feeding, and with body drag acting in opposition to propulsive fluking, rorquals and balaenids alike are exploiting food resources by managing both fluking and body drag in ways that ensure large prey capture efficiencies (energy in/energy out >10). As discussed in earlier sections, propulsive fluking carries important penalties

in drag arising from the dorsoventral motion of the tail- and flukes-generating vortices and other time-dependent wake structures (Fish 1999; Fish et al. 2014). Moreover, high drag can be found with bodies lacking streamlining and tapering, owning to the fact that body drag is proportional to body surface area, degree of streamlining (via its drag coefficient; Vogel 1994; Potvin et al. 2021) and body speed (squared). Thus, adopting either highly streamlined body forms while fluking at high speed (>2 m s^{-1}), or swimming slowly while fluking with a bluff and blunt body, becomes drag management strategies that enable low enough locomotor expenditures and high-feeding efficiencies. For examples, slow swimming with a "draggier" body is used extensively by balaenid whales, which consistently sweep through prey patches at speeds of 1 m s^{-1} or less, with an open and draggy mouth, and exposed oropharyngeal cavity and baleen racks (Potvin and Werth 2017). Slow-swimming speed is also exhibited by fish-feeding rorquals fluking not only during prey approach but also during engulfment ("powered engulfment"), given the high drag of an open mouth and expanding ventral pouch.

Interestingly, krill-feeding rorquals exhibit not just one but several energy-saving strategies, applied sequentially during the various stages of the same lunge. At first, a typical rorqual approaches its prey in acceleration, to reach speeds as high as 5 m s^{-1} (Gough et al. 2021, 2022). Here the animals keep their mouth closed to achieve a high degree of body streamlining (and lower drag), an action that necessitates lower levels of propulsive fluking to reach high velocities. This is followed by engulfment while coasting (thus without fluking), a stage marked by the high drag incurred by the unstreamlined expanding buccal cavity (i.e., growing into a tadpole shape) (**Figure 1.8**). Here extra metabolic expenditures are incurred, not from the use of locomotor musculature but by the eccentric pull of muscle embedded in the VGB to entrain (and accelerate) the large and initially immobile masses of engulfed prey and

Figure 1.8 Photograph of sei whale with an inflated throat pouch showing the "tadpole" shape. (Courtesy of Paolo Segre)

seawater to the current motion state of the body (Potvin et al. 2021). With the VGB musculature being in direct contact with the engulfed mass, Newtonian dynamics entails its corresponding force as proportional to the mass and acceleration of the engulfed mass which, in other words, translates into another source of drag. This, in turns, makes VGB expenditures as important as the locomotor expenditures incurred during prey approach (Potvin, Goldbogen and Shadwick, 2009, 2021).

TOOLBOX

A number of innovative technologies have been applied to study the hydrodynamics of marine mammals. Although some of these techniques have been around for some time, the application to large marine mammals and their study in open water environments has only been recently used.

Bubble particle image velocimetry

BPIV is a variant of standard PIV analysis, where microbubbles (<1 mm) are used as small particles to be tracked to visualize water movements so that water velocities and pressures can be measured. This technique avoids use of a laser and small silver-coated glass sphere which may be logistically impossible due to the large volume of water occupied by a marine mammal and be injurious to the animal. Microbubbles are generated from a compressed air source through a porous hose to provide a curtain of bubble through which the animal swims (Fish et al. 2014). A high-speed video camera records the natural light reflected from bubbles in the region of around and in the wake of the propulsor. The particle (microbubble) displacements divided by the time between successive video frames yields temporally and spatially resolved flow velocity field information. The video sequence is then digitized and used as an input to a PIV-processing program (e.g., DaVis) using standard cross-correlation techniques.

Hydrodynamic modeling

When experimental approaches are not available, computer modeling (or simulation) of the flows about a swimming animal, a technique known as Computational Fluid Dynamics (CFD), can provide important insights into how morphology can influence hydrodynamic performance.

CFD approaches exploit the capacity of modern computer-processing power to calculate the motions fluid particles down to scales of about one-hundredth millimeters, over body lengths spanning those up to the largest whales. This computational technique enables the visualization of fluid structures that cannot be resolved in the laboratory due to their extreme scale or ephemeral nature. An important example is the boundary layer found very close to the body, or the trail of vortices produced by the sweeping motions of a fluke. One can also simulate, although imperfectly, the turbulence trailing the body. These data allow the estimation of the force applied on and resulting motions by a swimmer. Tag design has also involved the use of CFD, to identify the flow characteristics moving past the sensors and the added drag associated with its attachment (Shorter et al. 2014). Such data can help inform tag placement-dependent hydrodynamic effects on the energetic cost of swimming (van der Hoop et al. 2014).

Tags

Many chapters in this book discuss how the use of instrumented bio-logging and satellite tags has significantly advanced knowledge of the movements and habits of marine mammals. Bio-logging tags (e.g., DTAG, CATS) are equipped with depth (pressure) and temperature sensors, three-axis accelerometers, three-axis magnetometers, and three-axis gyroscopes (Gough et al. 2021). Auxiliary sensors may include positional information from satellite systems or video from cameras integrated into the tag (Gough et al. 2021). Physio-logging includes the incorporation of medical grade components to measure physiological parameters. Accelerometers sampling at high rates (i.e., 400 Hz) can detect a wide range of important information including calling behavior, swimming speed, and heart rate. Magnetometers and gyroscopes provide orientation and rotational information that is important for understanding complex accelerometer signals related to animal movement (Segre et al. 2016, 2022).

Video analysis

The use of digital video cameras offers a means to examine detail of movement within small time intervals. Movements of the animal's trajectory and its propulsive appendages can be recorded. The time interval is determined by the framing rate of the camera. Typical video cameras record 30 frames per second (intervals of 0.33 s). Higher-frame rates can be obtained with high-speed video cameras. For detailed analysis, it is best to have the camera stationary so that swimming motions are recorded within the field of view with a fixed frame of reference (Fish 1998). This also allows for a scale (e.g., meter stick) to be put into the field of view for absolute measurements. Multiple video cameras held in stationary positions with overlapping fields of view can be used for three-dimensional imaging. In this case, a three-dimensional calibration device needs to be placed in the field of video to coordinate the images from each camera, and a means to synchronize the records is required. The cameras can be hard-wired for synchronization or an electronic underwater strobe can be periodically discharged to synchronize the video frames. Specialized software is required for two-dimensional analysis (e.g., Tracker, Deep Lab Cut) and three-dimensional analysis (e.g., Proanalyst).

CONCLUSIONS

Marine mammals must obey the physical laws that dictate their swimming and foraging performance to successfully exist in the marine environment. The prominent forces are thrust and drag. These forces are derived from the types of flow that develop over the body surface. The balance between these forces determines the speed of swimming and ability to procure food. Marine mammals have adapted to the aqueous environment of the marine realm through a variety of morphological, physiology, and behavioral adaptations. These adaptations that evolved over millions of years have allowed for highly efficient swimming modes in the most derived marine mammals. Similarly, feeding mechanisms have evolved in the largest of whales with respect to the increased drag as a result of entrapment and engulfment of small fish and planktonic organisms. Despite the detriment of increased drag and energetic effort of feeding using baleen, these baleen whales are able to maximize food intake and maintain a positive energy balance. Marine mammals have been able to control the flow and physical forces to effectively move and feed in the aquatic environment.

REVIEW QUESTIONS

1 Why do marine mammals have drag?
2 How is thrust in water produced in the swimming of marine mammals?
3 What are the different types of drag?
4 How are laminar and turbulent flows defined?
5 What is the most efficient swimming mode for thrust production?
6 How are the oscillations of the propulsive appendages of otariid seals, phocid seals, and cetaceans different?
7 What behavioral mechanisms do dolphins use to reduce the energy expended when swimming?
8 How does lift and drag generate a thrust force for propulsion during swimming?
9 What is the difference between the feeding modes of the rorqual (Balaenopteridae) and balaenid whales (Balaenidae)?
10 Why does drag increase with rorquals during lunge feeding?

CRITICAL THINKING

1 What specific shape would you expect to find in the body of marine mammals for rapid swimming?
2 What mechanisms can help to increase forging efficiency during lunge feeding?
3 Why do pinnipeds have a different means of generating propulsive thrust from that of cetaceans and sirenians?

FURTHER READING

1. Berta, A. 2012. *The Rise of Marine Mammals: 50 Million Years of Evolution*. Baltimore: Johns Hopkins University Press.
2. Fish, F.E. 2016. Secondary evolution of aquatic propulsion in higher vertebrates: Validation and prospect. *Integrative and Comparative Biology* 56(6): 1285–1297.
3. Fish, F.E., L.E. Howle, and M.M. Murray. 2008. Hydrodynamic flow control in marine mammals. *Integrative and Comparative Biology* 211: 1859–1867.
4. Gough, W.T., P.S. Segre, K.C. Bierlich, D.E. Cade, J. Potvin, F.E. Fish, J. Dale, J. di Clemente, A.S. Friedlaender, D.W. Johnston, S.R. Kahane-Rapport, J. Kennedy, J.H. Long, M. Oudejans, G. Penry, M.S. Savoca, K.A. Simon, F. Visser, D.N. Wiley, and J.A. Goldbogen. 2019. Scaling of swimming performance in baleen whales. *Journal of Experimental Biology* 222: jeb.204172.
5. Howell, A.B. 1930. *Aquatic Mammals*. Springfield, IL: Charles C. Thomas.
6. Weber, P.W., L.E. Howle, M.M. Murray, and F.E. Fish. 2009. Lift and drag performance of odontocete cetacean flippers. *Journal of Experimental Biology* 212: 2149–2158.
7. Williams, T.M. 1999. The evolution of cost efficient swimming in marine mammals: Limits to energetic optimization. *Philosophical*

Transactions Royal Society of London B Biological Sciences 353: 1–9.

8. Woodward, B.L., J.P. Winn, and F.E. Fish. 2006. Morphological specialization of baleen whales according to ecological niche. *Journal of Morphology* 267: 1284–1294.

GLOSSARY

Acceleration reaction: Changes in the kinetic energy of water accelerated by the action of a propulsive body structure where the force is dependent on the added mass.

Added mass drag: Extra drag due to acceleration of the mass of the body and the mass of fluid entrained with the body.

Adverse pressure gradients: Pressure in the area posterior of the maximum thickness of a streamlined body increases posteriorly. This impedes the flow and causes flow separation.

Angle of attack: Incident angle between the flow and a lifting surface.

Aspect ratio: Ratio of the square of the length of an appendage to the planar surface area of the appendage.

Baleen: Triangular thin plates made of keratinous tissue suspended from the palatal gingiva on the roof of the mouth that are found in mysticete whales.

Bluff body: A body where the length is smaller than the maximum girth.

Boundary layer: Region of flow next to a body that is slower than the outer freestream. The slower flow is due to viscosity.

Drag: A hydrodynamic force that is the resistance to forward motion caused by frictional (shear), pressure differences, and other body-generated forces that may generate large-scale vorticity.

Eddies: Small-scale rotating parcels of flow.

Fineness ratio: Ratio of length to maximum width of an elongate body.

Flow separation: When the boundary layer detaches from the surface of the body and mixes with the freestream flow.

Free-riding: Using external energy sources to lower the energetic cost of swimming.

Freestream: Freestream represents the flows **upstream**, and about to interact with the object moving in the fluid. Flows of fluid particles moving along smooth curvilinear paths; usually traced by continuous lines.

Friction drag: The drag force arising from the viscous friction among adjacent parcels of fluids.

Fusiform: Shape of a slender and tapered body (forward and aft).

Hydrodynamics: Scientific discipline devoted to the study fluid flows, either stand-alone or while moving past solid objects.

Hydrofoil: A lifting surface that operates in water.

Induced drag: A form of drag created by the generation of large-scale and (longitudinally) stretched vortices at the tip of flukes, flippers, and wings. Caused by the pressure difference atop and below the tips.

Laminar: Smooth flow where the particles of fluid follow along streamlines without crossing.

Laminar-to-turbulent transition: The change from laminar to turbulent flow either in the boundary layer or outer freestream flow. The change is associated with an increase in friction drag.

Lift: A hydrodynamic force that is perpendicular to the drag.

No-slip condition: Where the fluid in direct contact with a solid object is not moving. Created by the cohesive force(s) between the atoms of a solid surface and those of the fluid moving past and/or against the surface.

Particle image velocimetry: A laboratory technique used to visualize water movements and to measure water velocities.

Pelagic prey: Prey organisms that live in the open ocean.

Pressure drag: The drag force generated by the lower pressure in the near wake, in relation to the (higher) pressure applied at the front.

Strouhal number: A dimensionless number describing flow oscillation performance.

Thunniform: Swimming motions like a tuna where thrust is generated by an oscillating wing-like caudal propulsor (e.g., flukes).

Turbulent: Chaotic flow with mixing where the fluid is continually undergoing changes in the direction and magnitude.

Ventilation drag: The drag force created by body parts breaking the water surface and resulting in the lifting and depression of the mean water surface ahead and aft of the body parts.

Vortices: Flows moving along circular paths; usually of larger sizes than eddies with a cyclonic flow.

Viscosity: A cohesive force among same-type fluid particles; causes friction among adjacent fluid parcels moving at different speeds and directions.

Wave drag: The drag force created by body parts breaking the sea surface and resulting in waves attached to said parts; a wave-making resistance.

REFERENCES

Abrahms, B., E.L. Hazen, E.O. Aikens, et al. 2019. Memory and resource tracking drive blue whale migrations. *Proceedings of the National Academy of Sciences* 116(12): 5582–5587.

Ahlborn, B.K., R.W. Blake, and K.H. Chan. 2009. Optimal fineness ratio for minimum drag in large whales. *Canadian Journal of Zoology* 87(2): 124–131.

Bose, N. and J. Lien. 1990. Energy absorption from ocean waves: A free ride for cetaceans. *Proceedings of the Royal Society of London B* 240: 591–605.

Cade, D.E., N. Carey, P. Domenici, et al. 2020. Predator-informed looming stimulus experiments reveal how large filter feeding whales capture highly maneuverable forage fish. *Proceedings of the National Academy of Sciences* 117(1): 472–478.

Feldkamp, S.D. 1987a. Swimming in the California sea lion: Morphometrics, drag and energetics. *Journal of Experimental Biology* 131: 117–135.

Feldkamp, S.D. 1987b. Foreflipper propulsion in the California sea lion, *Zalophus californianus*. *Journal of Zoology, London* 212: 43–57.

Fish, F. E. 1984. Mechanics, power output, and efficiency of the swimming muskrat (*Ondatra zibethicus*). *Journal of Experimental Biology* 110: 183–201.

Fish, F.E. 1993. Influence of hydrodynamic design and propulsive mode on mammalian swimming energetics. *Australian Journal of Zoology* 42: 79–101

Fish, F.E. 1996. Transitions from drag-based to lift-based propulsion in mammalian aquatic swimming. *American Zoologist* 36(5): 628–641.

Fish, F.E. 1998. Comparative kinematics and hydrodynamics of odontocete cetaceans: Morphological and ecological correlates with swimming performance. *Journal of Experimental Biology* 201(20): 2867–2877.

Fish, F.E. 2004. Structure and mechanics of nonpiscine control surfaces. *IEEE Journal of Oceanic Engineering* 28(3): 605–621.

Fish, F.E. 2006. The myth and reality of Gray's paradox: Implication of dolphin drag reduction for technology. *Bioinspiration & Biomimetics* 1: R17–R25.

Fish, F.E., K.T. Goetz, D.J. Rugh, et al. 2013. Hydrodynamic patterns associated with echelon formation swimming by feeding bowhead whales (*Balaena mysticetus*). *Marine Mammal Science* 29(4): E498–E507.

Fish, F.E. and C.A. Hui. 1991. Dolphin swimming: A review. *Mammal Review* 21: 181–196.

Fish, F.E., S. Innes, and K. Ronald. 1988. Kinematics and estimated thrust production of swimming harp and ringed seals. *Journal of Experimental Biology* 137: 157–173.

Fish, F.E., P. Legac, T.M. Williams, et al. 2014. Measurement of hydrodynamic force generation by swimming dolphins using bubble DPIV. *Journal of Experimental Biology* 217(2): 252–260.

Fish, F.E. and J. Rohr. 1999. Review of dolphin hydrodynamics and swimming performance. *SPAWARS System Center Technical Report* 1801, San Diego, CA.

Friedman, C. and M.C. Leftwich. 2014. The kinematics of the California sea lion foreflipper during forward swimming. *Bioinspiration & Biomimetics* 9(4): 046010.

Goldbogen, J.A. 2010. The ultimate mouthful: Lunge feeding in rorqual whales: The ocean's depths have long shrouded the biomechanics behind the largest marine mammals' eating methods, but new devices have brought them to light. *American Scientist* 98(2): 124–131.

Goldbogen, J.A., D.E. Cade, J. Calambokidis, et al. 2017. How baleen whales feed: The biomechanics of engulfment and filtration. *Annual Review of Marine Science* 9(1): 367–386

Goldbogen, J.A., D.E. Cade, D.M. Wisniewska, et al. 2019. Why whales are big but not bigger: Physiological drivers and ecological limits in the age of ocean giants. *Science* 366(6471): 1367–1372.

Goldbogen, J.A., F.E. Fish, and J. Potvin. 2016. Hydrodynamics. In *Marine Mammal Physiology: Requisites for Ocean Living*, ed. M.A. Castellini and J-A. Mellish. Boca Raton: CRC Press.

Goldbogen, J.A. and P.T. Madsen. 2018. The evolution of foraging capacity and gigantism in cetaceans. *Journal of Experimental Biology* 221(11): jeb166033.

Goldbogen, J.A., J. Potvin, and R.E. Shadwick. 2010. Skull and buccal cavity allometry increase mass-specific engulfment capacity in fin whales. *Proceedings of the Royal Society B: Biological Sciences* 277(1683): 861–868.

Gough, W.T., D.E. Cade, M.F. Czapanskiy, et al. 2022. Fast and furious: Energetic tradeoffs and scaling of high-speed foraging in rorqual whales. *Integrative Organismal Biology* 4(1): https://doi.org/10.1093/iob/obac038.

Gough, W.T., H. Smith, M. Savoca, et al. 2021. Scaling of oscillatory kinematics and Froude efficiency in baleen whales. *Journal of Experimental Biology* 224: jeb237586.

Gray, J. 1936. Studies in animal locomotion VI. The propulsive powers of the dolphin. *Journal of Experimental Biology* 13: 192–199.

Hertel, H. 1966. *Structure, Form, Movement.* New York: Reinhold.

Innes, H.S. 1984. Swimming energetics, metabolic rates and hind limb muscle anatomy of some phocid seals. Ph.D. Dissertation, University of Guelph, Ontario, Canada.

Kojeszewski, T. and F.E. Fish. 2007. Swimming kinematics of the Florida manatee (*Trichechus manatus latirostris*): Hydrodynamic analysis of an undulatory mammalian swimmer. *Journal of Experimental Biology* 210: 2411–2418.

Kooyman, G.L. 1973. Respiratory adaptations in marine mammals. *American Zoologist* 13: 457–468.

Lang, T.G. and K. Pryor. 1966. Hydrodynamic performance of porpoises (*Stenella attenuate*). *Science* 152: 531–533.

Leahy, A.M., F.E. Fish, S.J. Kerr, et al. 2021. The role of California sea lion (*Zalophus californianus*) hindflippers as aquatic control surfaces for maneuverability. *Journal of Experimental Biology* 224: jeb243020.

Lighthill, J. 1975. *Mathematical Biofluiddynamics.* Philadelphia: Society for Industrial and Applied Mathematics.

Moore, M.J., T. Hammar, J. Arruda, et al. 2011. Hyperbaric computed tomographic measurement of lung compression in seals and dolphins. *Journal of Experimental Biology* 214: 2390–2397.

Noren, S.R. 2008. Infant carrying behaviour in dolphins? Costly parental care in an aquatic environment. *Functional Ecology* 22: 284–288.

Norris, K.S. and J.H. Prescott. 1961. Observations on Pacific cetaceans of California and Mexican waters. *University of California Publications in Zoology* 63: 291–402.

Pagano, A.M., A. Cutting, N. Nicassio-Hiskey, et al. 2019. Energetic costs of aquatic locomotion in a subadult polar bear. *Marine Mammal Science* 35(2): 649–659.

Potvin, J., D.E. Cade, A.J. Werth, et al. 2021. Rorqual lunge-feeding energetics near and away from the kinematic threshold of optimal efficiency. *Integrative Organismal Biology* 3(1): obab005.

Potvin, J., J.A. Goldbogen, and R.E. Shadwick. 2009. Passive versus active engulfment: Verdict from trajectory simulations of lunge-feeding fin whales Balaenoptera physalus. *Journal of the Royal Society Interface* 6(40): 1005–1025.

Potvin, J. and A.J. Werth. 2017. Oral cavity hydrodynamics and drag production in Balaenid whale suspension feeding. *PloS One* 12(4): e0175220.

Riedman, M. 1990. *The Pinnipeds: Seals, Sea Lions, and Walruses.* Los Angeles: University of California Press.

Rohr, J.J. and F.E. Fish. 2004. Strouhal numbers and optimization of swimming by odontocete cetaceans. *Journal of Experimental Biology* 207(10): 1633–1642.

Segre, P.S., D.E. Cade, F.E. Fish, et al. 2016. Hydrodynamic properties of fin whale flippers predict maximum rolling performance. *Journal of Experimental Biology* 219(21): 3315–3320.

Segre, P.S., W.T. Gough, E.A. Roualdes, et al. 2022. Scaling of maneuvering performance in baleen whales: Larger whales outperform expectations. *Journal of Experimental Biology* 225(5): jeb243224.

Shorter, A.K., M.M. Murray, M. Johnson, et al. 2014. Drag of suction cup tags on swimming animals: Modeling and measurement. *Marine Mammal Science* 30(2): 726–746.

Simon, M., M. Johnson, and P.T. Madsen. 2012. Keeping momentum with a mouthful of water: Behavior and kinematics of humpback whale lunge feeding. *Journal of Experimental Biology* 215(21): 3786–3798.

Tarasoff, F.J., A. Bisaillon, J. Pierard, et al. 1972. Locomotory patterns and external morphology of the river otter, sea otter, and harp seal (Mammalia). *Canadian Journal of Zoology* 50: 915–929.

van der Hoop, J.M, A. Fahlman, T. Hurst, et al. 2014. Bottlenose dolphins modify behavior to reduce metabolic effect of tag attachment. *Journal of Experimental Biology* 217(23): 4229–4236.

Vennell, R., D. Pease, and B. Wilson. 2006. Wave drag on human swimmers. *Journal of Biomechanics* 39(4): 664–671.

Vogel, S. 1994. *Life in Moving Fluids: The Physical Biology of Flow*, 2nd ed. Princeton,: Princeton University Press.

Wainwright, D.K., F.E. Fish, S. Ingersoll, et al. 2019. How smooth is a dolphin? The ridged skin of odontocetes. *Biology Letters* 15: 20190103.

Webb, P.W. 1975. Hydrodynamics and energetics of fish propulsion. *Bulletin of the Fisheries Research Board of Canada* 190: 1–158.

Weihs, D. 2004. The hydrodynamics of dolphin drafting. *Journal of Biology* 3(2): 1–16.

Werth, A.J. and H. Ito. 2017. Sling, scoop, and squirter: Anatomical features facilitating prey transport, processing, and swallowing in rorqual whales (Mammalia: Balaenopteridae). *The Anatomical Record* 300(11): 2070–2086.

Werth, A.J., J. Potvin, R.E. Shadwick, et al. 2018. Filtration area scaling and evolution in mysticetes: Trophic niche partitioning and the curious cases of sei and pygmy right whales. *Biological Journal of the Linnean Society* 125(2): 264–279.

Williams, T.M. 1989. Swimming by sea otters: Adaptations for low energetic cost locomotion. *Journal of Comparative Physiology A* 164: 815–824.

Williams, T.M. 2001. Intermittent swimming by mammals: A strategy for increasing energetic efficiency during diving. *American Zoologist* 41(2): 166–176.

Williams, T.M., R.W. Davis, L.A. Fuiman, et al. 2000. Sink or swim: Strategies for cost-efficient diving by marine mammals. *Science* 288: 133–136.

Williams, T.M., W.A. Friedl, M.L. Fong, et al. 1992. Travel at low energetic cost by swimming and wave-riding bottlenose dolphins. *Nature* 355: 821–823.

Williams, T.M. and G.L. Kooyman. 1985. Swimming performance and hydrodynamic characteristics of harbor seals *Phoca vitulina*. *Physiological Zoology* 58: 576–589.

Shawn R. Noren
Institute of Marine Science, University of California, Santa Cruz, CA

Jennifer L. Maresh
Department of Biology, West Chester University, West Chester, PA

Terrie M. Williams
Department of Ecology and Evolutionary Biology, University of California, Santa Cruz, CA

Shawn R. Noren

Dr. Noren is a marine mammal physiological ecologist with 25 years of experience. She has studied numerous dolphin and seal species, as well as belugas and walruses. She pioneered research on the development of diving physiology in cetaceans, and her research makes connections between morphology, biochemistry, and physiology to larger-scale functions and performance. To address these types of questions, she combines experimental and theoretical techniques in the laboratory, field, and at zoological parks. Ultimately, she aims to understand how marine mammal populations are impacted by rapid changes in the ocean associated with global climate change and other anthropogenic disturbances.

Jennifer L. Maresh

Dr. Maresh is a vertebrate biologist with 15 years of experience studying the intersection between animal ecology, physiology, and behavior. She has worked primarily with marine mammals, including elephant seals, Weddell seals, sea lions, bottlenose dolphins, and right whales, seeking to understand the energetic underpinnings behind why animals make the decisions they do regarding foraging and reproduction. Her fieldwork has brought her on an incredible journey from her humble beginnings in the suburbs of Philadelphia to field sites around the world, including coastal New England, North Carolina, Florida, California, and Antarctica. Today, you can find Jen and her students seeking to uncover and explain patterns of maternal energy investment in groups as diverse as marine mammals and lizards!

Terrie M. Williams

Dr. Williams is a wildlife ecophysiologist with 40 years of research experience studying terrestrial and aquatic mammals including sea otters, narwhals, polar bears, African lions, and Weddell seals. She directs the Integrative and Comparative Energetics (ICE) Lab and the Marine Mammal Physiology Project at UCSC. Her research expeditions have taken her from Arctic and Antarctica polar environments to

DOI: 10.1201/9781003297468-2

African savannahs and California mountains. She developed many of the instruments used in her studies. Terrie and her students strive to understand the ecological role of large mammals and the physiological adaptations necessary for species survival in a world changed by human activities.

Driving Question: How do marine mammals exercise while holding their breath?

INTRODUCTION

Anyone who has swam in a pool or in the ocean is well aware that moving quickly through water is more difficult than moving through air. It is one reason that water aerobics are so popular; humans tend to burn a lot of calories when exercising in water (thus, assisting in rapid weight loss). Now imagine trying to perform this exercise while holding your breath. That is the major challenge faced by marine mammals, how to swim fast and dive deep when the timing between breaths may range from a few minutes to over three hours. We learned in **Chapter 1** that marine mammals have streamlined bodies, which create low-drag profiles; this promotes cost-efficient swimming, especially when compared to surface-swimming human athletes and dog-paddling canines. Such cost-efficiency is critical for saving limited on-board oxygen stores that support cellular processes during prolonged breath-holds when diving.

Ultimately, the metabolic rates of marine mammals will determine the pace of oxygen utilization during rest and exercise, and thus, dictate how long each species can hold its breath. In this chapter, we will explore the various levels of metabolism that allow marine mammals to sleep, forage, and travel across the oceans. Key questions that we will address include (1) What are the resting metabolic rates (RMRs) of marine mammals, and are they similar to or different from those of terrestrial mammals?, (2) How does the energy expenditure of marine mammals change with swimming speed and

dive duration?, and (3) How much energy do marine mammals expend in a day to survive in today's changing oceans?

THE CHALLENGE OF EXERCISING WHILE DIVING

All marine mammals are air-breathing vertebrates, whose ancient ancestors were terrestrial mammals (Berta, Sumich, and Kovacs 2015). Thus, like us, marine mammals breathe air through their lungs, transfer oxygen from the air into the cardiovascular system, which then moves the oxygen into the tissues to support cellular functions (**Figure 2.1**). As marine mammal ancestors became better divers, there were associated adaptations in this pathway that allowed them to increase their breath-hold capacity. Adaptations included modifications to morphological, physiological, biochemical, and molecular mechanisms that originally supported locomotion on land (Williams 1999). Over time, these modifications enabled marine mammals to efficiently use oxygen stored in the lungs, blood, and muscles to support dives to remarkable depths that can never be approached by other mammalian groups (see **Chapter 3**).

This transition from land to sea in mammals creates a paradox concerning exercise while diving. This is because two seemingly conflicting physiological responses occur simultaneously when a marine mammal submerges (Williams et al. 2017a). The first is the **dive response**, which is characterized by the cessation of breathing (**apnea**), a slowing of the heart (**bradycardia**), peripheral vasoconstriction, and metabolic down-regulation of non-critical tissues. **Chapter 3** has detailed information about this dive response. The second physiological response involved is the **exercise response** of mammals, which in active terrestrial mammals promotes an increase in **metabolism** (oxygen consumption), heart rates (**tachycardia**), and respiratory rates. How can both responses occur when a submerged marine mammal is actively swimming to migrate, pursue underwater prey,

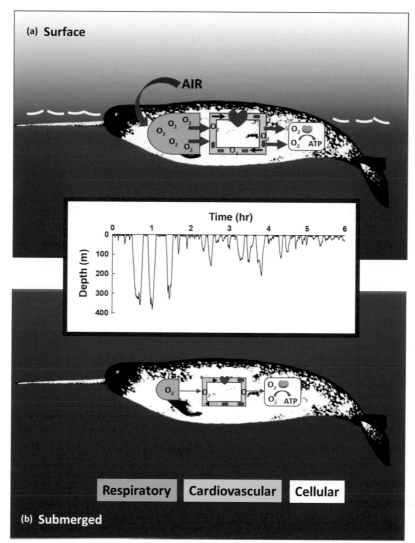

Figure 2.1 The pathway for oxygen for marine mammals. The top illustration (a) presents the pathway when a narwhal is breathing on the water surface, while the lower panel (b) shows the relative changes when the animal is holding its breath during a dive. The simplified model indicates the movement of oxygen as it travels sequentially through three major compartments: the respiratory system, the cardiovascular system, and into the cells of the tissues where O_2 is taken up by the mitochondria to produce ATP. Here the size of each compartment and the arrows indicate the relative change in oxygen transport between breathing and breath-holding. When at sea, marine mammals typically alternate between these states as the animals perform sequential dives and post-dive recovery periods, as shown by a typical time-depth record for a free-ranging narwhal (middle panel).

or avoid predation? We might expect that marine mammals would avoid energetically costly underwater behaviors to prevent depleting their valuable onboard oxygen reserves. Yet, we find that many marine mammal species exhibit exceptional athletic performances. One such underwater athlete is the short-finned pilot whale (*Globicephala macrorhynchus*), a cetacean species that can swim at speeds of up to 9 meters per second and has been named the cheetahs of the deep sea (Aguilar de Soto et al. 2008). Some of the largest whales, including humpback whales (*Megaptera novaeangliae*) and blue whales (*Balaenoptera musculus*), also perform complex, energetically costly maneuvers when opening their enormous mouths to forage underwater (**Chapter 1**; Goldbogen et al. 2008).

Studies that have simultaneously measured heart rate, swimming performance, and **stroke frequency** (how quickly flippers move back and forth, or flukes move up and down) in diving pinnipeds and cetaceans provide clues about the ability of marine mammals to balance the different physiological demands of the dive and exercise responses (Noren et al. 2012; Williams et al. 2022). Rather than a single dominant response, there is an interplay between these two responses, where both the depth of the dive and the intensity of exercise alter the level of bradycardia. The deeper marine mammals dive, the lower the minimum heart rate achieved (**Figure 2.2a, b and c**). However, the intensity of swimming exercise as the animal dives is also a factor that can modify its heart rate. Periods of high-stroke frequency result in a subtle relaxation of the diving bradycardia; in other words, diving heart rate can shift upward as the level of underwater exercise increases.

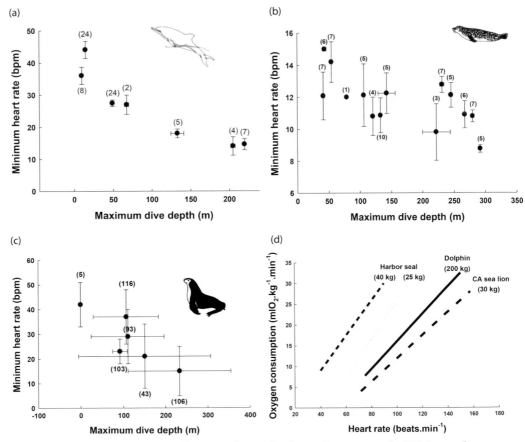

Figure 2.2 Cardiac responses to diving and exercise in marine mammals. Minimum heart rate (level of bradycardia achieved) is plotted in relation to maximum dive depth for (a) bottlenose dolphins, (b) Weddell seals, and (c) California sea lions. The number of dives is indicated in the parentheses. Panel (d) shows the relationship between heart rate and oxygen consumption during exercise for pinnipeds swimming in a flume and dolphins pushing on a load cell. (Data are from sources in the text and the reference section.)

Conversely, periods of low-stroke frequency, as occur during low energy, cruise swimming, are associated with the most intense levels of bradycardia; here the dive response dominates and heart rate is low. Such a physiological interplay has been demonstrated in a wide range of species, including bottlenose dolphins (*Tursiops truncatus*), narwhals (*Monodon monoceros*), Weddell seals (*Leptonychotes weddellii*), gray seals (*Halichoerus grypus*), California sea lions (*Zalophus californianus*), and even human breath-hold divers.

The variation in heart rate throughout a dive will reflect the overall energetic cost of exercise. This is because heart rate is directly correlated with oxygen consumption in marine mammals (**Figure 2.2d**), as is true of terrestrial mammals.

ENERGETIC COSTS OF MARINE MAMMALS

Determination of how much energy an animal needs to survive requires an understanding of the various factors that drive metabolism. The total **energetic costs** of a marine mammal can be determined by measuring its oxygen consumption, or by monitoring the total number of calories it consumes. These costs are important for (1) defining the animal's **dive capacity** (how long it can hold its breath, and thus, how deep an individual marine mammal can dive) and (2) delimiting the amount of food that is required within foraging grounds to ensure an individual's survival and successful reproduction, to promote population sustainability.

Typically, total energetic costs are partitioned among four main expenditures. First are the costs associated with supporting maintenance functions (collectively encompassing the **basal metabolic rate**) that represent the energy baseline for survival. Second, additional energy is needed to support activity costs for moving through the water. The third expenditure includes the additional energy required to support a variety of acute physiological functions.

These include energy used in the digestion and assimilation of food, growth, reproduction, and thermoregulation if the animal is outside of its **thermal neutral zone** (defined as the range of temperatures over which an animal can maintain a normal body temperature without requiring energy beyond its normal basal metabolic rate). For marine mammals with hair, there is a fourth component to account for in their total energetic costs, and this is associated with an annual molt.

Specific energetic costs for each of these four functions, and the proportion of total energy allocated to each, varies across age, class, and species. Some of these topics are discussed in detail in subsequent chapters. Here we will only address the first two major costs, that is, the energetic cost of resting and of activity (swimming and diving). With this foundation, we then discuss the total daily costs incurred by wild marine mammals, termed its **field metabolic rate (FMR)**, as well as behavioral strategies that marine mammals may use to help reduce total energetic costs while swimming and diving.

RESTING METABOLIC RATES AND COSTS

RMR, sometimes called the **basal metabolic rate**, is a measure of the minimum energy requirements for an animal that is awake and alert but otherwise not engaged in any activity that might elevate metabolism above baseline survival levels (Kleiber 1975). This metabolic rate represents the maintenance costs of the animal. To accurately measure this cost, "Kleiber standards" must be met, wherein the animal is a nonreproductive adult (i.e., no growth, pregnancy, or lactation costs), fasted, relaxed but not asleep, and in a thermally neutral environment, meaning it does not need to increase its metabolic rate to account for the extra energy required to warm-up or cool down its body. These "Kleiber standards" were originally developed for terrestrial mammals whose metabolic rates are measured in air; applying them to marine mammals has led to the questions, "What represents 'resting' metabolism for marine mammals?" Should the "resting" metabolism for marine mammals be measured in air or water?

Several factors make answering these questions difficult for animals built to live in water. Marine mammals have a thick blubber layer that enables them to stay warm when immersed but may also result in them overheating while on land (discussed in **Chapter 5**). As a result, measurements of marine mammal metabolism in air may not meet "Kleiber standards", especially if the animals are not in a thermally neutral environment. However, measurements of metabolism in water may be confounded by the dive response since face immersion alone is enough to trigger the dive response in marine mammals (Ridgway, Carder, and Clark 1975). In this case, metabolic rate may be decreased. Because marine mammals spend most (if not all) of their lives at sea, is it possible that metabolism measured during submergence may represent the true baseline energetic cost for them. The RMR measured in a marine mammal is also confounded by other factors, such as whether or not the animal is molting or if the animal is in a season where they naturally fast for a prolonged period of time. For these reasons, comparisons of the baseline resting costs between the various marine mammal groups, and between marine and terrestrial mammals, can be challenging.

Despite these limitations we have attempted to identify some general trends in the RMRs of the major marine mammal groups (**Figure 2.3a**). The relationship between body mass and RMR for marine mammals is best described by a log-log function (see Equations section for more details.) A recent review by Noren of the RMR of marine mammals noted that the mean RMR rate for odontocetes (toothed whales) was 2.18 ± 0.78 times Kleiber, which was intermediate in value compared to that of otariids (sea lions and fur seals, 3.01 ± 0.37 times Kleiber) and phocids (seals, 1.32 ± 0.31 times Kleiber), respectively. The lowest relative mean RMR

has been found in manatees (0.35 ± 0.19 times Kleiber), which is undoubtedly related to their herbivorous and sedentary lifestyle (John 2020). What this means is that most marine mammals have about 1.5–3 times the RMR of similar-sized terrestrial mammals, with exceptions found for some slow-moving species like the manatee. It is important to note, however, that the RMR of an adult mysticete has never been measured, so the effects of extreme body size on the metabolic rate of an enormous animal like the blue whale at more than 180,000 kg (Folkens et al. 2002) are unknown.

Many explanations have been proposed for the observed elevation in RMRs of marine mammals compared to terrestrial mammals. One widely accepted hypothesis suggests that the metabolic rates of marine mammals are elevated to offset the high cost of endothermy while living in water (e.g., Hudson, Isaac, and Reuman 2013). Alternatively, others suggest that maintaining homeothermy in water is only a problem for the smallest marine mammals (Porter and Kearney 2009) so that other traits in marine mammals set metabolism (e.g., Heim et al. 2015). In these later studies, high-metabolic rates were considered an **exaptation** rather than an adaptation; their ancestral carnivores that invaded the marine environment simply had an elevated metabolism that led to a competitive advantage over other aquatic species with low metabolic rates. Additional data that are collected under "Kleiber standards" are needed from a wider range of marine mammals to rectify this debate.

SWIMMING AND DIVING METABOLIC RATES AND COSTS

As observed for most vertebrate animals, the metabolic rate of marine mammals increases in response to exercise. However, the pattern of change in energy expenditure as physical exertion becomes progressively more difficult differs significantly when comparing mammals adapted for swimming to those adapted for

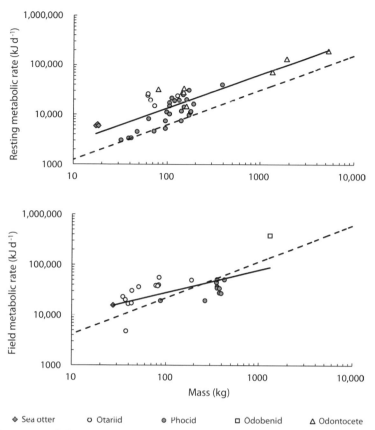

Figure 2.3 **(a) Resting and (b) field metabolic rates of marine mammals (solid lines) compared to terrestrial mammals (dashed lines). Each point denotes a separate metabolic measurement for an individual marine mammal. Symbols representing the major marine mammal groups are shown on the bottom. (Data from Maresh 2014)**

running (**Figure 2.4**). In addition, the energetic demands incurred by mammals like sea otters (*Enhydra lutris*), humans, and polar bears (*Ursus maritimus*) that swim on the water surface, differ from those mammals that swim fully submerged. These energetic differences are primarily due to the unique physical forces experienced when trying to move through water, air, or at the interface between the two.

For swimmers, hydrodynamic drag in its many forms is the dominant force to be overcome during exercise; for runners, gravity and friction dominate. In **Chapter 1**, we learned that the drag of a swimmer increases exponentially with its locomotor speed. Based on physics, we would expect the shape of the relationship describing oxygen consumption and

swim speed to show an exponential rise as performance level increases. In fact, such curvilinear increases in the rate of oxygen consumption (**VO₂**, that is the volume of oxygen consumed by an animal over a unit time) as the animals swim faster have been reported for phocid seals (Davis, Williams, and Kooyman 1985), sea lions (Feldkamp 1987), dolphins (Yazdi, Kilian, and Culik 1999), and killer whales (Kreite 1994; **Figure 2.4a**). In contrast, the VO₂ of runners typically increases linearly with speed until the aerobic limits of the animal are reached (Taylor, Heglund, and Maloiy 1982).

The shape of these oxygen consumption-speed relationships has a marked effect on the energy that is expended to move a set distance through water or air, and determines

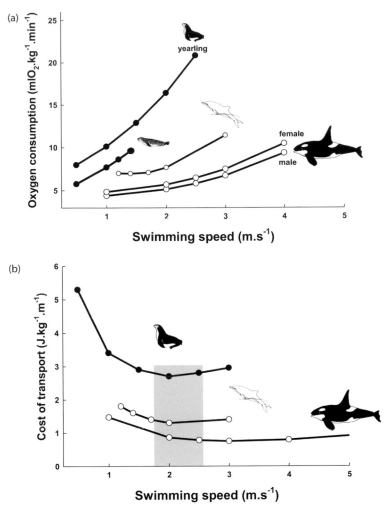

Figure 2.4 (a) Oxygen consumption and (b) cost of transport in relation to swimming speed for marine mammals. Panel (a) shows the characteristic curvilinear increase in oxygen consumption with swimming speed in pinnipeds and cetaceans. This results in a non-linear relationship between cost of transport (COT) and speed as shown in panel (b). Note that the minimum COT occurs in the mid-range of speeds as denoted by the gray bar.

the preferred locomotor speeds of each species (**Figure 2.4b**). For example, metabolic costs are low when marine mammals swim at slow or routine speeds. In fact, the swimming VO_2 measured for oceanic bottlenose dolphins swimming at routine speeds of 2.0 m.s^{-1} are only slightly greater than the resting metabolic costs of the animal. Some species, such as northern elephant seals (*Mirounga angustirostra*) may be hypo-metabolic (more energy efficient than while resting on land) during their long-

distance, at-sea foraging migrations (Maresh et al. 2015). As a result, these routine speeds can be performed for prolonged periods of time, enabling marine mammals to transit for hours between foraging areas or migrate for months with minimal energetic cost.

Another effect of the curvilinear relationship between oxygen consumption and speed is that swimming at very low or high speeds is enormously expensive in terms of the amount of energy needed to move a unit of body mass

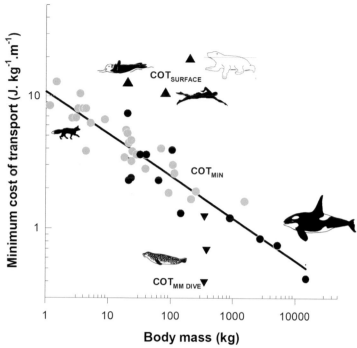

Figure 2.5 Cost of transport in relation to body mass for mammals. Minimum COT values are compared for swimming marine (black circles) and running terrestrial (gray circles) mammals. The black line denotes the least squares regression for the relationship (presented in the text as Equation 2.3 for submerged marine mammal swimmers). Note the comparatively high-minimum costs for surface swimmers (upright triangles) including sea otters, humans, and polar bears. Conversely, diving Weddell seals (downward triangles) often show lower transport costs when submerged.

a unit distance (called the **cost of transport, COT**; Schmidt-Nielsen 1972; **Figures 2.4b** and **2.5**). Equivalent to the miles per gallon (MPG) fuel economy rating of automobiles, COT indicates the energy expended per meter traveled. COT increases rapidly with small changes in speed as marine mammals move outside of their preferred ranges. Swimming slow costs little energy but the animals gain little distance, while swimming fast incurs exponentially rising energetic costs relative to the amount of distance gained. Both result in an elevated COT. This U-shaped COT with speed for swimming marine mammals (**Figure 2.4b**) is in marked contrast to the progressive decline in COT with speed that occurs for running mammals (Taylor, Schmidt-Nielsen, and Raab 1970). As a result, high-speed swimming by most marine mammals, even small, athletic dolphins,

are of comparatively short duration unless the animals employ behavioral maneuvers that help them circumvent the exceptionally high energetic transport costs encountered during fast swimming (see section Behavioral Strategies). For example, short-finned pilot whales can sustain their unusual 9.0 m.s⁻¹ sprints for chasing prey for only 20 to 80 seconds (Aguilar de Soto et al. 2008).

Importantly, for each species, there is a **minimum COT (COT$_{MIN}$)** that defines the range of speeds across which the animals move the furthest for the least investment of energy (Schmidt-Nielsen 1972). Moving at **COT$_{MIN}$** is energetically efficient for marine mammals. From **Figure 2.4b**, it is clear that COT$_{MIN}$ occurs within the trough of the U-shaped curve relating COT to swimming speed, and that it decreases with body mass. Interestingly, for many of the marine mammals measured to date, this minimum occurs at approximately 1.5 – 2.5 m.s^{-1}, regardless of the size of the animal. Not surprisingly, the preferred or routine speeds measured for many marine mammals fall within or near this cost-efficient range of swimming speeds (see Chapter 1).

Like RMR, **COT$_{MIN}$** for marine mammals varies predictably with body mass (*Table 2.1*; Williams 1999). Remarkably, this relationship is indistinguishable from that describing the total cost of transport for running mammals, a testament to the shared ancestral lineages of highly active mammalian specialists. Terrestrial mammals, phocid seals, otariids, large and small odontocetes, and an estimate for a mysticete, the gray whale (*Eschrichtius*

robustus), all follow the same equation line with a few notable exceptions (**Figure 2.5**). Phocid seals tend to have lower transport costs compared to other marine mammals, especially when diving. This difference has been attributed to metabolic changes associated with the dive response as well as to the energy savings associated with extended periods of gliding that occur during deep dives. Conversely, sea otters and polar bears demonstrate COT$_{MIN}$ levels that are approximately three times predicted for other similar-sized marine mammals, when swimming on the water surface (Williams 1999; Pagano et al. 2018). The inefficient paddling swimming style of these animals, as well as wave drag associated with surface swimming instigate these high costs. For the same reasons,

Table 2.1 Energetic costs and swimming speeds of marine mammals

SPECIES	BODY MASS (kg)	VO$_{2REST}$ (mlO$_2$.kg^{-1}.min^{-1})	VO$_{2SWIM}$ (mlO$_2$.kg^{-1}.min^{-1})	COT$_{MIN}$ (J.kg^{-1}.m^{-1})	SPEED (m.s^{-1})	METHOD
Sea otter (surface)	20	13.5	29.6	12.6	0.8	Flume
Sea otter (submerged)	20	13.5	17.6	7.4	0.8	Flume
California sea lion	21		13.7	2.3	2.0	Flume
	23	6.3	22.0	2.8	2.6	Flume
	23	6.6	13.0	2.4	1.8	Flume
Harbor seal	32		23.6	3.6	2.2	Flume
	33	5.1	15.2	3.6	1.4	Flume
	63	4.6	9.6	2.3	1.4	Flume
Gray seal	104	7.7	15.0	3.9	1.3	Flume
Steller sea lion	116		24.9	4.3	2.0	Flume
	139		22.6	3.5	2.0	Flume
Bottlenose dolphin	145	4.6	8.1	1.3	2.1	Ocean swim
Bottlenose dolphin (pregnant)	162	6.7		1.2	2.5	Pool transit
Polar bear (surface)	200			19.1	<1.0	Flume
Killer whale	2738			0.84	3.1	Field respiration rate
	5153			0.75	3.1	Field respiration rate
Gray whale	15,000			0.4	2.1	Field respiration rate

Note: Oxygen consumption at rest (VO$_{2rest}$) was determined for animals resting on the water surface prior to exercise. Swimming values (VO$_{2swim}$) were determined during exercise in flumes or open water as indicated. VO$_2$ in mlO$_2$.kg^{-1}.min^{-1} was converted to metabolic energy (joules, J) assuming a caloric equivalent of 4.8 kcal per liter of O$_2$ and a conversion factor of 4.187×10^3 J.kcal^{-1}.

even the best Olympic swimming athletes have comparatively high swimming COT_{MIN}. These comparisons also demonstrate the challenges that ancestral marine mammals must have encountered when transitioning from land to sea (Williams 1999).

FIELD METABOLIC RATES AND COSTS

As might be expected, measuring the energetic costs of free-ranging marine mammals that spend more than 90% of their lives submerged is challenging. This has required scientists to develop and utilize a wide variety of methods to tackle the question, "What are the FMRs of marine mammals?" (see Toolbox section). Because energy expenditure is related to body size for wild mammals (Nagy 2005), the FMR of marine mammals (FMR_{MM}) can be estimated by using the allometric equation 4 at the end of this chapter. Here, FMR_{MM} has the units kJ per day and body mass is in kg (Maresh 2014). Based on this equation developed from measurements of field energetic costs for marine mammal ranging in body mass from 27 kg sea otters to 1310 kg walruses (*Odobenus rosmarus*), we find that field energy expenditure in the smallest marine mammals is higher than would be predicted for a terrestrial mammal. Conversely, FMR is lower than would be predicted in the largest marine mammals (**Figure 2.3b**). When the comparison is made between marine mammals and only terrestrial carnivores, the differences in FMRs often disappear (Maresh 2014). Although the intercept for the FMR to body mass relationships are not significantly different between the groups, the slope as indicated by the exponent is shallower for marine mammals (exponent = 0.45) than that for terrestrial mammals (exponent = 0.72). Based on this, it appears that mass-specific FMRs decline with increasing body size at a faster rate in marine mammals than in terrestrial mammals, as proposed by Boyd (2002). The implication is that energy expenditure in marine mammals up to approximately

250 kg will be elevated above that of a similar-sized terrestrial mammals. Yet, for marine mammals above 250 kg, the field metabolic rate is more economical than that of a similar-sized terrestrial mammal. Importantly, the diversity of marine mammal groups comprising the FMR relationship may be a critical underlying factor driving this trend. Thus, equation 4, while useful as a first approximation of FMR, should be considered an estimate of the diverse energy needs of the many marine species that make up this mammalian group.

Based on their taxonomic and ecological diversity, we might expect different metabolic adaptations for aquatic living by mustelids (otters), otariids, phocids, odontocetes, and mysticetes. Indeed, phylogeny does seem to explain much of the residual variation of FMR after body size is considered. Among marine mammals, phylogeny is highly correlated with activity levels, reproductive strategies, locomotor mechanics, and other physiological and ecological drivers of overall energy expenditure. For example, among the pinnipedia, phocid seals tend to swim at more economical speeds and engage in less energetically costly acrobatic maneuvers than otariids (**Chapter 1**), which is reflected in their FMR. Both athletic otariids and odonotocetes demonstrate comparatively high FMRs relative to the general mammalian trend. Mysticete whales are more likely to swim at low-cost cruising speeds rather than engage in costly sprints typical of small odontocetes. FMRs have not been directly measured in adult mysticetes. However, based on shared economical swimming patterns, it is reasonable to predict that the FMRs of this cetacean group may follow the trends of phocid seals rather than those of otariids. Using the same logic, we might predict that sirenians (manatees and dugongs), as slow swimmers, will have low FMRs compared to other marine mammal groups (John 2020). Not surprisingly, sea otters, as small-bodied surface swimmers, show higher than predicted FMRs among the marine mammals (Thometz et al. 2014).

Because body size and phylogeny are correlated, it is possible to assign the major marine mammal taxonomic groups to one of two distinct "pace of life" categories. Species with a fast pace of life tend to be relatively small, highly active swimmers, and shallow-diving, with a reproductive strategy that involves long lactation periods and little-to-no separation of foraging from lactation (i.e., **income breeders**). Membership in this group includes otariids, most odontocetes, and sea otters. Species with a slow pace of life tend to be relatively large, economical swimmers, and deep-diving, with a reproductive strategy that involves short lactation periods and geographical and temporal separation of foraging from lactation (i.e., **capital breeders**). Membership in this group includes phocids, and possibly some of the larger but inaccessible odontocetes and mysticetes. There are exceptions that fall outside these groupings. The walrus (an odobenid) has a long lactation period with no separation of foraging from lactation yet is an economical swimmer and thus has a low metabolic rate (as reviewed by Noren). Also, as the only herbivorous marine mammal group, sirenians represent a special case of a taxonomic group with exceptionally low energy costs below those of any similar-sized marine or terrestrial mammal. Tropical living, as exemplified by both sirenians and the Hawaiian monk seal (*Neomonachus schauinslandi*), may also contribute to a slower life pace.

BEHAVIORAL STRATEGIES TO REDUCE THE ENERGETIC COSTS OF SWIMMING AND DIVING

Marine mammals, unlike terrestrial mammals, are adapted to lower their cost of locomotion in water. Their torpedo-shaped body streamlines them, resulting in lower-drag costs in water compared to a similar-sized terrestrial mammal. In addition, marine mammals use several energy-saving behaviors, such as maintaining a submerged swimming position, employing burst-and-glide performance (unlike the steady-state stroking of swimming humans), and using changes in buoyancy for energetic advantages while diving (Williams, Fuiman, and Davis 2015).

Swimming three body diameters below the water surface is the most economical position since swimming near the surface incurs additional drag associated with the generation of waves (**Chapter 1**). For example, a sea otter swimming completely submerged has a 40% reduction in VO_2 compared to when it swims at the same speed at the water surface (Williams 1999). This impacts the minimum cost of transport, which is 69% higher at the surface compared to the cost while fully submerged (**Figure 2.5**). The necessity of breathing, however, requires marine mammals to periodically surface, which will intermittently preclude them from using sub-surface energy-saving behavior. Consequently, the surface intervals of marine mammals tend to be brief, ranging from <1 second for fast-swimming dolphins and otariids to 3–5 seconds and longer for harbor seals (*Phoca vitulina*) and large whales. Indeed, most marine mammals, with the exception of sea otters and polar bears, spend 90–95% of their time below the water surface when migrating across ocean basins, moving between prey patches, or even transiting short distances.

Despite the limited time marine mammals spend at the surface, there are behavioral "tricks" that marine mammals can employ to lower their energetics costs when in this position. These include **porpoising** (leaping out of the water) and **wave-riding** (like surfing). As the former name implies, this behavior is most often associated with small cetaceans, but some pinnipeds (e.g., sea lions, harbor seals, and fur seals) have also been observed performing this energy-saving strategy. At slower speeds, the cost of leaping out of the water is greater than the energy expended to overcome **wave drag** at the surface, so slow-swimming dolphins generally remain in the water when breathing. Conversely, at high speeds the relative energetic costs are reversed since drag increases exponentially with speed. Thus, the most energetically efficient strategy for fast-swimming marine mammals is to "porpoise" when taking a breath to avoid the problem of elevated drag near the water surface (Au and Weihs 1980). Wave-riding, a behavior described by sailors since Greek mythology times, is when dolphins position their bodies in the bow or stern wakes of boats and match vessel speed without moving their flukes (Scholander 1959). Marine mammals in this position benefit because the vessel sets the water around it in motion, which can then be taken advantage of to decrease the energy required for locomotion. Wave-riding bottlenose dolphins have lower heart rates, lower respiration rates, and by inference lower metabolic rates, and move twice as fast as the dolphin swimming near the surface on its own. Marine mammals have also been observed taking advantage of waves generated by other sources, such as wind, currents, and even larger whales (Würsig and Würsig 1979).

Cetacean calves employ an energy-saving trick similar to that of "wave-riding" marine mammals. When maintaining close proximity to their mothers in **echelon position (Figure 2.6)**, calves are able to increase their swim speed by 8% while reducing their swim effort as evident by gliding over a third of the time and by reducing fluke stroke amplitude by 22% when they did stroke (Noren et al. 2008). By drafting off a larger body, the calf is afforded an energetic savings. However, this does come with a cost to the mother (Noren 2008). Compared to periods of solitary swimming, a mother with a calf in echelon position has a 13% reduction in distance covered per stroke. The extra energetic burden for mothers swimming alongside their calf must be considered when quantifying energetic costs for mother cetaceans, but is often overlooked in ignored in bioenergetic models.

The interaction between buoyancy and hydrostatic pressure as marine mammals move through the water column while diving affords many opportunities for saving energy. Changes in buoyancy associated with hydrostatically induced **lung compression** at depth allow

Figure 2.6 Echelon swimming position for dolphin moms and calves. Here lateral (a) and overhead (b) views are presented for bottlenose dolphin mother-calf pairs. Echelon position is described as a calf in very close proximity to its mother's mid-lateral flank in the region near her dorsal fin. (Photo © and permission courtesy of Dolphin Quest Hawaii)

marine mammals to "turn off the motor" such that, during certain periods of the dive they can maintain forward motion without the energetic cost of active stroking (Skrovan et al. 1999). In this way, stroke frequency can be decoupled from speed during the descent portion of the dive as the animal incorporates prolonged (>12 s) periods of gliding (**Figure 2.7**), a variation of **burst-and-glide swimming** (Williams et al. 2000). For one elite diver, the Weddell seal, over 78% of the dive descent can be spent gliding rather than actively stroking. By incorporating this intermittent mode of swimming during the dive, Weddell seals have reductions of 9.2–59.6% in energetic costs, depending on depth. Many deep-diving pinnipeds, including Weddell seals (Williams et al. 2004) and elephant seals (Adachi et al. 2014), as well as diving cetaceans including bottlenose dolphins and blue whales (Williams et al. 2000), beaked

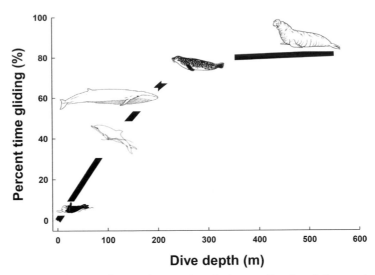

Figure 2.7 Percentage glide time during descent in relation to dive depth for marine mammals. Each animal symbol represents an individual species. The data were described by the non-linear function, percentage glide time = 85.9 – (2820.3/depth). Except for the dolphins, the range of depths was determined by the free-ranging behavior of instrumented animals in the wild. (Data from Williams, Fuiman, and Davis 2015)

whales (*Ziphius cavirostris* and *Mesoplodon densirostris*; Tyack et al. 2006), sperm whales (*Physeter macrocephalus*, Miller et al. 2004), pilot whales (Aguilar de Soto et al. 2008), and right whales (*Eubalaena glacialis*: Nowacek et al. 2001) also save energy on the ascent portion of the dive by using burst-and-glide swimming due to lung reinflation as the water pressure decreases when the animal approaches the surface. Regardless of the direction (descent versus ascent), the energetic savings associated with decreased stroking translates into reductions in energetic costs, which conserves on-board oxygen stores and prolongs dive duration, allowing for more foraging time underwater. Nonetheless, it should be noted that marked changes in buoyancy and body drag that occur as the percentage of body fat changes with season (e.g., Adachi et al. 2014), pregnancy (e.g., Noren, Redfern, and Edwards 2011), lactation and fasting (e.g., Costa et al. 1986; Crocker, Le Boeuf, and Costa 1997) will alter the ability of marine mammals to take advantage of these sub-surface energy-saving behaviors.

HIGH COSTS OF FORAGING ACTIVITIES

For active predators like most marine mammals, behaviors associated with foraging represent a major component of daily energy costs. In particular, high-speed chases, prey handling, and prey consumption can entail high-stroke frequencies and a large investment of energy when submerged (Williams et al. 2004; Aguilar de Soto et al. 2008; Maresh 2014). However, some of these costs are mitigated as marine mammals often switch between several swimming modes during a foraging dive as described before. Such variability and use of unsteady swimming modes are typical of swimming vertebrates from sharks to pinnipeds and have been shown to provide an energetic advantage over continuous locomotion (Williams et al. 2000; Gleiss, Wilson, and Shepard 2011).

In addition to foraging dives, both large-scale migrations of birds and mammals (Gleiss, Wilson, and Shepard 2011) and the intra-dive foraging periods of pinnipeds often incorporate a roller-coaster series of powered and non-powered phases that result in performance, behavioral, and energetic benefits depending on the context. For example, active foraging by Weddell seals feeding in an aggregation of Antarctic silverfish (*Pleuragramma antarcticum*) involves a series of roller-coaster dips and rises that are associated with low-frequency (7.2 ± 0.7 strokes.min^{-1}) stroking descents followed by moderate stroke frequency (28.5 ± 0.8 strokes. min^{-1}) ascents and fish encounters. Only rarely do the seals feed on descent. Average instantaneous energetic costs are dictated by the stroking patterns and alternate between 17.2 ± 1.6 J.kg^{-1}.min^{-1} and 68.2 ± 2.0 J.kg^{-1}.min^{-1} on each gliding dip and powered rise of the foraging period, respectively (Williams, Fuiman, and Davis 2015).

One of the most energetically costly feeding behaviors is displayed by lunge-feeding rorqual whales (balaenopterids such as blue and humpback whales), which has been described as "the largest biomechanical event on Earth and one of the most extreme feeding methods among aquatic vertebrates" (Brodie 1993). Balaenids are bulk filter feeders that capture prey by engulfing large volumes of water containing dense aggregations of plankton or nekton (Goldbogen et al. 2011). This lunge-feeding behavior requires acceleration to high speeds toward a prey patch, and inflation of the accordion-like buccal cavity to an 80° gape angle. During engulfment, the whale presents the equivalent of a massive, flat plate to oncoming water flow. The animal must overcome exceptionally high-drag forces due to a reduction in body streamlining and engulfment drag that rapidly decelerates the whale to a near halt (**Chapter 1**). The exceptionally high level of biomechanical work required to lunge feed underlies the relatively short maximum dive durations observed for foraging balaenids (Goldbogen et al. 2012), which are often shorter than predicted based on allometric calculations of aerobic dive limits

(**Chapter 3**). Moreover, there are thermoregulatory costs to consider as the whale engulfs freezing water into its mouth, which can be somewhat mitigated by counter-current heat exchangers in the tongue (Heyning and Mead 1997; **Chapter 5**). Despite these high costs, the energetic payoff of the lunge-feeding foraging strategy is high, with each lunge providing 6 to 237 times more energy consumed than expended by foraging blue whales (Goldbogen et al. 2011).

For marine mammals that have most recently invaded the aquatic environment (in an evolutionary sense), foraging activities are more energetically costly than for species from lineages that have maintained an aquatic existence for much longer. Sea otters and polar bears, whose ancestors re-entered the aquatic environment only 1–3 million years ago, lack many of the cost-saving adaptations associated with a fully aquatic lifestyle that are typical of cetaceans and pinnipeds, whose ancestors re-invaded the aquatic environment 25–50 million years ago. Due to elevated locomotor costs associated with buoyancy and an inefficient swimming style (described previously), it is estimated that a sea otter spends twice the amount of energy on a foraging dive as a phocid seal (Yeates, Williams, and Fink, 2007). Polar bears do not engage in foraging dives *per se* while hunting, but instead typically prey on ice seals and other marine mammals from the sea ice; however, the cost of movement across the sea ice, whether by walking or swimming, is substantial, such that movement and activity rates are the primary drivers of total energy costs in this species (Pagano et al. 2018). To remain in positive energy balance, both sea otters and polar bears must forage often on high-energy prey, counterbalanced with prolonged periods of rest. As a consequence, both species are particularly vulnerable to fluctuations in food resource availability, such as those associated with human overfishing (sea otters) and sea ice retreat (polar bears).

IMPORTANCE OF HAVING RELIABLE ENERGETICS DATA AND FUTURE DIRECTIONS

Movement of marine mammals is important for foraging, predator evasion, and migration to foraging grounds and mating grounds. Therefore, the ability to move is directly tied to **fitness** and **fecundity** (reproductive success). Although the motivations and cues for movement vary across species, one constant unifies the movements of all active animals; it comes at an highenergetic cost despite the energy-saving strategies that we have discussed in this chapter. Meanwhile, survival and successful reproduction depend on individuals achieving positive energy balance, whereby individuals acquire ample energy from foraging to fuel critical life processes. One can only imagine the total energetic costs that supported the remarkable movements of a female gray whale that migrated 22,511 km (nearly 14,000 miles) roundtrip between Russia and Baja, Mexico, to set the record for the longest migratory movement of any mammal (Mate et al. 2015). Yet, a sexually mature female gray whale will not produce a calf if her energy intake is reduced by just 4% during pregnancy, since her own survival could be impacted (Villegas-Amtmann et al. 2015). Similarly, the highenergetic demands of a female sea otter raising a pup superimposed on the daily energetic costs of foraging and activity can push the mother into a state of starvation. If this occurs the pup will be abandoned (Thometz et al. 2014). Thus, energy balance not only has consequences for individual survival, but it also has consequences for reproductive fitness and ultimately population demograpics.

Estimates of energy expenditure and, by extension, the prey requirements of free-ranging marine mammals, are fundamental to questions that must be answered for the successful management and conservation of marine mammal populations. With today's cumulative **anthropogenic disturbances** of marine mammals, especially during foraging activities that impact their energetic balance, it is imperative that we focus research efforts in this realm. Some species are taking longer, more circuitous transit routes to avoid acute anthropogenic disturbances, and others are changing diving behaviors. At the same time, some individuals are traveling farther to find prey due to changes in prey distribution patterns associated with climate change. The energetic consequences of increased movement are largely unknown. See **Chapter 15** for more discussion concerning this issue.

As evident from this chapter, our understanding of the energetics of marine mammals is limited and based primarily on a few well-studied species, most notably, harbor, gray, elephant and Weddell seals, California sea lions and fur seals, sea otters, and bottlenose dolphins. Much less is known about the energetic costs in other species, particularly large-bodied cetaceans that spend their entire lives at sea. In addition, the offshore lifestyles and prolonged submergence periods of some species, like beaked whales, make them extremely cryptic, making it impossible for researchers to measure their metabolism. Indeed, of the nearly 90 cetacean species (dolphins, whales, and porpoises), RMR measured via oxygen consumption has only been quantified for a few, including the harbor porpoise (*Phocoena phocoena*), bottlenose dolphin, Pacific white-sided dolphin (*Lagenorhynchus obliquidens*), beluga whale (*Delphinapterus leucas*), and killer whale (*Orcinus orca*). With over 120 species of marine mammals worldwide, living in tropical to polar habitats, there are clearly many questions regarding the unique energetic costs and resource demands of all marine mammal species. It is encouraging that the energetics of marine mammals has received growing attention in the scientific community, where predictive models are incorporating movement energetics to examine the cumulative impacts of these disturbances from individuals up through populations. However, these models are only as good as their inputs, so additional empirical data from both the field and laboratory on marine mammal metabolism are needed. Ultimately, our aim with improved models is to be able to predict the vulnerability of marine mammal populations to rapidly changing environments and other anthropogenic disturbances.

TOOLBOX

The methods for assessing energetic costs in active marine mammals are changing quickly as advances in microprocessor technology afford new opportunities and finer scale monitoring of free-ranging animals. Combined with traditional methods of direct and indirect calorimetry (the measurement of heat production or absorption by metabolic reactions in animals), it is now possible to evaluate the energetics of marine mammals on many scales, from instantaneous to prolonged intervals that range from seconds to daily, seasonal, and annual periods of time.

Measuring oxygen consumption

One of the most common methods for evaluating the energetic cost of rest and exercise in marine mammals is the determination of the rate of oxygen consumption (VO_2) using open-flow respirometry. This is considered to be a "gold standard" approach. Details of the methods and calibrations are provided in Fedak, Rome, and Seeherman (1981) and Davis, Williams, and Kooyman (1985) for marine mammals as modified from Withers (1977), who developed many of the techniques for open-flow respirometry on terrestrial species. VO_2 during rest, swimming, and diving has been directly measured by training or directing animals to breathe into an appropriately-sized metabolic chamber to capture exhalations. A Plexiglas skylight floating on the water surface often suffices as a metabolic chamber and has been used to measure the VO_2 of a wide variety of marine mammals, from sea otters to killer whales (*Table 2.1*). Creative placement of the metabolic chamber above load cells (Williams et al. 2017b), pools (Yazdi, Kilian, and Culik 1999; Thometz et al. 2014), flumes (Davis, Williams, and Kooyman 1985; Feldkamp 1987; Rosen and Trites 2002; Pagano et al. 2018), open

water (Fahlman et al. 2008), and even breathing holes in the polar ice (Castellini, Kooyman, and Ponganis 1992; Williams et al. 2004) have allowed measurement of the metabolic rates of resting and exercising marine mammals.

The open-flow respirometry technique involves drawing air through the metabolic chamber with a vacuum pump at flow rates that maintain the fractional concentration of oxygen in the metabolic chamber above 0.2000 to avoid hypoxic (low oxygen) conditions for the animals. Samples of expired air from the exhaust port of the chamber are usually dried with Drierite and scrubbed of carbon dioxide with Sodasorb before entering an oxygen analyzer (e.g., Sable Systems International, Inc., Las Vegas, NV). VO_2 is calculated based on the air flow rate (VI), and the difference between the fractional concentration of oxygen entering the chamber (FIO_2) and the expired air (FEO_2) using equations from Fedak, Rome, and Seeherman (1981) and an assumed respiratory quotient (RQ, the ratio of oxygen consumption to carbon dioxide production) of 0.77 (Williams et al. 2004). Readers are directed to Withers (1977) for typical equations used for determining VO_2 of marine mammals from open-flow respirometry. Note that the specific equation used as well as the interpretation of the results will depend on the exact experimental setup and the physiological status of the animals.

A variation of this technique has been used for cetaceans. In this case, a pneumotachometer respirometer is placed directly over the blowhole and enables breath-by-breath analyses (for details, see Fahlman et al. 2018; Allen et al. 2022). Most often, trained cetaceans at rest or following a submerged swim are measured as the animals return to a pneumotach station for measurement of oxygen content of the breath.

Indirect methods for determing energetic costs

Except in the few unique circumstances mentioned before, it is not possible to directly measure the oxygen consumption of most wild marine mammals. However, by pairing direct measures of metabolism in controlled circumstances with other physiological or biomechanical parameters measured with microprocessor tags, it is possible to estimate nearly instantaneous to long-term energetic costs for a wide variety of behaviors. For these measurements, calibrated instrumentation is deployed on wild marine mammals, which records respiration rates, heart rates, acceleration, swimming stroke rates, and/or speed. Because each of these parameters is correlated with the rate of oxygen consumption (e.g., **Figures 2.2** and **2.4**), they can be used to determine energetic costs as the animals move freely through the environment. Swimming, diving, migrating, and foraging energetics of pinnipeds (Butler 1993; Boyd 2002; Williams et al. 2004) as well as small (Williams et al. 2017b, 2022) and large (Goldbogen et al. 2008, 2011, 2012) cetaceans have been estimated using this indirect method.

Recently, the use of tri-axial accelerometers has enabled energetics to be measured on individual behaviors, and even on an instantaneous basis. Accelerometry methods rely on a quantified relationship between movement in three dimensions and the metabolic power required to fuel that movement. One method, termed overall dynamic body acceleration (ODBA), integrates the dynamic acceleration of the animal's body in each of the three movement vectors, with higher levels of acceleration indicating movements requiring more metabolic power (e.g., Fahlman et al. 2008; John 2020). A second method involving accelerometry uses information from only one movement axis to identify individual swim strokes, the total number of which can be used to estimate overall energy expenditure on a cost-per-stroke basis (e.g., Williams et al. 2004, 2017b; Maresh et al. 2015). To be successful, both of these methods require species-specific calibration of the acceleration-energetics relationship, for which respirometry or doubly labeled water (defined in the next section) can

be useful. While there remains some uncertainty regarding the use of accelerometry to measure the field energetics of free-ranging marine mammals (e.g., Dalton, Rosen, and Trites 2014), these methods are increasingly popular as they can provide data at a high temporal and behavioral resolution over long periods of time (Fahlman et al. 2008).

Measuring field metabolic rates

While the measurement of oxygen consumption makes respirometry the "gold standard" for estimating metabolic rates and energetic costs, it is generally limited to captive settings. Of the various methods available for measuring FMR, the doubly-labeled water (DLW) method has been used most often in free-ranging, foraging pinnipeds (see Boyd 2002). The DLW method is based on the principle that the bodies of living animals are always in a state of change, constantly exchanging materials such as oxygen, carbon dioxide, fuels, and water with the external environment. The rate at which these materials turnover in the body is proportional to the animal's metabolic rate (Speakman 1997). DLW specifically measures the rate of carbon dioxide (CO_2) produced in expired gases as a proxy for metabolism and energy expenditure. In this method, water labeled with heavy isotopes of oxygen (O^{18}) and hydrogen (deuterium D^2, or tritium H^3) is injected into the animal, and the rates of isotope dilution in the body over time are used to determine CO_2 production. Doubly-labeled water performs best over prolonged periods, from hours to days, depending on the size of the animal, and ultimately provides a single value that represents energy expenditure summed over the entire measurement period (Costa 1987). Researchers interested in measuring the energy costs of specific, discreet behaviors should consider other methods.

For marine mammals, the use of doubly-labeled water has three major limitations. First, it is prohibitively expensive for exceptionally large animals, most notably mysticete whales. For example, as of this writing, an average-sized, adult blue whale would require a 30-liter injection of D^2O^{18}, at a minimum cost of over \$13 million! Second, because the DLW method requires blood samples at both the beginning and end of the measurement period, it is of limited utility when the chance of recapture is low, as is the case for many far-ranging and elusive marine mammals, particularly cetaceans. Lastly, calculations of energy expenditure using the DLW method are sensitive to the estimates of the animal's mass; this limits the utility of the method except in instances where precise body mass measurements are possible – a rare luxury for biologists studying free-ranging marine mammals, and impossible for those studying mysticetes. Despite these limitations, DLW is still considered one of the best methods for estimating FMR when logistically feasible (Speakman 1997; Sparling et al. 2008).

CONCLUSIONS

The ability of a marine mammal to forage successfully during a dive, transit long distances across the world's oceans, and move with stealth deep beneath the water's surface to avoid predators – in other words, to exercise while in breath-hold – is ultimately determined by how long it can hold its breath. Breath-hold duration is itself determined by two factors: the size of an individual's on-board oxygen supply (its "scuba tank"), and the rate at which the individual uses up that oxygen during a dive (its metabolic rate). Because dive times impact a marine mammal's foraging success, survival, and, ultimately, reproductive fitness, adaptations that increase oxygen stores, decrease metabolic rate, or both, have been under strong selective pressure since the terrestrial ancestors of modern-day cetaceans and pinnipeds re-entered the aquatic environment 25–50 million years ago.

In this chapter, we have described some of the many morphological, physiological, and behavioral adaptations seen in modern-day marine mammals that serve as cost-saving mechanisms for decreasing oxygen utilization rates. Most of these adaptations – from convergence on a streamlined body plan to the use of behavioral "tricks" while swimming and diving – serve to reduce locomotion costs, one of the four main contributors to an animal's total energy (and thus oxygen) needs. But to really understand how quickly an animal burns through its oxygen stores during a dive requires a knowledge of the other contributors to an animal's total energy needs. To accommodate both diving and exercise responses when submerged, marine mammals require some minimum amount of energy for survival (measured as basal, or resting metabolism), and must also balance respiratory, cardiovascular, and metabolic physiological functions. Together with locomotion costs, the sum of these energy expenditures determines an animal's total energy needs, which we measure as its FMR.

The importance of quantifying these costs in our quest to understand how marine mammals are able to exercise while holding their breath has been long-recognized by researchers. Despite this recognition, as well as the hundreds of studies seeking to illuminate answers to these questions, the sheer logistical challenges of acquiring physiological measurements from marine mammals in the wild limits our understanding of marine mammal energetics to a few species. Additionally, disagreement among researchers as to when and how to take these measurements – as in, for example, the debate surrounding what constitutes true resting metabolism in a fully aquatic mammal experiencing a dive response – also leaves many questions unanswered.

Despite these limitations, it is important to keep trying. Quantifying the energetic costs of individual marine mammals, from rest to high activity, would enable us to predict their dive capacity and food requirements – critical, missing knowledge as we attempt to forecast how global climate change and other human perturbations may impact marine mammal populations.

STUDY QUESTIONS

1 For a diving marine mammal, what two physiological responses are in conflict with each other and what are the characteristics that define each of these responses?
2 What components make up the total energetic costs of marine mammals?
3 What are the two hypotheses as to why marine mammals might have higher resting metabolic rates than terrestrial mammals?
4 How is the relationship for metabolic rate and speed different between terrestrial and marine mammals?
5 Define the cost of transport and what range of speeds is typical for marine mammals.
6 How does field metabolic rate (FMR) relate to body size and describe how FMR in marine mammals compares to terrestrial mammals of similar size.
7 How is lung compression related to saving energy while swimming on a dive?
8 What is porpoising?
9 What is wave-riding and how is this behavior similar to dolphin calves swimming in echelon position with their mothers?
10 Why is it important to have reliable energetics data and what are our limitations in measuring this when it comes to marine mammals?

CRITICAL THINKING

1 What factors have made it difficult to know what truly defines a resting metabolic rate measurement in a marine mammal?
2 How is the relationship for the cost of running different than the relationship for the cost of swimming, and what factor(s) make these relationships look different?
3 What behavioral tricks can lower the cost of swimming and diving in a marine mammal, and how do they lower the energetic costs of swimming and diving?

EQUATIONS

Useful equations for predicting the metabolic rates and energetic costs of marine mammals:

1 **Resting Metabolic Rate (RMR).** The relationship between body mass and resting metabolic rate for marine mammals is best described by

$$RMR_{MM} = 581 mass^{0.68}$$
$$(n = 12 \text{ species}, r^2 = 0.66) \tag{2.1}$$

where metabolic rate is in kJ d^{-1} and body mass is in kg (Maresh 2014).

2 **Cost of Transport (COT).** The energy expended (J.kg^{-1}.m^{-1}) to move a body one meter in distance is calculated from

$$COT = VO_2/Speed \tag{2.2}$$

where VO$_2$ is in J.kg^{-1}.s^{-1} and speed is in m.s^{-1}. The predictive equation for minimum COT for **submerged swimming marine mammals** is

$$COT_{MIN} = 7.79 M^{-0.29} \tag{2.3}$$

where COT is in J.kg^{-1}.m^{-1} and M is mass in kg.

3 **Field Metabolic Rate (FMR).** The predictive equation for FMR of marine mammals (FMR$_{MM}$ in kJ d^{-1}) is described as

$$FMR_{MM} = 3511 mass^{0.45}$$
$$(n = 10 \text{ species}, r^2 = 0.43) \tag{2.4}$$

where body mass is in kg (Maresh 2014).

FURTHER READING

Costa, D.P., and J.L. Maresh. 2018. Energetics. *Encyclopedia of Marine Mammals* 2nd Edition: W. Perrin, B. Wursig, and J.G.M. Thewissen, eds. Amsterdam: Academic Press (pp. 329–335).
Kooyman, G.L., M.A. Castellini, and R.W. Davis. 1981. Physiology of diving in marine mammals. *Annual Review of Physiology* 43 (1): 343–356.

Lavigne, D.M., S. Innes, G.A.J. Worthy, K.M. Kovacs, O.J. Schmitz, and J.P. Hickie. 1986. Metabolic rates of seals and whales. *Canadian Journal of Zoology* 64 (2): 279–284.

McNab, B.K. 1988. Complications inherent in scaling the basal rate of metabolism in mammals. *The Quarterly Review of Biology* 63 (1): 25–54.

Stephens, D.W., and J.R. Krebs. 1986. *Foraging Theory* (Vol. 1). Princeton,: Princeton University Press, 248 pages.

GLOSSARY

Apnea: Temporary cessation of breathing.

Basal metabolic rate: Baseline energy requirements for survival per some unit time (aka, maintenance energy requirements).

Bradycardia: Slowed heart rate.

Burst-and-glide swimming: Swimming behavior characterized by the incorporation of prolonged periods of passive gliding (unpowered) in between periods of active propulsion (powered by muscles).

Calorimetry: The measurement of heat production or absorption by metabolic reactions in animals.

Capital breeder: Reproductive strategy characterized by the separation of feeding and breeding grounds, whereby fasting mothers rely on endogenous energy stores acquired before parturition to support lactation; extremely rare among mammals.

Cost of transport (COT): Amount of energy needed to move a unit of body mass some unit of distance; equivalent to the miles per gallon (MPG) fuel economy rating of automobiles.

Cumulative anthropogenic disturbances: Changes caused by the combined impact of numerous past, present, and future human activities.

Dive capacity: How deep and for how long an individual can dive; in marine mammals, determined by how long an individual can hold its breath.

Dive response: Series of temporary physiological changes that take place in the mammalian body in response to breath-hold, in order to conserve oxygen.

Echelon position: Swimming formation used by cetacean mother-calf dyads to increase calf swim performance at reduced locomotor effort; the dependent calf is positioned in very close proximity to its mother's mid-lateral flank in the region near her dorsal fin.

Energetic costs: Energy required by an animal to perform physiological work, including homeostasis, physical activity, growth, and reproduction.

Exaptation: A morphological or physiological feature that predisposes an organism to adapt to a different environment or lifestyle, resulting in a shift in the function of a trait during evolution.

Exercise response: Series of temporary physiological changes that take place in the mammalian body in response to physical exertion, in order to increase oxygen transport and use.

Fecundity: Physiological maximum potential reproductive output of an individual over its lifetime.

Field metabolic rate: The total daily energy costs incurred by a wild animal.

Fitness: An animal's ability to survive and reproduce in a particular environment.

Income breeder: Reproductive strategy characterized by the overlap of feeding and breeding grounds, whereby mothers continue to feed post-parturition to support lactation.

Lung compression at depth: Progressive decrease in the volume of the lungs and thoracic (chest) cavity that occurs during a breath-hold dive underwater due to an increase in hydrostatic pressure during descent.

Metabolism (aerobic): A measure of the energy required to fuel the biochemical processes that occur within the cells of living organisms (oxygen consumption).

Minimum COT (COT$_{MIN}$): Amount of energy needed to move a unit of body mass some unit of distance at a speed or range of speeds that enable the individual to move the furthest for the least investment of energy; provides an index of the energetic efficiency of marine mammals.

Porpoising: Continuous rapid swimming with rhythmic aerial leaps; in marine mammals, promotes rapid gas exchange at the water-air interface while minimizing time spent at the surface, and minimizes the added energetic costs of wave drag.

Stroke frequency: In swimming animals, the number of times the flippers or flukes move back and forth per some unit time during propulsion; a measure of swimming effort.

Tachycardia: Elevated heart rate.

Thermal neutral zone: The range of environmental temperatures over which an animal can maintain a normal body temperature without requiring additional energy beyond its normal basal or resting metabolic rate.

VO₂: A measure of the volume of oxygen that is consumed by an animal over a unit time. Often reported in units of mlO_2 per minute, or mlO_2 per kg body mass per minute.

Wave drag: A resisting force on swimmers traveling on the water's surface, created by the generation of waves that oppose the direction of motion.

Wave-riding: Similar to surfing, the exploitation by swimming animals of a pressure wave caused by displaced water at the front of a vessel, wave, or larger animal to swim faster with greatly reduced physical exertion.

REFERENCES

Adachi, T., J.L. Maresh, P.W. Robinson, et al. 2014. The foraging benefits of being fat in a highly migratory marine mammal. *Proceedings of the Royal Society B: Biological Sciences* 281 (1797): 20142120.

Aguilar de Soto, N., M.P. Johnson, P.T. Madsen, et al. 2008. Cheetahs of the deep sea: Deep foraging sprints in short-finned pilot whales off Tenerife (Canary Islands). *Journal of Animal Ecology* 77 (5): 936–947.

Allen, A.S., A.J. Read, K.A. Shorter, et al. 2022. Dynamic body acceleration as a proxy to predict the cost of locomotion in bottlenose dolphins. *Journal of Experimental Biology* 225 (4): jeb243121.

Au, D., and D. Weihs. 1980. At high speeds dolphins save energy by leaping. *Nature* 284 (5756): 548–550.

Berta, A., J.L. Sumich, and K.M. Kovacs. 2015. *Marine Mammals: Evolutionary Biology*, 3rd ed. San Diego, CA: Academic Press, 738pp.

Boyd, I.L. 2002. Energetics: Consequences for fitness. Marine Mammal Biology – An Evolutionary Approach. A.R. Hoezel, ed. Oxford: Blackwell Science (pp. 247–277).

Brodie, P.F. 1993. Noise generated by the jaw actions of feeding fin whales. *Canadian Journal of Zoology* 71 (12): 2546–2550.

Butler, P.J. 1993. To what extent can heart rate be used as an indicator of metabolic rate in free-living marine mammals. Paper read at Symposia of the Zoological Society, London.

Castellini, M.A., G.L. Kooyman, and P.J. Ponganis. 1992. Metabolic rates of freely diving Weddell seals: Correlations with oxygen stores, swim velocity and diving duration. *Journal of Experimental Biology* 165 (1): 181–194.

Costa, D.P. 1987. Isotopic methods for quantifying material and energy intake of free-ranging marine mammals. *Approaches to Marine Mammal Energetics* 1: 43–66.

Costa, D.P., B.J. Le Boeuf, A.C. Huntley, et al. 1986. The energetics of lactation in the northern elephant seal, *Mirounga angustirostris*. *Journal of Zoology* 209 (1): 21–33.

Crocker, D.E., B.J. Le Boeuf, and D.P. Costa. 1997. Drift diving in female northern elephant seals: Implications for food processing. *Canadian Journal of Zoology* 75 (1): 27–39.

Dalton, A.J.M., D.A.S. Rosen, and A.W. Trites. 2014. Season and time of day affect the ability of accelerometry and the doubly labeled water methods to measure energy expenditure in northern fur seals (*Callorhinus ursinus*). *Journal of Experimental Marine Biology and Ecology* 452: 125–136.

Davis, R.W., T.M. Williams, and G.L. Kooyman. 1985. Swimming metabolism of yearling and adult harbor seals *Phoca vitulina*. *Physiological Zoology* 58: 590–596.

Fahlman, A., M. Brodsky, R. Wells, et al. 2018. Field energetics and lung function in wild bottlenose dolphins, Tursiops truncatus, in Sarasota Bay Florida. *Royal Society Open Science* 5 (1): 171280.

Fahlman, A., R. Wilson, C. Svärd, et al. 2008. Activity and diving metabolism correlate in Steller sea lion *Eumetopias jubatus*. *Aquatic Biology* 2: 75–84.

Fedak, M.A., L. Rome, and H.J. Seeherman. 1981. One-step N2-dilution technique for calibrating open-circuit VO2 measuring systems. *Journal of Applied Physiology* 51 (3): 772–776.

Feldkamp, S.D. 1987. Swimming in the California sea lion: Morphometrics, drag and energetics. *Journal of Experimental Biology* 131 (1): 117–135.

Folkens, P.A., R.R. Reeves, B.S. Stewart, et al. 2002. *Guide to Marine Mammals of the World*. New York: AA Knopf.

Gleiss, A.C., R.P. Wilson, and E.L. Shepard. 2011. Making overall dynamic body acceleration work: On the theory of acceleration as a proxy for energy expenditure. *Methods in Ecology and Evolution* 2 (1): 23–33.

Goldbogen, J.A., J. Calambokidis, D.A. Croll, et al. 2008. Foraging behavior of humpback whales: Kinematic and respiratory patterns suggest a high cost for a lunge. *Journal of Experimental Biology* 211 (23): 3712–3719.

Goldbogen, J.A., J. Calambokidis, D.A. Croll, et al. 2012. Scaling of lunge-feeding performance in rorqual whales: Mass-specific energy expenditure increases with body size and progressively limits diving capacity. *Functional Ecology* 26 (1): 216–226.

Goldbogen, J.A., J. Calambokidis, E. Oleson, et al. 2011. Mechanics, hydrodynamics and energetics of blue whale lunge feeding: Efficiency dependence on krill density. *Journal of Experimental Biology* 214 (1): 131–146.

Heim, N.A., M.L. Knope, E.K. Schaal, et al. 2015. Cope's rule in the evolution of marine animals. *Science* 347 (6224): 867–870.

Heyning, J.E., and J.G. Mead. 1997. Thermoregulation in the mouths of feeding gray whales. *Science* 278 (5340): 1138–1140.

Hudson, L.N., N.J. Isaac, and D.C. Reuman. 2013. The relationship between body mass and field metabolic rate among individual birds and mammals. *Journal of Animal Ecology* 82 (5): 1009–1020.

John, J.S. 2020. Energetics of rest and locomotion in diving marine mammals: Novel metrics for predicting the vulnerability of threatened

cetacean, pinniped and sirenian species. PhD Thesis, Santa Cruz, CA: University of California.

Kleiber, M. 1975. *The Fire of Life: An Introduction to Animal Energetics*, 2nd ed. New York: Kreiger.

Kreite, B. 1994. Bioenergetics of the killer whale, *Orcinus orca*. Ph.D. Thesis, Vancouver, British Columbia, Canada: University of British Columbia.

Maresh, J.L. 2014. Bioenergetics of marine mammals: the influence of body size, reproductive status, locomotion and phylogeny on metabolism. Ph.D. Thesis, Santa Cruz: University of California.

Maresh, J.L., T. Adachi, A. Takahashi, et al. 2015. Summing the strokes: Energy economy in northern elephant seals during large-scale foraging migrations. *Movement Ecology* 3 (1):1–16

Mate, B.R., V.Y. Ilyashenko, A.L. Bradford, et al. 2015. Critically endangered western gray whales migrate to the eastern North Pacific. *Biology Letters* 11. DOI: 10.1098/rsbl.2015.0071.

Miller, P.J., M.P. Johnson, P.L. Tyack, et al. 2004. Swimming gaits, passive drag and buoyancy of diving sperm whales *Physeter macrocephalus*. *Journal of Experimental Biology* 207 (11): 1953–1967.

Nagy, K.A. 2005. Field metabolic rate and body size. *Journal of Experimental Biology* 208 (9): 1621–1625.

Noren, S.R. 2008. Infant carrying behaviour in dolphins? Costly parental care in an aquatic environment. *Functional Ecology* 22: 284–288.

Noren, S.R., G. Biedenbach, J.V. Redfern, et al. 2008. Hitching a ride: The formation locomotion strategy of dolphin calves. *Functional Ecology* 22: 278–283.

Noren, S.R., T. Kendall, V. Cuccurullo, et al. 2012. The dive response redefined: Underwater behavior influences cardiac variability in freely diving dolphins. *Journal of Experimental Biology* 215 (16): 2735–2741.

Noren, S.R., J.V. Redfern, and E.F. Edwards. 2011. Pregnancy is a drag: Hydrodynamics, kinematics, and performance in pre- and post-parturition bottlenose dolphins (*Tursiops truncatus*). *Journal of Experimental Biology* 214: 4151–4159.

Nowacek, D.P., M.P. Johnson, P.L. Tyack, et al. 2001. Buoyant balaenids: The ups and downs of buoyancy in right whales. *Proceedings of the Royal Society of London B: Biological Sciences* 268 (1478): 1811–1816.

Pagano, A.M, A. Cutting, N. Nicassio-Hiskey, et al. 2018. Energetic costs of aquatic locomotion in a subadult polar bear. *Marine Mammal Science* 35. DOI: 10.1111/MMS.12556.

Porter, W.P., and M. Kearney. 2009. Size, shape, and the thermal niche of endotherms. *Proceedings of the National Academy of Sciences* 106 (Supplement 2): 19666–19672.

Ridgway, S.H., D.A. Redfern, and W. Clark. 1975. Conditioned bradycardia in the sea lion *Zalophus californianus*. *Nature* 256: 37–38.

Rosen, D.A.S., and A.W. Trites. 2002. Cost of transport in Steller sea lions, *Eumetopias jubatus*. *Marine Mammal Science* 18 (2): 513–524.

Schmidt-Nielsen, K. 1972. Locomotion: Energy cost of swimming, flying, and running. *Science* 177 (4045): 222–228.

Scholander, P.F. 1959. Wave-riding dolphins: How do they do it? At present only the dolphin knows the answer to this free-for-all in hydrodynamics. *Science* 129 (3356): 1085–1087.

Skrovan, R.C., T.M. Williams, P.S. Berry, et al. 1999. The diving physiology of bottlenose dolphins (*Tursiops truncatus*). II. Biomechanics and changes in buoyancy at depth. *Journal of Experimental Biology* 202 (20): 2749–2761.

Sparling, C.E., D. Thompson, M.A. Fedak, et al. 2008. Estimating field metabolic rates of pinnipeds: Doubly labelled water gets the seal of approval. *Functional Ecology* 22 (2): 245–254.

Speakman, J.R. 1997. *Doubly Labelled Water: Theory and Practice*. London: Chapman and Hall. 416pp.

Taylor, C.R., K. Schmidt-Nielsen, and J.L. Raab. 1970. Scaling of energetic cost of running to body size in mammals. *American Journal of Physiology–Legacy Content* 219 (4): 1104–1107.

Taylor, C.R., N.C. Heglund, and G.M. Maloi. 1982. Energetics and mechanics of terrestrial locomotion. I. Metabolic energy consumption as a function of speed and body size in birds and mammals. *Journal of Experimental Biology* 97(1): 1–21.

Thometz, N.M., M.T. Tinker, M.M. Staedler, et al. 2014. Energetic demands of immature sea otters from birth to weaning: Implications for maternal costs, reproductive behavior and population-level trends. *Journal of Experimental Biology* 217 (12): 2053–2061.

Tyack, P.L., M.P. Johnson, N.A. Soto, et al. 2006. Extreme diving of beaked whales. *Journal of Experimental Biology* 209 (21): 4238–4253.

Villegas-Amtmann, S., L.K. Schwarz, J.L. Sumich, et al. 2015. Population consequences of lost foraging opportunity in eastern female gray whales. *Ecosphere* 6(10):1–19..

Williams. 1999. The evolution of cost efficient swimming in marine mammals: Limits to energetic optimization. *Philosophical Transactions of the Royal Society B: Biological Sciences* 354 (1380): 193–201.

Williams, T.M., S.B. Blackwell, B. Richter, et al. 2017a. Paradoxical escape responses by narwhals (*Monodon monoceros*). *Science* 358: 1328–1331.

Williams, T.M., S.B. Blackwell, O. Tervo, et al. 2022. Physiological responses of narwhals to anthropogenic noise: A case study with seismic airguns and vessel traffic in the Arctic. *Functional Ecology*. DOI: 10.1111/1365-2435.14119.

Williams, T.M., R.W. Davis, L.A. Fuiman et al. 2000. Sink or swim: Strategies for cost-efficient diving by marine mammals. *Science* 288 (5463): 133–136.

Williams, T.M., L.A. Fuiman, and R.W. Davis. 2015. Locomotion and the cost of hunting in large, stealthy marine carnivores. *Integrative and Comparative Biology* 55: 673–682.

Williams, T.M., L.A. Fuiman, M. Horning, et al. 2004. The cost of foraging by a marine predator, the Weddell seal *Leptonychotes weddellii*: Pricing by the stroke. *Journal of Experimental Biology* 207 (6): 973–982.

Williams, T.M., T. Kendall, P. Richter, et al. 2017b. Swimming and diving energetics in dolphins: A stroke-by-stroke analysis for predicting the cost of flight responses in wild odontocetes. *Journal of Experimental Biology* 220: 1135–1145. DOI: 10.1242/jeb.154245.

Withers, P.C. 1977. Measurement of VO2, VCO2, and evaporative water loss with a flow-through mask. *Journal of Applied Physiology* 42 (1): 120–123.

Würsig, B., and M. Würsig. 1979. Behavior and ecology of the bottlenose dolphin, *Tursiops truncatus*, in the South Atlantic. *Fishery Bulletin* 77 (2): 399–412.

Yazdi, P., A. Kilian, and B.M. Culik. 1999. Energy expenditure of swimming bottlenose dolphins (*Tursiops truncatus*). *Marine Biology* 134 (4): 601–607.

Yeates, L.C., T.M. Williams, and T.L. Fink. 2007. Diving and foraging energetics of the smallest marine mammal, the sea otter (*Enhydra lutris*). *Journal of Experimental Biology* 210 (11):1960–1970.

Cassondra L. Williams

National Marine Mammal Foundation,
San Diego, CA

Paul J. Ponganis

Center for Biotechnology and Biomedicine,
Scripps Institution of Oceanography, University of California,
San Diego, CA

Cassondra L. Williams

Cassondra Williams, a research scientist with the National Marine Mammal Foundation, is a comparative physiologist who studies the diving physiology of air-breathing vertebrates, including elephant seals, emperor penguins and marine turtles. Dr. Williams' research is focused on understanding oxygen management and cardiovascular mechanisms in diving mammals, birds and reptiles, as well as developing novel instruments to measure physiological parameters in freely diving animals. Her recent work includes examining cardiovascular response to routine activities and disturbances in diving seals and marine turtles. Dr. Williams' NSF-funded research has been featured by National Geographic, NBC News and LiveScience. She was also featured in the "Women Working in Antarctica" series, which highlights the research contributions of the select group of women who have lived, worked and studied in Antarctica.

Paul J. Ponganis

Paul Ponganis is a research physiologist at Scripps Institution of Oceanography, as well as an anesthesiologist. He has over 35 years' experience both in clinical anesthesiology and in investigations of the diving physiology of marine mammals and seabirds. Research subjects have included whales, dolphins, seals, sea lions, fur seals, and king and emperor penguins. He pioneered the development of many physiological instruments to measure blood oxygen, heart rate and temperature in freely diving marine mammals and birds.

Driving Question: How do marine mammals stay underwater for so long?

INTRODUCTION

Marine mammals are able to dive for incredibly long durations. Elephant seals and beaked whales can dive for as long as two hours. One of the central questions in marine mammal

DOI: 10.1201/9781003297468-3

physiology is how these air-breathing vertebrates are able to remain underwater for so long on a single breath (Ponganis 2015). The keys to such long dive durations are found in the amount of oxygen (O_2) marine mammals take with them, the rate of O_2 consumed and how much O_2 is usable during dives. In general, marine mammals store much greater amounts of O_2 than terrestrial mammals. But extra O_2 storage is not enough to explain the routine, let alone longest, dive durations of these animals. How quickly or slowly the O_2 is used is also essential to diving ability. Marine mammals have several mechanisms to reduce O_2 consumption during dives. Finally, while terrestrial mammals are not able to function with low levels of blood O_2, some marine mammals have demonstrated the ability to continue a dive with very little O_2 remaining in the blood (**hypoxemic tolerance**). Thus, marine mammals can extend dive durations by (1) bringing more O_2 with them, (2) using it more efficiently and (3) depleting it to very low levels during dives.

However, most dives are much shorter than the animals' record (longest) dive durations. During shorter dives, marine mammals rely primarily on aerobic metabolism, which uses O_2 in the production of biochemical energy (ATP; Adenosine triphosphate). By using aerobic metabolism, marine mammals can dive repetitively with very short surface intervals because O_2 stores are replenished quickly when they surface. This efficiency has led to the concept that most dives are aerobic in many species. Once dives extend beyond a certain duration, anaerobic (without oxygen) metabolism increases. This increase in anaerobic metabolism results in an accumulation of blood lactate, which defines a concept called the **aerobic dive limit** (ADL). Dives longer than the ADL are not as efficient because these dives tend to have longer surface intervals.

This chapter will review (1) how O_2 is stored in marine mammals and what adaptations they have to increase O_2 stores, (2) the dive response and how it reduces O_2 consumption and (3) the ADL concept.

Total body oxygen stores

Oxygen is stored in three main places in the body of marine mammals (and all mammals): the lungs, blood system and muscle. We refer to these as O_2 stores: the respiratory (or lung) O_2 store, the blood O_2 store and the muscle O_2 store. The amount of O_2 found in each store depends on many different variables. To determine the magnitude of O_2 stores, a number of anatomical or physiological measurements must be made. However, these measurements can be difficult (or impossible) depending on the measurement technique and the species involved. As a result, assumptions about O_2 stores have been made for some marine mammals, which have yet to be verified in freely diving animals.

The amount of O_2 stored in each location also varies among species; the best divers typically store more O_2 in the muscle and blood than in the lungs. All marine mammals except the manatee have mass-specific body O_2 stores 1.5 to almost 5 times higher than non-divers (e.g., humans) (**Figure 3.1**). Of the mammals where total body O_2 stores have been measured, deep-diving phocid seals and beaked whales have the largest mass-specific O_2 stores. For example, the northern elephant seal (*Mirounga angustirostris*), with some of the longest routine dive durations of any pinniped, has the highest mass-specific O_2 store at 94 ml O_2 kg^{-1} (**Figure 3.1**) (Ponganis 2015). The manatee (*Trichechus manutus*), a short duration, shallow diver, has the lowest O_2 store at 21 ml O_2 kg^{-1} (Ponganis 2015), even lower than the human mass-specific O_2 store (24 ml O_2 kg^{-1}) (**Figure 3.1**). As a percentage of the total body O_2 store in diving mammals, the blood and muscle O_2 stores range from 55% in the sea otter (*Enhydra lutris*) to 97% in the northern elephant seal (**Figure 3.1**). In some deeper divers, there is a decreased dependence on the respiratory O_2 store. This is especially evident in the sperm whale, beaked whale, Weddell seal and elephant seal (**Figure 3.1**).

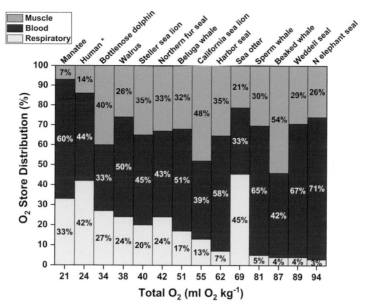

Figure 3.1 Total body O₂ stores in humans and marine mammals with the distribution of O₂ between the three stores illustrated by color. The muscle O₂ store is shaded gray, the blood O₂ store is shaded red and the respiratory O₂ store is shaded light blue. The species and percentages of contribution from each of the three O₂ stores are noted. (Data from Ponganis 2015) * for comparison

Respiratory oxygen store

In simple terms, the respiratory O₂ store is the amount of O₂ in the lungs that the animal can use during a dive. As the animal breathes before a dive, the lungs fill with air from the environment. This air contains approximately 21% O₂ but is lower by the time it reaches the alveoli (17–19%). The percentage of O₂ in the lungs declines throughout a dive. However, not all of the O₂ in the lungs can be used. Determining the usable amount of O₂ in the lungs (i.e., the respiratory O₂ store) is difficult. The respiratory O₂ store is dependent on the volume of air in the lungs at the start of a dive (diving lung volume) and the amount of O₂ that can be extracted from the lung during a dive (the decline from initial values of 17–19% O₂).

The diving lung volume is based on the total lung capacity, which is the volume of air in the lungs after maximal inspiration. The total lung capacities of diving mammals are in the general range of terrestrial mammals (Piscitelli et al. 2013; Ponganis 2015). Notable exceptions

are the small lungs of some of the deep-diving whales and the large lungs of the shallow-diving sea otter (**Figure 3.2**). A smaller lung volume, as in northern bottlenose whale, may allow for **lung collapse** (alveolar collapse) at a shallower depth and decrease the risk of nitrogen absorption and decompression sickness (see **Chapter 4**). For the otter, the high-lung volume contributes to its buoyancy at the surface, where it feeds, grooms and cares for its young. Such buoyancy also elevates more of the otter's body out of the water while the animal is at the surface; this should also reduce body heat loss due to conduction in water (see **Chapter 5**). Determining the total lung capacity is challenging in freely diving animals. While a variety of techniques have been used to determine total lung capacity, including helium dilution techniques, nitrogen washout, tidal volume measurements and inspiratory capacity, most total lung capacity measurement are made by inflating excised lungs (Piscitelli et al. 2010; Piscitelli et al. 2013; Ponganis 2015).

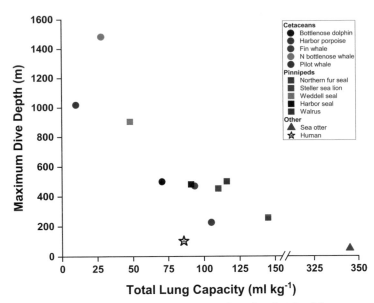

Figure 3.2 Total lung capacity in relation to maximum dive depth. Total lung capacity was lowest in marine mammals with higher maximum dive depths. Data points represent values for cetaceans (circles), pinnipeds (squares), other marine mammals and (triangles) humans (star). Species within the groups are represented by different colors. (Data from Ponganis 2015)

The level of inspiration (how deep a breath) the animal takes before a dive is the main reason that the diving lung volume is not equal to the total lung capacity. Some animals start a dive after inhaling and others after exhaling. Most cetaceans, manatees and otariids (e.g., sea lions) dive after inhalation and at full or near full total lung capacity. Although sea otters have the largest total lung capacity, they dive with about 60% of total lung capacity (Ponganis 2015). Phocid seals (e.g., elephant seals) typically dive after exhalation with diving lung volumes closer to 50% of total lung capacity (Gentry and Kooyman 1986; Kooyman 1989; Ponganis 2011). However, diving lung volumes in pinnipeds are likely quite variable depending on the circumstances. For example, research on diving sea lions suggests that sea lions inspire deeper and have larger diving lung volumes when they are about to do a deeper dive (McDonald and Ponganis 2012). These differences mean that the respiratory O_2 store can change depending on the level of inspiration (or type of dive).

Diving lung volumes are even more difficult to measure in freely diving mammals. In simulated dives in pressure chambers, the diving lung volumes of phocid seals and sea lions have been estimated to be about 50% of total lung capacity (Kooyman et al. 1973; Kooyman and Sinnett 1982). Such values, obtained from restrained animals, may not be representative of the initial lung volumes of freely diving animals. However, these values have been used extensively in calculations of O_2 stores. Since cetaceans dive on inspiration, it is often assumed that their diving lung volume is near total lung capacity. This assumption of full diving lung volume is supported by buoyancy-swim velocity calculations in freely diving cetaceans (Miller 2004).

As mentioned before, the diving lung volume is not equivalent to the total lung capacity. The diving lung volume gives us an idea of how much O_2 is in the lungs at the start of a dive if we multiply that volume by the initial percentage of O_2 in the lungs (typically 17–19%). However, the true, usable, respiratory O_2 store is based on the amount of O_2 that can be *extracted* from the diving lung volume during dives. The amount of O_2 extracted during dives is the difference between the percentage of O_2 in the lungs at the start of the dive (17–19%) and the lowest percentage of O_2 at the end of dives multiplied by the diving lung volume. It is unlikely that the O_2 in the lungs can be completely depleted. This difference is difficult to measure but is assumed to be 15% in diving mammals (Kooyman 1989). This assumption is supported by expiratory O_2 fractions measured at the end of dives or forced submersions in seals and other marine mammals, which were approximately 2–4%, consistent with the assumption of 15% net O_2 extraction (Scholander 1940; Ridgway, Scronce, and Kanwisher 1969; Ponganis, Kooyman, and Castellini 1993).

Blood oxygen store

The blood O_2 store is the amount of O_2 in the blood that can be used during a dive. It is vital, not only because it can store large amounts of O_2 but also because it distributes O_2 to the rest of the body. O_2 obtained from the lungs binds to the protein hemoglobin (Hb) in the blood and the O_2-bound Hb is then transported to the rest of the body through the arterial side of the circulatory (blood) system. As blood flows to organs and tissues, O_2 molecules are released from Hb and then the (deoxygenated) blood returns, via the venous side, to the lungs to be replenished with O_2.

The blood O_2 store is the largest O_2 store in many marine mammals (**Figure 3.1**). Almost all pinnipeds store 50% or more of their O_2 in the blood (**Figure 3.1**). In cetaceans, blood is the largest O_2 store in the deep, long duration divers, such as the sperm whale (*Physeter macrocephalus*). The blood O_2 store is also highest in both walruses and manatees; however, in sea otters, the respiratory O_2 store is the highest at 45% (**Figure 3.1**).

Blood O_2 stores are calculated using three main parameters: (1) blood volume, (2) Hb concentration and Hb-O_2 carrying capacity and (3) the amount of O_2 extracted from Hb during a dive. We have a good idea about the blood volume of

many marine mammals because it can often be measured using standard methods. The blood volume, assumed to be one-third arterial and two-thirds venous, is determined by dividing the total plasma volume by one minus the hematocrit. Plasma volume measurements are obtained using an injected dye (Evans blue dye) and the resulting dilution of the dye (El-Sayed, Goodall, and Hainsworth 1995) and are reported as ml of plasma/kg body mass. Hematocrit values, reported as a percentage, are obtained by centrifuging blood in a calibrated tube. Despite these standard methods, there are still limitations to our understanding of blood volume in some marine mammals. Most significantly, hematocrit can vary under different conditions (e.g., under anesthesia versus during a dive). This is because some divers, such as phocid seals, can store red blood cells in their spleens when on the shore or surface, releasing the cells during a dive, which increases hematocrit (Qvist et al. 1986; Ponganis, Kooyman, and Castellini 1993; Hurford et al. 1996; Thornton et al. 2001). In addition to blood volume, Hb concentration and carrying capacity are also needed to calculate the blood O_2 store. Hb concentration, although it does vary with hematocrit, is the most straightforward to determine as it can be measured from a blood sample using a standard lab assay, which is the same method used on humans during a medical exam. Each gram of Hb can carry 1.34 ml O_2. This value, the Hb-O_2 carrying capacity, is a constant and does not need to be measured.

The third component of the blood O_2 store is the amount of O_2 that can be extracted (is usable) during a dive. The measurement involves a number of variables. We can determine the amount of O_2 extracted from Hb as the difference between the initial Hb saturation and the end-of-dive Hb saturation on both the arterial side and venous side (Lenfant, Johansen, and Torrance 1970; Kooyman 1989). The Hb saturation is the percentage of Hb that is bound with O_2. When all Hb proteins are bound with O_2, Hb is 100% saturated. Hb saturation declines as Hb releases O$_2$ throughout the body during a dive. Typically, it is assumed that on the arterial side, Hb saturation can decrease during a dive from 95 to 20%. The venous side may start between 60 and 80% and can be completely depleted (Lenfant, Johansen, and Torrance 1970). These assumptions are based on expected resting values. However, recent technological advances have allowed researchers to measure the initial Hb-O$_2$ saturation values and the end-of-dive Hb-O$_2$ saturation values in some freely diving marine mammals (Meir et al. 2009; McDonald and Ponganis 2013). See **Toolbox**.

Marine mammal adaptations of the blood oxygen store

Most marine mammals store more O$_2$ in the blood than terrestrial mammals. This is in part because their Hb concentration and blood volume are elevated when compared to terrestrial mammals (**Figure 3.3**). The typical human Hb concentration of 15 g dl^{-1} and blood volume of 70 ml kg^{-1} are 50–70% lower than those of many marine mammals (**Figure 3.3**). The greatest elevations in both Hb and blood volume are typically found in the longest-duration divers and in highly active species (Ridgway and Johnston 1966; Ponganis 2011). This is exemplified in the phocid seals. The northern elephant seal and the Weddell seal have blood volumes above 200 ml kg^{-1} and Hb concentrations at or over 20 g dl^{-1} (**Figure 3.3**). In general, Hb and blood volume are not as high in otariids as in phocid seals (**Figure 3.3**). Hb concentration and blood volume span a wide range in cetaceans, from the shallow-diving bottlenose dolphin with values similar to standard human values to the sperm whale with values similar to the Weddell seal (**Figure 3.3**) (Sleet, Sumich, and Weber 1981; Ridgway 1986). Consequently, the bottlenose dolphin only stores 33% of total body O$_2$ in the blood store, while the sperm whale's blood store accounts for 64% of the total O$_2$ store (**Figure 3.1**). Walrus and manatee values are not particularly high, but the sea otter's values are within the otariid range (**Figure 3.3**) (Thometz, Murray, and Williams 2015).

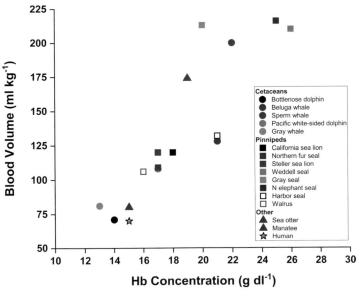

Figure 3.3 Blood volume in relation to hemoglobin (Hb) concentration in marine mammals and humans. Marine mammals that have high blood O$_2$ stores, such as the northern elephant seal or Weddell seal, have both elevated blood volume and high hemoglobin concentrations. Data points represent values for cetaceans (circles), pinnipeds (squares), other marine mammals (triangles) and humans (star). Species within the groups are represented by different colors. (Data from Ponganis 2015)

In addition to having increased blood volume and Hb concentrations, marine mammals make incredible use of their large blood O_2 stores. They are able to increase the usable O_2 in two ways. First, marine mammals may increase the Hb saturation of the venous side, called **arterialization** of venous blood. This arterialization has been observed both prior to and during dives of California sea lions, and during dives of elephant seals (Meir et al. 2009; Ponganis, Meir, and Williams 2011; McDonald and Ponganis 2013). As you may recall, venous Hb saturation has been assumed to be 60–80% at the beginning of a dive. In free-diving sea lions, venous blood at the beginning of some dives was increased to above 95% saturation (McDonald and Ponganis 2013). This could be an increase of O_2 by 15–35% in venous blood (which accounts for two-thirds of the blood volume). This venous arterialization would greatly increase the amount of O_2 that can be used during a dive. Since measurements of blood O_2 during diving have been made in very few species, it is unknown if arterialization of venous blood occurs in other marine mammals. However, as a mechanism to increase O_2 storage, and ultimately enhance dive duration, it would seem to be a highly beneficial mechanism.

Second, some marine mammals can use almost all of the O_2 stored in the blood. In most mammals, the blood O_2 store cannot come close to being completely depleted. A low level of O_2 in the blood (hypoxemia) of terrestrial mammals can cause blackouts. The limit of human hypoxemic tolerance is exemplified by a measurement of 34% arterial Hb saturation from a climber hyperventilating near the peak of Mt. Everest and an extreme free diver reaching 24% arterial Hb saturation (Grocott et al. 2009; McKnight, Mulder et al. 2021). These extreme values are not survivable by most terrestrial mammals. However, some marine mammals can tolerate even lower arterial saturations. While there have been few studies of blood O_2 store depletion during diving, the lowest arterial saturation is generally assumed to be 20%. Forced submersion studies demonstrated that seals were tolerant of arterial Hb saturation values down to about 20% (Elsner et al. 1970; Kerem and Elsner 1973). However, more recent studies using new technology (see Toolbox) have demonstrated northern elephant seals have extreme hypoxemic tolerance with arterial Hb saturation reaching values less than 5% (Meir et al. 2009). Similarly, in studies on freely diving northern elephant seals and California sea lions, venous Hb saturation values were near zero at times (Meir et al. 2009; McDonald and Ponganis 2013). Thus, the ability to increase O_2 stores by arterialization of the venous Hb saturation and to deplete O_2 stores to values below 20% on the arterial side and near 0% on the venous side provide increased capacity for longer duration dives in marine mammals.

Muscle oxygen store

The muscle O_2 store is the amount of O_2 in the muscle that can be used during a dive. Like the blood, muscle has a protein to which O_2 molecules bind: myoglobin (Mb). When all Mb proteins are bound with O_2, Mb is 100% saturated. Measurement of the muscle O_2 store is not complicated; it is based on two measurable components: (1) muscle mass and (2) Mb concentration. A third component, the Mb-O_2 carrying capacity, is a constant and the same value as the Hb-O_2 carrying capacity (1.34 ml O_2 g^{-1} Mb). Although not complicated, determining the two measured components can be challenging. Most measurements of muscle mass have been made by directly weighing muscle tissue during anatomical dissections. Thus, for a number of species, particularly endangered mammals or large cetaceans, directly measured muscle mass is still unknown. In many marine mammals where muscle mass has been measured, it is close to 30% of the body mass (Ponganis 2015). This value is often assumed for muscle O_2 store calculations. However, this assumption does not hold for some pelagic cetaceans, beaked whales and balaenopterid whales (e.g., minke whales), in which muscle mass measured during necropsies was 45–62% of body mass (Ponganis 2015).

Mb concentrations have most often been determined using a spectrophotometric (optical measurements) method (Reynafarje 1963). This technique involves determining the difference in tissue optical absorbances at two wavelengths (538 and 568 nm). There may be limitations to this approach (Masuda et al. 2008); nonetheless, it remains the most common method for measuring Mb concentration. A potentially more significant limitation may be the assumption that Mb concentration is constant throughout a muscle or among different muscles. This assumption may be unfounded in at least some species. In several studies of seals, Mb concentrations have been shown to differ: (1) between individual muscles, (2) between locations within a single muscle, and (3) with the season of the year (Neshumova and Cherepanova 1984; Davis 2014; Ponganis 2015). However, at least in the Baikal seal (*Phoca sibirica*), the mean Mb concentration of all muscles was not that different from the Mb content of the primary locomotory muscle (Neshumova and Cherepanova 1984).

Thus, most researchers consider the use of Mb concentration measurements from a single locomotory muscle provides a reasonable estimate of muscle O_2 stores.

Marine mammal adaptations of the muscle oxygen store

One of the hallmarks of deep and long duration divers is a high concentration of Mb in muscle. In all marine mammals except the manatee, muscle is a significant component of the total O_2 store in comparison to their non-diving counterparts (**Figure 3.1**). Marine mammal Mb concentrations vary almost 100-fold, from 0.1 g 100 g^{-1} muscle in manatees to 9.5 g 100 g^{-1} muscle in hooded seals. The Mb concentration in marine mammals is often indicative of their diving capacity and dive patterns. Mb concentration is highest in the deep, long duration divers such as the hooded, northern elephant, harp and Weddell seals (**Figure 3.4**) (Ponganis, Kooyman, and Castellini 1993; Burns et al. 2007; Hassrick et al. 2010). Among otariids,

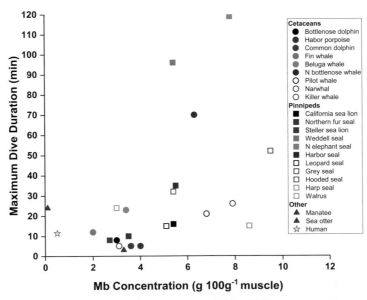

Figure 3.4 Maximin dive durations in relation to myoglobin (Mb) concentrations in marine mammals and humans. Those with the longest dive durations have the highest myoglobin concentrations. Data points represent values for cetaceans (circles), pinnipeds (squares), other marine mammals (triangles) and humans (star). Species within the groups are represented by different colors. (Data from Ponganis 2015; Keith-Diagne et al. 2022)

California (*Z. californianus*) sea lions have some of the highest Mb concentrations, but not as high as phocid seals (**Figure 3.4**) (Weise and Costa 2007).

A large muscle O_2 store, with a particular emphasis on the O_2 stored in the primary underwater locomotory muscles, will have implications for an effective dive response and conservation of O_2 during dives. A high-Mb concentration and large muscle O_2 store decrease the need for blood O_2 delivery to muscle, and increase the duration of aerobic muscle metabolism, especially if muscle work effort is low during a dive. As will be reviewed in the next section of the chapter, a large muscle O_2 store allows for regulation of blood O_2 distribution to other tissues and organs in the body through adjustments in heart rate and degree of constriction of blood vessels.

Dive response

There are several ways marine mammals reduce their O_2 use during dives, including having a lower heart rate or **bradycardia** and **redistributing** blood flow to tissues. These mechanisms occur while diving and are collectively referred to as the **dive response**. Marine mammals also have efficient underwater locomotion, meaning it takes less energy (or O_2 use) for them to move underwater compared to most terrestrial animals moving on land. However, it is the dive response that forms the basis for the management of O_2 stores, and, indeed, for all of diving physiology. While all mammals have some form of a dive response (even humans), the response is strongest in diving mammals. Early studies examined the dive response using forced submersions, where animals were forced underwater for time periods unknown to the animal and the response was extreme. For example, in harbor seals, resting, non-diving heart rate was approximately 110 beats min^{-1}; however, during forced submersions, heart rate dropped to 18 beats min^{-1} (Elsner 1965). Under these conditions, the amount of blood pumping out of the heart (**cardiac output**) is severely reduced.

In order to maintain normal blood pressure, there is widespread **peripheral vasoconstriction** (reducing the diameter) of arterial vessels. This reduced blood flow is directed away from most organs, including working muscles, and primarily distributed to the brain and heart (Blix, Elsner, and Kjekshus 1983). This reduces O_2 consumption, in part, because the metabolic rate of some organs is based on how much blood flow they receive. These organs' metabolic rates are perfusion-dependent and include the kidneys and stomach. Thus, by reducing blood flow to these organs, blood O_2 consumption is decreased. The muscle, which typically consumes much O_2 during exercise, is isolated from the circulation, meaning it receives no blood flow during a forced submersion. Even though the animal is not exercising (swimming), this requires the muscle to rely on its enhanced myoglobin-bound O_2 store for O_2 and, once that O_2 is depleted, to turn to anaerobic glycolysis for ATP production. For these reasons, this severe reduction in heart rate and peripheral blood flow in forced diving (see **Figure 3.5**) results in a very slow rate of blood O_2 depletion that conserves the blood O_2 store for use by the brain and heart (Elsner et al. 1966; Kerem and Elsner 1973).

In free dives and spontaneous breath holds, however, the dive response (reduction in heart rate and peripheral blood flow) is often variable and not as great as during forced submersion (Elsner 1965; Thompson and Fedak 1993; Andrews et al. 1997). This was well illustrated in Elsner's studies of trained breath holds of harbor seals in the 1960s (**Figure 3.5**). In free dives, diving heart rates can be quite variable but are typically in the range of 20–40 beats min^{-1} (about 2–5 times higher than during forced submersions). In addition, blood flow to perfusion-dependent organs (e.g., stomach and kidneys), as well as some muscles, appear to be maintained in short-duration dives of Weddell seals (Davis et al. 1983; Guyton et al. 1995). However, even during free dives, heart rate can drop to low levels at times. For example, during

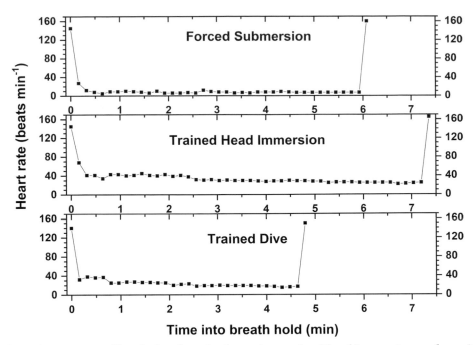

Figure 3.5 Heart rate profiles during forced submersion, trained head immersion, and a trained dive of a young harbor seal (Phoca vitulina) illustrate the differences in the dive response under these three conditions. During forced submersion, heart rates were 5–10 beats min^{-1} (the classic dive reflex with extreme bradycardia). In contrast, heart rates were more variable and higher during the voluntary head immersion and dive, with initial heart rates near 40 beats min^{-1} that later decreased to about 20 beats min^{-1}. The breath holds started at 0 min, and were approximately 6, 7.5, and 4.7 min in duration. Heart rate data are at 10-sec intervals. (Data from Elsner 1965)

a 14-min dive of a grey seal at sea, average dive heart rate was near 4 beats min^{-1} (Thompson and Fedak 1993). Similarly, during deep dives of California sea lions at sea (**Figure 3.6**), heart rate can be less than 10 beats min^{-1} at maximum depth (McDonald and Ponganis 2014). These examples demonstrate that the degrees of bradycardia and tissue blood flow reduction during free dives are variable and probably depend on the nature and circumstances of a given dive. Due to the variability of the dive response, the depletion rate of the blood O_2 store and the duration of aerobic metabolism (before resorting to anaerobic metabolism) can vary in different dives. While during forced submersions, the muscle appears completely cut-off from the blood supply, this may not be true during free dives. This means the depletion rate of the muscle O_2 store will also be variable.

This variability depends on the muscle workload (e.g., stroke rate) during a given dive (see **Chapters 1** and **2**) as well as any additional O_2 delivered by the blood during the dive. The dive response and rates of O_2 store depletion are further detailed in recent reviews (Ponganis, Meir, and Williams 2011; Davis 2014; Ponganis 2015). The range and complexity of the dive response during free dives is illustrated for shallow versus deep dives of a California sea lion in **Figure 3.6**.

Aerobic dive limit

The enhanced O_2 stores, the dive response and the reduced workload of muscle all combine to contribute to the duration of primarily aerobic metabolism during a dive. At some dive duration, O_2 will be depleted in one or more tissues and significant increases in anaerobic metabolism will occur, resulting in lactate accumulation.

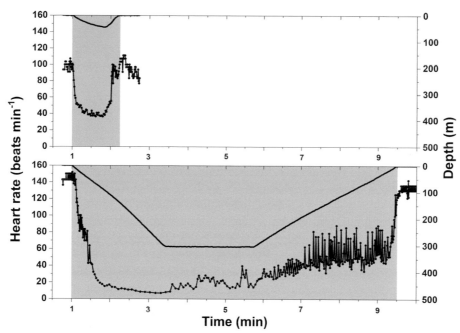

Figure 3.6 Heart rate profiles of a California sea lion during a 1.3-min dive to 45 meters and an 8.5-min dive to 305 meters illustrate the range and variability of the dive response during different types of dives. Note that pre- and post-dive heart rates were higher for the deep dive and that minimum heart rates during the deep dive were near 10 beats min^{-1}, in the same range as observed during forced submersions. Although heart rates during the short dive were higher, minimum heart rates were still less than that at rest on land (54 beats min^{-1}). Gray-shaded area indicates dive period. (Data from McDonald and Ponganis 2014)

This duration is called the aerobic dive limit (ADL) with units of minutes. In dives longer than the ADL, the animal is believed to have increased its use of anaerobic metabolism to produce ATP. Originally, the ADL (**Figure 3.7**) was determined by measuring the dive times that corresponded to increased post-dive blood lactate levels in freely Weddell seals (Kooyman et al. 1980; Kooyman et al. 1983). Increased blood lactate is a signal that anaerobic metabolism had increased during the dive. Importantly, it was noted that most free dives (90–97%) of Weddell seals were less than their ADL, and that dives beyond the ADL were associated with longer surface intervals. The surface interval is the time period during which the animal, while breathing at the surface, replenishes its O$_2$ stores and removes CO$_2$. In dives longer than the ADL, it was believed that additional

time at the surface was needed to process the accumulated blood lactate (Kooyman et al. 1980; Kooyman et al. 1983). Hence, the concept developed that most dives were aerobic, and that it was the efficiency of aerobic metabolism that allowed animals to dive frequently and with short surface intervals.

Although an ADL has only been determined in a few other marine mammals besides the Weddell seal (Ponganis 2015), the concept of an ADL and aerobic diving has been applied to multiple species and has become fundamental in the interpretation of diving behavior and foraging ecology. This widespread application of the ADL concept to animals where the ADL could not be measured has been based on three observations from Kooyman's original papers. These three observations have been used to develop alternate ways to estimate the ADL. Kooyman

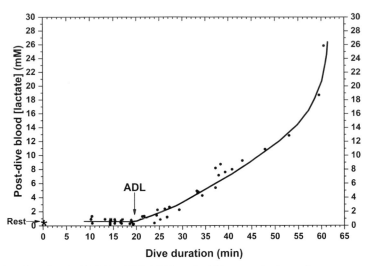

Figure 3.7 Aerobic dive limit (ADL) of Weddell seals. The graph shows lactate concentrations measured from blood samples taken from seals after dives of various durations. Initially, post-dive blood lactate values remain near resting values and then begin to increase at a dive duration near 20 min, the ADL. (Data from Kooyman et al. 1980)

and co-workers noted two behavioral observations: (1) about 95% of dives were less than the blood-lactate determined ADL of the Weddell seal, and (2) after dives in which blood lactate had accumulated, post-dive surface intervals increased (Kooyman et al. 1980; Kooyman et al. 1983). These first two observations led to the use of a **behavioral ADL** (ADL$_b$), which is determined on the basis of either a dive duration distribution with a 95–97% cut-off threshold for the ADL$_b$ (Burns and Castellini 1996; Hindle, Mellish, and Horning 2011) or the dive duration at which surface intervals begin to increase (Kooyman and Kooyman 1995; Horning 2012). The determination of an ADL$_b$ can be difficult because not all species have clear dive duration distributions and some marine mammals will undertake short (recovery) dives even with high post-dive blood lactate levels (Kooyman et al. 1980).

The third observation led to a **calculated aerobic dive limit, ADL$_c$**. Kooyman and co-workers noted that the blood lactate-determined ADL could be reasonably estimated by dividing O_2 stores by a diving metabolic rate. The diving

metabolic rate was the average metabolic rate when the seal surfaced and breathed under a metabolic dome (Castellini, Kooyman, and Ponganis 1992; Ponganis, Kooyman, and Castellini 1993). The metabolic dome was used to measure the total amount of O_2 consumed by the seal during the post-dive surface interval. To calculate the diving metabolic rate, the total O_2 consumed was divided by total minutes of the dive duration plus the surface interval. To obtain the ADL$_c$, this metabolic rate is then divided into the total body O_2 stores.

However, there are important limitations to the ADL$_c$. The diving metabolic rate described before is not the actual O_2 consumption rate during a dive, but rather the O_2 consumption rate of the dive plus the post-dive surface interval. Further, although this approach has been widely used to predict the ADL (dive duration beyond which there is post-dive blood lactate accumulation), this calculation does not reflect the actual status of body O_2 stores at the ADL. As emphasized in many reviews, body O_2 stores are not completely depleted at the ADL (Ponganis, Meir, and Williams 2011; Ponganis 2015), which is often assumed under the ADL_c. Finally, the ADL_c calculation does not describe physiological processes during the dive. The lactate accumulated at and beyond the ADL is considered to be a result of muscle O_2 depletion in the primary locomotory muscles, with subsequent anaerobic metabolism-producing lactate during the dive, which is then washed out into blood after the dive.

Thus, the ADL is an important concept that describes physiological processes and can be difficult to estimate. The ADL and its implications have been most completely demonstrated in an avian diver, the emperor penguin (*Aptenodytes forsteri*), where the ADL and depletion of all three O_2 stores have been measured in freely diving penguins (Ponganis et al. 1997; Williams, Meir, and Ponganis 2011).

TOOLBOX: TECHNOLOGY INNOVATIONS

1   Measuring Oxygen in the Blood
2   Measuring ECG and Heart Rate
3   Measuring Tissue or Blood Oxygen Using Near-infrared Spectroscopy

The measurement of physiological parameters is challenging in a number of ways. As mentioned earlier, the results from early studies of forcibly submerged animals can be quite different from how animals respond while freely diving. Yet, making similar measurements in freely diving animals is much more difficult. Beginning in the 1980s and expanding in the 1990s and 2000s, data loggers were developed which could record and store physiological data. The early instruments could only store a few hours' worth of data, but with technological advances, data from multiple parameters can now be collected and stored for weeks at a time. These small microprocessor-based data recorders can be attached to free-diving animals. However, because these instruments do not transmit data, the need to recapture the animal and remove the physiological sensors is still a major impediment in increasing our understanding of the diving physiology of many marine mammals. Here we review technological devices that measure blood O_2, heart rate, and tissue/blood O_2.

Measuring oxygen in the blood

The amount of O_2 carried by Hb in the blood is primarily determined by the partial pressure of O_2 (P_{O2}). If one knows the P_{O2}, the Hb saturation can be calculated from the relationship between P_{O2} and the percentage of bound Hb (the O_2-Hb dissociation curve). This relationship is explained in more detail in most basic physiology textbooks and reviewed and illustrated in Ponganis, 2015. Continuous P_{O2} and Hb saturation profiles in the blood of diving marine mammals have been obtained using indwelling O_2 electrodes and backpack recorders (Meir et al. 2009; McDonald and Ponganis 2013). The O_2 electrode is an electrochemical sensor that has an O_2-reducing cathode, a reference anode and a membrane permeable to O_2. When the electrode is inserted in a vessel, O_2 in the blood passes through the membrane to the cathode where it reacts and creates an electric current that is correlated with the blood P_{O2}. These studies require multiple steps: electrode calibration, water- and pressure-proofing the entire system, anesthesia of the animal, catheter/electrode insertion, recorder attachment and documentation of the O_2-Hb dissociation curve. To date, such measurements have only been recorded in two species of marine mammals (northern elephant seals and California sea lions) (Meir et al. 2009; McDonald and Ponganis 2013). Blood P_{O2} and Hb saturation have also been measured in one species intermittently, but again with a very complex technique – a programmable, backpack blood sampler, which provided blood samples that could be later evaluated with clinical blood gas analyzers (Hill 1986; Qvist et al. 1986). As will be reviewed in the final section of the toolbox, new near-infrared spectroscopy techniques also hold promise for documentation of Hb saturation.

Measuring ECG and heart rate

Heart rates of diving animals are usually determined from the electrocardiogram (ECG) signal. The typical ECG signal that is associated with a heartbeat consists of a P wave (atrial contraction), QRS complex (ventricular contraction) and T wave (ventricular relaxation). Beat-to-beat heart rates are usually calculated from the time intervals between successive R waves (the R-R interval). There have been two common types of recorders used in heart rate studies of marine mammals (Williams and Ponganis 2021). The first technique consists of an electronic R wave detector that recognizes the R wave based on its height, width and polarity. The second technique is the actual recording of the ECG signal.

The R-wave detector is usually programmed to count the number of detected R waves over a given time interval, and then stores that value in memory. Such recorders provide a record of heart rate at fixed intervals throughout the dive. The advantages of such recorders are that signal processing is automatic, memory usage is minimal and heart rates can be recorded over long time periods. Disadvantages are that, without the ECG, potential error due to artifact cannot be identified and beat-to-beat heart rates are not recorded.

In contrast, an ECG recorder provides a continuous ECG record, from which each R-R interval can be calculated so that the instantaneous heart rate profiles can be constructed. The advantages of this technique are that accurate R-wave detection can be verified, beat-to-beat changes in heart rate are recorded, and variability in heart rate profiles can be assessed. Disadvantages include high memory storage requirements (typically the ECG is recorded at 50–100 Hz), the need for peak detection programs to recognize the R waves and calculate the heart rate, and the time to process the ECG records with such programs.

Another challenge in collecting heart rate data is secure placement of the ECG electrodes in the proper position so that a good signal is detected and there is no muscle or movement artifact in the record. The best position varies with species. Both surface and subcutaneous electrodes have been used in marine mammals (Hill et al. 1987; Thompson and Fedak 1993; Hindle et al. 2010; McDonald and Ponganis 2014). Surface electrodes have usually been attached with epoxy glues in pinnipeds and with suction cups in cetaceans.

Measuring tissue or blood oxygen using near-infrared spectroscopy

The O_2 saturation of Hb in blood or of myoglobin (Mb) in muscle can also be determined using near-infrared spectroscopy (NIR). These techniques typically use light-emitting diodes (LEDs) of different NIR wavelengths and photodiodes to measure absorption of the NIR light by the tissue or blood. How much of the light is absorbed is proportional to Hb or Mb saturation.

A backpack recorder and custom NIR probe are required for study in a diving marine mammal. Implanted NIR probes (and sensors) have been utilized to examine Mb saturations in two species, harbor seals and Weddell seals (Guyton et al. 1995; Jobsis, Ponganis, and Kooyman 2001). In a recent advancement, oxygenation in the brain and intracranial arterial Hb saturations have been estimated non-invasively with surface-mounted NIR probes (McKnight, Ruesch et al. 2021). Although anesthesia is required for attachment of the surface probes and recorder, further evaluation and development of this non-invasive technique hold promise for investigation of blood and tissue O_2 levels.

CONCLUSIONS

The primary challenge in understanding the physiology underlying extended dive durations remains making measurements in freely diving mammals. There are significant differences in the physiological responses to mammals undergoing forced submersions and animals freely diving in their natural environment. Measurements in free-diving cetaceans are especially difficult due to their completely aquatic lifestyle and the complexity of recorder attachment.

As more physiological data are collected in freely diving marine mammals, we are learning how the dive response and O_2 store management vary during different types of dives and between species. We now know that heart rates and O_2 depletion rates can vary considerably depending on the nature and circumstances of a given dive. Yet, the how and why these parameters vary remains a challenge for future studies and new technological advances. Additional physiological parameters discussed before may also vary depending on the type of dive or even within dives; these include diving lung volume, O_2 extraction from lung or blood, hematocrit and initial/final blood O_2 saturations. How these parameters change during or between dives also remains an unanswered question. Finally, the ADL has been defined by the elevation of blood lactate after dives, yet we still have not conquered the challenge of making continuous blood lactate measurements. Such measurements are essential to interpretations of diving/foraging behavior and to evaluation of the ADL hypothesis that most dives are aerobic.

It should now be clear that this area of diving physiology is still an active area of research as new measurement techniques are developed and new measurements are made. Consequently, what we know about O_2 storage in marine mammals is subject to frequent revision and our understanding of how marine mammals can make extended dives continues to improve.

REVIEW QUESTIONS

1 What are the three places where marine mammals store oxygen?
2 How can total lung capacities be similar to terrestrial mammals, yet diving lung volumes may be quite different?
3 Why do some marine mammals exhale before a dive? What are the benefits?

4 How does having high-hypoxemic tolerance help extend dive durations?
5 How do you calculate the amount of O_2 extracted from the blood O_2 store during a dive?
6 Is a high Mb concentration more important for marine mammals that do short or long duration dives? Why?
7 Is it appropriate to use one Mb concentration for all muscles of a marine mammal when calculating the muscle O_2 store? Why or why not?
8 How does the dive response conserve O_2?
9 What is the difference between the dive response in forcibly submerged versus freely diving marine mammals? What might explain this difference?
10 What is the difference between the ADL, ADL_c, ADL_b? Which one gives us information about physiological processes?

CRITICAL THINKING

1 Which body oxygen store would you argue might be the least important for a deep-diving marine mammal and why?
2 Under what circumstances might a marine mammal dive beyond its ADL? What are the benefits and disadvantages?
3 Why are physiological measurements in most freely diving cetaceans so difficult to obtain? How might we overcome those challenges?
4 What are potential applications for human medicine based on marine mammal diving adaptations?

FURTHER READING

Davis, R.W. 2019. *Marine Mammals: Adaptations for an Aquatic Life*. Cham: Springer Nature.
Kooyman, G.L. 1981. *Weddell Seal: Consummate Diver*. Cambridge: Cambridge University Press.
Kooyman, G.L., B.I. McDonald, C.L. Williams, J.U. Meir, and P.J. Ponganis. 2021. The aerobic dive limit: After 40 years, still rarely measured but commonly used. *Comparative Biochemistry and*

Physiology Part A: Molecular & Integrative Physiology 252(110841). https://doi.org/10.1016/j.cbpa.2020.110841.

Williams, C.L., and A.G. Hindle. 2021. Field Physiology: Studying organismal function in the natural environment. *Comprehensive Physiology* 11(3): 1979–2015. https://doi.org/10.1002/cphy.c200005.

GLOSSARY

Aerobic dive limit (ADL): The shortest dive duration associated with an elevation in post-dive blood lactate concentration. This is determined by lactate measurements in post-dive blood samples. The ADL is important because it indicates an increased reliance on anaerobic metabolism, the result of which is lactate accumulation. In the Weddell seal, the accumulation of lactate has been associated with increased time at the surface, which results in less time spent underwater for important activities, such as foraging.

Arterialization: The arterialization of venous blood means that more O_2 than normal is found in the venous circulation. For a diving mammal, such an increase in pre-dive venous O_2 saturation (arterialization) would result in an increased blood O_2 store, allowing these mammals to dive aerobically for longer periods.

Behavioral aerobic dive limit (ADL$_b$): The behavioral ADL is often considered the dive duration below which are 90–97% of all dives of a given animal. The behavioral ADL has also been estimated as the dive duration after which the time spent at the surface (post-dive surface interval) is increased significantly.

Bradycardia: A bradycardia is a heart rate that is below resting levels. In diving mammals, a bradycardia during diving is part of the dive response and is considered to reduce O_2 consumption during the dive.

Calculated aerobic dive limit, ADL$_c$: The total sum of all three oxygen stores in a species divided by the diving metabolic rate. The ADL$_c$ is subject to many assumptions and is often used when the aerobic dive limit cannot be determined.

Cardiac output: The volume of blood that is pumped by the ventricles to the circulation, usually per minute. Cardiac output is calculated from the stroke volume (blood volume per beat) and heart rate (beats per minute).

Dive response: The cardiovascular response during a breath hold or dive. This response involves both a decrease in heart rate and an increase in peripheral vasoconstriction via activation of the parasympathetic and sympathetic nervous systems. **Dive reflex** usually refers to a most extreme dive response as elicited by forced submersion.

Forced submersion: An experimental technique in which an animal is constrained and held underwater for a duration unknown to it. This approach was utilized in early studies on diving physiology.

Hypoxemia: A low blood O_2 level; in humans, arterial hemoglobin saturations less than 90% are considered hypoxemic; **hypoxemic tolerance** refers to the ability of an animal or tissue to tolerate and survive such low O_2 levels.

Lung collapse: The process of alveolar collapse and lack of gas exchange at depth secondary to the increase in ambient pressure during a dive; see Chapter 4.

O_2-Hb dissociation curve: The curve that describes the relationship between hemoglobin saturation and the partial pressure of O_2 (P_{O_2}).

Peripheral vasoconstriction: The constriction of arterial blood vessels. Such constriction maintains blood pressure when heart rate and cardiac output decrease.

REFERENCES

Andrews, R.D., D.R. Jones, J.D. Williams, et al. 1997. Heart rates of northern elephant seals diving at sea and resting on the beach. *Journal of Experimental Biology* 200(15): 2083–2095.

Blix, A.S., R. Elsner, and J.K. Kjekshus. 1983. Cardiac output and its distribution through capillaries and A-V shunts in diving seals. *Acta Physiologica Scandinavica* 118(2): 109–116.

Burns, J.M., and M.A. Castellini. 1996. Physiological and behavioral determinants of the aerobic dive limit in Weddell seal (*Letonychotes weddellii*) pups. *Journal of Comparative Physiology B-Biochemical Systemic and Environmental Physiology* 166: 473–483.

Burns, J.M., K.C. Lestyk, L.P. Folkow, et al. 2007. Size and distribution of oxygen stores in harp and hooded seals from birth to maturity. *Journal of Comparative Physiology B-Biochemical Systemic and Environmental Physiology* 177(6): 687–700.

Castellini, M.A., G.L. Kooyman, and P.J. Ponganis. 1992. Metabolic rates of freely diving Weddell seals: Correlations with oxygen stores, swim velocity, and diving duration. *Journal of Experimental Biology* 165: 181–194.

Davis, R.W. 2014. A review of the multi-level adaptations for maximizing aerobic dive duration in marine mammals: From biochemistry to behavior. *Journal of Comparative Physiology B-Biochemical Systemic and Environmental Physiology* 184(1): 23–53.

Davis, R.W., M.A. Castellini, G.L. Kooyman, et al. 1983. Renal GFR and hepatic blood flow during voluntary diving in Weddell seals. *American Journal of Physiology* 245: 743–748.

El-Sayed, H., S. Goodall, and R. Hainsworth. 1995. Re-evaluation of Evans blue dye dilution method of plasma volume measurement. *Clinical & Laboratory Haematology* 17(2): 189–194.

Elsner, R. 1965. Heart rate response in forced versus trained experimental dives of pinnipeds. *Hvalrådets Skrifter* 48: 24–29.

Elsner, R., D.L. Franklin, R.L. Van Citters, and D.W. Kenney. 1966. Cardiovascular defense against asphyxia. *Science* 153: 941–949.

Elsner, R., J.T. Shurley, D.D. Hammond, et al. 1970. Cerebral tolerance to hypoxemia in asphyxiated Weddell seals. *Respiration Physiology* 9(2): 287–297.

Gentry, R.L., and G.L. Kooyman, eds. 1986. *Fur Seals: Maternal Strategies on Land and at Sea.* Princeton: Princeton University Press.

Grocott, M.P., D.S. Martin, D.Z. Levett, et al. 2009. Arterial blood gases and oxygen content in climbers on Mount Everest. *New England Journal of Medicine* 360(2): 140–149.

Guyton, G.P., K.S. Stanek, R.C. Schneider, et al. 1995. Myoglobin-saturation in free-diving Weddell seals. *Journal of Applied Physiology* 79: 1148–1155.

Hassrick, J.L., D.E. Crocker, N.M. Teutschel, et al. 2010. Condition and mass impact oxygen stores and dive duration in adult female northern elephant seals. *Journal of Experimental Biology* 213(4): 585–592.

Hill, R.D. 1986. Microcomputer monitor and blood sampler for free-diving Weddell seals. *Journal of Applied Physiology* 61: 1570–1576.

Hill, R.D., R.C. Schneider, G.C. Liggins, et al. 1987. Heart rate and body temperature during free diving of Weddell seals. *American Journal of Physiology* 253: R344–R351.

Hindle, A.G., J.A. Mellish, and M. Horning. 2011. Aerobic dive limit does not decline in an aging pinniped. *Journal of Experimental Zoology Part A: Ecological Genetics and Physiology* 315(9): 544–552.

Hindle, A.G., B.L. Young, D.A.S. Rosen, et al. 2010. Dive response differs between shallow- and deep-diving Steller sea lions (*Eumetopias jubatus*). *Journal of Experimental Marine Biology and Ecology* 394(1–2): 141–148.

Horning, M. 2012. Constraint lines and performance envelopes in behavioral physiology: The case of the aerobic dive limit. *Frontiers in Physiology* 3: 381.

Hurford, W.E., P.W. Hochachka, R.C. Schneider, et al. 1996. Splenic contraction, catecholamine release, and blood volume redistribution during diving in the Weddell seal. *Journal of Applied Physiology* 80(1): 298–306.

Jobsis, P.D., P.J. Ponganis, and G.L. Kooyman. 2001. Effects of training on forced submersion responses in harbor seals. *Journal of Experimental Biology* 204(22): 3877–3885.

Kerem, D., and R. Elsner. 1973. Cerebral tolerance to asphyxial hypoxia in harbor seal. *Respiration Physiology* 19(2): 188–200.

Keith-Diagne, L.W., M.E. Barlas, J. P. Reid, et al. 2022. Diving and Foraging Behaviors. In *Ethology and Behavioral Ecology of Sirenia* (pp. 67–100). Cham: Springer.

Kooyman, G.L. 1989. Diverse Divers: Physiology and Behavior. In *Zoophysiology*, edited by D.S. Farner, vol. 23. Berlin: Springer-Verlag.

Kooyman, G.L., M.A. Castellini, R.W. Davis, et al. 1983. Aerobic diving limits of immature Weddell seals. *Journal of Comparative Physiology B-Biochemical Systemic and Environmental Physiology* 151(2): 171–174.

Kooyman, G.L., D.H. Kerem, W.B. Campbell, et al. 1973. Pulmonary gas exchange in freely diving Weddell seals (*Leptonychotes weddellii*). *Respiration Physiology* 17: 283–290.

Kooyman, G.L., and T.G. Kooyman. 1995. Diving behavior of emperor penguins nurturing chicks at Coulman Island, Antarctica. *The Condor* 97(2): 536–549.

Kooyman, G.L., and E.E. Sinnett. 1982. Pulmonary shunts in harbor seals *Phoca vitulina* and sea lions *Zalophus californianus* during simulated dives to depth. *Physiological Zoology* 55(1): 105–111.

Kooyman, G.L., E.A. Wahrenbrock, M.A. Castellini, et al. 1980. Aerobic and anaerobic metabolism during voluntary diving in Weddell seals: Evidence of preferred pathways from blood chemistry and behavior. *Journal of Comparative Physiology B-Biochemical Systemic and Environmental Physiology* 138(4): 335–346.

Lenfant, C., K. Johansen, and J.D. Torrance. 1970. Gas transport and oxygen storage capacity in some pinnipeds and the sea otter. *Respiration Physiology* 9: 277–286.

Masuda, K., K. Truscott, P.-C. Lin, et al. 2008. Determination of myoglobin concentration in blood-perfused tissue. *European Journal of Applied Physiology* 104(1): 41.

McDonald, B.I., and P.J. Ponganis. 2012. Lung collapse in the diving sea lion: Hold the nitrogen and save the oxygen. *Biology Letters* 8(6): 1047–1049.

McDonald, B.I., and P.J. Ponganis. 2013. Insights from venous oxygen profiles: Oxygen utilization and management in diving California sea lions. *Journal of Experimental Biology* 216(17): 3332–3341.

McDonald, B.I., and P.J. Ponganis. 2014. Deep-diving sea lions exhibit extreme bradycardia in long-duration dives. *Journal of Experimental Biology* 217(9): 1525–1534.

McKnight, J.C., E. Mulder, A. Ruesch, et al. 2021. When the human brain goes diving: Using near-infrared spectroscopy to measure cerebral and systemic cardiovascular responses to deep, breath-hold diving in elite freedivers. *Philosophical Transactions of the Royal Society B: Biological Sciences* 376(1831): 20200349.

McKnight, J.C., A. Ruesch, K. Bennett, et al. 2021. Shining new light on sensory brain activation and physiological measurement in seals using wearable optical technology. *Philosophical Transactions of the Royal Society B: Biological Sciences* 376(1830): 20200224.

Meir, J.U., C.D. Champagne, D.P. Costa, et al. 2009. Extreme hypoxemic tolerance and blood oxygen depletion in diving elephant seals. *American Journal of Physiology-Regulatory, Integrative and Comparative Physiology* 297(4): R927–R939.

Miller, P.J.O. 2004. Swimming gaits, passive drag and buoyancy of diving sperm whales *Physeter macrocephalus. Journal of Experimental Biology* 207(11): 1953–1967.

Neshumova, T.V., and V.A. Cherepanova. 1984. Blood supply and myoglobin stocks in muscles of the seal *Pusa siberica* and muskrat *Ondatra zibethica. Journal of Evolutionary Biochemistry and Physiology* 20: 282–287.

Piscitelli, M.A., W.A. McLellan, S.A. Rommel, et al. 2010. Lung size and thoracic morphology in shallow- and deep-diving cetaceans. *Journal of Morphology* 271(6): 654–673.

Piscitelli, M.A., S.A. Raverty, M.A. Lillie, et al. 2013. A review of cetacean lung morphology and mechanics. *Journal of Morphology* 274(12): 1425–1440.

Ponganis, P.J. 2011. Diving mammals. *Comprehensive Physiology* 1: 447–465.

Ponganis, P.J. 2015. *Diving Physiology of Marine Mammals and Seabirds*. Cambridge: Cambridge University Press.

Ponganis, P.J., G.L. Kooyman, and M.A. Castellini. 1993. Determinants of the aerobic dive limit of Weddell seals: Analysis of diving metabolic rates, postdive end tidal PO_2's, and blood and muscle oxygen stores. *Physiological Zoology* 66(5): 732–749.

Ponganis, P.J., G.L. Kooyman, L.N. Starke, et al. 1997. Post-dive blood lactate concentrations in emperor penguins, *Aptenodytes forsteri. Journal of Experimental Biology* 200(11): 1623–1626.

Ponganis, P.J., J.U. Meir, and C.L. Williams. 2011. In pursuit of Irving and Scholander: A review of oxygen store management in seals and penguins. *Journal of Experimental Biology* 214(20): 3325–3339.

Qvist, J., R.D. Hill, R.C. Schneider, et al. 1986. Hemoglobin concentrations and blood gas tensions of free-diving Weddell seals. *Journal of Applied Physiology* 61: 1560–1569.

Reynafarje, B. 1963. Simplified method for the determination of myoglobin. *Journal of Laboratory and Clinical Medicine* 61: 139–145.

Ridgway, S.H. 1986. Diving by cetaceans. In *Diving in Animals and Man*, edited by A.O. Brubakk, J.W. Kanwisher and G. Sundnes, 33–62. Trondheim: Royal Norwegian Society of Science and Letters.

Ridgway, S.H., and D.G. Johnston. 1966. Blood oxygen and ecology of porpoises of three genera. *Science* 151: 456–458.

Ridgway, S.H., B.L. Scronce, and J. Kanwisher. 1969. Respiration and deep diving in the bottlenose porpoise. *Science* 166: 1651–1654.

Scholander, P.F. 1940. Experimental investigations on the respiratory function in diving mammals and birds. *Hvalrådets Skrifter* 22: 1–131.

Sleet, R.B., J.L. Sumich, and L.J. Weber. 1981. Estimates of total blood volume and total body weight of a sperm whale (*Physeter catodon*). *Canadian Journal of Zoology-Revue Canadienne De Zoologie* 59(3): 567–570.

Thometz, N.M., M.J. Murray, and T.M. Williams. 2015. Ontogeny of oxygen storage capacity and diving ability in the southern sea otter (*Enhydra lutris nereis*): Costs and benefits of large lungs. *Physiological and Biochemical Zoology* 88(3): 311–327.

Thompson, D., and M.A. Fedak. 1993. Cardiac responses of grey seals during diving at sea. *Journal of Experimental Biology* 174: 139–164.

Thornton, S.J., D.M. Spielman, N.J. Pelc, et al. 2001. Effects of forced diving on the spleen and hepatic sinus in northern elephant seal pups. *Proceedings of the National Academy of Sciences of the United States of America* 98(16): 9413–9418.

Weise, M.J., and D.P. Costa. 2007. Total body oxygen stores and physiological diving capacity of California sea lions as a function of sex and age. *Journal of Experimental Biology* 210(Pt 2): 278–289.

Williams, C.L., J.U. Meir, and P.J. Ponganis. 2011. What triggers the aerobic dive limit? Patterns of muscle oxygen depletion during dives of emperor penguins. *Journal of Experimental Biology* 214(11): 1802–1812.

Williams, C.L., and P.J. Ponganis. 2021. Diving physiology of marine mammals and birds: The development of biologging techniques. *Philosophical Transactions of the Royal Society B: Biological Sciences* 376(1830): 20200211.

CHAPTER 4

UNDER PRESSURE

Sascha K. Hooker and Chris McKnight
Sea Mammal Research Unit, University of St. Andrews, Scotland, UK

Andreas Fahlman
Fundacion Oceanografic, Spain and Kolmården Wildlife Park, Sweden

Sascha K. Hooker

Sascha has been studying the diving and movements of marine mammals for over three decades. She was the first person to tag a beaked whale, showing that northern bottlenose whales regularly dive to a mile deep. She has maintained a strong interest in the physiology underlying the diving behavior of both whales and seals. She has been a faculty member at the University of St. Andrews since 2004.

Chris McKnight

Chris is a marine mammal biologist who has spearheaded efforts to incorporate medical wearable optical technologies into marine mammal tags. He primarily focuses on the comparative aspects of diving physiology and studies both marine mammal and human free-divers. His particular interests are in cerebrovascular and cardiovascular function.

Andreas Fahlman

Andreas is a comparative physiologist with experience estimating field metabolic rate of marine vertebrates, in particular helping test ethically and logistically viable methods to estimate cetacean energy requirements. He is interested in marine mammal cardiorespiratory physiology, especially the effect of pressure on lung function. He uses 'modern physiology', incorporating tools from other disciplines such as biochemistry, molecular biology and mathematics to complement physiological data and solve central physiological questions.

Driving Question: How do marine mammals cope with huge pressures at depth?

Human free-diving record-holders can perform a single dive to 214 m, but this sport is notoriously dangerous and prone to fatalities caused by the effect of pressure and lack of air on the body. In contrast, marine mammals like elephant seals and beaked whales routinely catch prey at depths of 2 or 3 km. To all extents and purposes, they live at depth with occasional visits to the surface. How do they cope with the enormous pressure,

DOI: 10.1201/9781003297468-4

hundreds of times greater than air pressure, found at those depths? How do they avoid the pressure-related diving diseases that limit human divers: 'the bends', high-pressure nervous syndrome, shallow-water blackout or nitrogen narcosis?

MARINE MAMMAL-DIVING BEHAVIOR

Marine mammal species show very different diving patterns, driven by where they live and the depth of their prey (*Table 4.1*, **Figure 4.1**). Elephant seals, beaked whales and sperm whales dive repeatedly to great depths, while others, such as dugongs, spend much of their lives in relatively shallow sea-grass environments. Polar bears might be thought of as surface dwellers but often swim daily in the summer and reach depths of 14 m. The smallest marine mammal, the sea otter, often feeds at 10 m depth but can dive to 100 m. Even among the deep divers, some are relatively slow-moving, while others descend and ascend at high speeds (**Figure 4.1**). With the advent of sophisticated microelectronic-monitoring devices, we now have a better understanding of

Table 4.1 **Overview of marine mammal genera and their approximate diving depths (shallow <200 m; moderate 200–500 m; deep 500–1000 m; very deep >1000 m, those shown in bold have been confirmed by time-depth recorder, those not in bold are speculated)**

ORDER Suborder	FAMILY	COMMON NAME (NUMBER OF SPECIES)	APPROXIMATE DIVING DEPTHS*	SHOWN IN FIGURE 4.1
CETACEA				
Mysticeti	Balaenidae	Right and bowhead whales (4)	**shallow/moderate**	
	Neobalaenidae	Pygmy right whale (1)	shallow	
	Balaenopteridae	Rorquals (8)	**shallow/moderate**	
	Eschrichtiidae	Gray whale (1)	**shallow**	
Odontoceti	Physeteridae	Sperm whale (1)	**very deep**	Physeter
	Kogiidae	Pygmy and dwarf sperm (2)	very deep	
	Monodontidae	Narwhal and beluga (2)	**deep**	
	Ziphiidae	Beaked whales (21)	**very deep**	Ziphius
	Delphinidae	Oceanic dolphins (36)	**moderate/deep**	Globicephala
	Phocoenidae	Porpoises (6)	**shallow/moderate**	Phocoena
	Platanistidae	South Asia river dolphin (1)	shallow	
	Iniidae	Boto (1)	shallow	
	Lipotidae	Baiji (extinct)	shallow	
	Pontoporiidae	Franciscana (1)	**shallow**	
SIRENIA	Trichechidae	Manatees (3)	**shallow**	
	Dugongidae	Dugong (1)	**shallow**	
CARNIVORA	Mustelidae	Otters (2)	**shallow**	
	Ursidae	Polar bear (1)	**shallow**	
Pinnipedia	Otariidae	Fur seals and sea lions (16)	**shallow/moderate**	Arctocephalus
	Phocidae	True seals (18)	**moderate/deep/ very deep**	Mirounga
	Odobenidae	Walrus (1)	**moderate**	

* We suggest the online Penguinness database for updates on diving depths listed by species (Ropert-Coudert et al. 2018).

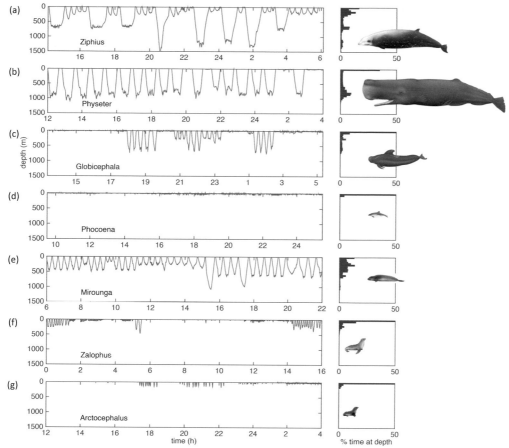

Figure 4.1 Dive traces (left) and frequency histograms (right) showing time at depth (in 50 m depth intervals) for various marine mammal species. Icons show the relative size of these species © Uko Gorter. Many marine mammal species spend over 50% of their time in the upper 50 m of the water, but elephant seals and beaked whales spend less than 20% of their time shallower than 50 m. Dive traces are plotted to identical scales: 1500 m depth over a 16 h time period, to illustrate the differences in use of depth and patterns of diving between species. (a) Cuvier's beaked whale (Ziphiidae), Mediterranean, Sept-03, (b) sperm whale (Physeteridae), Azores, Aug-10, (c) short-finned pilot whale (Delphinidae), Canary Islands, Oct-04, (d) harbor porpoise (Phocoenidae), Jutland Peninsula, Denmark, Oct-12, (e) northern elephant seal (Phocidae), eastern Pacific, Mar-14, (f) California sea lion (Otariidae), San Nicholas Island, California USA, Nov-06, and (g) Antarctic fur seal (Otariidae), South Georgia, Dec-01. (Reproduced from Hooker and Fahlman (2016). Data courtesy of (a) M. Johnson, (b) C. Oliveira, (c) N. Aguilar and M. Johnson, (d) D. Wisniewska (e) D. Costa, (f) D. Costa, and (g) S. Hooker)

the diving behavior of many marine mammal species, but we still do not fully comprehend how their physiology may limit their diving behavior.

Body mass and dive depth

The deepest divers tend to be relatively large (*Table 4.1*). This may be more related to dive duration than depth, but since marine mammals tend to ascend and descend at similar speeds, the deepest dives are by default the longest.

The tendency for greater body size to allow greater dive durations is largely due to the differential scaling. Oxygen stores scale similarly to mass but metabolic rate scales to the 3/4 power of the animal's mass (Kleiber's law, see also **Chapter 2**). This means that larger species have larger oxygen stores relative to the rate at which they use oxygen, and they can breath-hold for longer. Of course, this is somewhat of an oversimplification, and there is also considerable variation in diving ability that is not explained by body mass alone. Many other factors can affect both oxygen stores and metabolic rate (see **Chapter 10**). For example, energetic-foraging methods such as the lunges performed by blue and fin whales to engulf swarms of krill cost a lot of energy, increase metabolic rate and so limit their diving capacity.

As animals grow, their diving capacity increases due to this relative difference in scaling of oxygen stores and metabolism. Older animals can dive deeper, and this has been observed among both cetacean and pinniped species. It is thought to underlie the babysitting behavior seen in sperm whales because young animals are unable to dive as deep or for as long as adults. Preliminary work on the mechanical properties of the lung in pinnipeds has also shown higher and more variable lung **compliance** in wild pinnipeds compared with those

raised in captivity, possibly indicating that lung conditioning is important for diving animals and that repeated diving helps protect against **lung squeeze** and maintain healthy lung function (Fahlman, Loring et al. 2014).

PRESSURE, GAS VOLUME AND SOLUBILITY

Why is diving problematic? We can take a step back and look at some of the physics. Pressure is a measure of the force exerted over an area. Atmospheric pressure is the force exerted on a surface by the air above it as gravity pulls it toward the center of the Earth. If we go up a mountain, the air pressure becomes less. In the ocean, hydrostatic pressure is the pressure exerted by water due to the force of gravity. Pressure, therefore, increases as you go deeper because of the increasing weight of water exerting downward force from above. From atmospheric pressure (1 **ATA**) at the surface, pressure increases by 1 ATA for every 10 m descended. At 1000 m depth, the pressure is one hundred and one times greater than at the surface (**Figure 4.2**).

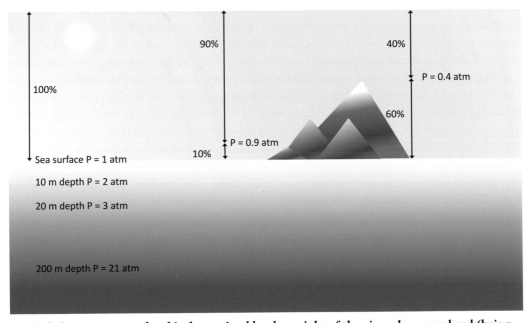

Figure 4.2 Pressure at any level is determined by the weight of the air and sea overhead (being pulled down by gravity).

Partial pressure is where things get a little more complicated. Partial pressure is the pressure exerted by one among a mixture of gasses if it occupies the same volume. The volume of 1 mole of any dry gas (6.02×10^{23} molecules) under standard temperature (0°C) and pressure (1 ATA) is the same (22.7 l) for all gases. In a mixture of gasses, each gas exerts its own pressure (partial pressure) and the total gas pressure is the sum of all partial pressures (Dalton's law; **Figure 4.3**). So, partial pressure will be related to the proportions of each gas. In air at the surface (1 ATA), nitrogen (80%) will exert 0.8 ATA, oxygen (15%) will exert 0.15 ATA and carbon dioxide (4%) will exert 0.04 ATA. At 10 m depth (2 ATA), these will double.

Pressure exerts a large influence on the body via its effect on gas spaces and solubility.

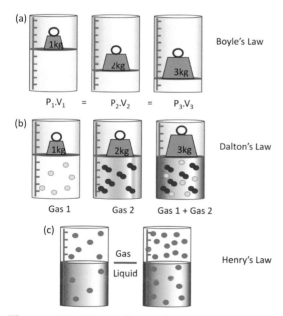

Figure 4.3 The gas laws relating to pressure: **(a)** Boyle's law states that volume will decrease in inverse proportion to the increase in pressure; **(b)** Dalton's law states that total gas pressure is the sum of all partial pressures so that all gases remain in the same proportions; **(c)** Henry's law states that the solubility of gas in a liquid is directly proportional to the partial pressure of the gas above the liquid.

Gas volume changes in a diving animal in an inverse relationship with pressure (Boyle's law, **Figure 4.3**). A popular experiment to show this is to attach a Styrofoam cup to deep-ocean sampling equipment. Such cups return from depth thimble sized as the pressure has compressed the air in them, forcing this structural change. Air spaces in animals similarly become compressed at depth, potentially causing damage to any surrounding rigid structures, e.g., **barotrauma** or lung squeeze.

Gas solubility is also affected by pressure. Increasing pressure increases the solubility of dissolved gases (Henry's law, **Figure 4.3**). This is well illustrated by the gas dissolved in a carbonated drink. Unopened drinks contain carbon dioxide at slight pressure, causing high solubility of carbon dioxide in the drink. When the lid is opened, the pressurized CO_2 escapes and the partial pressure of CO_2 drops to match that in the air. The carbon dioxide in solution is 'released' forming gas bubbles as its solubility decreases (and after a while your drink sadly goes flat). Hyperbaric (increased pressure) therapy uses the same idea. Exposure to oxygen at higher than atmospheric pressure is used to treat hypoxia (low oxygen supply in the tissues), by increasing the oxygen partial pressure in the lungs, which then increases dissolved oxygen in the blood.

Variation in solubility also affects the relationship between pressure and dissolved gas. Gases vary in their solubility: N_2 is less soluble than O_2, but CO_2 has high solubility (and dissociates into carbonic acid). Consequently, the amount of gas dissolved will be related to both the partial pressure and the solubility. Human divers sometimes breathe an artificial mixture called heliox (helium and oxygen) rather than air, because helium is only one-fifth as soluble in blood as nitrogen. As a result, less helium is dissolved in the blood at depth, and divers are less likely to get the bends (see the **Diving Diseases** section).

During descent, the pressure the animal is exposed to increases, and the partial pressure of the gases in the lung increases in proportion (Dalton's law). While the alveoli are open and there is gas exchange, the gas tension in the **arterial** blood increases as gas diffuses from high to low pressure. This leads to many of the gas diseases that all divers face.

BIOCHEMISTRY AND BLOOD FLOW

Pressure can also alter biochemical reactions and cellular structure (Somero 1992). Lipid chains become extended under pressure resulting in increased thickness of the lipid membrane layer around cells and impacts on membrane transport reactions. Protein denaturation (changing their 3D structure and affecting their function) is only caused at pressures >4000 ATA, which is much deeper than marine mammals would ever experience. However, lower pressures that marine mammals face (up to 300 ATA) can still affect the changes in volume caused by chemical reactions, altering the biochemical reactions in the cells. Under higher pressure, the position of equilibrium in a chemical reaction will move in the direction of fewest moles of gas, so when a process occurs with an increase in volume, high pressure inhibits that process, and when a process occurs with a decrease in volume, pressure enhances that process.

Pressure sensitivities of enzymes, structural proteins, and membrane-based systems appear to differ between shallow- and deep-living species. Some fish that live in shallow waters up to 500 m already show biochemical adaptations that allow their enzymatic reactions to be pressure-tolerant (Somero 1992). Although many marine mammal species stay in relatively shallow waters for much of the time, others visit depths well beyond 500 m (*Table 4.1*). In general, marine deep-sea fish and other deep-water species are adapted to their specific depth range and optimize function for a relatively narrow range of pressure. In contrast, marine mammals need to travel through and function in the entire range of pressures experienced between the surface and their maximum dive depth.

The few studies of biochemical tolerance to pressure in marine mammals have shown adaptations among tissue enzymes and red blood cells, which either show no reaction to pressure changes, or function better under pressure conditions (Castellini et al. 2002). Immune responses investigated in beluga blood plasma showed either no or decreased response under pressure, hypothesized to assist with ensuring no damage or aberrant immune reaction to inflammation or bubbles (see the **Diving Diseases** section). Normal clotting tendency was found to be reduced in northern elephant seal platelets, thought to be a protective mechanism developed in response to rapid pressure changes and cold temperatures. Cetaceans in general appear to lack blood-clotting factors (e.g., Hageman factor) common to terrestrial mammals, so their blood is described as **hypocoagulable**. This may help to improve microcirculation at depth or reduce **venous** thrombosis.

Pressure in the body can have a detrimental effect on both blood pressure and blood flow to the brain. Marine mammals possess extensive networks of blood vessels (called venous **plexuses**). Some, such as the retia mirabilia found in cetaceans and sirenians, might be related to diving ability. In cetaceans, these are a series of vascular networks of densely looped arteries along the base of the brain case, along and within the vertebral column, and in the thorax. It is the only path of arterial blood to the brain in adult cetaceans and may be an adaptation allowing effective drainage of blood from the central nervous system during periods of elevated pressure. The rete appears to dampen pressure variation and although blood pressure is maintained, the pressure is non-pulsatile: there is no systolic peak or diastolic trough. Alternatively, or additionally, the venous portion of these retia may allow expansion of these blood vessels to fill some of the reduced thoracic volume caused by pressure increase, i.e., helping to prevent lung squeeze (see the **Lungs and Alveolar Collapse** subsection). The arterial portion of the rete could also potentially act as a filter (neuroprotective effect) for arterial gas emboli preventing decompression sickness (see the **Diving Diseases** section) (Nagel et al. 1968).

AIR SPACES

It is the airspaces that cause the most problem to diving animals, and for this reason many of the airspaces present in terrestrial mammals have been lost in the evolution of marine mammals. There are three major airspaces that are a liability for divers – the lung, the facial **sinuses** and the middle ear.

The middle ear is an air-filled rigid cavity with little or no compressibility. In pinnipeds, a pressure differential is prevented by a complex vascular sinus that can fill with blood to line the wall of the middle ear cavity (Odend'hal and Poulter 1966). Increasing pressure would reduce the air volume, but instead the negative pressure that develops in the middle ear during diving pulls blood into the venous sinus and helps fill the gap left by air volume reduction, balancing the pressure differential. Marine mammals have lost the paranasal facial sinuses found in terrestrial mammals and so avoid problems with these. Nevertheless, they do have a series of gas-filled sinuses in the head and possess elaborate plexiform **veins** in these. Bigger and more elaborate sinus vasculature is found in deep divers (e.g., physeterids, kogiids, ziphiids) compared to shallow-diving delphinids (Fraser and Purves 1960). These gas-filled sinuses are associated with the middle ear and upper respiratory system.

Lungs and alveolar collapse

The major air space that is affected by pressure during diving is the lungs. Lung squeeze occurs when pressure reduces the air volume within an incompressible rib cage. Human divers have largely incompressible ribs and chest wall and so historically it was thought that they could not physically breath-hold dive deeper than 40 m. In fact, they can, but this is achieved by shifting blood into these spaces – expanding blood vessels in the chest area to counteract the compression of the lungs. In pinnipeds, and possibly cetaceans, the chest is much less rigid and offers little resistance to compression regardless of the depth to which they dive (Fahlman, Loring et al. 2014).

The venous thoracic rete that exists in many cetaceans may also help pool blood in the thoracic cavity and prevent excessive negative intrathoracic pressures from developing (Hui 1975). In addition, the inner wall of the trachea of dolphins and sperm whales may also engorge with blood and so off-set the limited compression of a stiff trachea. Blood engorgement may help with any reduction in gas volume beyond the compliance available in both chest and trachea, preventing barotrauma.

The second problem relates to gas uptake by the pressurized lung. As depth increases, the lung volume reduces and the pulmonary pressure and the partial pressure of the gases in the respiratory system increase (Boyle's and Dalton's laws). This increases the solubility of these gases within the blood (Henry's law), potentially causing a problem with uptake of N_2 and the risk of forming inert gas bubbles during ascent and decompression (DCS) symptoms (see the **Diving Diseases** section).

Marine mammals show modifications to their lungs compared to terrestrial mammals. In general, marine mammals have reinforced upper airways and a lack of smaller respiratory bronchii compared to terrestrial mammals. This reinforced lung structure may facilitate high-ventilation rates at the surface (Denison and Kooyman 1973). There is variation in tracheal stiffness observed between several phocid and odontocete species (Moore et al. 2014), perhaps related to variation in life history and diving abilities. Dolphins show the most extreme modifications including the presence of a series of bronchial **sphincter** muscles found in the terminal segments of the airways. The function of these is largely unknown, but it is hypothesized to relate to management of lung air. In combination with chest wall musculature, compression/expansion of the chest could be used to help control volume (and thus buoyancy) without the need to exhale. Otariids (fur seals and sea lions) have robust cartilaginous airway reinforcement extending to the alveolar sac, whereas phocids have no cartilage in the last airways before the alveolar sacs, but the walls are thickened by connective tissue and smooth muscle (Kooyman 1973).

During diving, these more compliant alveoli and stiffer trachea have been likened to a balloon (the alveoli) on the end of a pipe (the trachea). As the pressure increases with depth, the thin-walled alveoli compress. Effective gas exchange between lungs and blood ceases when all alveoli are collapsed (also called **atelectasis** and often referred to as lung collapse). Alveolar compression results in a gradual reduction of the alveolar surface area and increasing alveolar thickness both of which reduce diffusion (i.e., causing a pulmonary shunt) and eventually lead to a termination of gas exchange (**Figure 4.4**). Alveolar collapse has been estimated at 70 m for bottlenose dolphins based on N_2 washout rates observed (Ridgway and Howard 1979), at 30 m in Weddell seals based on an arterial N_2 blood sampler (Falke et al. 1985) and at 225 m for California sea lions based on continuous arterial O_2 sensors (McDonald and Ponganis 2012). **Hyperbaric chambers**, sealed chambers within which the pressure can be increased, are usually used to treat diving and other injuries (often combined with oxygen inhalation). Experimental work has used these chambers to investigate alveolar collapse depths and showed depths closer to 170 m for sea lions (Kooyman and Sinnett 1982).

But what if the pipe is not completely rigid? The trachea is certainly more robust than the alveoli, but hyperbaric chamber studies have

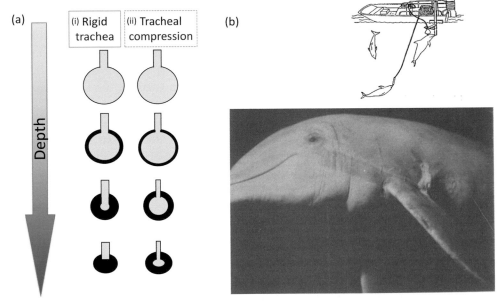

Figure 4.4 Compression of lung and thorax during diving results in changes in gas/diffusion between lungs and blood. **(a)** With increasing depth and pressure, the lungs show a graded compression as the air volume reduces according to Boyle's law. Two models are shown: (i) balloon-and-pipe model of a rigid trachea and compliant alveoli; (ii) tracheal compression at depths less than alveolar collapse. **(b)** Chest compression of a bottlenose dolphin (Tuffy), photographed at 300 m depth, with experimental setup shown above. Thoracic collapse is visible behind the left flipper. (Image copyright US Navy; details published in Ridgway, Scronce, and Kanwisher (1969). Reproduced from Hooker and Fahlman 2016)

shown that it can still compress under pressure (Kooyman et al. 1970). If the trachea were to begin to compress prior to full alveolar collapse, the predicted alveolar collapse depths would be deeper than predicted by Scholander's balloon-pipe model (**Figure 4.4**). A model with a compliant trachea predicts alveolar collapse depths of 110m for bottlenose dolphins based on the (Ridgway and Howard 1979) study. It is possible that the peak arterial N_2 seen at 30 m depth by (Falke et al. 1985) might correspond to changes caused to the diffusion rate and that alveolar collapse had not yet occurred (Bostrom et al. 2008). We still have very little information about these physiological responses and so this is still relatively unknown.

Management of gas exchange probably varies both within and between species due to differences in respiratory anatomy and behavior, e.g., shallow versus deep divers. Deep divers seem to have smaller lungs as compared with shallow-diving species (Piscitelli et al. 2010). The diving lung volume, i.e., the air volume at the outset of the dive, is likely to be the most important predictor of the depth of alveolar collapse. Different species also have different tactics in terms of inspired air volume. Although phocid seals are thought to exhale prior to diving, the important question is to what extent does this occur, i.e., what is the diving lung volume? They may in fact still dive with as much as 60% of their inspiratory volume (Kooyman et al. 1971). Fur seals, sea lions and cetaceans are thought to dive after inhalation (Kooyman 1973).

Whatever the depth of alveolar collapse, many marine mammal species will reach depths sufficient to cause a period of alveolar collapse followed by re-inflation. That they do this repeatedly and without any apparent side-effects as the animal resurfaces is remarkable. The lungs produce a pulmonary **surfactant**, which acts as a lubricant. It lines the alveolar air-water interface and varies the surface tension to reduce the work of breathing, alter compliance, and prevent adhesion of respiratory surfaces. An anti-adhesive surfactant with greater fluidity

and rapid expansion capabilities to cope with repeated collapse and reinflation has been reported in pinnipeds. These functional adaptations are caused by molecular modifications in key protein and lipid compositional changes, as well as adaptations in the secretory mechanisms of the cells (Foot et al. 2006). The surfactant compositions in pinnipeds may have been selected to help recruitment of closed alveoli following deep dives. Surfactant production may also be triggered by diving, supporting the idea that pressure is the driving force behind observed differences in surfactant levels (Foot et al. 2006). While selection at the molecular level has been investigated between different marine mammal groups, analysis of surfactant composition has been primarily on samples from pinnipeds and much less is known about cetacean surfactants. Preliminary data suggest that the fluidizing lipids are not increased in odontocetes (Gutierrez et al. 2015).

PRODUCING SOUNDS

Whales need air to produce sounds at depth. For toothed whales this is particularly crucial since they hunt their prey using echolocation clicks. As we have already seen, some species of toothed whales may dive very deeply. This means that the air available for sound production will be vastly diminished in volume (due to Boyle's law) when they need it at depth. As discussed above, lung air is sequestered into the thick-walled trachea and distensible nasal sacs as the lungs collapse with increasing pressure. Air is passed through structures called phonic lips to produce a click. This air does not exit the blowhole but collects in sacs on the other side of the phonic lips. Particularly at depth, where air is limited, the number of clicks produced before air needs to be recycled will be limited. Echolocating whales must recycle air from one side of the sound-producing lips back to the other, entailing a pause in click production (**Figure 4.5**). To produce their continuous series of clicks (allowing homing-in on

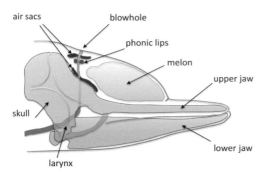

Figure 4.5 Toothed whales (odontocetes) need air at depth for sound production. Lack of blood vessels around such air sacs will minimize risk of gas absorption, but the problem for deep divers arises from the reduction in volume of these sacs as the whale or dolphin dives. Whales need to recycle air from anterior to posterior sacs to continue producing sound at great depths.

escaping prey), very small air volumes as little as 50 microliters are needed for each click. This is the volume of a drop of water. Furthermore, animals need to adjust click intensity to help ensure continuity of clicks when needed, and likely manage recycling pauses during times when sensory input from echolocation is less critical (Foskolos et al. 2019). See also **Chapter 6** for more on this topic.

DIVING DISEASES

There are several diving diseases brought about by the effect of pressure on the body (**Figure 4.6**). These may be separated into three types:

Those caused by high pressure:
- **high-pressure nervous syndrome (HPNS)** which is caused by hydrostatic pressure-inducing tremors and convulsions.

Those caused by elevated concentrations of gas under pressure:
- **nitrogen narcosis** is caused by the anesthetic effect of nitrogen at high

pressure ultimately leading to loss of consciousness at deeper depths.
- **oxygen toxicity** is caused by high partial pressure of O_2 for which short exposure causes central nervous system toxicity, and longer exposure causes toxicity to the lungs and eyes.

Those caused by the reduction in pressure during ascent:
- **shallow-water blackout** is a loss of consciousness due to cerebral **hypoxia** caused by depletion of blood O_2 associated with a rapid drop in partial pressure of lung O_2 toward the end of a breath-hold dive.
- **decompression sickness** (or 'the bends') is caused by dissolved gas coming out of solution and forming bubbles during a reduction in pressure. Bubbles may cause damage to tissues or organs or may **embolize** blood vessels leading to ischemic damage. It was called the bends because in human divers, bubbles would cause joint and muscle pain, resulting in a characteristic bent shape of the patient.

We are familiar with these syndromes due to their sometimes dramatic effect on human divers. Diseases caused by elevated concentrations of gases are more of a problem for scuba divers than for breath-hold divers because breathing compressed air exposes them to the damaging effects of increased blood gas concentrations. At 30 m depth, they breathe four times the gas concentrations as someone breathing at the surface. In contrast, breath-hold divers bring down only what is contained in the lungs, which mixes with gases previously absorbed in the blood that were not completely removed during their time at the surface. Scuba divers absorb significantly more N_2 compared to breath-hold divers. Scholander (1940) demonstrated this by pressurizing a container holding two frogs. One frog was held underwater, and one frog was allowed to breathe under pressure.

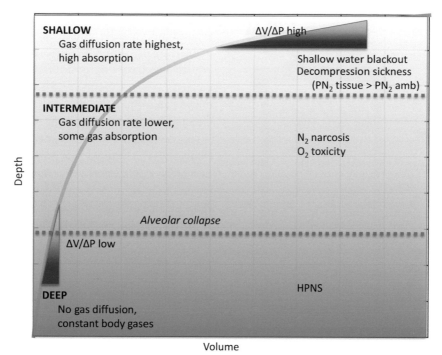

Figure 4.6 Risk of different diving related problems are related to pressure and gas diffusion. The water column can be divided into a shallow, intermediate and deep region. In the shallow region, the rate of change of volume ($\Delta V/\Delta P$) is high, and there is high gas diffusion (see Figure 4.4). In this region, gases are exchanged and animals may be at risk of gas bubble disease when blood and tissue PN_2 exceeds ambient pressure. Shallow-water blackout may also occur in this region due to the rapid changes in gas volume. In the intermediate region, a reduction in alveolar surface area and thickening of the alveolar membrane reduces gas exchange. The N_2 and O_2 taken up may cause nitrogen narcosis and increase the risk for O_2 toxicity. Once the alveoli collapse in the deep region, no further gas is exchanged and as pressure increases, animals may be more at risk of HPNS. (Reproduced from Hooker and Fahlman 2016)

The frog breathing under pressure died while that held underwater (breath-holding) survived (**Figure 4.7**).

Pressure and elevated gas concentrations

In humans, the symptoms of HPNS appear at pressures exceeding 10 ATA, although individual differences make it difficult to give an absolute pressure (or depth) at which the symptoms first appear (Bennett and Rostain 2003). Although species show differences in their susceptibility to elevated pressures, this variability appears to be related to the complexity of the CNS in that organisms with a less complex CNS seem to have a higher tolerance. Many marine mammals would, therefore, appear to be at risk of HPNS since they have a more complex CNS, they have fast descent rates (1–2 m/s) and spend time at pressures far exceeding 11 ATA.

We still know very little about why marine mammals are apparently unaffected by HPNS, and whether they possess specific neuroanatomical and physiological adaptations to protect them. The exact mechanism is currently unknown, but compression is thought to change the structure and function of the neural membranes, with secondary effects resulting in

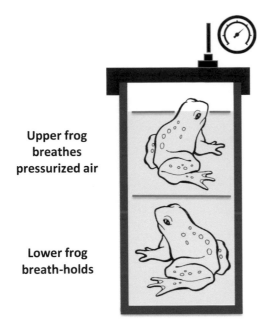

Upper frog breathes pressurized air

Lower frog breath-holds

Figure 4.7 Scholander (1940) demonstrated the greater danger from breathing air under pressure compared to breath-holding at the same pressure. Placing two frogs in a pressure bomb with one able to breathe, but the other breath-holding, the breathing frog would die, while the breath-holding frog would be unharmed.

HPNS. Secondary effects could involve changes in axonal conduction, synaptic transmission, or changes in the release of neurotransmitters. Generally, membranes of the nervous system function over only a small range of pressures – so the deep-diving marine mammals are all the more remarkable for their tolerance to a high range of pressures. Recent work has suggested that there are genetic differences between nervous system receptors of marine mammals and terrestrial mammals and that these may help prevent HPNS.

Interestingly, anesthetic gases appear to ease symptoms and provide increased pressure tolerance in terrestrial mammals (Hunter and Bennett 1974). As N_2 behaves like an anesthetic gas at elevated pressure, it has been suggested that high N_2 levels could help. The myelin sheath that surrounds nerve cell axons is made up of lipids, and since N_2 has a higher solubility in lipids, this might alter the structure around nerve cells and help ensure effective neurotransmission. Of course, high pressures of N_2 can also have a narcotic effect (causing nitrogen narcosis in humans), and it is not known why marine mammals do not suffer from the same effect. The critical PN_2 for prevention of HPNS is less than that causing N_2 narcosis in humans (Halsey 1982), and so it is possible that regulation of N_2 could ensure that levels are enough to ensure that animals do not get HPNS but that they are low enough to avoid N_2 narcosis.

In terrestrial mammals, there is a relationship between O_2 tension (PO_2), exposure time and O_2 toxicity (Harabin et al. 1995). During breath-holding, however, O_2 toxicity is unlikely to be a problem as the PO_2 will only transiently reach high pressures and the continuous uptake of O_2 will not last for an extended period. Oxygen toxicity is, therefore, not considered a problem for marine mammals.

Shallow-water blackout

Shallow-water blackout has been documented for human breath-hold divers, and therefore presents a potential problem for marine mammals. In humans, it is often linked with hyperventilation before the dive, which helps reduce the vascular CO_2 tension (PCO_2) and thus the urge to breathe. During the dive, the arterial PO_2 (PaO_2) drops as O_2 is consumed. On ascent, re-expanding lung volume reduces the partial pressure of O_2, and potentially reverses the O_2 gradient across the lung, causing a rapid drop in alveolar PO_2 and resulting **hypoxemia** (**Figure 4.8**). Marine mammals hyperventilate before and after deep dives (Kooyman 1989), and could therefore be susceptible to shallow-water blackout if they dive on **hypocapnic** blood PCO_2 levels.

Some phocid species, e.g., Weddell seal, harbor seal and elephant seal have been documented to be extremely hypoxia-tolerant. Unlike phocids, otariid seals are thought to be more reliant on their lung O_2 stores (Kooyman 1973). Continuous arterial PO_2 loggers attached to California sea lions have shown a relatively

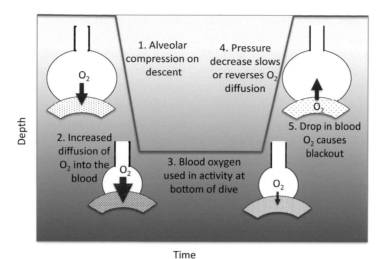

Figure 4.8 Mechanism underlying shallow-water blackout. Green line shows hypothetical dive. Lung O_2 pressure increases on descent to depth, increasing diffusion into blood. Blood O_2 then decreases (due to metabolism) over the course of the dive. During ascent, lung O_2 pressure drops and lung volume increases. This can cause a reversal of the diffusion gradient, pulling O_2 from the blood into the lungs, causing a transient decrease in blood O_2 and leading to blackout. Darker vessel shading indicates higher blood gas tension. Arrow shows direction of net diffusion and greater arrow thickness indicates greater diffusion rate.

deep depth of alveolar collapse (225 m). Alveolar collapse mitigated the potential for shallow-water blackout as cessation of gas exchange preserved a reservoir of O_2 that supplemented blood O_2 during ascent (McDonald and Ponganis 2012). In contrast, during the relatively shallow (less than 160 m) dives of fur seals, animals were found to exhale during the ascent portion of dives (Hooker et al. 2021). This could function to reduce buoyancy during ascent, but it would also impact susceptibility to shallow-water blackout. Since these dives are shallower than the likely depth of alveolar collapse, they could potentially result in extreme lung O_2 depletion and severe hypoxemia. Exhaling on ascent would maintain compressed alveoli and minimize diffusion between blood and lungs during ascent (Hooker et al. 2005). A similar effect was previously observed during measurement of blood N_2 levels in a forced dive of a harbor seal. The seal exhaled during decompression from a simulated dive to 130 m in a hyperbaric chamber, which resulted in delaying

the removal of N_2 from its blood until it surfaced and breathed (Kooyman et al. 1972). Exhalation can, therefore, maintain blood gas concentrations during ascent and help prevent drops in arterial PO_2 that may cause unconsciousness.

Whether shallow-water blackout is a problem for cetaceans and whether they have methods to mitigate this are unknown. However, given that shallow-water blackout is more likely to be a problem for active inhalation divers, diving to depths shallower than alveolar collapse, we would expect that some species of cetaceans would be vulnerable (**Figure 4.8**).

Decompression sickness

It was widely believed until recently that freedivers, and particularly those with shallow depths of alveolar collapse (helped by exhalation prior to diving), would be relatively immune to decompression sickness. Alveolar collapse at shallow depths (**Figure 4.4**) would reduce the amount of inert gas absorbed and minimize the likelihood of supersaturation and bubble formation during ascent. However, recent work looking at human free-divers has suggested otherwise. In fact, rapid, repetitive breath-hold diving in humans can result in decompression sickness, and modelling work suggests that it may even be possible to develop DCS after a single deep breath-hold dive (Fitz-Clarke 2009). It seems that in free-diving animals, tissues can become highly saturated under certain circumstances depending on the process of gas loading during diving and wash-out time at the surface. There is ample evidence that marine mammals are living with blood and tissue N_2 tensions that exceed ambient air levels (Hooker et al. 2012).

Theoretical models have attempted to estimate blood and tissue levels of N_2, O_2 and CO_2 in diving vertebrates (**Figure 4.9**). Most results suggest that accumulation of both N_2 and CO_2 may reach levels that would cause DCS symptoms in similar-sized terrestrial mammals (Parraga et al. 2018).

Decompression sickness may explain why beaked whales have mass stranded at the same time as military exercises deploying sonar (Bernaldo de Quiros et al. 2019). Necropsies of these dead whales have shown intravascular and major organ damage from gas **emboli**. There has been some controversy about whether the emboli caused death or were a side-effect of something else (Hooker et al. 2012) although it is widely agreed that anthropogenic disturbance was the root of the problem. These types of tissue damage have also been reported in some single-stranded cetaceans for which they do not appear to have been immediately fatal.

Osteonecrosis is the death of bone tissue caused by disruption to the blood supply and leads to weakening of the bone. Sperm whale bones have been found to have osteonecrosis-type surface injury, hypothesized to have been caused by repetitive formation of N_2 emboli over time. This suggests sperm whales may live with non-lethal decompression-induced bubbles on a regular basis, but with long-term impacts on bone health. A low level of bubble incidence was also detected in stranded common and

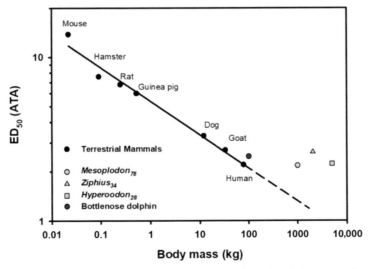

Figure 4.9 Estimated whole body N_2 saturation pressure (ED50, ATA) that would result in 50% of individuals experiencing severe decompression sickness (DCS) in terrestrial mammals (black) after a rapid reduction in pressure (decompression). Also shown are theoretical model estimates for end-of-dive mixed venous N_2 levels (which represent whole body N_2 saturation) for bottlenose dolphins (red) and beaked whales (gray), based on recorded dive profiles. This suggests marine mammals regularly experience gas levels that would cause decompression sickness for similarly sized terrestrial mammals.

white-sided dolphins using ultrasound. Once released back to sea, these dolphins showed normal behavior and did not re-strand, suggesting some tolerance to bubble formation. Bubbles were also observed in marine mammals caught in fishing nets, which died at depth. The fact that bubbles have been found in stranded animals suggests the animals' tissues were supersaturated such that bubbles either formed as animals were brought to the surface or were held stranded and unable to dive. California sea lions admitted to a rehabilitation facility with neural deficiencies underwent magnetic resonance imaging (MRI) which showed cerebral gas lesions. These cases suggest that bubble-related injuries, while not common, do occur (Hooker et al. 2012).

In terrestrial mammals, high CO_2 levels have been associated with a higher likelihood of DCS (Behnke 1951). Since carbon dioxide is produced following a burst of activity, it was suggested that its much higher diffusion rate could trigger the initiation of a bubble which would then grow with diffusion of N_2 into the bubble as supersaturation increases close to the surface. The gas composition in bubbles from stranded animals believed to have experienced gas bubble disease was found to contain high levels of CO_2 (Bernaldo de Quirós et al. 2012). This hypothesis could explain how an aversion response to sonar causes a burst of activity increasing anaerobic metabolism and increasing muscle and vascular CO_2 levels which initiates bubble growth (Fahlman, Tyack et al. 2014).

The working hypothesis for avoidance of excessive blood and tissue N_2 tensions in marine mammals has been that compression of the alveolar space results in alveolar collapse. The collapse would limit N_2 uptake during the dive and thereby the risk of gas bubble formation. However, if the depth of passive alveolar collapse is more than 100 m, considerable uptake of N_2 is possible before the animal reaches this depth (**Figure 4.4**). Theoretic models, even using the most conservative estimates to limit N_2 uptake, indicate that passive alveolar collapse should result in considerable blood and tissue N_2

tensions, and would result in a high likelihood for gas emboli formation.

A new mechanism, called the *selective gas exchange* hypothesis proposes a mechanism allowing marine mammals to regulate gas exchange (Parraga et al. 2018; Fahlman et al. 2021). The unusual lung architecture, which permits routes of airflow in addition to the main airways (known as collateral ventilation), means that some parts of the lung could be collapsed while other parts are not. The unusual hypoxic pulmonary vasodilatation seen in marine mammals maintains blood flow even under hypoxic conditions, and together with the conditioned heart rate response would regulate blood flow to the open/closed regions. The key, however, is use of the ventilation-perfusion mismatch to promote selective exchange of gases based on differences in their solubility. A high ratio of alveolar ventilation to cardiac output/lung perfusion would promote exchange of oxygen and carbon dioxide while reducing or even reversing exchange of nitrogen. This mechanism potentially provides an explanation of how marine mammals can manage gas exchange during normal diving to avoid excessive uptake of N_2 but still exchange O_2 and CO_2. It also suggests that unusual stress, such as exposure to man-made sound, might result in failure of this mechanism, uptake of N_2, and ultimately symptomatic gas bubble formation.

TOOLBOX

Animal-attached dive recorders

The development of the animal-attached dive recorder has been crucial to our understanding of the exposure of marine mammals to pressure. Although some information about diving behavior is available through direct observation – noting dive times or following animals subsurface by sonar, long time-series information on diving behavior necessitates the use of animal-attached instruments (Ropert-Coudert and Wilson 2005). The earliest of these instruments were simple capillary tube manometers

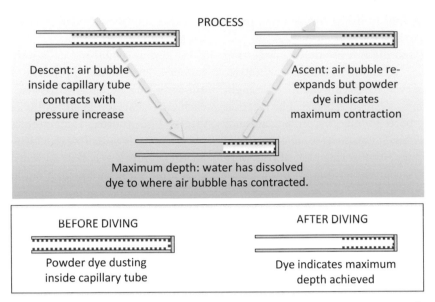

PROCESS

Descent: air bubble inside capillary tube contracts with pressure increase

Ascent: air bubble re-expands but powder dye indicates maximum contraction

Maximum depth: water has dissolved dye to where air bubble has contracted.

BEFORE DIVING

Powder dye dusting inside capillary tube

AFTER DIVING

Dye indicates maximum depth achieved

Figure 4.10 The earliest measurements of maximum depth were via Boyle's law – marking the position of a compressed air bubble inside a capillary tube manometer.

(**Figure 4.10**). The tubes were welded closed at one end and dusted internally with dye that was easily soluble in water. As the air within the tube was compressed, the intrusion of water would show the maximum pressure achieved. In some cases, multiple rings could be identified showing dives to several depths. Addition of a radioactive bead that followed the water-air boundary could record time spent at depth. However, such instruments were unable to provide the resolution achieved by time-depth recorders.

The earliest time-depth recorders linked measurement of pressure to a moving needle which recorded directly as a trace on a 60-minute kitchen timer with a smoked glass disc mounted on it. Later models used carbon-coated paper with a quartz motor allowing recording durations of 25 days. These analogue systems were superseded by digital solid-state recorders. Early limitations were memory capacity and temporal resolution of data sampling, number of parallel sensors which could be supported, 8-bit resolution (allowing only 255 measurements, e.g., 1000 m depth could only be recorded to the nearest 4 m), and instrument size (Ropert-Coudert and Wilson 2005).

Most of these impediments have become vastly improved, and tags can now combine hydrophone, video, accelerometer, speed sensor, and oceanographic measurements alongside records of animal depth. These instruments not only record a detailed diary of the behavior and physiology of the animal but can also provide a record of their environment.

Observation of lung structure and dynamics

For the semi-aquatic pinnipeds, several aspects of the dive response have been effectively studied in lab-based settings. These have included sub-cellular-based studies but also whole animal studies using hyperbaric chambers. They have examined anatomical changes, such as the flexibility of the trachea under pressure (Kooyman et al. 1970) but also changes in pulmonary shunt and its effects on blood gas content under breath-hold and pressure (Kooyman et al. 1972; Kooyman and Sinnett 1982).

The field of medical imaging (ultrasound, CT and MRI) is becoming ever more refined, and there is great scope for its application to investigations of diving animals. A hyperbaric chamber inside a CT scanner was used to obtain 3-D images of the changes in lung compression of dead marine mammals as pressure was increased (Moore et al. 2011). In future, perhaps an MRI-compatible hyperbaric chamber, combined with trained animals that dive on cue, would provide great scope for investigations of the respiratory alterations caused by pressure.

While imaging studies may become feasible in the future, recent work has investigated the structural properties of the respiratory system in anesthetized pinnipeds (Fahlman et al. 2017). The static compliance (pressure-volume relationship) of the lung and chest have been estimated in anesthetized individuals during manual ventilation (manual compression of a gas-filled reservoir bag to push gases into the lungs). The results agreed with Scholander's suggestion that the chest provides little resistance and can compress to very low volumes. In cetaceans, anesthesia is more difficult and studies on respiratory physiology and lung mechanics have been performed in conscious animals using breath-by-breath spirometry (measurement of amount and speed of inhalation and exhalation). A custom-made pneumotachometer (converting the flow of gases into a pressure signal) can be placed over the blowhole of dolphins. Both anatomical studies and work on live animals has suggested that exhalation during voluntary breaths is passive, while inhalation requires substantial work from the respiratory muscles. The elastic recoil of the chest may help empty the lung to achieve low volumes to help avoid negative pressures (lung squeeze) developing inside the lung during diving. Work on excised lungs and on live dolphins has suggested that the respiratory architecture may play a role in their unusual ability to respire. For example, respiratory flow rates are as much as three times higher than in a horse. The **vital capacity** can be as much as 80–90% of the total lung capacity and exchanged in as little as 200–300 msec.

Measurement of blood flow, gases and bubble formation

The cardiac output and the blood flow distribution are, in addition to pulmonary gas exchange, possibly the most important variables that diving animals can govern to help manage gases. Blood flow measurements are few and difficult to perform. Dye or thermal dilution experiments can be made to determine cardiac output. To study blood flow distribution, studies have either provided a snap-shot of blood flow distribution at one time point using micro-labelled spheres during forced dives (Zapol et al. 1979) or continuous measurements using doppler probes from a single blood vessel (Ponganis et al. 1990). While the micro-sphere studies provide global distribution of blood flow, these studies are done during forced-diving, which most likely alters the physiology of the animal and questions how valid the response may be as compared with natural dives. More recent work has used echocardiography to evaluate how cardiac output varies before, during and following both exercise and breath-holding, and also attempts to separate the confounding effect of respiration on cardiac function. These studies have suggested that the cardiorespiratory systems are linked, and to understand the physiological adaptations to diving, these need to be separated to avoid the confounding effect of breathing on cardiac function, i.e., the respiratory sinus arrythmia (Blawas et al. 2021). Future alternatives may be to use animals trained to dive on command and short-lived radioactive isotopes that can be imaged in a PET-CT, or ultrasound blood flow probes at selected blood vessels. While these require invasive surgery, technological advances may minimize impact and provide an interesting alternative for use on wild animals.

Optical technologies are currently showing great promise, and simple optical sensors are widely used for heart-rate measurement (you can even use a mobile phone app for this!). Medical technology such as diffuse correlation spectroscopy and broadband continuous-wave near-infrared spectroscopy (CW-NIRS) are currently static technologies used to provide measures of blood flow, intracranial pressure and metabolic rate changes. However, wearable near-infrared spectroscopy has been tested on seals and humans and can provide data on relative measures of oxyhaemoglobin and deoxyhaemoglobin (McKnight et al. 2019). Calculation of total haemoglobin can be used as a proxy for changes in blood volume while calculation of relative haemoglobin difference can be used as a proxy for oxygenation changes.

Blood gases (O_2 and N_2) have been measured in captive animals during forced dives (Kooyman et al. 1972; Kooyman and Sinnett 1982) and from wild animals with a computerized device that sampled arterial gas for later processing in the lab (Falke et al. 1985). One clever study inserted a needle covered in a gas permeable silicone sleeve into the muscle of dolphins following a dive. The needle was attached to a mass-spectrometer which pulled gases out of the muscle according to the partial pressure gradients, enabling the N_2 washout to be measured, and providing the estimate of alveolar collapse depth of 70 m for bottlenose dolphins (Ridgway and Howard 1979). In another study, blood samples were collected following a dive and a Van Slyke manometer was used to derive volumetric blood gas nitrogen levels (Houser et al. 2010). Use of intravascular O_2 electrodes has generated some very interesting results in California sea lions and elephant seals. In the latter, evidence of extreme hypoxia tolerance was shown where arterial and venous PO_2 levels of 15 mmHg and 3 mmHg, respectively, were observed during extended dives (Meir et al. 2009). In the sea lion, the arterial data showed evidence of alveolar collapse and the depth of where gas exchange stopped was much deeper than formerly believed (McDonald and Ponganis 2012). Together these studies continue to show the physiological plasticity that appears possible in marine mammals.

Other studies beyond blood gas measurements have shown that during certain circumstances the blood gas levels may become supersaturated and result in intravascular bubbles (Hooker et al. 2012). When and if bubbles are formed is of particular interest in establishing preconditions for diving diseases. Audible Doppler ultrasound and imaging ultrasound have been used to study intravascular bubbles in humans. Emerging technologies, such as dual frequency ultrasound, should enable the study of extravascular bubbles (see **Chapter 5** for other uses of ultrasound). Development of technology to continuously monitor free-ranging animals will be needed to test whether free-ranging animals do experience gas bubbles while foraging.

CONCLUSIONS

We do not fully understand how marine mammals cope with repeated exposure to high pressure and need a better understanding of both their physiology and their physiological plasticity. Information from beach-cast animals has been helpful but knowing what an organ looks like is not the same as understanding how it works. Unusual observations have raised questions but rarely given answers about the cause of death. Experimental physiological research on marine mammals is difficult, particularly for species which never come ashore, although advances in electronics are beginning to offer greater insights. Most diving physiology studies on wild individuals have been done on seals and trained bottlenose dolphins. These will not have the N_2 load that wild animals routinely deal with and so may not be useful models for some questions. Caution is also needed in extrapolating results broadly across all marine mammals. For instance, the often-quoted 'seals dive on exhalation' appears to relate only to phocid seals. Alveolar collapse is the major mechanism governing animals' uptake of gases at depth, but there are only a handful of studies suggesting the depth that occurs, and even fewer documenting changes in blood flow with diving. These are crucial to our ability to build a model framework to understand gas uptake and circulation during dives. Carefully structured and ethical studies on captive individuals may help provide vital information to direct studies and conservation efforts for wild animals. However, in the wild, managing the effects of pressure is likely to occur simultaneously with managing several other issues (e.g., buoyancy, thermoregulation, energy balance, muscle oxygenation).

REVIEW QUESTIONS

1 Why are marine mammals better than humans at diving deep?
2 How does body size affect the ability to breath-hold?
3 How can pressure affect biochemical reactions?
4 How does alveolar collapse limit the harmful effects of pressure in breath-hold divers?
5 Why do toothed whales need to recycle air to echolocate at depth?
6 How do marine mammals avoid decompression sickness?
7 How might nitrogen help avoid high-pressure nervous syndrome?
8 Why was decompression sickness called the bends?
9 What do hyperbaric chambers do?
10 How did a capillary manometer measure pressure?

CRITICAL THINKING

1 Is there a depth limit for marine mammal dives?
2 How can understanding the adaptations of marine mammals help provide novel medical treatments for humans?
3 Are captive animals good model systems to help understand diving physiology?

EQUATIONS

Boyle's law	$P_1 \cdot V_1 = P_2 \cdot V_2$
Dalton's law	$p(1) \propto P$
Henry's law	$p = k_H M$

where V = volume; P = pressure; p = partial pressure; M = molar concentration; k_H = Henry's law constant (this varies for different gases, e.g., N_2 less soluble [k_H = 1640 L.atm/mol] than O_2 [k_H = 770 L.atm/mol] or CO_2 [k_H = 29 L.atm/mol])

FURTHER READING

Kooyman, G.L. 1989. *Diverse divers*. Berlin: Springer-Verlag.

Kooyman, G.L. 2006. Mysteries of adaptation to hypoxia and pressure in marine mammals. The Kenneth S. Norris lifetime achievement award lecture presented on 12 December 2005 San Diego, California. *Marine Mammal Science* 22: 507–526.

Ponganis, P.J. 2015. *Diving physiology of marine mammals and seabirds*. Cambridge: Cambridge University Press, pp. 246.

Scholander, P.F. 1940. Experimental investigations on the respiratory function in diving mammals and birds. *Hvalradets Skrifter* 22: 1–131.

GLOSSARY

ATA: Atmospheres Absolute, a pressure unit that includes surface pressure.

Artery/arterial: A blood vessel carrying blood away from the heart. Systemic arteries carry oxygenated blood from the heart to the body and the pulmonary artery carries deoxygenated blood from the heart to the lungs.

Atelectasis: Alveolar closure when no gas exchange occurs.

Barotrauma: Trauma caused by pressure.

Compliance: A measure of the ease of expansion of a structure.

Decompression sickness (or 'the bends'): This is caused by dissolved gas coming out of solution and forming bubbles during a reduction in pressure. Bubbles may cause damage to tissues or organs or may **embolize** blood vessels leading to ischemic damage. It was called the bends because in human divers, bubbles would cause joint and muscle pain, resulting in a characteristic bent shape of the patient.

Emboli/Embolism: The lodging of an embolus (blood clot, fat globule or gas bubble) in a blood vessel.

High-pressure nervous syndrome (HPNS): This is caused by hydrostatic pressure-inducing tremors and convulsions.

Hyperbaric: Higher pressure.

Hypocapnia: CO_2 levels lower than normal.

Hypocoagulable: Less prone to coagulate (form clots).

Hypoxemia: Arterial O_2 levels lower than normal.

Hypoxia: Tissue O_2 levels lower than normal.

Lung squeeze: When the chest is exposed to a pressure that reduces the lung volume below the functional residual capacity and a negative pressure develops inside the lung.

Nitrogen narcosis: This is caused by the anesthetic effect of nitrogen at high pressure ultimately leading to loss of consciousness at deeper depths.

Oxygen toxicity: This is caused by high partial pressure of O_2 for which short exposure causes central nervous system toxicity, and longer exposure causes toxicity to the lungs and eyes.

Partial pressure: Pressure exerted by one among a mixture of gasses if it occupies the same volume.

Shallow-water blackout: This is a loss of consciousness due to cerebral **hypoxia** caused by depletion of blood O_2 associated with a rapid drop in partial pressure of lung O_2 toward the end of a breath-hold dive.

Sinus: A cavity within a bone or tissue.

Sphincter: A muscle that can close off a body cavity.

Surfactant: Compound which lowers the surface tension.

Vein/venous: A blood vessel carrying blood toward the heart. Systemic veins carry deoxygenated blood from the body back to the heart and the pulmonary vein carries oxygenated blood from the lungs to the heart.

Vital capacity: The amount of air that can be exhaled after a maximum inhalation.

REFERENCES

Behnke, A.R. 1951. Decompression sickness following exposure to high pressures. In *Decompression sickness*, ed. J.F. Fulton. Philadelphia: Saunders.

Bennett, P.B., and J.C. Rostain. 2003. The high pressure nervous syndrome. In *Bennett and Elliott's physiology and medicine of diving*, ed. A.O. Brubakk and T.S. Neuman. Philadelphia: Saunders.

Bernaldo de Quiros, Y., A. Fernandez, R.W. Baird, et al. 2019. Advances in research on the impacts of anti-submarine sonar on beaked whales. *Proceedings of the Royal Society B-Biological Sciences* 286(1895): 20182533.

Bernaldo de Quirós, Y., Ó. González-Díaz, M. Arbelo, et al. 2012. Decompression versus decomposition: Distribution, quantity and gas composition of bubbles in stranded marine mammals. *Frontiers in Physiology* 3: 177.

Blawas, A.M., D.P. Nowacek, J. Rocho-Levine, et al. 2021. Scaling of heart rate with breathing frequency and body mass in cetaceans. *Philosophical Transactions of the Royal Society B-Biological Sciences* 376(1830): 20200223.

Bostrom, B.L., A. Fahlman, and D.R. Jones. 2008. Tracheal compression delays alveolar collapse during deep diving in marine mammals. *Respiratory Physiology & Neurobiology* 161: 298–305.

Castellini, M.A., P.M. Rivera, and J.M. Castellini. 2002. Biochemical aspects of pressure tolerance in marine mammals. *Comparative Biochemistry and Physiology A* 133: 893–899.

Denison, D.M., and G.L. Kooyman. 1973. The structure and function of the small airways in pinniped and sea otter lungs. *Respiration Physiology* 17: 1–10.

Fahlman, A., S.H. Loring, S.P. Johnson, et al. 2014. Inflation and deflation pressure-volume loops in anesthetized pinnipeds confirms compliant chest and lungs. *Frontiers in Physiology* 5: 433.

Fahlman, A., M.J. Moore, and D. Garcia-Parraga. 2017. Respiratory function and mechanics in pinnipeds and cetaceans. *Journal of Experimental Biology* 220(10): 1761–1773.

Fahlman, A., M.J. Moore, and R.S. Wells. 2021. How do marine mammals manage and usually avoid gas emboli formation and gas embolic pathology? Critical clues from studies of wild dolphins. *Frontiers in Marine Science* 8: 598633. doi: 10.3389/fmars.2021.598633.

Fahlman, A., P.L. Tyack, P.J.O. Miller, et al. 2014. How man-made interference might cause gas bubble emboli in deep diving whales. *Frontiers in Physiology* 5: 13.

Falke, K.J., R.D. Hill, J. Qvist, et al. 1985. Seal lungs collapse during free diving: Evidence from arterial nitrogen tensions. *Science* 229: 556–558.

Fitz-Clarke, J.R. 2009. Risk of decompression sickness in extreme human breath-hold diving. *Undersea Hyperbaric Medicine* 36: 83–91.

Foot, N.J., S. Orgeig, and C.B. Daniels. 2006. The evolution of a physiological system: The pulmonary surfactant system in diving mammals. *Respiratory Physiology & Neurobiology* 154(1–2): 118–138.

Foskolos, I., N.A. de Soto, P.T. Madsen, et al. 2019. Deep-diving pilot whales make cheap, but powerful, echolocation clicks with 50 µL of air. *Scientific Reports* 9(1): 15720.

Fraser, F.C., and P.E. Purves. 1960. Hearing in cetaceans – Evolution of the accessory air sacs and the structure and function of the outer and middle ear in recent cetaceans. *Bulletin of the British Museum (Natural History) Zoology* 7(1): 1–140.

Gutierrez, D.B., A. Fahlman, M. Gardner, et al. 2015. Phosphatidylcholine composition of pulmonary surfactant from terrestrial and marine diving mammals. *Respiratory Physiology & Neurobiology* 211: 29–36.

Halsey, M.J. 1982. Effects of high pressure on the central nervous system. *Physiological Reviews* 62(4): 1341–1377.

Harabin, A.L., S.S. Survanshi, and L.D. Homer. 1995. A model for predicting central nervous system oxygen toxicity from hyperbaric oxygen exposures in humans. *Toxicology and Applied Pharmacology* 132(1): 19–26.

Hooker, S.K., R.D. Andrews, J.P.Y. Arnould, et al. 2021. Fur seals do, but sea lions don't - cross taxa insights into exhalation during ascent from dives. *Philosophical Transactions of the Royal Society B-Biological Sciences* 376(1830): 20200219.

Hooker, S.K., A. Fahlman, M.J. Moore, et al. 2012. Deadly diving? Physiological and behavioural management of decompression stress in diving mammals. *Proceedings of the Royal Society B-Biological Sciences* 279(1731): 1041–1050.

Hooker, S.K., and A. Fahlman. 2016. Pressure regulation. In *Marine Mammal Physiology: Requisites for Ocean Living*, eds. M.A. Castellini and J.E. Mellish. Boca Raton: CRC Press.

Hooker, S.K., P.J.O. Miller, M.P. Johnson, et al. 2005. Ascent exhalations of Antarctic fur seals: A behavioural adaptation for breath-hold diving? *Proceedings of the Royal Society of London B* 272: 355–363.

Houser, D.S., L.A. Dankiewicz-Talmadge, T.K. Stockard, et al. 2010. Investigation of the potential for vascular bubble formation in a repetitively diving dolphin. *Journal of Experimental Biology* 213: 52–62.

Hui, C.A. 1975. Thoracic collapse as affected by the *retia thoracica* in the dolphin. *Respiration Physiology* 25(1): 63–70.

Hunter, W.L., Jr., and P.B. Bennett. 1974. The causes, mechanisms and prevention of the high pressure nervous syndrome. *Undersea Biomedical Research* 1(1): 1–28.

Kooyman, G.L. 1973. Respiratory adaptations in marine mammals. *American Zoologist* 13: 457–468.

Kooyman, G.L. 1989. *Diverse divers*. Berlin: Springer-Verlag.

Kooyman, G.L., D.D. Hammond, and J.P. Schroeder. 1970. Bronchograms and tracheograms of seals under pressure. *Science* 169: 82–84.

Kooyman, G.L., D.H. Kerem, W.B. Campbell, et al. 1971. Pulmonary function in freely diving Weddell seals, *Leptonychotes weddelli*. *Respiration Physiology* 12: 271–282.

Kooyman, G.L., J.P. Schroeder, D.M. Denison, et al. 1972. Blood nitrogen tensions of seals during simulated deep dives. *American Journal of Physiology* 223: 1016–1020.

Kooyman, G.L., and E.E. Sinnett. 1982. Pulmonary shunts in harbor seals and sea lions during simulated dives to depth. *Physiological Zoology* 55: 105–111.

McDonald, B.I., and P.J. Ponganis. 2012. Lung collapse in the diving sea lion: Hold the nitrogen and save the oxygen. *Biology Letters* 8(6): 1047–1049.

McKnight, J.C., K.A. Bennett, M. Bronkhorst, et al. 2019. Shining new light on mammalian diving physiology using wearable near-infrared spectroscopy. *Plos Biology* 17(6): e3000306.

Meir, J.U., C.D. Champagne, D.P. Costa, et al. 2009. Extreme hypoxemic tolerance and blood oxygen depletion in diving elephant seals. *American Journal of Physiology-Regulatory Integrative and Comparative Physiology* 297(4): R927–R939.

Moore, C., M. Moore, S. Trumble, et al. 2014. A comparative analysis of marine mammal tracheas. *Journal of Experimental Biology* 217(7): 1154–1166.

Moore, M.J., T. Hammar, J. Arruda, et al. 2011. Hyperbaric computed tomographic measurement of lung compression in seals and dolphins. *Journal of Experimental Biology* 214: 2390–2397.

Nagel, E.L., P.J. Morgane, W.L. McFarlane, et al. 1968. Rete mirabile of dolphin: Its pressure-damping effect on cerebral circulation. *Science* 161(844): 898–900.

Odend'hal, S., and T.C. Poulter. 1966. Pressure regulation in the middle ear cavity of sea lions: A possible mechanism. *Science* 153(737): 768–769.

Parraga, D.G., M. Moore, and A. Fahlman. 2018. Pulmonary ventilation-perfusion mismatch: A novel hypothesis for how diving vertebrates may avoid the bends. *Proceedings of the Royal Society B-Biological Sciences* 285(1877): 20180482.

Piscitelli, M.A., W.A. McLellan, S.A. Rommel, et al. 2010. Lung size and thoracic morphology in shallow- and deep-diving cetaceans. *Journal of Morphology* 271(6): 654–673.

Ponganis, P.J., G.L. Kooyman, M.H. Zornow, et al. 1990. Cardiac-output and stroke volume in swimming harbour seals. *Journal of Comparative Physiology B* 160(5): 473–482.

Ridgway, S.H., and R. Howard. 1979. Dolphin lung collapse and intramuscular circulation during free diving: Evidence from nitrogen washout. *Science* 206: 1182–1183.

Ridgway, S.H., B.L. Scronce, and J. Kanwisher. 1969. Respiration and deep diving in the bottlenose porpoise. *Science* 166: 1651–1654.

Ropert-Coudert, Y., A. Kato, A. Robbins, et al. 2018. *The Penguiness book*. World Wide Web electronic publication (http://www.penguiness.net) version 3.0, October 2018.

Ropert-Coudert, Y., and R.P. Wilson. 2005. Trends and perspectives in animal-attached remote sensing. *Frontiers in Ecology and the Environment* 3: 437–444.

Scholander, P.F. 1940. Experimental investigations on the respiratory function in diving mammals and birds. *Hvalradets Skrifter* 22: 1–131.

Somero, G.N. 1992. Adaptations to high hydrostatic pressure. *Annual Review of Physiology* 54: 557–577.

Zapol, W.M., G.C. Liggins, R.C. Schneider, et al. 1979. Regional blood flow during simulated diving in the conscious Weddell seal. *Journal of Applied Physiology* 47(5): 968–973.

Michael Castellini
University of Alaska, Fairbanks, AK

Jo-Ann Mellish
North Pacific Research Board, Anchorage, AK

Michael Castellini

Mike earned his Ph.D. in 1981 in marine biology on the diving and medical physiology of marine mammals. He recently retired from the University of Alaska, Fairbanks, where his students worked around the world on marine mammal biology. Mike has been on more than a dozen scientific marine mammal expeditions to the Antarctic and lived over 3 years on the sea ice, including winter. He was part of the team that defined the aerobic diving limit in seals.

Jo-Ann Mellish

Jo-Ann is a recovering academic with 20 years of adventures in tracking, gluing, prodding, evaluating, and happily releasing seals and sea lions in various cold parts of the world. Now she uses her vast knowledge of just how many things can and will go wrong in the field to guide her current role as a funding program manager to help other scientists get out there and get it done.

Driving Question: How do marine mammals stay warm in the cold ocean?

INTRODUCTION

Humans cannot survive for long in the ocean outside of equatorial regions because the water is much cooler than our core body temperature and carries heat away from body surfaces faster than our normal air environment. Marine mammals often have very thick layers of fur or blubber that can protect them from water as much as 40°C cooler than their body temperature, but this has mixed consequences. What happens if animals are in warm water, swimming hard, or lounging on land under the sun? Marine mammals must be able to keep cool in warm waters and stay warm in the cold ocean. How do they do both?

THERMOREGULATORY SYSTEMS

The answer is a range of highly evolved **thermoregulatory** systems that can both conserve and dissipate heat. Of course, a species' ability to do either varies widely given their global distribution. Some cetaceans spend their entire lives in ice-laden waters while others migrate regularly between polar and tropical waters. Many temperate pinnipeds spend considerable time in cold water but haul out onto hot beaches, while polar pinnipeds haul out on sea ice or land in air temperatures that are dramatically lower than

DOI: 10.1201/9781003297468-5

the water. A combination of physical, biochemical, and behavioral tools allows these animals to move effectively between temperature conditions that would be challenging for many terrestrial mammals.

If we back up for a moment to put this into a larger perspective, both terrestrial and marine mammals are considered **endothermic homeotherms**. They generate their own heat and can maintain a constant body temperature well above that of the environment. It is one of the main characteristics that define them as a "mammal". These groups were called "warm-blooded" in older literature, to distinguish them from the "cold-blooded" reptiles, amphibians, etc. In modern terminology, lizards, snakes, fish, and other groups are considered **ectothermic heterotherms**. This means that these animals are dependent on external sources to regulate their body temperature.

Hypothermia (body temperature too cold) and **hyperthermia** (too hot) can be lethal to all mammals, and they regulate their body temperature carefully to avoid these problems. The key to regulating their body temperature is the balance between heat production and heat loss. How your body balances those two needs vary depending on the situation. For instance, your body will react very differently when exercising in a cold environment while skiing or snowshoeing compared to running a race in the heat of the summer. Heat balance equations help us understand the process of how energy in the form of heat is conserved, gained, or lost.

What does the term "heat energy" mean, and how do we use it? Throughout this chapter, we refer to heat, energy, and thermal balance. A warm object has more heat energy than a cold object and that energy can flow from the warm to the cold body by a variety of pathways as described in the following sections. At the most basic level, the molecules of matter in the warm object are vibrating more than in the cold object and this is defined as heat energy. At absolute zero (0 K or −273°C), all molecular movement stops, and the object has no heat energy. Heat balance is a constant flow of energy through the combined effects of **conduction**, **convection**, **radiation**, and **evaporation**.

> Conduction occurs when heat energy moves from a warm body in contact with a colder body.
> Convection is heat lost to air or water moving over a surface.
> Radiation is the release of heat as electromagnetic energy, usually as infrared energy.
> Evaporation is the loss of heat through the evaporation of surface water (sweat).

Conduction

Conduction is the direct contact transfer of heat energy from a warm body to a cooler body. A common misperception is that cold flows from the cold to the warm body. Energy can only move down a gradient from higher temperature to lower temperature, not up from cold to warm. The rate of conductive heat transfer is determined by the area of contact between the two bodies, the thickness of the surface that the heat must travel through, the insulating nature of the material in contact with the object, and the temperature difference between the warm and colder body. One can visualize this by thinking about being outside on a cold day with a warm down jacket. We know that our down jacket with its great insulation capacity works better than a thin, cotton jacket. Similarly, if we can curl up into a smaller size, we will stay warmer. Heat transfer follows a formula known as the **Fick Equation** where the rate of heat flow is directly proportional to the thermal conductivity of the material (down coat), the temperature difference between the two bodies (outside temperature versus body temperature), and inversely proportional to the thickness of the material (thick coat versus thin). For marine mammals, heat will be lost through conduction to any contact material (land, sea ice, water, air) that is cooler than their external body temperature.

For example, a polar bear (*Ursus maritimus*) that is sleeping on sea ice will lose heat to both the ice and the cold air around it. The full Fick Equation and its variables are provided in the **Equations (1)** section of this chapter.

Convection

Convection is the loss of heat through the combined effect of conduction and the movement of the surrounding air or water molecules. We commonly refer to this in air as "wind chill". Because it is the result of two variables, wind chill cannot be measured simply on a thermometer. When the air temperature is −30°C, the absolute temperature will be the same whether the wind is blowing or not, despite how much colder it may feel if windy. This function is important not only for marine mammals that haul out on land or ice, but also when animals encounter strong water currents or are swimming rapidly through the water, which can drive the cooling power up to almost 100 times that of air.

Radiation

Radiation is the release of heat energy to a cooler body, which does not require surface-to-surface contact, air, or water for transfer. In biology, this is usually in the form of infrared energy. One example of radiation is how the sun heats the earth even though there is no air in the vacuum of space. A seal warming on the ice surface by orienting to the sun is an excellent example of radiative heat gain.

Evaporation

Generated heat can also be lost by using its energy to evaporate water on the surface of an object or animal. Convection can speed up the process by pulling away the moisture on the skin surface. Humans experience this phenomenon as sweating. Because marine mammals spend most of their time in water, evaporation to offset excess body heat is impossible. There is visual and chemical evidence of sweat glands in some marine species (Khamas et al. 2012) but there has been no direct observation of sweating. Instead, many species will employ behavioral tactics on land to compensate when cooling is necessary, such as flipping sand over their backs to reduce radiative heat gain on hot and sunny days, or returning to the water (Beentjes 2006).

Other variables

Body size and the *ratio of surface area to volume* are also important to the Fick Equation. The greater the surface area of the animal compared to its body size (volume) will influence how much relative area there is to transfer heat from one body to another. All else being equal, a smaller animal will experience greater heat loss than a larger animal because its surface area is proportionately larger. Consider a sea otter (*Enhydra lutris*) compared to a bowhead whale (*Baleana mysticetus*). The otter has a proportionately larger surface area than the whale, losing more heat per unit of volume than the larger animal, and may struggle to stay warm in cold water. The bowhead whale has a comparatively very small surface area to body size ratio and might instead struggle with too much body heat and have difficulty cooling down in warmer waters or during exertion. The formulas for how surface area and volume relate to each other are found in the **Equations (2)** section.

The difference between the temperature of the animal and its environment (*temperature differential*) is also a factor in the transfer of heat. A biologist sitting outside in the warm tropical air of Hawaii watching humpback whales (*Megaptera novaeangliae*) off the coast can stay warm in shorts and a t-shirt. A biologist counting those same humpback whales after they migrate to Alaska would quickly regret sitting outside in shorts and a t-shirt. Most likely the Alaskan biologist will use layers of cold-weather gear to insulate themselves from the large temperature differential between their body temperature and the cold Alaska air.

Finally, when comparing heat loss in air or water, an essential point is the **heat capacity** of water. A mass of water can hold about 25 times more heat than the same mass of air. Applied to the Fick Equation, this means that if all other factors are held constant (temperature differential, surface area, etc.), heat will flow out of a warm body 20–25 times faster in water than in air. For a human, you notice this when you are in water that is 18–19°C. While that temperature would seem pleasant outside in the air, you will feel very cold in the water as heat flows out of you much faster. In fact, to be thermal-neutral in water, you would need to be in water between 33°C and 35°C. For a marine mammal, being in the water implies a heat challenge significantly greater than compared to a terrestrial mammal, or even to itself if on land.

PHYSICAL CHARACTERISTICS

So how do the principles of heat balance apply to marine mammals? We started with the knowledge that insulation by fur or blubber is a well-known mechanism for effective thermoregulation in the cold ocean. However, the details of how fur and blubber work are interestingly complex.

How well insulation works depends on the thickness and **thermal conductivity** (**k**) of the material that is doing the insulating. As you increase thickness of an insulator, you also increase the distance the energy needs to travel. For example, 5 cm of an insulator is better than 2 cm for reducing heat flow. Thermal conductivity (k) of a material describes how well heat moves through it and is the inverse of insulation (*Table 5.1*). Returning to our winter jacket analogy, feather down traps large amounts of air and is a far superior insulator than thin cotton. This is because the air-filled feather down has a very low thermal conductivity, whereas thin cotton is an excellent conductor. Several textbooks and review articles discuss the many intricacies of thermoregulation, fur, and blubber and are included in the **Further Reading** section.

Fur

Fur traps air between hairs that makes it an effective insulator and a poor heat conductor.

Table 5.1 **Thermal conductance of various materials (units are cal s⁻¹ cm⁻¹°C⁻¹)**

Silver	1.0181
Water	0.0013
Organic tissue	0.0011
Leather	0.0004
Crude oil	0.00025
Blubber (as pelt)	0.0001
Fur	0.00009
Air	0.00005
Eider down	0.00001

Source: Data from *CRC Handbook of Chemistry and Physics* and from Bryden (1964), Elam et al. (1989), Schmidt-Nielsen (1997), Folk, Riedesel, and Thrift (1998), and Bagge et al. (2012).

In this case, the thermal gradient defined by the Fick Equation follows a path from the warm skin through the fur, to the external environment (**Figure 5.1**). The common rat has a hair density of 95 hairs/mm², while most fur seals have about 400 hairs/mm² and a sea otter has 1188 hairs/mm² (Liwanag et al. 2012a). By contrast, the hair densities of phocid seals and sea lions are less than 25 hairs/mm² and cetaceans may only have a few vestigial hairs on their bodies (Liwanag et al. 2012a). Young sea otters have hair that is so dense and full of air that they are often too buoyant to dive. The thick and luxurious fur of otters and northern fur seals (*Callorhinus ursinus*) inspired the Russian exploration of Alaska, and fur trading fueled the regional economy for many years. Fish et al. 2002 (**Further Reading**) provide an excellent review of the buoyancy and hydrodynamic characteristics of fur and hair.

The temperature gradient for marine mammals with water-resistant fur like sea otters and fur seals occurs primarily across the thickness of the fur and the cold water does not reach the skin. The costly trade-off is that these animals must maintain their fur in prime condition to keep the protective fur-air barrier intact. Sea otters can spend up to 20% of their **daily metabolic rate** (DMR) on fur

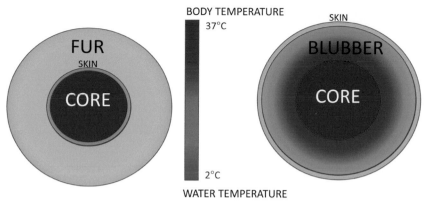

Figure 5.1 A comparison of thermal gradients from core body temperature to cold water environments with fur or blubber as the primary insulator. Left: Species with water-impermeable fur have very warm skin next to the core and are protected from cold water by the air trapped in the hair fibers. Most of the thermal gradient occurs within the fur layer. Right: Species with blubber typically have little or no fur and most of the thermal gradient is within the blubber layer. The skin is often only a few degrees above water temperature.

grooming (Walker, Davis, and Duffield 2008). **Chapters 2**, **9**, and **11** have detailed discussions on DMR values and measurements. A fur coat fouled with oil loses its protective air barrier and rapidly becomes a poor insulator (Costa and Kooyman 1982; *Table 5.1*), which was the unfortunate situation for sea otters impacted by the 1989 Exxon Valdez oil spill in Alaska. A large rescue effort included cleaning the fur of oiled otters followed by extended rehabilitation to allow time for grooming and replacement of the natural water repellents that were unavoidably stripped by the cleaning process.

Pagophilic (ice-loving) species like harp (*Pagophilus groenlandicus*) and gray seals (*Halichoerus grypus*) spend significant amount of time on or around sea ice, including during reproduction and pupping seasons. Pagophilic pups stay mostly on the sea-ice surface until they accumulate sufficient blubber reserves from the lipid-rich milk provided by their mothers to provide insulation. These species typically are born with a very thick **lanugo fur coat** that is excellent at keeping the pup warm in air (**Figure 5.2**). However, lanugo is only temporary. It is not water repellent and is mostly

useless as a thermal barrier in the water (Pearson et al. 2019). These pups shed their lanugo as they pack on insulating fat layers and begin to take on adult coloration and appearance with thinner, coarser hair as they are weaned. This process is not universal for ice seals and there are almost as many variations as there are species. In a mixed thermoregulatory strategy, the hooded seal (*Cystophora cristata*) is born with a lighter lanugo and a pre-loaded fat layer. Pups are weaned in just under four days due to a phenomenal efficiency of fat transfer (Frisch, Øritsland, and Krog 1974; Mellish et al. 1999).

Some seals and sea lions have a very rough, short, and relatively low-density fur that affords little insulation because it allows the skin to become wet. Their fur provides protection from the rough surfaces of the sea floor, rocky beach or sea ice but does little to contribute to insulation. The skin surface may be only a few degrees warmer than the environment, if at all (**Figure 5.1**) and the thermal gradient is internal to the animal. In these pinniped species, and all cetaceans, the insulating layer is usually found in the form of blubber. An interesting variant on this pattern is the heavily furred polar bear, where the fur allows water to permeate to the skin.

Figure 5.2 A Weddell seal (*Leptonychotes weddellii*) mother and her pup. Note the lanugo coat of the pup that is beginning to molt. (Photo: J. Mellish, Marine Mammal Protection Act 15748, Antarctic Conservation Act 2012-003)

For these bears, the fur acts mainly to keep them warm in air, but fat layers beneath the skin are the primary thermal barrier when in water (Frisch, Øritsland, and Krog 1974).

Marine mammal hair is shorter, denser, and flatter than other carnivores, but those species that rely heavily on fur for heat retention have longer external layers to their hair shaft than those that do not (Liwanag et al. 2012a). For short-furred marine mammal species, characteristics like color variation related to sex or species are likely more important than the small amount of heat retention provided.

Blubber

Blubber is not the same as fat deposits found in terrestrial mammals. Blubber is so varied in structure and function it is referred to as an organ. For starters, it is just built differently. Blubber begins with a protein matrix, mixed with **adipocytes** that are specialized fat storage cells. While a solid block of beef fat will melt entirely into liquid when heated, blubber can be heated and still maintain its shape. Blubber in general is an effective insulator against the cold because it has low-conductivity capacity (*Table 5.1*). The Fick thermal gradient occurs through the blubber from the warm muscle to the cold skin. The thicker the blubber, greater the insulation (*Table 5.2*). How we measure blubber thickness is discussed in the **Toolbox** section.

Blubber varies in depth, location, and lipid content depending on the type of marine mammal and the species (**Figure 5.3**). You can find blubber just about anywhere on a marine mammal except for the fins, flukes, and flippers. As one might expect, large whales have thick and complex blubber (e.g., bowhead whales, 50 cm thick; Haldiman and Tarpley 1993). Seals are generally smaller than whales but they can have proportionally very thick blubber layers that are found across most of the body (Weddell

Table 5.2 **Blubber depths (cm) in marine mammals**	
Bowhead whale	Average ~35
Sperm whale	15–20
Pygmy sperm whale	3
Orca	7–10
Common dolphin	1.5
Weddell seal	3–4.5
Harbor seal	2–2.5
California sea lion	<1
Manatee	3.5

Source: Data from Haldiman and Tarpley (1993), Pabst, Rommel, and McLellan (1999), Luque and Aurioles-Gamboa (2001), Smith and Worthy (2006), Castellini et al. (2009), George (2009), Bagge et al. (2012), Liwanag et al. (2012b), and Ashley (2014).

In general, blubber near the surface tends to be metabolically stable, while the blubber nearer to the muscle is used for fuel and water production. In cetaceans, there is a distinctively higher protein content near the skin surface to provide structure and influence hydrodynamics. The collagen protein content drops as blubber nears the warmer muscle layer eventually becoming extremely flexible and oily. **Chapter 14** describes how the high-lipid content of blubber makes it ideal for measuring **lipophilic** (dissolve easily in lipids) **persistent organic pollutants** (POPS; e.g., DDT, PCBs, dioxins). Blubber samples for any kind of contaminants, diet, or genetics study must have inner and outer edges identified as well as a recording of depth at which the sample was taken.

Blubber is the product of evolutionary and adaptation pressures that go beyond the need to keep an animal warm. It can be stored in bulk to provide fuel and energy during fasting (see **Chapters 9** and **11**). Significant amounts of water are produced in the process of metabolizing blubber fat for energy, which keeps marine

seals; *Leptonychotes weddellii*, 8 cm thick). Sea lions and fur seals have much thinner blubber layers that can vary by a factor of 3 depending on the part of the body (e.g., Steller sea lions; *Eumetopias jubatus*, 1–3 cm thick).

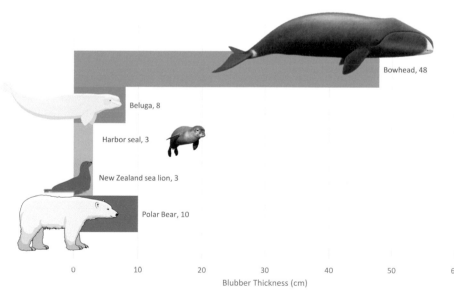

Figure 5.3 **Comparison of blubber thickness in a variety of marine mammals. See also *Table 5.2* for additional species and data references. (Bowhead whale (*Balaena mysticetus*): https://www.fisheries. noaa.gov/species/bowhead-whale; Beluga whale (*Delphinapterus leucas*): Cornick et al. 2016; Harbor seal (*Phoca vitulina*): Mellish et al. 2007; New Zealand sea lion (*Phocarctos hookeri*): Liwanag et al. 2012; Polar bear (*Ursus maritimus*): Øritsland 1970)**

mammals hydrated without food intake (see **Chapters 8** and **11**). Blubber influences hydrodynamics through both the shape of the protein matrix and the buoyancy provided by 90% lipid content (see **Chapter 1**). It also provides a thick shield against injury in the equivalent of marine mammal bar brawls (e.g., fights over breeding females).

Our own research has investigated the roles of blubber relative to thermal conditions for Antarctic ice seals. In 2009, Castellini and others used ultrasound on all four Antarctic ice seals (Weddell, leopard *Hydrurga leptonyx*, crabeater *Lobodon carcinophagus*, and Ross *Ommatophoca rossii*) and found that Weddell seals have far more blubber than they need just for insulation (Castellini et al. 2009). They thought the extra blubber might be necessary to provide energy for the breeding season and nursing pups. Mellish et al. (2015) later explored this idea using both ultrasound and **thermal imaging** (see **Toolbox**). Confirming Castellini's earlier work, adult females without pups had 60% greater fat reserves and reduced heat loss due to conduction compared to their breeding counterparts.

We started this section recognizing that thermal balance in marine mammals is not simply a matter of reducing heat loss. These animals will encounter situations where they must deal with an excess of heat. How do they lose heat if they are covered in blubber? The answer is that blubber has blood vessels that run through it to the skin surface. These veins and arterioles can be opened or shut to control blood flow to the skin surface. When the animal is too hot, it opens circulation, "blushes," and warm blood flows to the skin surface to dump heat. Imagine seals "steaming" on the sea-ice surface in subzero temperatures and melting their "shadow" into the ground. Crazy as it sounds, it happens. On the other hand, a cold animal will shut down the blood flow through the blubber and reduce heat loss enough to not even melt falling snow.

This effect of opening and closing blood vessels to control heat at the skin surface is often described as a "thermal window." In both cetaceans and pinnipeds, thermal windows are also seen when controlling blood flow to the relatively blubber-free flippers and flukes (**Figure 5.4**). These appendages have almost no insulation of any kind and, following the Fick Equation, allow large amounts of heat to be released or conserved by directing blood to or from these areas. The arteries in the flukes and flippers transporting warm blood are usually surrounded by a series of veins bringing the cold blood back from the outer extremities. This sets up a "counter-current" heat exchanger system that helps transfer the heat (via conduction) from the arteries to the veins. These circulatory exchangers can be opened or closed by the animal to either dump or conserve heat.

Field biologists can confirm how it is virtually impossible to collect a blood sample from a cold seal or sea lion flipper regardless of the skill of the collector. Shunting of blood to provide cooling is not limited to the extremities. Male cetaceans have internal testes, completely enclosed inside the blubber layer. There is anatomical evidence that cool blood can be routed from dorsal fins to the testes, presumably to keep them from over-heating (Rommel et al. 1992). Pinnipeds may keep the face, snout, and the base of the vibrissae warm to facilitate high-resolution neural sensitivity of the vibrissae for feeding or navigation (see **Chapter 7**). For an in-depth review of how peripheral perfusion plays a leading role in thermoregulation, see Favilla, Horning, and Costa (2021).

BIOCHEMICAL VARIABLES

To balance the thermal equation, we need to consider the production of heat. Terrestrial and marine endothermic homeotherms alike create most of their heat through metabolic processes. Like all mammals, the normal body temperature of most marine species is around 37°C. There is nothing particularly special about 37°C as a temperature, but what is special is its

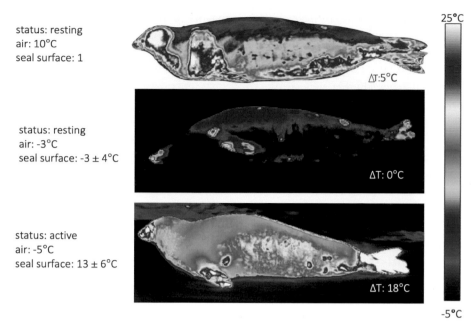

status: resting
air: 10°C
seal surface: 1

ΔT:5°C

status: resting
air: -3°C
seal surface: -3 ± 4°C

ΔT: 0°C

status: active
air: -5°C
seal surface: 13 ± 6°C

ΔT: 18°C

25°C

-5°C

Figure 5.4 **Infrared images of Weddell seals (*Leptonychotes weddellii*) in different air temperatures and activity states demonstrate the thermoregulatory abilities of these animals. Top: At rest in warm air, the skin surface temperature will be close to ambient. Middle: When the air is cool and the animal is at rest, there is also very little heat lost to the air thanks to the generous insulating blubber layer. Bottom: Active seals will lose heat to the cooler surrounding air from body parts that are uninsulated as well as from areas with working muscles. (Photo: J. Mellish, Marine Mammal Protection Act 15748, Antarctic Conservation Act 2012-003)**

consistency (homeostasis). Homeotherms hold body temperature constant because it is biochemically more efficient than having a body temperature that fluctuates widely (Hochachka and Somero 1984). There is a chemical principle known as the **Q_{10} concept** that describes the change in metabolic rate driven by changes in temperature. For mammals, a 10°C increase in tissue temperature generally induces a doubling in the speed of a metabolic reaction. Because biochemical homeostasis is important for effective metabolic control, it is advantageous to not allow body temperature to fluctuate too widely. Imagine driving a car with a speed that varies by itself from 50 to 100 mph. This car would be much more difficult to control than a car that stays at a consistent 60 mph. **Hyperthermia**, or when the body temperature rises to dangerous levels, is a problem because it causes significant

disruption of the delicate biochemical reaction balances in the body. Hyperthermic animals do not die because tissues are breaking down from the temperature but because of Q_{10} imbalances in biochemical reactions.

Surprisingly, heat generation by all mammals is a waste product of the basic cellular metabolism that keeps ion balances across cell walls in check. Mammals have relatively "leaky" cell walls compared to ectotherms, and the cellular ion pumps must constantly work to keep the correct ion balance (Hochachka and Guppy 1987), consuming large amounts of **ATP** in the process. The body burns ATP at a 25% efficiency, with the remaining 75% producing waste heat. Even more heat is generated during exercise and digestion (specific dynamic action, **SDA**).

The **Kleiber principle (Chapter 2)** describes the relationship between metabolic rate and body mass such that small animals have a much higher metabolic rate per body mass than larger animals. The sea otter has a very high Kleiber value compared to a larger seal or whale. The smaller sea otter also has a higher surface area to body volume ratio and loses heat more quickly because of that ratio. The combined higher relative metabolism and SA:V ratio means that otters must consume a huge amount of food to maintain thermal homeostasis, including the contributions of SDA (Costa and Kooyman 1982; Yeates, Williams, and Fink 2007).

Despite the long-held theory that marine mammal metabolic rates were about double that of terrestrial mammals to deal with the added thermal challenge of being in water, "thermoregulation may not be a problem for all but the smallest of marine mammals and other traits in marine mammals set metabolism" (**Chapter 2**). Smaller marine mammals may still struggle to maintain enough heat generation while the larger species with reduced SA:V ratios may have more issues in creating too much heat. This leads to an essential point about the thermal biology of marine mammals: they are not warmer than terrestrial mammals, yet they appear to generate more heat. It seems that the fundamental difference between marine and terrestrial mammal heat loss lies almost entirely in their control of the processes, rather than in specialized systems to generate heat.

BEHAVIORAL IMPACTS

Intuitively, we expect an exercising animal to generate heat. This assumption becomes more complex when that exercising animal is diving and swimming in cold water (see **Chapter 2**). For many deep-diving mammals, the water temperature can change dramatically from the sea surface to the deepest locations. Short-finned pilot whales (*Globicephala macrorhynchus*) off the coast of North Carolina regularly forage at 500 m and can dive as deep as 1500 m, experiencing temperature differences of up to 18°C (Adamczak et al. 2020). Consider also that a swimming

marine mammal will likely be moving quickly (usually ~2 m/sec) and experience significant convective heat loss. In the 1970s–1980s, freely diving Weddell seals in the Antarctic were fitted with central arterial temperature devices, showing that deep circulating body temperature decreased by 1–2°C (Kooyman et al. 1980; Hill et al. 1987). Since that time, a suite of seal (and penguin) studies demonstrated that even a few degrees drop can provide reduction in metabolic demand as defined by the Q_{10} relationship (Meir and Ponganis 2010), allowing these animals to extend their dive time. While this works for some, other species such as northern elephant seals (*Mirounga angustirostris*) that remain at sea for months and spend 90% of their time at sea diving, a continued drop in body temperature would be a problem. This takes us back to the discussions in **Chapters 1**, **2**, and **3** and others about swimming patterns, muscle use, and the energetics of diving. The relationships between thermal biology, physical insulation, metabolic rate, and diving physiology for marine mammals are all connected.

CLIMATE CHANGE

Climate change is at the forefront of biology, economics, and world politics, but how has an increasing environmental temperature affected marine mammals and how might that change if not addressed? From the basic and fundamental aspects of heat balance discussed here, the answer is that it likely has little impact on their physiology, biochemistry, or thermal biology (Huntington and Moore 2008; Castellini 2012; see also **Chapter 15**). Marine mammals maintain a constant internal body temperature in the face of massive temperature swings seasonally, on migrations, off and onto beaches, and in and out of freezing seawater. A 1–2°C water temperature change due to climate shift is easily accommodated by their thermoregulatory physiology. However, these oceanic temperature changes will noticeably impact the distribution and abundance of their prey and might influence the migration routes of large whales as ice, prey, or other oceanic patterns shift. Warming temperatures can and will continue to alter the ice pack that is relied upon by polar seals, walrus, and bears for rest and breeding. It is on these bases of predicted changes in ice platform that the recent listings of Endangered under the US Endangered Species Act for polar bears and some of the Arctic seals have been enacted. See **Chapter 15** for a more in-depth discussion of climate change and its potential impacts on marine mammals.

CASE STUDIES

Bowhead whale: behemoth ice breakers

The 100-ton Bowhead whale lives in Arctic and subarctic waters and is the second largest mammal on earth. Bowheads have extremely thick blubber (up to 50 cm) and transit through ice-laden waters. Some believe the thick blubber is necessary as a fuel source for their long migrations rather than for thermal needs. Others propose that the thick blubber acts as a physical shield against the heavy Arctic ice. It turns out that deep body temperatures of bowheads are cooler than expected (33–34°C; George 2009). One explanation for this is whales have thick blubber to protect their bodies from ice damage, making them over-insulated (**Figure 5.5**).

Sirenia: life in the slow lane

All the marine mammals we have discussed so far have been carnivores that are active swimmers and strong divers, with a high-metabolic rate. The sirenia (dugongs; *Dugong dugon*, and manatees; *Trichechus* spp.) are warm water vegetarians with a low-metabolic rate. They are not phylogenetically related to any of the pinnipeds or cetaceans and are most closely related to the elephants. They are "marine" but in a group by themselves. However, because they live in a marine environment, they must be able to dive, hold their breath, swim, and have many of the same evolutionary adaptation pressures that are faced by other marine mammals.

Figure 5.5 Bowhead whale (*Balaena mysticetus*) blubber from a whale taken in the subsistence hunt near Barrow (now Utqiagvik), Alaska, in fall 2003. (Photo credit: Michael Castellini. Access to animals through courtesy and cooperative agreement with Alaska Eskimo Whaling Commission, NOAA, whaling boat captains, and North Slope Borough Department of Wildlife Management)

Returning to Fick, sirenia are relatively large (small SA:V ratio), tend to inhabit warm water (reduced convective heat loss), and will even move into warm water springs during the colder winter months (behavioral thermoregulation) (**Figure 5.6**). They have a thick, fat layer but no fur or hair. Their fat layer is not homogenous blubber as seen in cetaceans or phocid seals. It is multi-layered with muscle, similar to the fat/blubber of Steller sea lions. Their body temperature is low compared to other marine mammals, and the temperature gradient from body core to water is reduced (Gallivan, Best, and Kanwisher 1983). Because they lack robust insulation mechanisms, they must compensate for their low-metabolic rate and have a very limited environmental thermal window. That is,

Figure 5.6 A manatee (*Trichechus manatus*) floats in the warm coastal waters near Crystal River, Florida. Manatees have a low metabolic rate and no fur, relying on their blubber to provide insulation. However, their blubber is not as thermally protective as in whales or seals, and the manatees seek warm water to keep from getting cold. (Photo courtesy of Robert Bonde, USFWS Wildlife Research Permit MA791721)

they must stay in warmer waters. The endangered status of many sirenian populations is a result of human interference and loss of essential warm water habitat.

Steller's sea cow: extinct but not forgotten

The extinct Steller's sea cow (*Hydrodamalis gigas*) is a fascinating case study in thermal biology. The sea cow was a vegetarian like dugongs and manatees; however, unlike modern sirenia, the Steller's sea cow lived in the cold, shallow intertidal areas of southeast Alaska to Japan. It was hunted to extinction by 1768 by the same fur traders mentioned earlier in search of otter and seal pelts.

How did the sea cow, a low-metabolic rate vegetarian, survive in sub-Arctic waters? We know from historical records and recovered bones that the sea cow was the largest of all the non-cetacean marine mammals. It approached 8 m in length, reached 10 mt in mass, had very thick skin and may have had such a significant fat layer that it was unable to dive because of its high buoyancy. The only reasonable answer to Fick requirements is that their overwhelming mass and blubber combined with a relatively small surface area afforded them a much smaller heat loss than their much smaller current relatives.

Polar bear: life in perpetual cold

Polar bears are not the usual marine mammals like whales or seals, but they are well adapted to a far north marine and sea-ice environment. They utilize a thick fur coat to stay warm in the air, and even have fur on the bottom of their paws that helps to make them quieter when stalking their seal prey on the ice surface (**Figure 5.7**).

Figure 5.7 Polar bear (*Ursus maritimus*) showing heavy fur on the bottom of the feet and covering the belly and genital regions. (Photo: Kathy Crane, NOAA Arctic Research Program https://www.pmel.noaa.gov/arctic-zone/gallery_bear.html accessed August 2022)

They have a thick fat layer to help with insulation and utilize circulatory adjustments to control blood flow to the extremities. Their blubber is layered with well-vascularized muscular sheets (for control of blood flow to the skin surface) and is not as homogenous as cetacean and phocid blubber. They might be over-insulated while in the air, and rely on behavioral thermoregulation when heat stressed (for example, laying down with their legs spread out on the sea-ice surface). The dense under-fur works well in the air but allows water to penetrate to the skin surface, which is actually black. The outer larger guard hairs are hollow and transparent which makes their fur appear white. The inner dense hair is finer and transparent but not hollow.

Interestingly, these characteristics of their fur have been used to help create synthetic commercial insulation material.

Polar bears have become the "poster child" of climate change advocacy because of their critical dependence on sea ice for hunting and raising their cubs, their public appeal and their low population numbers.

Sea otters: eating to stay warm

As noted in **Chapter 2**, sea otters have an elevated metabolic rate, high-dietary demand, a large surface area to volume ratio, lose heat easily, and live in cold water environments. They use extremely dense fur found everywhere except on their paws to insulate themselves. When they need to conserve heat, they float on the surface with their paws out of the water. If they need to dump heat, they submerge their paws. However, it is costly to maintain their fur coat and upward of 25% of their daily metabolic energy goes to maintaining and grooming the fur. Like fur seals, they must keep their inner fur coats dry and coated with natural oils to repel water.

Because they rely on their fur coat to stay warm, they are very susceptible to oil spills. If their fur is fouled, then it can no longer repel water, clumps together, and loses the air layer that keeps their skin warm (**Figure 5.8**). They are extremely

Figure 5.8 Oiled sea otter (*Enhydra lutris*) on beach in Prince William Sound after the Exxon Valdez Oil Spill (EVOS) in March 1989. (Photo courtesy of the Exxon Valdez Oil Spill Trustee Council)

dependent on high quality and high volume nearshore food resources, which can place them in direct conflict with humans for nearshore resource use.

Sea otters and northern fur seals were heavily hunted in the 1700s for their valuable and luxurious fur coats. There is evidence suggesting sea otters are a primary prey of Orca.

Orca: fast with a bad reputation

Orca (*Orcinus orca*) are not the largest of the cetaceans, but they are apex predators that have earned several distinctive common names such as killer whale (English), assassin whale (Portuguese), and blubber chopper (Norwegian). They combine speed, hydrodynamics, and a fast metabolic rate into the marine mammal equivalent of a sports car. Swim speeds can reach up to 30 mph. They are above the Kleiber curve for predicting metabolic rate based on body mass, which means they have a higher metabolic rate than predicted. Yet, they have a thick 6-inch blubber layer, with higher levels of membrane-specific lipids than other cetaceans, which may help them maintain their streamlined shape. Their surface area-to-volume ratio is small compared to terrestrial mammals and almost a third of that available surface area is claimed by the uninsulated but vascular flukes. Not only can they shunt warm blood to or from the extremities as needed, but they also have the countercurrent circulatory heat exchangers discussed above, which means warmer arteries are surrounded by veins and heat can be conserved much longer under cold conditions. Unlike the Arctic polar bear, they have a global distribution from pole to pole, which demonstrates the thermoregulatory adaptability of this remarkable combination of attributes.

Steller's sea lion: on the edge of survival

Steller sea lions are the largest of the sea lions, with adult males over 10 feet long and 1000 kg in mass. They have the unfortunate distinction of also following in the footsteps of the other species we have already discussed that was named by Steller as being endangered in their North Pacific home range. Where there once were 300,000 or more animals, best estimates are now closer to 40,000. There was a dramatic population crash in the 1970s–1980s, which drew a lot of attention in the 2000s. The annual decline seems to have stabilized but the cause(s) are unknown. Aside from their size, Steller's are squarely in the middle of the road for most other characteristics. They do not have particularly thick blubber or fur to help with thermoregulation. They do have very large, hairless flippers that not only make them very maneuverable under water but also function as heat radiators. It is not uncommon to see sea lions waving their hind flippers in the air as they lounge on the beach or float in the water on a warm day. They fast from food at various points during their annual cycle but not for very long. They swim fast in water but do not usually dive deep (usually less than 50 m) or for very long (5–10 min). The males do have substantially heavier fur and blubber around their necks to provide protection during fighting in the breeding season (**Figure 5.9**). Ultrasound studies have found that the blubber in Steller's has layers rather than a single defined border as in seals and cetaceans. Thermal imaging has shown that they have consistent thermal windows along their trunk. While their physiological attributes may help explain why they do not range as far into the cold poles as phocids, whether it contributed their dramatic decline is unknown.

Gray seals: fat and fluffy pups

Gray seals (*Halichoerus grypus*) are found on exposed coasts of both North America and the United Kingdom. They are a mid-sized seal (adults 150–250 kg) with about one-third of their body mass as fat or blubber. Aside from their occasional cranky attitude, one of their more defining features is the grayish-white fur called lanugo found on newborn pups. This temporary fur coat provides a thermoregulatory solution to the land-born pups until they

Figure 5.9 Male Steller's sea lion (*Eumetopias jubatus*) along with pups and females on Ugamak Island, Aleutians, Alaska. (Photo credit: Michael Castellini. MMPA permit # 809)

are ready to enter the water. Pups are born with very limited fat stores (approx. 5% of body mass) but nursing pups do nothing apart from consuming large volumes of high-fat milk (2–3 kg of 40% fat milk per day) and sleep. Pups are 30% body fat and have shed their lanugo before being abruptly weaned and venturing into the water at two weeks of age. Curiously enough, gray seals on the North American side of the Atlantic are born in the middle of winter, while UK gray seals pup in the summer. UK gray seals are approximately 25% smaller in body mass than the winter-born North American gray seals. Heat loss through surface area to body volume differences are consistent with the smaller (higher heat loss) summer-born UK pups, compared to the larger winter-born North American pups and their comparatively lower heat loss.

Weddell seals: southern exposure

The Weddell seal is mentioned many times in this chapter, due to their large size, accessibility during the breeding season, and their Antarctic habitat that provides an excellent tool for

thermoregulation modeling. Weddell seals are the southernmost breeding mammal, live only in the Southern Ocean and are closely associated with the sea ice. They do not tolerate even the relatively warmer waters of the sub-Antarctic. They are an excellent example of a **pagophilic** species that must stay warm in very cold water but cannot live in warmer waters. Mellish combined both thermal imaging and skin surface heat flux sensor deployments in Weddell seals on the sea-ice surface as a thermoregulatory model for phocid seals in a wide range of mass and condition states (Mellish et al. 2015). The overall body surface temperature of the seals was a surprising 14°C on average, despite the below-freezing air temperatures. However, on cold days (−18°C not including wind chill), there could be as little as a 2–4°C difference between the outer temperature of the skin and the environment. Unlike more temperate pinniped species, there was no regional difference in heat loss across the body. According to the Fick Equation, this would greatly limit the amount of heat lost to the environment.

TOOLBOX

Thermoregulation has so many components to it that there are an equally large number of technical and scientific tools to investigate questions about body temperature, heat loss, insulation and more. From relatively simple remote thermometers to complex heat flow devices, the field is one where much of the technical advancement comes from the fields of human medicine and engineering.

Ultrasound

Much of the early knowledge about blubber was limited to what could be collected from harvested, hunted, or culled animals. For seals, that initially meant taking a single measurement at the equivalent of the chest bone area to standardize the data. Advances in **ultrasound** technology in the 1980s dramatically changed our ability to measure blubber in live animals. Medical ultrasound functions on the principle that sound travels at different speeds through different types of tissues. In some cases, such as bone, the sound does not travel through at all but instead is refracted/returned. Portable medical ultrasound devices were used to estimate blubber depth in phocids in particular given their relatively uniform blubber layer and clear signal at the blubber-muscle interface. This led to the ability to perform time-series studies in individuals, and the important finding that not only does blubber depth vary by body site, it can vary seasonally, and is quite species-specific (Rosen and Renouf 1997; Mellish, Horning, and York 2007).

The first portable ultrasound devices provided a simple LED scale of blubber depth but not long after "advanced graphic display imaging" ultrasound revolutionized the ability to visualize and measure blubber depth both rapidly and effectively (Mellish, Tuomi, and Horning 2004). As these ultrasound units shrunk from the size of a wheeled cart to the size of a backpack, and now to the size of a smartphone, they have become a part of regular health exams for cetaceans and pinnipeds at aquaria around the world, and in the gear complement of most field physiologists (Noren et al. 2008; Mellish, Hindle, and Horning 2011; Hoopes et al. 2014).

In its simplest form, one can model the body shape of a marine mammal as a core of muscle, surrounded by a ring of blubber and build a series of volume cones to estimate body mass, blubber content, etc. (Castellini et al. 2009), allowing for areas with little or no blubber like the head, flippers, and tail. It is most accurate to obtain a suite of blubber thickness measurements around the body and then build a species-specific model of total blubber content (Shero et al. 2014; Shuert, Skinner, and Mellish 2015).

Photogrammetry

Ultrasound methods work well when an animal can be directly handled. However, there are many cases where the animals cannot be touched (almost all wild cetaceans) or where they are not restrained easily (many wild pinnipeds). In those cases, there have been great advances in digital imagery where measurements can be made of width, girth, etc. Using controls, studies of live and dead specimens and animal geometry, values of blubber depth and total body mass can be modeled using photographic techniques. The basic assumption in these methods is that a marine mammal with large amounts of blubber appears different from the same animal with less blubber. The concern here is that it is possible to remove a great deal of the lipid from blubber without changing its shape, because of the protein matrix. In this example, the same whale under "low blubber fat" and "high blubber fat" conditions may appear to have the same width and girth and the method can be controversial (Christiansen et al. 2020).

Infrared thermography

Thermal imaging cameras can measure the skin surface temperature profile of the entire animal in a single frame. One of the pioneering wildlife applications of this technology was by one of the authors of **Chapter 2**, where they defined differences in heat transfer in elephant ears (Williams 1990). The method has since been applied to numerous thermoregulatory studies (Tattersall and Cadena 2010). There have been many applications of thermal imaging specific to pinnipeds, ranging from tracking the energy budgets of newly weaned gray seals in comparison to their environment (McCafferty et al. 2005) to the variation in heat loss by territorial male northern elephant seals during battle (Norris, Houser, and Crocker 2010). The thermal windows mentioned previously provide evidence beyond the visual that marine mammals can and will use increased blood flow to the skin to enhance evaporative heat loss (Mauck et al. 2003). With some care, thermal images of individuals can be used in biophysical modeling to estimate metabolic heat loss (McCafferty et al. 2011; Mellish et al. 2015).

As discussed in the Weddell seal case study, their skin surface temperature on the ice surface is largely dictated by a combination of the mass of the animal, ambient air temperature, and wind speed. Thermographic imagining methods have provided fascinating new information on heat transfer modeling (**Figure 5.4**). For example, it was possible to estimate that radiation contributed more than half of the total heat loss in dry animals resting on ice, regardless of the size of the individual. Conduction to the ice surface accounted for about a third, and convection came in last place with about 15% of total heat loss. Further, the studies were able to estimate how environmental conditions could impact heat loss in these animals, providing an example of how drastic changes in energy budgets can occur with changes in an animal's environment. For example, a small 120 kg Weddell seal (or other similarly sized seal) in low wind conditions of 4 m/s will lose 52 W, however, if you increase the wind to a moderate (for Antarctica) wind speed of 17 m/s, the heat loss jumps to 124 W.

This is just a small sample of the species-specific information that we can now model due to the improvement of two types of imaging tools – infrared (thermal) and ultrasound. These technologies have a multitude of diverse applications that we only expect to grow, given their no or low-contact requirements, high portability, and increasingly affordable prices. However, caution must be exercised in that these images are providing the combined effect of the anatomy (insulation), physiology (metabolic state), and environment. A major limitation of thermal imaging cameras is that they can only be used in air.

Heat flux sensors

Throughout this section, we have explained that the heat capacity of water is so high that heat will flow out from a warm marine mammal to the cold water at elevated rates. We have also explained how diving will reduce the metabolic rate and reduce internal heat generation. However, we have mostly discussed the characteristics and measurements of heat flow in air. How can one measure heat flow in marine mammals while underwater?

The Fick Equation provides some background for this problem. In general, the surface area, inside to outside temperature differential, and the thickness of a material are usually the easiest values to obtain when underwater. However, measuring thermal conductivity is usually very difficult in living tissue (such as blubber) given the dynamic nature of blood flow. Computer-aided analyses of pelts using suites of thermocouples and standard reference materials with known heat conductance work well in the laboratory, but they are not able to be utilized for studies of live animals, either in air or in water.

The invention of electronic heat flow disks was a breakthrough in the field of thermoregulation for both cetaceans and pinnipeds. These small disks can measure the temperature difference between the skin surface where they are attached and the water temperature surrounding them, they are of a precise thickness and their thermal conductivity is accurately known.

Using the Fick Equation, they can be used to calculate the heat flow that is moving through them. Noren et al. were able to adapt this technology by holding heat-flow disks against the skin of dolphins accustomed to divers. In this way, they were able to measure heat flow from the surface of the dolphin skin at multiple sites around the body, after the animals had been resting, exercising, etc. They found that most of their "extra heat" from exercise was dissipated when they returned to the surface and that the diving responses over-rode the need to dump heat to the periphery. Evidence for this included that heat flow from the fins and flukes decreased when the animals were underwater, suggesting a reduction in blood flow which would be consistent with blood flow redistribution during diving.

A pair of highly trained Steller sea lions at the Vancouver Aquarium in British Columbia, Canada, provided the stage for application of this method to otariid seals (Willis and Horning 2005). Custom-made housings were designed for a temporary attachment of heat flux sensors for up to seven days during routine swim bouts. Pilot deployments of this method led to a recent full-scale effort with Weddell seals in McMurdo Sound, Antarctica. Hindle, Horning, and Mellish (2015) deployed skin surface heat flux sensors on free-ranging Weddell seals and recorded heat loss in both air and water over days to weeks (**Figure 5.4**). By analyzing additional baseline information about individuals (body size and condition, insulation, and surface temperature patterns from infrared thermograms, mentioned before), they were able to suggest a method to integrate point measurements of heat flux across the body into a total measurement of whole-animal heat loss.

While adult female Weddell seals in very good condition with high blubber insulation showed little need for additional thermoregulatory heat production in air or in cold Antarctic water (−1.9°C), the smallest-bodied pups consistently showed thermoregulatory heat costs. Juvenile Weddell seals (yearlings), having the least blubber insulation, lost heat at the fastest rate as their swim speed (convective water flow over their bodies) increased. This type of information will help scientists to understand when higher thermoregulatory heat production becomes necessary to maintain stable core body temperature, and when changes in the environment or in animal condition impose thermoregulatory costs for wild marine mammals.

Designer temperature telemetry

The development of ever more sophisticated dive recorders is discussed in the Toolbox sections of many chapters in this book. They can also be important for the study of thermoregulation because they can be designed to collect skin surface temperatures, stomach, deep muscle, and any other temperature from wherever the investigator is able to place a temperature probe. Temperature data from these recorders have been integral in the models of diving blood flow, identifying tissues that are metabolically depressed during diving, and in understanding the overall thermal balance of animals underwater (Andrews 1998; Kuhn and Costa 2006; Meir and Ponganis 2010). They can even be used to measure food ingestion by showing the temperature change in the stomach after a marine mammal has consumed a cold fish meal.

CONCLUSIONS

There are several concepts that are essential to the thermal biology of marine mammals. Animals not only have to stay warm when in cold water but must also be able to remove excess heat when needed. They do not have any specialized heat-generating organs and must be able to conserve or dump heat produced from normal metabolic activity, exercise, and food digestion. They do this with very fine control of blood and heat flow through blubber or through selective blood flow to skin areas without fur or blubber. Because marine mammals span such a large size range (from sea otters to blue whales), the relative heat flow issues differ based on mass, volume, and surface area. Water impermeable fur is used for insulation by a few species, but most use blubber for insulation. However, blubber is also used as a fuel, a water source, and for buoyancy control, and not solely for thermal needs. It is very likely that only the smallest individuals of a given species experience a cold stress situation, but rather the most common thermoregulatory challenge for marine mammals is how to prevent overheating.

REVIEW

1 Describe the four factors in the Fick Equation that determine how much heat can be lost from a warm marine mammal.
2 What is wind chill and why is that important for a seal pup resting on sea ice?
3 Why is it easier to lose heat in water than in air?
4 How does heat loss across fur differ from heat loss through blubber?
5 Why is blubber important for more than just thermal insulation?
6 What is the normal body temperature of most marine mammals?
7 What is the primary biochemical cellular process that generates heat in mammals and birds?
8 Why does diving complicate heat balance in marine mammals?
9 Why do smaller marine mammals lose more heat than larger ones?
10 How does ultrasound measure blubber depth?

CRITICAL THINKING

1 Which is more important, using blubber for thermoregulation or as fuel reserve?
2 If the Arctic Ocean warms by 2°C, will marine mammals be able to adapt? Will their thermoregulatory abilities be adaptive enough to handle the change? Why are we concerned about their survival?

3 Is there an optimal core body temperature for marine mammals?

EQUATIONS

1 Fick Equation

$$Rate\ of\ heat\ flow = \frac{kA(T2-T1)}{thickness}$$

where:

k is the thermal conductivity of the material (calories per s^{-1} cm^{-1} $°C^{-1}$, *Table 5.1*),
A is area of contact (cm^2),
$T1-T2$ (°C) is the temperature difference between the two bodies, and
thickness of the material (cm).

The final units are cal/s.
Example comparison of heat flow through water versus fur:

Area of contact = 150 cm^2
Thickness/depth = 5 cm
Temperature difference = 15°C
k (water) = 0.0013 calories per s^{-1} cm^{-1} $°C^{-1}$
k (fur) = 0.00009 calories per s^{-1} cm^{-1} $°C^{-1}$

$$Rate\ of\ heat\ flow\ for\ water = \frac{0.0013 \times 150 \times 15}{5}$$

$$Water\ solution = 0.585\ cal/s$$

$$Rate\ of\ heat\ flow\ for\ fur = \frac{0.00009 \times 150 \times 15}{5}$$

$$Fur\ solution = 0.0405\ cal/s$$

Water moves heat 14 times faster than fur (0.585/0.0405)

2 SA:Volume ratios

$$volume = \frac{4}{3}\Pi r^3$$

$$surface\ area = 4\ \Pi r^2$$

A sphere with a diameter of 10 cm has surface area of 314 cm^2, a volume of 524 cm^3, and a SA:V ratio of 324/524 = 0.60.

If we reduce the diameter to 5 cm, the surface area is 78.5 cm², the volume is 65.5 cm³, and the SA:V ratio is now increased to 78.5/65.6 = 1.2.

A quick shortcut to this calculation is that the SA:V ratio of a sphere is given by the value of 3/r.

FURTHER READING

Castellini, M. 2009. Thermoregulation. *Encyclopedia of Marine Mammals* (2nd Edition), 1166–1171. New York: Academic Press.

Elsner, R. 2015. *Diving Seals and Meditating Yogis: Strategic Metabolic Retreats*. Chicago, IL: University of Chicago Press.

Fish, F.E., J. Smelstoys, R.V. Baudinette, et al. 2002. Fur doesn't fly, it floats: Buoyancy of pelage in semi-aquatic mammals. *Aquatic Mammals* 28(2): 103–112.

Folk, G.E., M.L. Riedesel, and D.L. Thrift. 1998. *Principles of Integrative Environmental Physiology*. San Francisco: Austin & Winfield Publishers.

Kooyman, G.L. 1981. *Weddell Seal: Consummate Diver*. Cambridge: Cambridge University Press.

MacArthur, R.A. 1989. Aquatic mammals in cold. In *Animal Adaptation to Cold*. Berlin, Germany: Springer.

Schmidt-Nielsen, K. 1997. *Animal Physiology: Adaptation and Environment*. Cambridge, UK: Cambridge University Press.

GLOSSARY

Adipocytes: Cells that contain fat

ATP: Adenosine tri-phosphate. Biochemical energy compound for metabolism.

Conduction: Transfer of heat by direct contact between two bodies

Convection: "Wind chill." Heat loss by moving air or water over surface.

Daily metabolic rate: Energy cost of an animal over a 24-h period

Ectothermic: "Cold-blooded" animals whose body temperature depends on the environment

Endothermic: "Warm-blooded" animals whose body heat is generated internally

Evaporation: Loss of heat energy by changing liquid surface water to gas

Fick Equation: Defines variables for heat transfer between two bodies

Heat Capacity: Heat energy required to raise 1 gm of material by 1°C

Heterothermic: Animals with a variable internal body temperature

Homeostasis: Constancy of metabolism

Homeothermic: Animals with a constant body temperature

Hyperthermia: Body temperature high above normal

Hypothermia: Body temperature below normal

Kleiber principle: How metabolic rate changes with body mass in mammals

Lanugo: "Pup fur." Very fine and good insulation in air. Poor in water.

Lipophilic: Compounds that dissolve easily in lipids

Pagophilic: "Ice loving." Animals that live closely associated with ice.

POPS: Persistent organic pollutants, often containing chlorine.

Q_{10}: Ratio of reaction rate change with 10°C temperature difference

Radiation: Heat transfer through electromagnetic radiation

Specific dynamic action: Heat release through the process of food digestion

Thermal conductivity: Description of how well a material will conduct heat energy

Thermal imaging: Method to visualize surface temperatures

Thermoregulatory: Referring to the ability to control heat loss and gain

Ultrasound: Use of sound waves to visualize tissue type

REFERENCES

Adamczak, S.K., D.A. Pabst, W.A. McLellan, et al. 2020. Do bigger bodies require bigger radiators? Insights into thermal ecology from closely related marine mammal species and implications for ecogeographic rules. *Journal of Biogeography* 47(5): 1193–1206.

Andrews, R.D. 1998. Remotely releasable instruments for monitoring the foraging behaviour of pinnipeds. *Marine Ecology Progress Series* 175: 289–294.

Ashley, C. 2014. *The Yankee Whaler*. Newburyport: Dover Publications.

Bagge, L.E., H.N. Koopman, S.A. Rommel, et al. 2012. Lipid class and depth-specific thermal properties in the blubber of the short-finned pilot whale and the pygmy sperm whale. *Journal of Experimental Biology* 215(24): 4330–4339.

Beentjes, M.P. 2006. Behavioral thermoregulation of the New Zealand sea lion (*Phocarctos hookeri*). *Marine Mammal Science* 22(2): 311–325.

Bryden, M.M. 1964. Insulating capacity of the subcutaneous fat of the southern elephant seal. *Nature* 203: 1299–1300.

Castellini, M.A. 2012. On thin ice: Marine mammals and climate change. *Temperature Adaptation in a Changing Climate: Nature at Risk* 3: 131.

Castellini, M.A., S.J. Trumble, T.L. Mau, et al. 2009. Body and blubber relationships in Antarctic pack ice seals: Implications for blubber depth patterns. *Physiological and Biochemical Zoology* 82(2): 113–120.

Christiansen, F., K.R. Sprogis, J. Gross, et al. 2020. Variation in outer blubber lipid concentration does not reflect morphological body condition in humpback whales. *Journal of Experimental Biology* 223(8): jeb213769.

Cornick, L.A., L.T. Quakenbush, S.A. Norman, et al. 2016. Seasonal and developmental differences in blubber stores of beluga whales in Bristol Bay, Alaska using high-resolution ultrasound. *Journal of Mammalogy* 97(4): 1238–1248.

Costa, D.P., and G.L. Kooyman. 1982. Oxygen consumption, thermoregulation, and the effect of fur oiling and washing on the sea otter, *Enhydra lutris. Canadian Journal of Zoology* 60(11): 2761–2767.

Elam, S.K., I. Tokura, K. Saito, and R.A. Altenkirch. 1989. Thermal conductivity of crude oils. *Experimental Thermal and Fluid Science* 2(1): 1–6.

Favilla, A.B., M. Horning, and D.P. Costa. 2021. Advances in thermal physiology of diving marine mammals: The dual role of peripheral perfusion. *Temperature* 9(1): 46–66.

Folk, G.E., M.L. Riedesel, and D.L. Thrift. 1998. *Principles of Integrative Environmental Physiology.* San Francisco: Austin & Winfield.

Frisch, J., N.A. Øritsland, and J. Krog. 1974. Insulation of furs in water. *Comparative Biochemistry and Physiology Part A, Physiology* 47(2): 403–410.

Gallivan, G.J., R.C. Best, and J.W. Kanwisher. 1983. Temperature regulation in the Amazonian manatee *Trichechus inunguis. Physiological Zoology* 56: 255–262.

George, J.C. 2009. Growth, morphology and energetics of bowhead whales (*Balaena mysticetus*). Ph.D., Biology and Wildlife, University of Alaska Fairbanks.

Haldiman, J.T., and R.J. Tarpley. 1993. Anatomy and physiology. In *The Bowhead Whale* (pp 71–156), ed. J.J. Burns, J.J. Montague, and C.J. Cowles. Lawrence, KS: Allen Pr.

Hill, R.D., R.C. Schneider, G.C. Liggins, et al. 1987. Heart rate and body temperature during free diving of Weddell seals. *American Journal of Physiology* 253: R344–R351.

Hindle, A.G., M. Horning, and J. Mellish. 2015. Estimating total body heat dissipation in air and water from skin surface heat flux telemetry in Weddell seals. *Animal Biotelemetry* 3(1): 1–11.

Hochachka, P.W., and M. Guppy. 1987. *Metabolic Arrest and the Control of Biological Time.* Cambridge, UK: Harvard University Press.

Hochachka, P.W., and G.N. Somero. 1984. *Biochemical Adaptation*. Princeton: Princeton University Press.

Hoopes, L.A., L.D. Rea, A. Christ, et al. 2014. No evidence of metabolic depression in western Alaskan juvenile Steller sea lions (*Eumetopias jubatus*). *PLoS One* 9(1). doi.org 10.1371/journal.pone.0085339

Huntington, H.P., and S.E. Moore. 2008. Assessing the impacts of climate change on Arctic marine mammals. *Ecological Applications* 18(sp2): S1–S2.

Khamas, W.A., H. Smodlaka, J. Leach-Robinson, et al. 2012. Skin histology and its role in heat dissipation in three pinniped species. *Acta Veterinaria Scandinavica* 54(46).doi.org/10.1186/1751-0147-54-46

Kooyman, G.L., E.A. Wahrenbrock, M.A. Castellini, et al. 1980. Aerobic and anaerobic metabolism during voluntary diving in Weddell seals: Evidence of preferred pathways from blood chemistry and behavior. *Journal of Comparative Physiology* 138: 335–346.

Kuhn, C.E., and D.P. Costa. 2006. Identifying and quantifying prey consumption using stomach temperature change in pinnipeds. *Journal of Experimental Biology* 209(22): 4524–4532.

Liwanag, H.E.M., A. Berta, D.P. Costa, et al. 2012a. Morphological and thermal properties of mammalian insulation: The evolution of fur for aquatic living. *Biological Journal of the Linnean Society of London* 106(4): 926–939.

Liwanag, H.E.M., A. Berta, D.P. Costa, et al. 2012b. Morphological and thermal properties of mammalian insulation: The evolutionary transition to blubber in pinnipeds. *Biological Journal of the Linnean Society of London* 107(4): 774–787.

Luque, S.P., and D. Aurioles-Gamboa. 2001. Sex differences in body size and body condition of California sea lion (*Zalophus californianus*) pups from the Gulf of California. *Marine Mammal Science* 17(1): 147–160.

Mauck, B., K. Bilgmann, D.D. Jones, et al. 2003. Thermal windows on the trunk of hauled-out seals: Hot spots for thermoregulatory evaporation? *Journal of Experimental Biology* 206(10): 1727–1738.

McCafferty, D.J., C. Gilbert, W. Paterson, et al. 2011. Estimating metabolic heat loss in birds and mammals by combining infrared thermography with biophysical modelling. *Comparative Biochemistry and Physiology Part A, Molecular and Integrative Physiology* 158(3): 337–345.

McCafferty, D.J., S. Moss, K. Bennett, et al. 2005. Factors influencing the radiative surface temperature of grey seal (*Halichoerus grypus*) pups during early and late lactation. *Journal of Comparative Physiology B-Biochemical Systemic and Environmental Physiology* 175(6): 423–431.

Meir, J.U., and P.J. Ponganis. 2010. Blood temperature profiles of diving elephant seals. *Physiological and Biochemical Zoology* 83(3): 531–540.

Mellish, J.A., P.A. Tuomi, and M. Horning. 2004. Assessment of ultrasound imaging as a noninvasive measure of blubber thickness in pinnipeds. *Journal of Zoo and Wildlife Medicine* 35(1): 116–118.

Mellish, J.A.E., A.G. Hindle, and M. Horning. 2011. Health and condition in the adult Weddell seal of McMurdo Sound, Antarctica. *Zoology* 114(3): 177–183.

Mellish, J.A.E., M. Horning, and A.E. York. 2007. Seasonal and spatial blubber depth changes in captive harbor seals (*Phoca vitulina*) and Steller's sea lions (*Eumetopias jubatus*). *Journal of Mammalogy* 88(2): 408–414.

Mellish, J.E., A.G. Hindle, J.D. Skinner, et al. 2015. Heat loss in air of an Antarctic marine mammal, the Weddell seal. *Journal of Comparative Physiology B* 185(1): 143–152.

Mellish, J.E., S.J. Iverson, W.D. Bowen, et al. 1999. Fat transfer and energetics during lactation in the hooded seal: The roles of tissue lipoprotein lipase in milk fat secretion and pup blubber deposition. *Journal of Comparative Physiology B-Biochemical Systemic and Environmental Physiology* 169(6): 377–390.

Noren, S.R., L.E. Pearson, J. Davis, et al. 2008. Different thermoregulatory strategies in nearly weaned pup, yearling, and adult Weddell seals (*Leptonychotes weddellii*). *Physiological and Biochemical Zoology* 81(6): 868–879.

Norris, A.L., D.S. Houser, and D.E. Crocker. 2010. Environment and activity affect skin temperature in breeding adult male elephant seals (*Mirounga angustirostris*). *Journal of Experimental Biology* 213(24): 4205–4212.

Øritsland, N.A. 1970. Temperature regulation of the polar bear (*Thalarctos maritimus*). *Comparative Biochemistry and Physiology* 37(2): 225–233.

Pabst, D.A., S.A. Rommel, and W.A. McLellan. 1999. The functional morphology of marine mammals. In *Biology of Marine Mammals*, ed. J.E. Reynolds III and S. Rommel, 15–72. Washington, DC: Smithsonian Institution Press.

Pearson, L.E., E.L. Weitzner, J.M. Burns, et al. 2019. From ice to ocean: Changes in the thermal function of harp seal pelt with ontogeny. *Journal of Comparative Physiology B* 189(3–4): 501–511.

Rommel, S.A., D.A. Pabst, W.A. Mclellan, et al. 1992. Anatomical evidence for a countercurrent heat-exchanger associated with dolphin testes. *Anatomical Record* 232(1): 150–156.

Rosen, D.A.S., and D. Renouf. 1997. Seasonal changes in blubber distribution in Atlantic harbor seals: Indications of thermodynamic considerations. *Marine Mammal Science* 13(2): 229–240.

Schmidt-Nielsen, K. 1997. *Animal Physiology: Adaptation and Environment*. Cambridge, UK: Cambridge University Press

Shero, M.R., L.E. Pearson, D.P. Costa, et al. 2014. Improving the precision of our ecosystem calipers: A modified morphometric technique for estimating marine mammal mass and body composition. *PLoS One* 9(3). doi.org/10.1371/journal.pone.0091233

Shuert, C.R., J.P. Skinner, and J.E. Mellish. 2015. Weighing our measures: approach-appropriate modeling of body composition in juvenile Steller sea lions (*Eumetopias jubatus*). *Canadian Journal of Zoology* 93: 177–180.

Smith, H.R., and G.A.J. Worthy. 2006. Stratification and intra- and inter-specific differences in fatty acid composition of common dolphin (*Delphinus* sp.) blubber: Implications for dietary analysis. *Comparative Biochemistry and Physiology Part B: Biochemistry and Molecular Biology* 143(4): 486–499.

Tattersall, G.J., and V. Cadena. 2010. Insights into animal temperature adaptations revealed through thermal imaging. *Imaging Science Journal* 58(5): 261–268.

Walker, K.A., J.W. Davis, and D.A. Duffield. 2008. Activity budgets and prey consumption of sea otters (*Enhydra lutris kenyoni*) in Washington. *Aquatic Mammals* 34(4): 393–401.

Williams, T.M. 1990. Heat-transfer in elephants – Thermal partitioning based on skin temperature profiles. *Journal of Zoology* 222: 235–245.

Willis, K., and M. Horning. 2005. A novel approach to measuring heat flux in swimming animals. *Journal of Experimental Marine Biology and Ecology* 315(2): 147–162.

Yeates, L.C., T.M. Williams, and T.L. Fink. 2007. Diving and foraging energetics of the smallest marine mammal, the sea otter (*Enhydra lutris*). *Journal of Experimental Biology* 210(11): 1960–1970.

Dorian S. Houser and Jason Mulsow
National Marine Mammal Foundation, San Diego, CA

Dorian S. Houser

Dorian Houser has studied the physiology and bioacoustics of marine mammals for over 25 years, including marine mammal hearing, dolphin echolocation, behavioral and physiological responses to anthropogenic sound, and various aspects of diving, fasting, and metabolic physiology. Dorian is the director of Biologic and Bioacoustic Research and the director of Environmental Stewardship at the National Marine Mammal Foundation.

Jason Mulsow

Dr. Jason Mulsow is the deputy director of Biologic and Bioacoustic Research. He began working with Dorian Houser as a graduate student and joined the National Marine Mammal Foundation after his post-doc with the US Navy Marine Mammal Program. Jason is an expert in both psychoacoustic and neurophysiological methods and has led or collaborated on numerous studies of marine mammal hearing and dolphin echolocation.

Driving Question: How do marine mammals hear and produce sound underwater?

INTRODUCTION

Marine mammals live in a world where light is transmitted poorly and the range over which vision is useful is limited. Light is absorbed rapidly by sea water and little usable light exists beyond ~200 m depth in most oceans. Above this depth, called the euphotic zone, the usefulness of vision seldom extends more than tens of meters. In very murky water, visibility may be negligible. In contrast to light (and vision), sound is transmitted efficiently in sea water. Sound travels faster in water (~1500 m/s) than in air (~340 m/s), and sound energy decreases more slowly with the distance it travels in water than in air. So, how does an animal whose ancestors were land-walking, visual creatures adapt to an environment that is unfavorable to vision (see **Chapter 7**)? Like terrestrial mammals, marine mammals must hunt prey, find mates, navigate, and avoid predators, but over spatial scales where vision may be of limited use. Marine mammals need to detect and identify sound, localize sound sources, and efficiently produce and project sound. How strictly marine mammals are tied to the ocean for life history functions (e.g., mating, foraging, caring for young) directly relates to how specialized their hearing and mechanisms for producing sound have become. Before getting too deep into underwater hearing and sound production, we should start with a baseline understanding of hearing in humans and other terrestrial mammals.

DOI: 10.1201/9781003297468-6

SOUND AND THE EAR

Sound is generated when an object vibrates. The vibration causes changes within the pressure of the medium (e.g., air or water) in which the object exists. As the object vibrates back and forth, it creates both positive and negative pressure changes around the static pressure of the medium, which we might refer to as the air pressure or the hydrostatic pressure (for water). The changes from positive to negative create waves in the medium that travel away from the vibrating object. Based upon the structure of an animal's auditory system, an animal may be able to **transduce** these pressure waves into neural signals that are perceived as **pitch** and **loudness**, the perceptual correlates of **frequency** and **sound pressure level**.

A schematic of the human ear is provided to demonstrate the major components of the peripheral auditory system and how they work together to allow an animal to hear (**Figure 6.1a**). We use the human peripheral auditory system because it is both familiar and similar to the peripheral auditory systems of other terrestrial mammals. The peripheral auditory system roughly consists of the outer, middle, and inner ear, and the

auditory nerve. Sound waves traveling through air enter via the outer ear (or pinna) and through the external auditory canal. The sound waves strike the **tympanic membrane**, also called the eardrum, and cause it to vibrate. On the opposite side of the tympanic membrane, a small bone, called the **malleus**, is attached. The malleus is the first of three bones connected in series, the remaining two being the **incus** and the **stapes**. Collectively, these bones are called the **ossicles** and they are the primary component of the middle ear. The last of the ossicles in the chain, the stapes, has a plate that attaches to a membrane called the oval window. The oval window sits at the base of a coiled, bony structure called the **cochlea**, which is part of the inner ear. The cochlea is fluid-filled and has a membrane that runs along its length and splits the cochlea into upper and lower parts. This membrane is called the **basilar membrane** and it is the base upon which other critical hearing structures exist. The cochlea and middle ear bones are partially contained within the **auditory bulla**, a bony structure that is part of the skull in terrestrial mammals.

Vibrations caused by sound waves arriving at the tympanic membrane cause the ossicles to

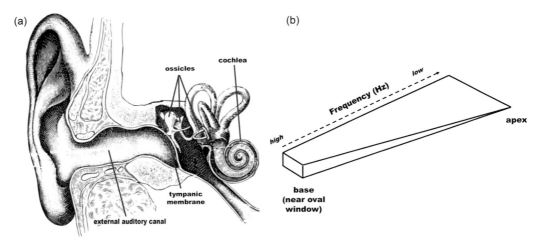

Figure 6.1 (a) Illustration of the human ear with important anatomical features identified. The ossicles consist of three bones; from left to right, these are the malleus, incus, and stapes. This image is reproduced and modified under CC0 1.0 Universal (CC0 1.0) license and is in the public domain. (b) Schematic of the basilar membrane showing the change in thickness and width from its base to its apex and how the thickness and width affect the frequencies at which the basilar membrane vibrates.

move back and forth, much like a piston. As the stapes moves back and forth on the oval window, pressure waves are created within the fluid of the cochlea. These waves travel along the length of the cochlea and cause the basilar membrane to vibrate. The basilar membrane serves as the frequency map of the auditory system. It is narrow and thick at its base near the oval window but widens and thins as it progresses toward the apex of the cochlea (**Figure 6.1b**). Differences in the thickness and width of the basilar membrane dictate where along the basilar membrane vibration to a particular frequency of stimulation occurs; where it is narrower and thicker at the base, it best responds to higher frequencies, and where it is wider and thinner at the apex it responds best to lower frequencies.

Set upon the basilar membrane is the **organ of Corti**, a specialized structure that holds sensory cells, known as hair cells, and a tectorial membrane that lays on top of projections (stereocilia) from the hair cells. The hair cells are aligned such that there is one row of inner hair cells and multiple rows of outer hair cells. As pressure waves cause the basilar membrane to vibrate, the stereocilia of the inner hair cells are deflected. This, in turn, causes the inner hair cells to release neurotransmitters into junctions that each cell shares with the auditory nerve. The auditory nerve then sends neural signals to the brain. Since the basilar membrane vibrates at a specific place along its length as a function of the frequency of sound received, the neural pulses initiated by the hair cells encode the frequency of the sound that is heard. In this manner, the auditory system transduces (sound) pressure waves from the environment into neural signals that are perceived as sound.

Sound waves underwater present a different set of challenges for hearing in marine mammals. Because the sound transmission properties of water and the tissues of the body are similar, sound passes easily through the body of an organism when it is submerged underwater. Sound waves in air mostly reflect off of the body of an organism leading to the need for the tympanic membrane and ossicles to transduce sound waves in air into pressure waves within the fluid of the cochlea. So then, what would be the use of the tympanic membrane for a fully aquatic marine mammal if underwater sound can simply pass through its body? How would underwater sound waves get into the cochlea of a submerged marine mammal so neural signals could be created? And what about marine mammals that live both on land and in the water, like the seals and sea lions? They have to be able to hear in both media. Marine mammals started with an ear built for hearing in air, but have since evolved adaptations to accommodate hearing underwater. The extent of these adaptations varies considerably by species, with the fully aquatic marine mammals (e.g., whales, dolphins) being more derived and specialized than their amphibious counterparts (e.g., pinnipeds).

HEARING IN CETACEANS

The most specialized marine mammal auditory systems belong to the cetaceans, as they are fully aquatic. Whales demonstrate adaptations to the auditory system and vestibular systems that are unique among mammals, even though they are believed to have evolved from hoofed terrestrial mammals (ungulates). There are some dramatic differences in the ear anatomy between the toothed whales and the baleen whales, the toothed whale ear being the more specialized of the two.

Toothed whales

All whales have a streamlined, fusiform body (see also **Chapter 1**) and a **telescoping** skull (**Figure 6.2a**), although telescoping is most prominent in the toothed whales. The process of telescoping resulted in the nares, the external opening to our respiratory passageways (or nostrils, in humans), migrating to the dorsal surface of the whale and a forward elongation of certain cranial bones. The fleshy external ear that is commonly found in terrestrial mammals is absent in toothed whales. The external opening

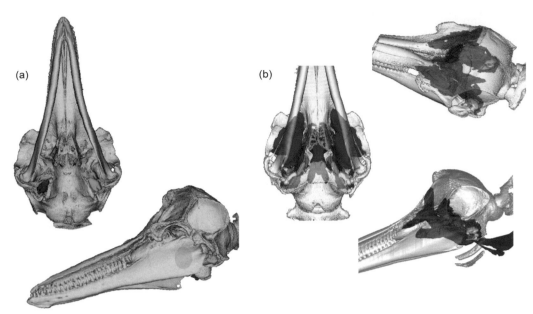

Figure 6.2 (a) Computed tomographic (CT) renderings of the skull of the bottlenose dolphin from ventral (upper image) and lateral perspectives (lower image). Note the elongation of the cranial bones prevalent as a result of "telescoping." The auditory bullae are colored red and green; both are completely separate from the dolphin skull. The "pan" region of the mandible is indicated by a red, transparent oval. (b) CT renderings of another dolphin with cranial air spaces colored in red and the auditory bullae colored in yellow. The air spaces are made semi-transparent so the relationship of the air covering to the auditory bullae can be observed. Perspectives are ventral (left), dorsal (top right), and lateral (bottom right).

of the ear, the external meatus, is significantly diminished and appears as a pin-sized hole in the side of a whale's head (**Figure 6.3**). The external auditory canal does not connect to the tympanic membrane, as it does in terrestrial mammals, and the ear canal is filled with wax and cellular debris (which likely come from sloughed cells of the canal wall). For these reasons, the outer ear is believed to contribute relatively little to sound reception in toothed whales (McCormick et al. 1970; Darlene R. Ketten 2000).

The middle and inner ears of the toothed whales are contained within the auditory bullae. The auditory bullae are the densest bones in the body of a toothed whale and are detached from the skull (**Figure 6.2**), which is a feature that is unique to the toothed whales. The bullae are suspended by ligaments and partially surrounded by air medially (toward the inward side) and dorsally (on the topside) (**Figure 6.2b**). The arrangement of air around the auditory bullae helps to acoustically isolate the ears, at least partially, because air is highly reflective to underwater sound. For example, for a sound coming from the right side of a toothed whale, the air on the medial, or internal, side of the left ear will partially shield the left ear from the arriving sound, but there will be a direct and uninterrupted path of sound to the right ear. The partial acoustic isolation contributes to differences in the level of sound received at each ear as it travels from one ear to the other, as well as in the **spectral content** that the ear receives. The difference in received sound level and spectral content at the two ears is one means by which toothed whales localize sound underwater.

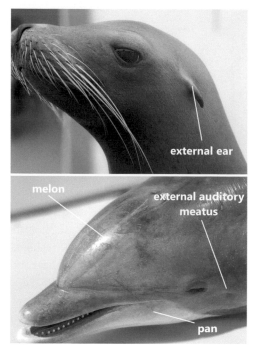

Figure 6.3 Picture of a (top) California sea lion (*Z. californianus*) and (bottom) bottlenose dolphin (*T. truncatus*). Note the reduction in external ear size in the sea lion and its absence in the dolphin. The opening to the external ear canal of the dolphin, the external auditory meatus, appears as a small indention several centimeters behind the eye. Additional arrows indicate other important anatomical modifications for sound production (melon) and reception (pan) in the dolphin.

The middle ear bones (the malleus, incus, and stapes) of toothed whales are more massive than in most terrestrial mammals and are stiffened by ligaments. Although the tympanic membrane has a membranous attachment to the malleus, the lack of a connection between the external auditory canal and the tympanic membrane has led to much debate over how the middle ear bones function in toothed whale hearing. Unlike in a terrestrial mammal, where the middle ear bones work like a piston to vibrate the oval window and generate pressure waves in the cochlea, research with dolphins

suggests that it is the motion of the auditory bulla relative to the middle ear bones that generate waves in the cochlea (McCormick et al. 1970). Having the auditory bulla free from the skull likely enhances such relative motion.

Figure 6.4 shows the **audiograms** for several toothed whales – the bottlenose dolphin (*Tursiops truncatus*), killer whale, and harbor porpoise (*Phocoena phocoena*). An audiogram shows the **threshold of hearing**, that is, the minimum level of sound an animal can hear, at a particular frequency. The hearing range of toothed whales can range from several hundred Hz to in excess of 160 kHz and is much broader than that observed in humans and most terrestrial mammals (**Figure 6.4**). The upper frequency limits of hearing in larger toothed whales (e.g., killer whales, beaked whales; 80–100 kHz) tend to be lower than that of the smaller-toothed whales. The broad frequency range of hearing is critical to the toothed whale's use of echolocation. Although the toothed whale inner ear generally looks like that of a terrestrial mammal, it shows some adaptations that enable ultrasonic hearing (Further Reading: Ketten [2000]). First, the base of the toothed whale basilar membrane is narrower and thicker than in terrestrial mammals and the ratio of the basilar membrane thickness to width (T/W) has been found to be consistent with the maximum high and low frequencies that can be heard by these species (Ketten and Wartzok 1990). Second, the basilar membrane is supported by outer and inner bony connections (or lamina) that stiffen the membrane in the areas associated with high-frequency hearing. The degree that the outer lamina supports the basilar membrane varies by species with more of the basilar membrane being supported in toothed whales producing echolocation signals with the highest-peak frequencies (e.g., >120 kHz for harbor porpoises, Ketten 2000).

The brain of modern toothed whales is large relative to its body size. Dolphin brains have an **encephalization quotient** (EQ), which is the ratio of the actual brain mass to that predicted

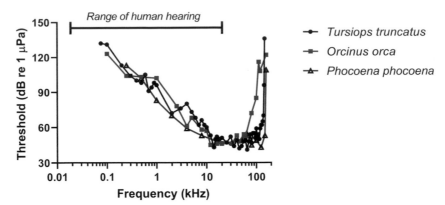

Figure 6.4 Audiograms for three cetacean species – the bottlenose dolphin (*T. truncatus*), killer whale (*O. orca*), and harbor porpoise (*P. phocoena*). The audiogram shows the frequency range of hearing and how sensitive an animal is to sound at a particular frequency. The lower the threshold, the more sensitive the animal is to the associated frequency of sound. The range of human hearing is indicated for comparative purposes. (Data from the bottlenose dolphin are from Johnson (1967) and data for the harbor porpoise are from Kastelein et al. (2010). The killer whale audiogram is a mean audiogram from multiple killer whales (Branstetter et al. 2017).)

based on body mass, as high as 6.3 (Manger 2006). The EQ of toothed whales is close to the EQ of humans and has often been used to argue for the intelligence of dolphins. However, it is the regions of the brain associated with the processing of sound – for example, the auditory cortex, cerebellum, and pons (in dolphins) – that are enlarged and contribute substantially to the overall increase in the size of the dolphin brain. The large size of the odontocete brain may have largely arisen during the evolution of sonar-guided navigation and communication (Oelschlager 2008).

Hearing pathways in echolocation

Toothed whales echolocate, which means they project sound into the water and listen for echoes to return to them from objects that were insonified with the sound (see Sound production in cetaceans, below). Toothed whales receive the echoes from targets through a thin region of their jaw, or mandible, referred to as the "pan" (**Figures 6.2a** and **6.3**). The mandible is hollow and filled with specialized fats that have sound transmission characteristics similar to water. These fats, referred to as **acoustic fats**, extend

from the pan backward to connect to the auditory bulla. Sound waves (echoes) pass into the jaw fat through the pan and are then transferred to the middle ear within the auditory bulla (Hemilä, Nummela, and Reuter 2010). Unlike terrestrial mammals, the outer ear and ossicles are bypassed by this. So, how then are pressure waves created within the cochlea? It is believed that this occurs through **bone conduction**, where vibrations in the bony structures surrounding the cochlea cause the formation of pressure waves in the cochlea. It is this type of sound conduction that is likely most important to whales, and also to amphibious marine mammals when underwater.

The sound path through the jaw of the toothed whales is sensitive to the ultrasonic frequencies used in echolocation (ultrasonic is here defined as frequencies >20 kHz, the approximate upper limit of human hearing). The peculiarity of the sound path makes hearing at ultrasonic frequencies very directional; toothed whales are sensitive to ultrasonic sounds that are received from directly in front of them, but sensitivity declines as the location of the sound source (or echoes) moves away from in front of the animal.

A narrow area for receiving sound is important to toothed whale biosonar because it enables the toothed whale to better localize the source of echoes while minimizing unwanted interference (e.g., echoes from other targets that are not of interest). In other words, the special path for receiving ultrasonic sounds allows toothed whales to narrow the field of view for things they inspect with echolocation. At frequencies below those used in echolocation, where communication signals such as dolphin "whistles" are more common (see Sound production in cetaceans), hearing is less directional.

Baleen whales

No one has ever measured hearing in a baleen whale, primarily because they are too big to keep under human care for the period needed to perform a behavioral hearing test (see **Toolbox**). Nevertheless, we can use the anatomy of their auditory systems to try and understand how they may receive and process sound. Unlike toothed whales, the external ear canal of baleen whales connects to the tympanic membrane, which suggests that it might have a role in sound reception (Ketten 2000). Also, unlike toothed whales in which the auditory bullae are completely separated from the skull, bony connections exist between the skull and the auditory bullae. This could potentially contribute to bone conduction of sound through the skull. Conversely, fatty sound reception pathways may not be unique to toothed whales but might also be present in baleen whales. For example, minke whales (*Balaenoptera acutorostrata*) have a fatty body that connects to the auditory bulla, as it does in odontocetes (Yamato et al. 2012). It extends from the bulla laterally to the blubber suggesting the possible presence of a lateral sound transmission pathway. However, the fat is not specialized as it is in toothed whales, but rather is similar in composition to the fats in blubber (Yamato et al. 2014).

The middle ear bones in baleen whales are massive and loosely coupled, which suggests they are adapted for the detection of low-frequency sound. Although the structure of the basilar membrane suggests that some baleen whales may have high-frequency hearing limits as high as humans (or higher) (Ketten 2000), the broadness and flimsiness of the membrane at its apex suggests that baleen whales are sensitive to low frequency sound (even possibly infrasonic sound, in some species).

The baleen whale brain does not have the same degree of hypertrophy observed in toothed whales. Acoustic processing centers appear to be in proportion to other brain structures. With an EQ < 1.0 for all baleen whales, brain size appears to be uncoupled from body size and deviates less from allometric predictions than it does in toothed whales (Boddy et al. 2012). As previously stated, the high EQ in toothed whales is likely because hypertrophied sound-processing centers are required to support echolocation. No definitive evidence for echolocation has yet been found in baleen whales.

SOUND PRODUCTION IN CETACEANS

It was once hotly debated as to how toothed whales produced sound and whether it occurred in the nasal passages or the larynx. One school argued that sounds must be generated in the larynx, whereas others argued for a sound source somewhere in the nasal passages. Collective evidence now points to a location within the nasal system of toothed whales as the source. Specialized structures located within the nasal passages and below the blowhole plug, termed the **phonic lips**, are believed to produce the most typical toothed whale acoustic signals, mainly whistles, burst-pulses, and echolocation clicks (Cranford 2000; Cranford et al. 2011). There are two sets of phonic lips, and it is debated as to whether one or both sets are involved in echolocation click production (Madsen et al. 2013). The nasal passages of toothed whales pressurize just prior to signal production (Ridgway et al. 1980; Amundin and Andersen 1983) and it is the passage of pressurized air across the phonic lips that is responsible for sound production. The phonic lips are also associated with a complex of air sacs that support the operation of signal production (see **Figure 6.2b**), although the exact role they play is not understood.

Whistles are tone-like signals in which the frequency can be changed, or modulated, over time. The primary (or fundamental) frequency of whistles is generally less than 20 kHz, but harmonics at higher frequencies are often observed. Whistles are used for communication and social interactions, but not all species of toothed whale produce whistles (e.g., sperm whale). Whistle sounds propagate almost omnidirectionally from a vocalizing whale; that is, in nearly all directions from the source. This distinguishes them from directional, high-frequency sounds like echolocation clicks (see below). Based upon how frequencies in a whistle change, whistles can roughly be grouped as constant frequency, upsweep, downsweep, concave, or convex (see Au and Hastings 2008 for review). The communication role of the different signal types remains largely unknown, although the identification of individually distinctive whistles produced by dolphins suggests a class of "signature whistles" that dolphins utilize for identification and individual localization (Janik and Sayigh 2013).

Burst pulses are short duration, **broadband** signals produced in rapid succession and which resemble squeaks, creaks, or groans. They are produced by all species of toothed whale studied to date, although there is some deviation and specialization (e.g., sperm whale production of codas). Burst pulses are much less studied than whistles and will not be discussed extensively here. However, it is worth noting that burst pulses may be the primary means by which some species communicate (e.g., harbor porpoises).

Echolocation clicks are short-duration (10s to 100s of μs), transient signals, the duration and frequencies of which vary by species. Bottlenose dolphins produce broadband clicks with more than half of the sound energy spread across frequency bandwidths greater than 85 kHz

(Houser, Helweg, and Moore 1999). Harbor porpoises produce narrower band signals with most sound energy between 120 and 140 kHz (Koblitz et al. 2012). It has been suggested that harbor porpoises produce high-frequency, narrowband signals to avoid detection by one of its primary predators, the killer whale (*Orcinus orca*, Koblitz et al. 2012), as the frequencies of the echolocation signal are above the killer whale's hearing range.

Echolocation clicks are projected through the melon, a specialized fat body that forms the forehead of the toothed whale (**Figure 6.3**). The fats of the melon are similar in composition to the acoustic fats that fill the lower jaw and which connect the pan to the auditory bullae. The melon has a low sound-velocity core but a gradient of increasing sound velocity toward the margins of the melon (Norris and Harvey 1974). This important feature of the melon allows it to **collimate** a projected echolocation signal, which means that as the sound moves through the melon it travels faster on the edge of the melon and slower towards the melon's center. This causes sound waves that are moving away from each other to become parallel and allows toothed whales to produce a forward-looking echolocation beam. In all toothed whales, the forward-looking echolocation beam is narrow and corresponds well with the directionality of hearing at echolocation frequencies.

Echolocation clicks are produced in a series, termed "trains," and for some species the frequency content of the clicks can be changed over the course of the train. The dimensions of the echolocation beam can also be varied to some extent and steered, much like one's vision can be changed by movement of the eyes but without movement of the head (Moore, Dankiewicz, and Houser 2008; Madsen, Wisniewska, and Beedholm 2010). The ability to manipulate the frequency content of clicks and the shape of the beam likely occurs by muscular manipulation of the melon and changes in the shape and air volume of associated air sacs, which would act as reflectors (Aroyan et al. 1992; Cranford 1992;

Moore, Dankiewicz, and Houser 2008; Madsen, Wisniewska, and Beedholm 2010). The complex arrangement of air spaces and air sacs can be seen in **Figure 6.2b**.

Exactly how baleen whales produce vocalizations remains undetermined, but unlike toothed whales where considerable evidence supports the phonic lips as the primary means of sound production, it is possible that sound production originates in the larynx where modified vocal folds have been identified (Reidenberg and Laitman 2007). Baleen whales produce a variety of calls that can be characterized as songs, simple calls, complex calls, and knocks, grunts, pulses, or clicks (Clark 1990). The duration of sounds produced by mysticetes can range from <100 ms (e.g., knocks and grunts) to several seconds (simple calls) and may contain dominant frequencies as low as 10 Hz and as high as 10 kHz. Because low-frequency sounds travel farther than high-frequency sounds underwater, communication sounds by some baleen whales (e.g., blue and fin whales) may travel hundreds of kilometers. Sounds produced by baleen whales might also be strung together in units and phrases to produce songs, as is probably most well known in the humpback whale. There is considerable species variability in sound production by mysticetes and a summary of the frequency, duration, and source levels can be found in Au and Hastings (2008).

HEARING IN PINNIPEDS

A defining characteristic of these marine carnivores is that they are amphibious; while all can swim in the ocean, each carries out some essential life history functions on land, such as breeding, birthing and nursing young, and molting. Since pinnipeds use hearing both underwater and in air (when on land), their hearing adaptations are a compromise that addresses the challenges of hearing both in air and underwater.

Auditory anatomy and processing capabilities are generally similar among otariids, or

eared seals. In contrast, the phocids, or true seals, display greater interspecies differences. Odobenids, who's only member is the walrus, are thought to be more closely related to otariids but possess an auditory system that appears in many ways to be intermediate between those of otariids and phocids (Nummela et al. 2007).

The outer ear of the pinnipeds is reduced in size relative to most terrestrial carnivores. In the otariids, the external ear has been reduced to small folds (**Figure 6.3**). In phocids, the external ear is altogether absent (Ramprashad, Corey, and Ronald 1972; Repenning 1972). Similar to cetaceans, the external auditory meatus appears as a small opening on the head (**Figure 6.5**). For this reason, phocids are also referred to as the "earless" seals. The reduction in the size of the outer ear is likely a result of selection for a more hydrodynamic shape when swimming underwater (see **Chapter 1**). In air, the outer ear directs and filters sound from the external environment into the ear canal towards the tympanic membrane and helps individuals localize sound sources. Underwater, the tissue properties of the outer ear are similar to water and likely contribute little to sound reception or localization.

The ear canals of otariids are generally similar to those of terrestrial mammals (Ramprashad, Corey, and Ronald 1972; Repenning 1972), and in air, they appear to have hearing that is generally as sensitive as terrestrial carnivores (**Figure 6.6a**). Otariid hearing sensitivity is also quite good underwater, although otariids are not as sensitive as toothed whales and they lack the broad frequency range of hearing used to support echolocation (**Figure 6.6b**). These findings have suggested that otariids are adapted primarily for aerial hearing (Reichmuth et al. 2013).

Phocid seal ear morphologies are more diverged from their terrestrial ancestors than are the otariids. In the extreme case of the elephant seal (*Mirounga* spp.), the opening to the ear canal is reduced to the size of a pinhole (**Figure 6.5**) and there does not appear to be an air space that connects

Figure 6.5 Picture of (top) harbor seals (*P. vitulina*) and (below) an adult male elephant seal (*Mirounga angustirostris*). (The image of the harbor seals is reproduced and modified under CC0 1.0 Universal [CC0 1.0] license and is in the public domain.) Both seals lack an external ear but differ in the structure of the external auditory meatus. In the harbor seal, a well-formed sphincter muscle enables the harbor seal to open and shut the meatus. The harbor seal has hearing in air that is comparable to other terrestrial carnivores. In contrast, the external auditory meatus of the elephant seal is pin-hole sized and seems to contribute little to the collection of airborne sound. The hearing abilities of the elephant seal in air are worse than the harbor seal (Figure 6.6). Differences in the anatomy and hearing ability reflect the degree to which the two species are amphibious; elephant seals are the most aquatic of the seals, whereas harbor seals spend much of their time on land.

to the tympanic membrane. This modification, which likely supports the deep diving behavior of this species, contributes to relatively poor aerial hearing sensitivity (**Figure 6.6a**). Other phocid species (e.g., harbor seals, *Phoca*

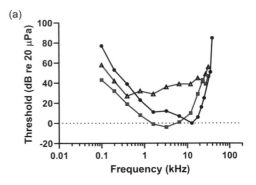

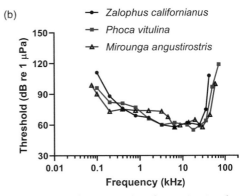

Figure 6.6 Audiograms for three pinniped species – the California sea lion (*Z. californianus*), the harbor seal (*P. vitulina*), and the northern elephant seal (*Mirounga angustirostris*). (a) In-air audiograms. (b) Underwater audiograms. (Data are from Reichmuth et al. 2013; Reichmuth, Sills, and Ghoul 2017; Kastak and Schusterman 1999).

vitulina; **Figure 6.5**) possess musculature that can constrict the outer ear canal, presumably as an adaptation for restricting the entrance of water into the ear canal when diving (Ramprashad, Corey, and Ronald 1972; Repenning 1972). However, in these seals, the ear canal can remain open in air and contribute to aerial hearing sensitivity that is comparable to that of terrestrial carnivores (Reichmuth et al. 2013) (**Figure 6.6a**). These differing degrees of adaptation reflect how different pinniped species spend their time in terrestrial and underwater environments: elephant seals are the "most aquatic" of the pinnipeds, otariids the "most terrestrial," while harbor seals exist somewhere in between (Kastak and Schusterman 1998; Kastak and Schusterman 1999).

The three middle ear bones of pinnipeds function like those of terrestrial mammals – the tympanic membrane vibrates when exposed to sound waves, the ossicles move with the tympanum and act like a piston that generates pressure waves in the fluid of the cochlea by moving against the oval window. While the mass of otariid ossicles is similar to that of terrestrial carnivores, that of phocids has greater mass and increased density. Additionally, present in the pinniped middle ear and outer ear canal is a specialized (cavernous) tissue that can engorge with blood during diving. This presumably serves to minimize or eliminate air spaces in the ear and minimizes pressure differences between the middle ear and the hydrostatic pressure of the ocean while diving (Odend'hal and Poulter 1966; Møhl 1967; 1968; Repenning 1972). Collectively, these modifications suggest that the primary means of generating pressure waves in the cochlea while underwater is through bone conduction.

The pinniped cochlea is similar to terrestrial carnivores but notably different from whales, particularly the toothed whales that possess adaptations for ultrasonic underwater hearing. As with all mammals, the shape of the basilar membrane dictates what frequencies pinnipeds hear. The high-frequency limits to hearing are approximately 40 and 100 kHz for the otariids and phocids, respectively (Reichmuth et al. 2013). In air, the greater mass of the middle ear bones in phocids results in an inertial constraint on the transfer of high-frequency sound into the inner ear, reducing the high-frequency limit to between 20 and 30 kHz (Hemilä et al. 2006) (**Figure 6.6a**). The only hearing data that exist for odobenids are underwater hearing thresholds for one individual, which shows a high-frequency limit of approximately 15 kHz (Kastelein et al. 2002). It is unknown whether this relatively low cut-off is representative of walruses in general, or just that particular walrus.

SOUND PRODUCTION IN PINNIPEDS

Pinnipeds produce a myriad of vocal sounds both in air and underwater. Vocalizations are typically associated with mother-pup pairing on land and reproductive displays on land and in water. Pinniped vocalizations can range from simple vocalizations, such as the bark of the California sea lion (*Zalophus californianus*), to complex calls, such as the underwater trills produced by bearded seals (*Erignathus barbatus*). Sound production in pinnipeds in air is similar to that of land mammals, but exactly how it occurs underwater remains uncertain and may vary by species, although the vocal tract is certainly involved. Some species of pinnipeds have a larynx that is very similar to terrestrial mammals (e.g., harbor seals), whereas others have laryngeal modifications, in some cases similar to baleen whales (e.g., elephant seals, Reidenberg and Laitman 2018). These modifications are suggestive of some form of adaptation to vocal production, but the exact mechanisms of how they work remain unknown.

HEARING IN SEA OTTERS

The sea otter has a much shorter evolutionary history in the marine environment than pinnipeds. Nevertheless, sea otters display some of the same adaptations of the outer and middle ear that are observed in pinnipeds. These include, reductions in the size of the outer ear, an increase in the size of the ossicles, and a thickening of the tympanic membrane (Solntseva 2007). Like the pinnipeds, some of these adaptations may have been driven by swimming and diving constraints, not hearing. Behavioral measurements of hearing sensitivity with the sea otter have shown that aerial hearing is similar to that of otariids, while underwater hearing is inferior to that of both pinnipeds and whales (Ghoul and Reichmuth 2014). Collectively, the findings suggest that a relatively recent transition to the marine environment has left the sea otter auditory system adapted primarily for hearing in air.

HEARING IN POLAR BEARS

Similar to sea otters, polar bears have a relatively short evolutionary history in the marine environment, having diverged from terrestrial bears less than 2 million years ago. Little research has been conducted on the auditory system of the polar bear outside of measuring the audiogram. The only modification to the auditory system distinct from terrestrial bears is the presence of prominent fur inside the external ear and a reduction in external ear size, which likely reduces heat loss in the Arctic cold (Stirling 1998).

Behavioral and electrophysiological measurements of hearing in air show that polar bears are sensitive to airborne sound over a frequency range similar to that of pinnipeds, mustelids, and other carnivores: approximately 125 Hz to 32 kHz (Nachtigall et al. 2007; Owen and Bowles 2011). There are currently no data on underwater hearing in polar bears, but their hearing ability underwater is probably poor given the absence of adaptations to underwater hearing. This is intuitive given that there are limited reasons to require sensitive underwater hearing – polar bears do not use underwater hearing for communication, foraging, or navigation.

OCEAN NOISE AND HEARING IMPACTS

The ocean is not necessarily a quiet place. Biological noise sources can significantly contribute to ocean noise in some ocean habitats (e.g., snapping shrimp), abiotic processes can add to ocean noise (e.g., cracking ice, wind, and rain), and marine mammals themselves may be significant contributors to ocean noise (e.g., song competition among whales). However, over the last 200 years or so, humans have industrialized and increased utilizing the ocean for commerce, developed the use of underwater sound for military purposes, and increased ocean exploration for scientific and commercial purposes. As a result, humans have introduced noise into the habitats of marine mammals that did not exist pre-industrialization. Sources of human-caused ocean noise include shipping, commercial and recreational boating, military activities, green energy construction and operations (e.g., wind turbines), and seismic exploration. Noise sources may be continuous (e.g., shipping) or might only occur over short periods of times (e.g., underwater explosions). Some noise sources are stationary (e.g., wind turbines), whereas others are mobile and potentially move great distances (e.g., seismic surveys). In short, there are many types of ocean noise sources, but of these, commercial shipping is likely the most persistent noise source in the ocean as commercial shipping occurs 24 hours a day on a global scale. Human-caused ocean noise has been observed or has been hypothesized to affect marine mammals in many ways: altering behavior (e.g., diving or vocalizations), causing habitat abandonment, inducing stress, and contributing to strandings (Cox et al. 2006; Harris et al. 2018; Erbe et al. 2019). However, given that our focus is on adaptations to underwater hearing, we will focus on impacts to hearing.

If a marine mammal is exposed to an intense underwater noise source, or is exposed to sound for a long duration, it can suffer hearing loss. Hearing loss is described as an increase in the hearing threshold. If the increase is temporary and eventually returns to normal, it is called a **temporary threshold shift**, or TTS (**Figure 6.7**). Most people are familiar with TTS as it is easily related to the so-called "rock concert effect," a phenomenon in which a person exposes themselves to loud music for a period of time and experiences a reduction in their hearing sensitivity that recovers over the ensuing hours and days. A TTS is a form of auditory fatigue indicative of the ear being overstimulated. At higher sound exposure levels and/or longer durations, physical damage can occur to the ear that causes a reduction in hearing sensitivity that is permanent. This is referred to as a **permanent threshold shift**, or PTS (**Figure 6.7**).

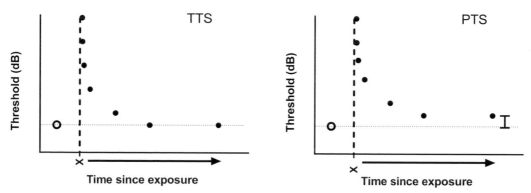

Figure 6.7 Figure showing the differences between a temporary threshold shift (TTS) and a permanent threshold shift (PTS). The normal threshold of the animal is shown by the open circle. The animal receives an intense noise exposure at time = x. Immediately after the noise exposure, there is an immediate increase in the measured threshold (filled circles). Over time, the thresholds gradually decrease toward the baseline. In the case of a TTS, the baseline value is eventually reached. In the case of a PTS, some degree of threshold shift remains permanently.

Toothed whales have been found to have a unique hearing adaptation for dealing with high-level sound exposures. Toothed whales have the ability to reduce their hearing sensitivity in the presence of noise; that is, they intentionally make their hearing worse in order that the noise is not perceived to be as loud. This ability was discovered by researchers studying echolocation. They noticed that the hearing sensitivity of a false killer whale changed depending on whether a target it was supposed to report after inspecting it with echolocation was placed in front of it or not. The researchers then devised a study that demonstrated that the whale could change its hearing at will. The phenomenon has now been demonstrated in several species of odontocetes. How the toothed whales accomplish this feat remains a mystery, but growing evidence suggests that it happens in the inner ear, probably under neural control and at the level of the cochlea. The extent that this adaptation can serve as a means of protecting toothed whales from high-level noise exposures remains to be determined.

Ocean noise also has the potential to interfere with a marine mammal's ability to detect and identify sounds important to their life functions. This interference is known as **masking** and it occurs when the presence of noise impedes the detection or interpretation of other sounds. Humans are familiar with this as the "cocktail party effect," which is when a person has difficulty understanding the speech of another person because of the presence of noise sources at the party (e.g., other conversations, band music, the clattering of plates and silverware). Marine mammals have different approaches to dealing with masking depending on whether they are trying to hear or to be heard.

When marine mammals face conditions in which noise limits their ability to communicate, they may change the characteristics of the signals they produce. The amplitude of the signal, the rate at which it is produced, or the signal duration may individually or collectively be increased, a phenomenon known as the **Lombard effect**. This phenomenon, named for the man who discovered it, is found to occur in humans and other mammals, as well as in birds (Brumm and Zollinger 2013). The Lombard effect has been observed in North Atlantic right whales (*Eubalaena glacialis*) and killer whales (*Orcinus orca*) in the presence of increased ambient noise (Holt et al. 2008; Parks et al. 2011). Increases in the rate of signal production, amplitude, and duration as well as shifts in the frequency of signals outside that of interfering noise, all serve to increase the detectability of the caller's signal. In this regard, evidence suggests that marine mammals deal with environmental noise in much the same manner that terrestrial mammals do. Marine mammals similarly utilize many of the same physiological and behavioral tools as terrestrial mammals when attempting to detect a sound in masking noise. This includes things such as changing their orientation relative to a noise source so it is less interfering (called spatial release from masking), or listening to how the frequencies of a complex sound change in level together over time (comodulation release from masking).

TOOLBOX

Sound transducers

The sound projector, microphone (for amphibious species), and hydrophone (underwater microphone) are the primary tools of the marine mammal bioacoustician. Projectors and hydrophones are transducers, which means they either change an electrical signal into motion that generates sound waves (e.g., projector) or they change pressure waves into an electrical signal (e.g., hydrophone) that can be captured and digitized. Many underwater transducers can do both, to some degree. There are different types of underwater projectors and receivers, and each has different sound projection capabilities or sound receiving sensitivities. These will vary as a function of frequency based upon their material composition and design. For this reason, proper calibration of a transducer is critical to any study that quantifies sound or needs to project a desired level of sound. In line with the hydrophone is generally an amplifier, for increasing the signal level, and a filter. Filtering is often required in order to eliminate unwanted signals or to prevent aliasing, which is a problem that occurs when high-frequency signals are not digitally sampled (recorded) at a high enough rate to adequately characterize them.

In passive acoustic monitoring, which is used to study, localize, or track free-ranging marine mammals, hydrophone arrays are used to record sounds produced by the marine mammals. Historically, as few as one hydrophone has been used to characterize sounds produced by marine mammals. However, arrays of hydrophones, appropriately placed in space, can be used for more advanced procedures such as localizing a vocalizing animal (e.g., hyperbolic fixing). These techniques are useful for tracking vocal animals at sea, monitoring their acoustic behavior, and in estimating population size (Marques et al. 2013; Stanistreet, Risch, and Van Parijs 2013; Durbach et al. 2021). With the increased awareness of the potential for anthropogenic sound to impact marine mammals, the techniques of localizing, tracking, and monitoring acoustic behavior have become important methods for relating noise exposure to changes in marine mammal behavior. For example, work with bottom-mounted hydrophone arrays has demonstrated that beaked whales cease the production of foraging-related echolocation signals when exposed to certain levels of sonar (Moretti et al. 2014).

Underwater sound transducers are also the mainstay of research performed with marine mammals under human care. Almost all of the research on underwater hearing using psychophysical approaches uses these tools. **Psychophysics** is the field of study related to how a physical quantity (e.g., sound frequency) translates to a subject's perception of the physical quantity (e.g., pitch). Psychophysical procedures in hearing studies involve training an animal to perform a behavior in response to a sound in which one or more of its physical properties are varied. The process is behavioral in that the animal is asked to make a decision, such as, do you hear the sound? For example, until ~2005 when electrophysiological techniques became more common for studying hearing in toothed whales (see below), information about the frequency range of hearing and hearing sensitivity were nearly always determined through psychophysical means. A simple form of a psychophysical hearing test relates the amplitude of a sound, typically measured as the sound pressure level, to an animal's ability to detect the sound. Procedures involve presenting a sound to an animal that is trained to station itself in a precise location within a calibrated sound field, or in the case of some pinnipeds, to wear headphones. If the animal hears the tone, it provides a response. The response could be the production of a sound (e.g., whistle) or the performance of a behavior (e.g., touching a paddle). If no sound is heard, the animal either provides a different response (e.g., touches a different paddle) or remains quiet. This process can be used to determine such things as the lowest level of sound an animal can hear. If the process is repeated for a broad range of frequencies, at least enough to cover the range of hearing, then an audiogram can be constructed. The audiogram provides the most fundamental piece of sensory information necessary to understanding marine mammal hearing, how hearing is intertwined into marine mammal ecology, and the potential for human-caused noise to affect marine mammals.

There are many standardized psychoacoustic procedures that can be performed and a detailed discussion of the psychophysical procedures used in marine mammal research is beyond the scope of this chapter. Nevertheless, it is important to be aware that the use of these procedures is critical to exploring both sensory capability and inferring physiological function. Examples of auditory system information obtained with psychophysical methods includes: hearing thresholds, time and intensity differences of sound received at both ears, angular discrimination in the location of sound source, minimum detectable frequency and amplitude differences, the filtering performed by the ear, how the duration of a sound affects its perceived loudness, directionality of hearing, masking, and other phenomena.

Electrophysiological recorders

Electrophysiological studies of marine mammal hearing date to the 1960s, when it was performed by a relatively small group of researchers. With modern advances in computational capabilities and the miniaturization of technological components utilized for electrophysiological research, the ability to perform electrophysiological research with marine mammals has become more common both in the laboratory and in the field. Electrophysiological approaches to studying hearing monitor the electrical response of nerves in the auditory system to the reception of sound. In marine mammals, responses are generally measured at the skin surface or through the use of subcutaneous needle electrodes inserted just under the skin. The presence and amplitude of neural responses can be correlated with some psychophysical functions; for example, one of the most common electrophysiological procedures performed in studying hearing sensitivity in marine mammals is the recording of the **auditory evoked potential** (AEP). Auditory evoked potentials are voltages generated by the brain in response to the detection of a sound, and the threshold of acoustic stimulation required to produce a measurable response has been shown to correlate with hearing sensitivity. The use of AEP methods to study hearing has greatly advanced the number of marine mammal species for which hearing tests have been performed and has permitted a number of species not maintained under human care to be tested when stranded or when undergoing rehabilitation for release (e.g., Nachtigall et al. 2005; Finneran et al. 2009; Schlundt et al. 2011; Mooney et al. 2015).

CONCLUSIONS

Sound travels efficiently through water and marine mammals have evolved to capitalize on this. Adaptations for hearing underwater vary across marine mammal species with the fully aquatic whales having the most derived auditory systems. The degree to which amphibious marine mammal auditory systems are derived correlates with the extent to which their life history functions occur in the ocean relative to the extent that it occurs on land. Most marine mammals rely on hearing as a critical component of some life history functions, particularly mating (e.g., vocal displays), foraging, and navigation. The introduction of noise into the oceans by humans is a relatively recent phenomenon but consists of many types of noise sources varying across a broad range of frequencies, with different durations, and having potentially very high sound levels. Given that most marine mammals rely on hearing to make a living in the ocean, how the introduction of human-caused noise might affect marine mammals is a critical conservation question, particularly for those marine mammals that are endangered or threatened.

REVIEW QUESTIONS

1 What anatomical adaptations support echolocation in toothed whales?
2 What anatomical adaptations support low-frequency hearing in baleen whales?
3 How does the degree of auditory system derivation relate to whether animals live only in the ocean or live on both land and in the sea (amphibious)?
4 In what ways are sounds used by marine mammals to live in the ocean?
5 Do all marine mammals hear as well underwater as they do in air?
6 What are the "mainstay" tools of the marine mammal bioacoustician?
7 Describe the two ways in which marine mammal hearing is typically studied?
8 Describe the information provided by an audiogram.
9 How might noise put into the ocean by humans affect marine mammal hearing?
10 How are hydrophones used to study marine mammals, both in the wild and under human care?

CRITICAL THINKING

1 Echolocating dolphins are known to scan back and forth across a target as they approach it. How might the scanning behavior help them localize the target?
2 If baleen whales are too large to keep from performing behavioral hearing tests, how might we determine what they hear?
3 Human-caused noise has the potential to directly affect a marine mammal's hearing by causing a temporary hearing loss (i.e., the "rock concert effect"), interfering with the detection or discrimination of sounds (i.e., masking), disrupting behavior (e.g., foraging, diving, mating), or by causing stress. Which of these will have the greatest impact to a population of animals?

FURTHER READING

Au, W.W.L. 1993. *The Sonar of Dolphins*. New York: Springer-Verlag.
Richardson, W.J., Greene, C.R., Malme, et al. 1995. *Marine Mammals and Noise*. San Deigo: Academic Press.
Supin, A.Y., Popov, V.V., and Mass, A.M. 2001. *The Sensory Physiology of Aquatic Mammals*. Boston: Kluwer Academic Publishers.

GLOSSARY

Acoustic fats: These are specialized fats that have sound transmission characteristics similar to water; they are found in the hollow mandible and melon of toothed whales.

Audiogram: This is a graph depicting the threshold of hearing at specific frequencies across the range of hearing.

Auditory bulla: It is a bony complex, consisting of the tympanic and periotic bones, that houses the middle and inner ear.

Auditory evoked potential: This is a voltage produced by the nervous system in response to the hearing of a sound.

Basilar membrane: It is a membrane that runs the length of the cochlea and divides it in two; the thickness and width of the membrane dictates frequencies at which it vibrates, and this subsequently causes associated hair cells to initiate neural pulses to the brain.

Bone conduction: This is the transmission of sound to the inner ear through cranial bones; bone conduction bypasses the tympanic membrane.

Broadband: This is a descriptive term used to describe a sound which encompasses a wide range of frequencies.

Cochlea: This is the coiled cavity of the inner ear that holds the basilar membrane and organ of Corti, which is responsible for transducing pressure waves into neural signals.

Collimate: It is to align sound waves; it is the process by which the toothed whale melon turns clicks produced by the phonic lips into a forward projected sonar beam.

Encephalization quotient (EQ): It is the ratio of the actual brain mass to that predicted based on an animal's body mass.

Frequency: It is the number of cycles of a sound pressure wave that occurs in a second; frequency is measured in Hertz (Hz).

Incus: This is the second bone in the chain of ossicles; it lies between and attaches to the malleus and stapes.

Lombard effect: This is an increase in the amplitude of an acoustic signal, the rate at which it is produced, or its duration, when the animal producing the signal is in the presence of masking noise.

Loudness: It is the perception of sound pressures received by the ear.

Malleus: It is the first bone in the chain of ossicles; it attaches to the tympanic membrane.

Masking: This is the phenomenon of a sound (or sounds) interfering with the detection or interpretation of another sound.

Organ of Corti: This is a structure in the inner ear containing hair cells and other structures used to create nerve impulses in response to sound vibration.

Ossicles: These are the bones of the middle ear, consisting of the malleus, incus, and stapes; the bones are connected in sequence and are important in stimulating pressure waves within the fluid of the cochlea.

Permanent threshold shift (PTS): This is a permanent increase in the threshold of hearing.

Phonic lips: These are the anatomical structures in the nasal passages of toothed whales responsible for the production of echolocation clicks, whistles, burst pulses, and other sounds.

Pitch: It is the perception of a sound's frequency.

Psychophysics: This is a subfield of psychology that studies the relationship between physical phenomenon and how they are perceived.

Sound pressure level: This is the magnitude of change in the ambient pressure caused by a sound wave; it is typically measured as a logarithm of the ratio of the measured pressure to a reference pressure (the reference pressure typically differs between air [gases] and water [liquids]).

Spectral content: Spectral content of a sound is the distribution of sound energy across the different frequencies that make up the sound.

Stapes: This is the third bone in the chain of ossicles; it has a foot plate that attaches to the oval window of the cochlea.

Telescoping: This is the evolutionary process by which cetacean cranial bones became elongated and the nares migrated to the dorsal surface.

Temporary threshold shift (TTS): This is a recoverable increase in the threshold of hearing.

Threshold of hearing: This is the minimum level of sound that an animal can detect, usually measured at a specific frequency.

Transduce: This means to change from one form to another; in acoustics, it typically refers to either converting sound pressure waves to electricity or vice versa.

Tympanic membrane: This is the eardrum; it separates the outer and middle ears and is responsible for transducing sound pressure waves into the mechanical motion of the ossicles.

REFERENCES

Amundin, M., and S.H. Andersen. 1983. Bony nares air pressure and nasal plug muscle activity during click production in the harbour porpoise, *Phocoena phocoena*, and the bottlenosed dolphin, *Tursiops truncatus*. *Journal of Experimental Biology* 105(1): 275–282.

Aroyan, J.L., T.W. Cranford, J. Kent, et al. 1992. Computer modeling of acoustic beam formation in *Delphinus delphis*. *Journal of the Acoustical Society of America* 92(5): 2539–2545.

Au, W.W.L., and M.C. Hastings. 2008. *Principles of Marine Bioacoustics*. New York: Springer.

Boddy, A.M., M.R. McGowen, C.C. Sherwood, et al. 2012. Comparative analysis of encephalization in mammals reveals relaxed constraints on anthropoid primate and cetacean brain scaling. *Journal of Evolutionary Biology* 25: 981–994.

Branstetter, B.K., J. St. Leger, D. Acton, et al. 2017. Killer whale (*Orcinus orca*) behavioral audiograms. *Journal of the Acoustical Society of America* 141(4): 2387–2398.

Brumm, H., and S.A. Zollinger. 2013. Avian vocal production in noise. In *Animal Communication and Noise*, ed. H. Brumm, 187–227. Berlin: Springer-Verlag.

Clark, C.W. 1990. Acoustic behavior of mysticete whales. In *Sensory Abilities of Cetaceans*, ed. J. Thomas and R. Kastelein, 571–583. New York: Plenum Press.

Cox, T.M., T.J. Ragen, A.J. Read. et al. 2006. Understanding the impacts of anthropogenic sound on beaked whales. *Journal of Cetacean Research and Management* 7(3): 177–187.

Cranford, T.W. 1992. Functional morphology of the odontocete forehead: Implications for sound generation. PhD Dissertation, Department of Biology, University of California Santa Cruz.

Cranford, T.W. 2000. In search of impulse sound sources in odontocetes. In *Hearing by Whales and Dolphins*, ed. W.W.L. Au, A.N. Popper, and R.R. Fay, 109–156. New York: Springer-Verlag.

Cranford, T.W., W.R. Elsberry, W.G. Van Bonn, et al. 2011. Observation and analysis of sonar signal generation in the bottlenose dolphin (*Tursiops truncatus*): Evidence for two sonar sources. *Journal of Experimental Marine Biology and Ecology* 407: 81–96.

Durbach, I.N., C.M. Harris, C. Martin, et al. 2021. Changes in the movement and calling behavior of minke whales (*Balaenoptera acutorostrata*) in response to navy training. *Frontiers in Marine Science* 8(880). https://doi.org/10.3389/fmars.2021.660122

Erbe, C., S.A. Marley, R.P. Schoeman, et al. 2019. The effects of ship noise on marine mammals – A review. *Frontiers in Marine Science* 6(606). https://doi.org/10.3389/fmars.2019.00606

Finneran, J.J., D.S. Houser, B. Mase-Guthrie, et al. 2009. Auditory evoked potentials in a stranded Gervais' Beaked whale (*Mesoplodon Europaeus*). *Journal of the Acoustical Society of America* 126(1): 484–490.

Ghoul, A., and C. Reichmuth. 2014. Hearing in the sea otter (*Enhydra lutris*): Auditory profiles for an amphibious marine carnivore. *Journal of Comparative Physiology A* 200(11): 967–981.

Harris, C., L. Thomas, E. Falcone, et al. 2018. Marine mammals and sonar: Dose-response studies, the risk-disturbance hypothesis and the role of exposure context. *Journal of Applied Ecology* 55: 396–404.

Hemilä, S., S. Nummela, A. Berta, et al. 2006. High-frequency hearing in phocid and otariid pinnipeds: An interpretation based on inertial and cochlear constraints (L). *Journal of the Acoustical Society of America* 120(6): 3463–3466.

Hemilä, S., S. Nummela, and T. Reuter. 2010. Anatomy and physics of the exceptional sensitivity of dolphin hearing (Odontoceti: Cetacea). *Journal of Comparative Physiology A* 196: 165–179.

Holt, M.M., D.P. Noren, V. Veirs, et al. 2008. Speaking up: Killer whales (*Orcinus orca*) increase their call amplitude in response to vessel noise. *Journal of the Acoustical Society of America* 125(1): EL27–EL32.

Houser, D.S., D.A. Helweg, and P.W.B. Moore. 1999. Classification of dolphin echolocation clicks by energy and frequency distributions. *Journal of the Acoustical Society of America* 106(3): 1579–1585.

Janik, V.M., and L.S. Sayigh. 2013. Communication in bottlenose dolphins: 50 years of signature whistle research. *Journal of Comparative Physiology A* 199(6): 479–489.

Johnson, C.S. 1967. Sound detection thresholds in marine mammals. In *Marine Bioacoustics*, ed. W.N. Tavolga, 247–260. Oxford: Pergamon Press.

Kastak, D., and R.J. Schusterman. 1998. Low-frequency amphibious hearing in pinnipeds: Methods, measurements, noise, and ecology. *Journal of the Acoustical Society of America* 103(4): 2216–2228.

Kastak, D., and R.J. Schusterman. 1999. In-air and underwater hearing sensitivity of a northern elephant seal (*Mirounga angustirostris*). *Canadian Journal of Zoology* 77(11): 1751–1758.

Kastelein, R.A., L. Hoek, C.A.F. de Jong, et al. 2010. The effect of signal duration on the underwater detection thresholds of a harbor porpoise (*Phocoena phocoena*) for single frequency-modulated tonal signals between 0.25 and 160 kHz. *Journal of the Acoustical Society of America* 128(5): 3211–3222.

Kastelein, R.A., P. Mosterd, B. van Santen, et al. 2002. Underwater audiogram of a Pacific walrus (*Odobenus rosmarus divergens*) measured with narrow-band frequency-modulated signals. *Journal of the Acoustical Society of America* 112(5): 2173–2182.

Ketten, D.R. 2000. Cetacean ears. In *Hearing by Whales and Dolphins*, ed. W. Au, A.N. Popper and R.R. Fay, 43–108. New York: Springer-Verlag.

Ketten, D.R., and D. Wartzok. 1990. Three-dimensional reconstructions of the dolphin ear. In *Sensory Abilities of Cetaceans: Laboratory and Field Evidence*, ed. J.A. Thomas and R.A. Kastelein, 81–105. New York: Plenum Press.

Koblitz, J.C., M. Wahlberg, P. Stilz, et al. 2012. Asymmetry and dynamics of a narrow sonar beam in an echolocating harbor porpoise. *Journal of the Acoustical Society of America* 131(3): 2315–2324.

Madsen, P.T., M. Lammers, D. Wisniewska, et al. 2013. Nasal sound production in echolocating delphinids (*Tursiops truncatus* and *Pseudorca crassidens*) is dynamic, but unilateral: Clicking on the right side and whistling on the left side. *Journal of Experimental Biology* 216(Pt 21): 4091–4102.

Madsen, P.T., D. Wisniewska, and K. Beedholm. 2010. Single source sound production and dynamic beam formation in echolocating harbour porpoises (*Phocoena phocoena*). *Journal of Experimental Biology* 213: 3105–3110.

Manger, P.R. 2006. An examination of cetacean brain structure with a novel hypothesis correlating thermogenesis to the evolution of a big brain. *Biological Reviews* 81: 293–338.

Marques, T.A., L. Thomas, S.W. Martin, et al. 2013. Estimating animal population density using passive acoustics. *Biological Review* 88: 287–309.

McCormick, J.G., E.G. Wever, J. Palin, et al. 1970. Sound conduction in the dolphin ear. *Journal of the Acoustical Society of America* 48(6): 1418–1428.

Møhl, B. 1967. Seal ears. *Science* 157(3784): 99.

Møhl, B. 1968. Hearing in seals. In *Behavior and Physiology of Pinnipeds*, ed. R.J. Harrison, R. Hubbard, C. Rice, and R.J. Schusterman, 172–195. New York: Appleton-Century.

Mooney, T.A., W.-C. Yang, H.-Y. Yu, et al. 2015. Hearing abilities and sound reception of broadband sounds in an adult Risso's dolphin (*Grampus griseus*). *Journal of Comparative Physiology A* 201(8): 751–761.

Moore, P.W., L.A. Dankiewicz, and D.S. Houser. 2008. Beamwidth control and angular target detection in an echolocating bottlenose dolphin (*Tursiops truncatus*). *Journal of the Acoustical Society of America* 124(5): 3324–3332.

Moretti, D., L. Thomas, T. Marques, et al. 2014. A risk function for behavioral disruption of Blainville's beaked whales (*Mesoplodon densirostris*) from mid-frequency active sonar. *PLoS One* 9(1): e85064.

Nachtigall, P.E., A.Y. Supin, M. Amundin, et al. 2007. Polar bear *Ursus maritimus* hearing measured with auditory evoked potentials. *Journal of Experimental Biology* 210(7): 1116–1122.

Nachtigall, P.E., M.M.L. Yuen, T.A. Mooney, et al. 2005. Hearing measurements from a stranded infant Risso's dolphin, *Grampus griseus*. *Journal of Experimental Biology* 208: 4181–4188.

Norris, K.S., and G.W. Harvey. 1974. Sound transmission in the porpoise head. *Journal of the Acoustical Society of America* 56(2): 659–664.

Nummela, S., J.G.M. Thewissen, S. Bajpai, et al. 2007. Sound transmission in archaic and modern whales: Anatomical adaptations for underwater hearing. *Anatomical Record* 290: 716–733.

Odend'hal, S., and T.C. Poulter. 1966. Pressure regulation in the middle ear cavity of sea lions: A possible mechanism. *Science* 153: 768–769.

Oelschlager, H.H.A. 2008. The dolphin brain – A challenge for synthetic neurobiology. *Brain Research Bulletin* 75: 450–459.

Owen, M.A., and A.E. Bowles. 2011. In-air auditory psychophysics and the management of a threatened carnivore, the polar bear (*Ursus maritimus*). *International Journal of Comparative Psychology* 24: 244–254.

Parks, S.E., M. Johnson, D. Nowacek, et al. 2011. Individual right whales call louder in increased environmental noise. *Biology Letters* 7: 33–35.

Ramprashad, F., S. Corey, and K. Ronald. 1972. Anatomy of the seal ear *Pagophilus groenlandicus* (Erxlebel 1777). In *Functional Anatomy of Marine Mammals*, ed. R.J. Harrison, 264–306. London: Academic Press.

Reichmuth, C., M.M. Holt, J. Mulsow, et al. 2013. Comparative assessment of amphibious hearing in pinnipeds. *Journal of Comparative Physiology A* 199(6): 491–507.

Reichmuth, C., J.M. Sills, and A. Ghoul. 2017. Psychophysical audiogram of a California sea lion listening for airborne tonal sounds in an acoustic chamber. *Proceedings of Meetings on Acoustics* 30(1): 010001.

Reidenberg, J.S., and J.T. Laitman. 2007. Discovery of a low frequency sound source in mysticeti (baleen whales): Anatomical establishment of a vocal fold homolog. *Anatomical Record* 290: 745–759.

Reidenberg, J.S., and J.T. Laitman. 2018. Comparative anatomy of the larynx in pinnipeds (seal, sea lion, walrus). *The FASEB Journal* 32(S1): 780.12

Repenning, C.A. 1972. Underwater hearing in seals: Functional morphology. In *Functional Anatomy of Marine Mammals*, ed. R.J. Harrison, 307–331. London: Academic Press.

Ridgway, S.H., D.A. Carder, R.F. Green, et al. 1980. Electromyographic and pressure events in the nasolaryngeal system of dolphins during sound production. In *Animal Sonar Systems*, ed. R.G. Busnel and J.F. Fish, 239–250. New York: Plenum Press.

Schlundt, C.E., R.L. Dear, D.S. Houser, et al. 2011. Auditory evoked potentials in two short-finned pilot whales (*Globicephala macrorhynchus*). *Journal of the Acoustical Society of America* 129(2): 1111–1116.

Solntseva, G.N. 2007. *Morphology of the Auditory and Vestibular Organs in Mammals, with Emphasis on Marine Species*, Vol. 4: Sofia, Bulgaria: Pensoft Publishers.

Stanistreet, J.E., D. Risch, and S.M. Van Parijs. 2013. Passive acoustic tracking of singing humpback whales (*Megaptera novaeangliae*) on a Northwest Atlantic feeding ground. *PLoS One* 8(4): e61263.

Stirling, I. 1998. *Polar Bears*. Ann Arbor: University of Michigan Press.

Yamato, M., D.R. Ketten, J. Arruda, et al. 2012. The auditory anatomy of the minke whale (*Balaenoptera acutorostrata*): A potential fatty sound reception pathway in a baleen whale. *The Anatomical Record*. 295(6): 991–998.

Yamato, M., H. Koopman, M. Niemeyer, et al. 2014. Characterization of lipids in adipose depots associated with minke and fin whale ears: Comparison with "acoustic fats" of toothed whales. *Marine Mammal Science* 30(4): 1549–1563.

CHAPTER 7

SENSING IN DARK AND MURKY WATERS

Kenneth Sørensen and Frederike Diana Hanke

Department of Neuroethology, University of Rostock,
Rostock, Germany

In this chapter, we will summarize how vision and touch/hydrodynamics help marine mammals to safely move, navigate, and detect prey in clear and bright but also in dark and murky waters where very little light is available. We will also learn how marine mammals use particle-rich water to orient themselves, and how they detect and discriminate prey items in the water column without any visual cues available. We will later discuss some methodological approaches to sensory research of marine mammals for future considerations.

Kenneth Sørensen

Kenneth took his first scientific steps at the University of Southern Denmark in Odense. He is very interested in sensory systems of semiaquatic organisms. While he tried to unravel how and what penguins can hear in air and underwater in his Ph.D., which he gained in 2021, he shifted his focus to vision in pinnipeds during his postdoctoral fellowship at the University of Rostock. In general, Kenneth has great passion for marine wildlife and the ocean which he prefers to discover during endless hours of diving – not surprisingly thus, he has already combined all his interests in a study on human underwater hearing.

Frederike Diana Hanke

Frederike recently became a professor of neuroethology at the University of Rostock. She is interested in vision in semiaquatic organisms from periphery to neuroanatomy and would also like to understand how seals navigate the ocean – maybe even with the support of a well-developed sense of time that her group has started to characterize. Although her interest lies in pinnipeds, Frederike still enjoys thinking back to different stages of her career during which she was working with honeybees, rats, or corvids – with the cephalopods, she had the impression that it was not her examining these wonderful creatures but the reverse!

Driving Question: How can marine mammals find their food and navigate the oceans even in dark and murky waters?

INTRODUCTION

Marine mammals are generally very mobile species. While some species undertake large-scale movements, many marine mammals move on small scales within, for example, a home range. These movements are required throughout the year for successful foraging and breeding.

DOI: 10.1201/9781003297468-7

Consequently, marine mammals can sometimes experience tremendous changes in the environment, and these changes occur with horizontal as well as vertical movements during dives. However, under all conditions, the animals have to rely on their senses as interfaces between environment and behavior. While the senses generally assist orientation, they also help to collect important information about general features of the environment, predators, prey, or mating partners. Although an individual's senses most likely interact to form a multimodal representation of the environment (Dehnhardt 2002), it may happen under certain conditions that only a few or, in the most extreme case, only one sensory system is providing reliable information.

Among the senses, **vision** is of paramount importance from a human perspective. However, in marine mammals, the role of the visual system for underwater orientation is controversial. Many authors speculate that vision cannot play a major role underwater due to low light levels that dominate the environment of most marine mammals (**Figure 7.1**).

Low light levels are encountered at night and also when diving into deep waters even during the day (**Figure 7.1a**). As a consequence, marine mammals move back and forth between high light conditions at the water surface and darkness at depth many times per hour. When diving, low light levels are caused due to more and more absorption and scattering of sunlight with increasing depth. **Absorption** and **scattering** increase when particles are dissolved in the water which especially characterizes coastal waters or rivers (**Figure 7.1b**). Are marine mammals able to use vision under these dim light conditions? Which role could vision nevertheless play in their environment? Which alternative senses could marine mammals rely on in dark and murky waters when trying to find food or when navigating the ocean?

ADAPTATIONS OF THE VISUAL SYSTEM TO LOW LIGHT CONDITIONS

The eyes of marine mammals show many adaptations for increased sensitivity which allow them to see even during low light levels; these adaptations can also be found in crepuscular (living in twilight habitat) and nocturnal animals. The eyes of manatees (*Trichechus* sp.), sea otters (*Enhydra lutris*.), and polar bears (*Ursus maritimus*) share only some of these adaptations, most likely because manatees and polar bears inhabit light-rich environments and sea otters are, in evolutionary terms, young semiaquatic animals with eyes still largely resembling the eyes of terrestrial mammals.

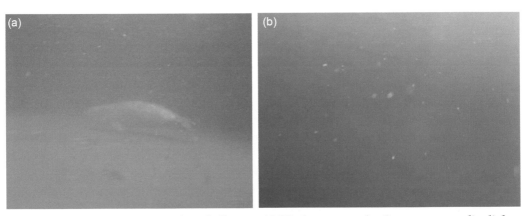

Figure 7.1 **Visual environment and its challenges. (a) Marine mammals often encounter dim light conditions even during the day, for example, when diving down to the bottom to hunt for benthic prey. (b) When the water is murky, particles are dissolved in the water that increase absorption and scattering.**

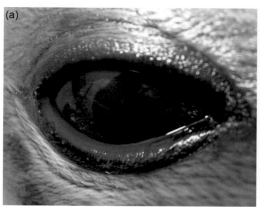

Figure 7.2 Marine mammal eyes. (a) Eye of a harbor seal. In the center of the eye, the pupil can be seen which is closed to a tiny slit under bright light conditions and finally constricts even more reaching the status of a small pinhole. (b) Eye of a bottlenose dolphin. In the center of the eye, the crescent-shaped pupil can be seen. While the dolphin's pupil is circular in darkness, it closes as a crescent and finally constricts to two tiny pinholes in bright light conditions. (Source: Photo by Yvonne Krüger)

First, if an animal wants to see when it is dim, it needs to collect as much of the small amount of light available in its environment as possible, for example by the means of a large **pupil**. However, to allow the pupil to dilate widely, the eye needs to be large too (**Figure 7.2**). And indeed, most marine mammals possess large eyes; the largest extant eyes in vertebrates can actually be found in cetaceans. In the pinnipeds, such as in northern elephant seals (*Mirounga angustirostris*), these large eyes allow the pupil to get more than five times larger than the fully dilated pupil of a young human (Levenson and Schusterman 1997). Generally, the pupillary range – from circular in darkness to (a) pinhole(s) in bright light – of the eyes of marine mammals is enormous reflecting the large range of ambient light that these animals encounter in their lives (**Figure 7.2**).

Second, let us assume that some light has entered the eye via the widely dilated pupil. This light will ultimately reach the retina in the back of the eye, where the elements that transduce light into electric signals, i.e., the photoreceptors, are located: the **rods** and **cones**. To assure that the small amount of light can be used for seeing, it is advantageous for the retina to be mainly equipped with rods. In comparison to cones, the rods contain a more light-sensitive pigment (**rhodopsin**); they collect light over longer time periods (temporal summation), and the information of numerous rods is pooled (spatial summation). For example, in the ringed seal (*Phoca hispida*), only around 1.5% of the photoreceptors are cones. The ringed seal eye has a mean cone to rod ratio of 1:64 (Peichl and Moutairou 1998) in comparison to the human eye which has approximately 5% cones and a cone to rod ratio of 1:19 (Jonas, Schneider, and Naumann 1992). Some baleen whales lack cones entirely and have an all-rod retina (Meredith et al. 2013). Spatial summation in marine mammal eyes happens when the visual information reaches the final layer of the retina, i.e., the ganglion cell layer, in which extremely large **retinal ganglion cells** can be found (for review see Mass and Supin 2018).

A third means to assure that the small amount of light that has entered the eye can be used for seeing is the presence of a **tapetum** (a reflective layer) behind the retina. The tapetum reflects light that was not absorbed by the photoreceptors during the first passage through the retina and enables light absorption during the second passage through the retina. Most marine mammals have well-developed tapeta; the tapeta of seals are the thickest tapeta found in the animal kingdom. In some marine mammals, the tapetum backs the whole back of the eye (ocular fundus) and is not restricted to the ventral fundus as in terrestrial carnivores (see Johnson 1901). This way the tapeta can increase the probability of photon absorption irrespective of the direction from which the photon is entering the eye. This "omnidirectionality" of the tapeta might have evolved in marine mammals as a response to the three-dimensional (3D) underwater world in which a marine mammal can not only move but also rotate in any orientation and in which photons thus might arrive at the eye from above, below, or from the sides. In contrast, in terrestrial animals, significant light comes from the sky above. Thus, the tapetum only needs to be present in the ventral fundus to support vision from dorsal viewing directions, from the ground below, from which less light is reaching the eye. Please note that you all might have observed the function of the tapetum when you encounter a cat on the streets at night and the light of your car is reflected from the tapetum, causing the cat's eyes to shine bright.

Marine mammal eyes not only possess adaptations to increase light sensitivity but their eyes can also quickly adapt to low light levels. Imagine as a marine mammal is departing from the potentially very bright sunny water surface diving down straight to deeper waters, its eyes need to quickly adapt to the light conditions encountered at depth, sometimes even to darkness, otherwise they will not have the required sensitivity to see anything at all. This **dark adaptation** process takes approximately 18 minutes in shallow diving pinnipeds and 20 minutes in the human eye; however, when a northern elephant seal (*M. angustirostris*) dives down to a depth of approximately 1,500 m within 6 minutes, its eyes will dark adapt within the same short period of time (Levenson and Schusterman 1999). But what do the deep diving marine mammals want to see at these depths?

The deep divers are particularly on the prowl for **bioluminescent** prey. Although sunlight never reaches these depths, the deep sea is not necessarily completely dark as organisms produce bioluminescent point-like flashes that contribute light for vision in the deep sea (Warrant and Locker 2004). Besides the adaptations mentioned above that make an eye highly light sensitive or that allow a quick adaptation to light levels encountered at foraging depth, the deep-diving marine mammals also have a rhodopsin pigment that is highly blue-shifted (see Lythgoe and Dartnall 1970). This blue-shifted pigment is maximally sensitive to light of short wavelengths that penetrates best into deep waters and to bluish bioluminescence, an advantage in deep water or even in the deep sea. Moreover, the blue pigments have a reduced noise level which helps to detect a signal from noise and is critical to gain a reasonably good image when light is scarce and consequently noise is high. Furthermore, the number of axons within the optic nerves of deep-diving animals is highest among the marine mammals. A high number of axons means that a lot of optic information is sent to the brain, and usually the more axons within a sensory nerve, the higher the resolution of the eye. This means that the likely high resolution of the eyes of deep-diving marine mammals might enable these animals to detect small bioluminescent sources even at distance; however behavioral visual acuity studies to confirm the eyes' high resolution are not yet available.

Altogether, the eyes of many marine mammals are adapted to dim light conditions. However, what happens when the water gets turbid; is vision then at its limits?

VISION IN TURBID WATERS

We have all experienced, while swimming in lakes or in the ocean, that the more turbid the water, the shorter the viewing distances. Also, seeing fine details is difficult. In harbor seals (*Phoca vitulina*), underwater visual acuity is very good in clear water (Weiffen et al. 2006); visual acuity of the carnivorous marine mammals in general compares well with the visual acuity of terrestrial carnivores. However, underwater visual acuity reduces drastically with increasing turbidity (Weiffen et al. 2006). Consequently, the detection of small prey items is limited. Under these conditions in turbid water, vision might only provide information at short viewing distances.

Short viewing distances occur when following a **pelagic** fish/prey item swimming in the free water column in the final stages of prey pursuit or when hunting **benthic** prey at the sea bottom. When swimming directly over the sea floor (**Figure 7.1a**) with short viewing distances to objects on the bottom, marine mammals might detect prey items visually. They might even be able to identify prey that is cryptic (**Figure 7.3a**), meaning hard to detect against the sea floor, when it is moving. Even slight movements of the prey cause all body parts to move in the same direction; that is, they move coherently (**Figure 7.3**). Harbor seals could use this visual cue as they are extremely sensitive to **coherent motion** (**Figure 7.3b**; Weiffen et al. 2014). It is important to note that high visual sensitivity to coherent motion would also enable seals to assess the motion direction of a school of pelagic fish. It could also be important to assess fish moving in a deviant manner that, consequently, might be easy prey objects.

While vision may support finding close-by prey in water, maybe even with many particles, there are some marine mammals, such as river dolphins (*Platanistidae* and *Lipopotidae*), that live in habitats with extreme amounts of particles constantly dissolved in the water. For these animals, vision might be almost useless. As expected, these species have "reduced" eyes (Pilleri 1974). In the most extreme case, such as the Indus river dolphin (*Platanista gangetica*), the eyes are very small and immobile with only a small eyelid opening, the lens is lacking, and the tapetum is almost absent. The optic nerves which carry the optic information to the brain contain a low number of axons as also found in the Amazon river dolphin (*Inia geoffrensis*) whose optic nerve possesses only approximately 15,000 axons (Mass and Supin 1989). This highly contrasts with the bottlenose dolphin (*Tursiops truncatus*), for example, with its well-developed eyes from which optical information is transmitted to the brain via more than 100,000 times more axons (for overview see Mass and Supin 2018). The only functions that river dolphin eyes probably can carry out underwater are the detection and gathering of light and the determination of the direction of light (Pilleri 1974).

Generally, as all marine mammals might sometimes face the problem that vision does not provide any reliable information about the

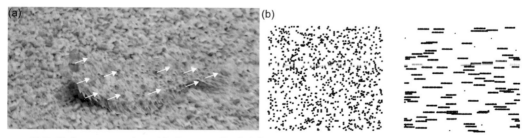

Figure 7.3 Coherent motion. (a) Cryptic flatfish lying on the sea floor. White arrows indicate that when the flatfish moves sideways, all body parts move coherently in the same direction. (b) Random dot stimuli used during the assessment of the harbor seal's sensitivity to coherent motion (modified from Weiffen et al. 2014). In the display on the right side, all dots move into the same direction which means that they move at 100% coherence. In contrast, in the display on the left side, all dots move randomly at 0% coherence. For the visualization of dot movement, the movement of the dots was averaged over 10 frames in order to obtain the figure.

environment, the question arises which alternative senses marine mammals could rely on to find their prey?

We will now consider alternative sensory systems that marine mammals utilize for finding prey. However, we will return to (1) vision in turbid waters with an astonishing new finding and novel hypotheses for visual orientation and navigation later in this chapter (see Optic Flow Perception), and (2) examine the use of vibrissae and hydrodynamic flow perception. Using acoustics for finding prey is covered extensively in **Chapter 6** of this volume. Readers interested in chemoreception in marine mammals can consult existing reviews covering this field of research (see Dehnhardt 2002).

HOW TO FIND PREY WITHOUT VISION

Let us return to benthic foraging. At the sea bottom, prey items sometimes cannot be detected visually even if enough light is available and the water is not turbid, as many benthic fish hide motionless in the mud. Under these circumstances, marine mammals have to adopt alternative strategies for prey detection. When hunting near the sea floor, marine mammals might come very close to their prey. For odontocetes, it was speculated that **electroreception** will help in prey detection when the animals are digging in the mud, such as during crater feeding (for review see Dehnhardt et al. 2020). Research has shown that at least two species of odontocetes are able to perceive low electric fields with their **electroreceptive** organs, the follicle crypts, that remain after the vibrissae on the rostrum are lost before or shortly after birth. For more information on vibrissae and their follicles, see the "The Vibrissal System and its Adaptations to the Aquatic Medium" section later in this chapter.

Benthic prey detection could be based on **active touch**, meaning that the animals directly come in contact with their prey, mainly with their vibrissae in the facial region. Although this opinion was commonly accepted for a long time, video recordings of harbor seals revealed

that the animals are not in direct contact with the sea bottom or with their prey. Is there any sensory information that a marine mammal could use to detect benthic prey when cruising over the sea bottom without direct contact and/ or under conditions in which visual detection is impaired, if not impossible?

There is sensory information generated by every organism, even when lying perfectly still on and camouflaged to the sea floor: every animal has to breathe and, during breathing, water is expelled from the gills – the so-called **breathing current**. Breathing causes water movements, which is hydrodynamic information. Hydrodynamic information is also generated by every organism while swimming through the water: it leaves a **hydrodynamic trail** behind itself (**Figure 7.4**; see Hanke and Bleckmann 2004). Because the vibrissae of marine mammals function as hydrodynamic sensors, they can sense breathing currents (Niesterok et al. 2017) as well as hydrodynamic trails (Dehnhardt et al. 2001). This means that with the help of their vibrissae, marine mammals possess sophisticated means to detect prey in dark and murky water where visual detection is impaired.

Now let us step back in time and have a look on how it was established that vibrissae can sense hydrodynamic information.

HYDRODYNAMIC FLOW PERCEPTION

The hydrodynamic function of vibrissae was established in harbor seals when a visually and acoustically masked seal was asked to sense water movements generated by a sinusoidally oscillating sphere (**Figure 7.5a;** Dehnhardt, Mauck, and Bleckmann 1998). The seal achieved a detection threshold (see "Psychophysics") for dipole water movements with velocities of 245 µm/s vibrating at 50 Hz (**Figure 7.5b;** Dehnhardt, Mauck, and Bleckmann 1998). This sensitivity compares well to the sensitivity of the **lateral line** of some decapods and marine teleosts (Bleckmann 1994). The detection thresholds of a California sea lion (*Zalophus californianus*)

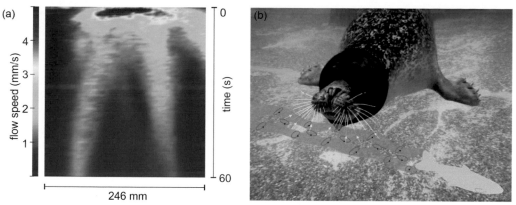

Figure 7.4 Hydrodynamic trail following (a) Velocity profile of the wake behind a fish (*Lepomis gibbosus*; modified from W. Hanke and Bleckmann 2004). The velocity profile was obtained from particle image velocimetry (PIV; see the "Particle Image Velocimetry" section). Flow speeds within a 246 mm region of interest are color coded from blue to red corresponding to water velocities from 0 to 5 mm/s. The fish swam from the bottom to the top of the figure, thus the flow speeds at the top are measured in the wake directly behind the fish, whereas the flow velocities at the bottom of the figure depict the flow speeds that persist after 60 s. (b) A visually masked harbor seal is encountering a hydrodynamic trail of a fish. The flow direction of the water particles in the hydrodynamic trail is indicated by arrows (modified from Bleckmann 1994). The seal can gain information on the 3D structure of the hydrodynamic trail by multiple point-to-point measurements (indicated by yellow arrows).

measured at 20 and 30 Hz were even higher than the highest sensitivity of the harbor seal (Dehnhardt and Mauck 2008), and the manatee's (*Trichechus manatus latirostris*) exceptional sensitivity reached 1 μm/s at 150 Hz (Gaspard et al. 2013). These data show that the vibrissae of the marine mammals tested so far function as very sensitive hydrodynamic receptors.

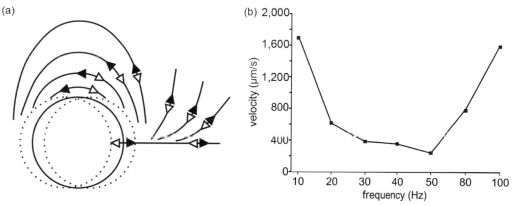

Figure 7.5 Hydrodynamic flow perception. (a) Water movements generated by a dipole as usually used to examine hydrodynamic receptors (modified from Tautz 1989). Movement vectors are displayed in the vicinity of a dipole (oscillating sphere), the amplitude of oscillation is indicated by the arrows and the dotted lines. (b) Performance of a harbor seal during hydrodynamic flow testing (modified from Dehnhardt, Mauck, and Bleckmann 1998). The seal's performance is depicted as velocity threshold (in μm/s) as a function of the frequency (in Hz) of the stimulus (see a). The seal achieved its best threshold performance of 245 μm/s at 50 Hz.

After these first studies characterized the vibrissae as hydrodynamic sensors, harbor seals and California sea lions were asked to follow hydrodynamic trails (**Figure 7.4**) – and they did this successfully. Hydrodynamic trails up to 40 m length were generated by remote-controlled miniature submarines (Dehnhardt et al. 2001) by conspecifics or artificial fish for comparison (for review see Dehnhardt et al. 2014). All experiments revealed the harbor seal's extraordinary trail-following abilities. They could follow the trails with high precision irrespective of which objects had generated the trail, and they did so even when they were allowed to follow the trail after the trail had aged by more than 20 s. The seals even performed well when the trail consisted of a phase in which the submarine was actively swimming and a phase in which it was only gliding (Wieskotten et al. 2010b) mimicking **burst-and-glide-swimming** of some fishes (see Blake 1983). However, burst-and-glide swimming might be an effective anti-predator strategy, as, if the seal was occasionally unsuccessful in following this kind of trail, it lost the trail mainly in the transition zone between active propelled and gliding phases.

In 2011, a comparative hydrodynamic trail-following study was conducted with a California sea lion (Gläser et al. 2011). The sea lion also followed the trails successfully; however, its performance broke down rapidly when a delay was introduced between trail generation and start of trail following. These differences might be due to different morphologies of the vibrissae of seals and sea lions (see the "The Vibrissal System and its Adaptation to the Aquatic Medium" section).

Just imagine a seal hitting a hydrodynamic trail: before actually starting trail following, it is essential to know in which direction the object that had generated the trail was swimming. In addition, if the marine mammal was able to analyze a hydrodynamic trail in detail, reading parameters such as the size or form of the trail generator out of the trail, it could optimize foraging: it could, for example, go for the "best" prey item only or it could also decide to swim into the opposite direction if the trail had been generated by a predator. These suggestions are based on data which show that the hydrodynamic trails generated by different fish species differ due to difference in size, shape, or swimming style of fish (Hanke and Bleckmann 2004). Wieskotten et al. (Wieskotten et al. 2010a, 2011) approached these questions with harbor seals and generated hydrodynamic trails by using many paddles. All these experiments were conducted in a box which enabled the generation of well-defined and measurable hydrodynamic events in otherwise calm waters. The seal determined the movement direction and trail, of less than 0.5s with high precision and with a contact time, of vibrissae with the hydrodynamic trail, less than 0.5 s (Wieskotten et al. 2010a). Again, the seal reliably did this even when the hydrodynamic event was up to 35 s old. In addition, the seal discriminated objects differing in size and shape with high precision (Wieskotten et al. 2011). In conclusion, harbor seals can gain a multitude of information from reading a hydrodynamic trail.

Which parameters could the seals potentially use to base their decisions on? The water vortices – looking similar to "smoke rings" in air when visualized – included in the trails seem to provide powerful information. This experimental evidence was gained from the seals' behavior during hydrodynamic experiments as well as from the generated trails, both of which were analyzed with PIV (for an example of PIV image see **Figure 7.5a** and the "Particle Image Velocimetry" section). Experimental work continues on what kind of information harbor seals gain from a single vortex (for review, see Dehnhardt et al. 2014). The results indicate that harbor seals are able to assess the travel direction of a single vortex ring irrespective of position of stimulation at the vibrissal pad. Furthermore, when presented with two vortex rings in succession, they can judge these vortices on the basis of size. In conclusion, the harbor seal's documented sensitivity to hydrodynamic events allows the seal to detect and follow objects.

Moreover, when encountering a hydrodynamic interpret the event, seals can interpret the events with high precision – this interpretation is comparable to experienced human hunters who read and interpret the traces left behind by animals.

But let us have a closer look at the hydrodynamic sensor: Where can these vibrissae be found and what do they look like?

THE VIBRISSAL SYSTEM AND ITS ADAPTATIONS TO THE AQUATIC MEDIUM

Vibrissae can be found in the facial region of many marine mammals (**Figure 7.6a;** see Bauer, Reep, and Marshall 2018 for detailed information on vibrissae in all groups of marine mammals). In sirenians, they additionally cover the whole body, and in baleen whales, vibrissae can also be found around the blowholes. Toothed whales possess vibrissae on the jaw; however, they lose them prenatally or shortly after birth with the remaining crypts functioning as electroreceptors (see the "How to Find Prey Without Vision" section). Amazon river dolphins are an exception as adults still have bristles on

the jaw. However, these hairs more resemble **guard hairs** that are found in the body fur of mammals. The river dolphin's bristles might also be richly innervated but lack the **follicle–sinus complex** (FSC) of vibrissae.

An FSC encompasses the vibrissal hair or sinus hair and the follicle in which it is anchored. Detailed anatomical investigations of the FSCs of marine mammals are available for some pinnipeds (see Hyvärinen et al. 2009) and sirenians (Reep et al. 2001). These detailed investigations revealed specific adaptations of the FSC of marine mammals which can be linked to their function in the aquatic environment as they differ from the FSC anatomy of terrestrial animals.

First the mystacial FSCs of pinnipeds are much larger in comparison to the FSCs of terrestrial animals. The pinnipeds' FSCs possess an additional blood sinus, the upper cavernous sinus, and its length accounts for 60% of the total FSC length. The existence of an additional cavernous sinus probably leads to elevated surface temperatures measured at the vibrissae of harbor seals (see thermogram in Dehnhardt, Mauck, and Hyvärinen 1998). This way the mechanoreceptors at the ring sinus that convert

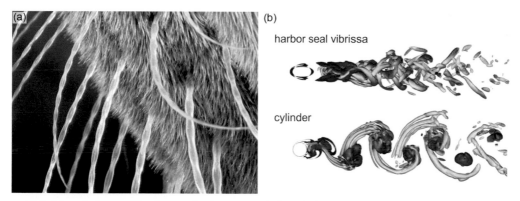

harbor seal vibrissa

cylinder

Figure 7.6 Morphology of harbor seal vibrissae and their behavior in flow. (a) Close-up view of a snout of a harbor seal showing the characteristic morphology of the vibrissae of most phocids. Most phocids possess vibrissae that are flattened and possess an undulatory shape. (b) Wake-flow behind a vibrissa of a harbor seal and behind a circular cylinder as obtained from numerical simulations (modified from W. Hanke et al. 2010). Behind a vibrissa, a complex 3D vortex structure generates downstream from the vibrissa leaving a gap between the vibrissa and the region with fluctuating vortex flows. Furthermore, the complex vortex structure is not stable over time. In contrast, behind a circular cylinder, primary vortices regularly shed directly from the cylinder (Kármán street) and largely persist over time.

the deflection of the hair into an electric signal to be sent to the brain are thermally shielded from low external temperatures, and, consequently, the vibrissae can retain high tactile sensitivity even in cold waters (Dehnhardt, Mauck, and Hyvärinen 1998). In contrast, we have all experienced that our fingers lose their accurate sense of touch in cold temperatures. Retaining function under low temperatures in marine mammals is also facilitated by a second adaptation of their FSC: low-melting-point monoenoic fatty acids in the adipose tissue around the vibrissae which render the tissue very flexible ensuring high vibrissal mobility even when it is cold (for review see Dehnhardt et al. 2014). And third, the FSCs of pinnipeds and sirenians have a higher density of innervation in comparison to terrestrial species. In ringed seals (*Phoca hispida*), e.g., 1,350 axons transmit the sensory information from one FSC to the brain. This is approximately 10 times higher as the innervation density of the vibrissae of terrestrial animals (see Hyvärinen et al. 2009).

Within the pinnipeds, two different vibrissal morphologies exist. The vibrissae of most *Phocidae* are flattened in one direction and undulated in the other (**Figure 7.6a**). Among the phocids, perfectly smooth vibrissae along the entire hair shaft can only be found in the leopard seal, the bearded seal, and in the Mediterranean monk seal (*Monachus monachus*). These species as well as all eared seals (*Otariidae*) and walruses possess vibrissae that are oval in diameter and smooth in outline. The undulated shape of harbor seal vibrissae seems to lead to an almost motionless sensor even when the seal is swimming fast (Hanke et al. 2010). This is unexpected as hydrodynamic vortices should shed from the vibrissae and ultimately lead to vibration of the vibrissae. However, measurements with PIV and **numerical simulations** revealed (**Figure 7.6b**) that in contrast to the smooth vibrissae of a sea lion, the vortices shedding from the vibrissae of a harbor seal are directly destroyed and thus fluctuating lift and drag forces acting on the vibrissae are maximally reduced (Hanke et al. 2010). With a motionless vibrissal sensor, the external hydrodynamic event can be directly measured without the need of extracting the signal from noise (Miersch et al. 2011). In contrast, noise is considerable when the vibrissae of sea lions detect water movements. However, the noise signal contains a dominant frequency that is modulated by a hydrodynamic event. As the dominant frequency was found to correlate with flow velocity, the sea lion might extract flow velocity from the dominant frequency contained in the signal measured by the vibrissae (Miersch et al. 2011). Although the flow performance and signal detection differ in the two vibrissal types, the harbor seal and the California sea lion can perform the same sensory flow tasks, albeit with some qualitative differences as already mentioned (see the "How to Find Food Without Vision" and "Hydrodynamic Flow Perception" sections).

Besides hydrodynamic trail following, what else can marine mammals achieve with their vibrissae? Marine mammals can use their vibrissae for active touch as already mentioned briefly (see the "How to Find Food Without Vision" section). Active touch means that the animals bring their vibrissae in contact with an object. During this haptic inspection process, the pinnipeds perform head movements which ultimately lead to the protruded but motionless vibrissae being moved over the object (Dehnhardt 1994). This process allows them to judge specific object parameters such as size, form, or texture. The semiaquatic harbor seals (for review see Dehnhardt and Mauck 2008) and sea otters (McKay Strobel et al. 2018) can examine objects with their vibrissae in air and underwater with the same precision; the sea otters however are even better in examining objects with the help of their very sensitive paws. Besides the already mentioned harbor seals and sea otters, active touch performance was investigated in numerous marine mammals including California sea lions, walruses (*Odobenus rosmarus divergens*), and manatees (*Trichechus manatus latirostris*) (for review see Dehnhardt and Mauck 2008; Bauer, Reep, and Marshall 2018). Altogether these studies reveal tactile abilities in marine mammals that compare to and sometimes even surpass the abilities of terrestrial species. For completeness, researchers speculated, mainly from the position of the hairs, that the sensory hairs of baleen whales might have a function during surfacing or the localization of prey fields (Murphy et al. 2022; Reichmuth, Casey, and Friedlaender 2022).

We have seen in the last paragraphs that some marine mammals are tactile specialists and experts in hydrodynamic flow perception. Are marine mammals also able to perceive optic flow and what is optic flow?

OPTIC FLOW PERCEPTION AND MOTION VISION

Optic flow is defined as the pattern of visual motion present on the retina of a moving observer (Gibson 1950). When moving forward, an optic flow pattern is generated in which all motion information emanates from the point to which the moving observer is heading/travelling, the so-called **focus of expansion** (FOE). The interpretation of optic flow patterns may allow an animal to avoid colliding with obstacles in its immediate surroundings, estimate its heading, locomote precisely toward a goal, or assess the distance that it has just travelled.

Recently, it was suggested that the particles (**Figure 7.1b**) which render vision in murky waters difficult, create a perfect optic flow environment when passed by an animal. The first and so far only optic flow experiment in a marine mammal was conducted with a harbor seal (Gläser et al. 2014) and was based on two preceding motion vision studies. Research in the field of marine mammal motion vision started in 2008. This "late" focus of research on motion vision is surprising, as the perception of moving objects such as prey, conspecifics, or predators seems of ultimate importance. In 2008, Hanke et al. investigated **optokinetic eye movements** in harbor seals with the optokinetic eye movements being a basic motion

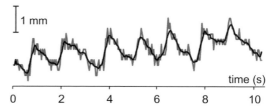

Figure 7.7 Optokinetic eye movements. Characteristic movement of the eye in the head over time during optokinetic stimulation (modified from F.D. Hanke et al. 2008). The eye makes a following movement to the right (downward movement in trace), during which the external motion is optimally nullified, succeeded by a fast saccade in the opposite direction to recenter the eye in the orbit. The gray line represents the changes in position of one point on the eye over time and noise is due to errors in tracing this specific point. The dark line was obtained by a moving average filter in order to approach the real movement of the eye. Scale 1 mm.

stabilizing reflex (**Figure 7.7**). To explain what happens in this reflex let us assume you are sitting in a train and watching the landscape outside the window. Your eyes will start "flipping" meaning they will first follow the moving landscape by making a **pursuit eye movement**. Optimally the visual motion is cancelled during the pursuit eye movement which is the prerequisite for seeing details. However, at some point, the eye needs to be recentered in the orbit, thus the eye performs a fast movement (saccade) against the stimulus movement direction. Pursuit eye movement and saccade alternate as long as there is a visual motion stimulus. Hanke et al. (2008) showed that harbor seals possess optokinetic eye movements and that the eye stabilizes image motion equally well irrespective of stimulus movement direction, up or down, left or right. The latter phenomenon has never been described before and might be an adaptation to the low structured, 3D underwater world in which harbor seals move and sense and in which important visual information might arrive at the eye from any direction. In conclusion, this study showed that harbor seals are indeed able to perceive and stabilize visual motion.

In a second experiment on motion vision, it was determined how sensitive harbor seals are with respect to global motion stimuli (Weiffen et al. 2014). For this purpose, the seal was presented with large **random dot displays** in which a specific number of dots are displayed in the test area. At the extremes, these dots move either randomly (0% coherence) or all dots move in one direction (100% coherence); you might remember a fish lying on the ground and moving which ultimately leads to all body parts moving incoherently or moving with 100% coherence (**Figure 7.3b**; see the "Vision in Turbid Waters" section). A threshold (see the "Psychophysics" section) of motion sensitivity is achieved by varying the coherence of the dots. The suite of experiments showed that harbor seals can achieve motion sensitivity at least as high as most vertebrates, if not better (Weiffen et al. 2014).

After these first motion vision studies, the first optic flow experiment (in 2014) was conducted in which a harbor seal was presented with an underwater optic flow stimulus which simulated a forward movement through a cloud of particles. These particles (**Figure 7.8a**) were represented by 500 white dots moving out of the FOE toward the seal on straight paths. The FOE could be on numerous positions which were, however, covered by a mask (see inset **Figure 7.8b**). In this experiment, the FOE and its vicinity were not visible to the seal which forced the animal to rely on the global motion pattern instead of on local motion signals at the FOE. In addition, a cross was superimposed on the flow pattern (see inset **Figure 7.8b**). It could either directly mark the position of the FOE or could deviate from the position of the FOE by a preset angle of deviation. In the first experimental condition, the seal correctly responded if it touched the cross on the projection; in the second condition, it was required to turn away

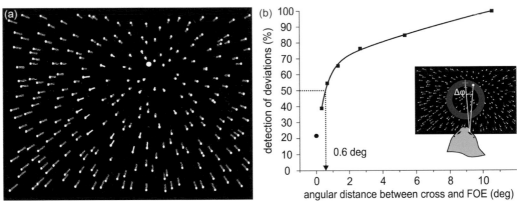

Figure 7.8 Optic flow perception. (a) Optic flow simulation mimicking a forward movement on a straight path through a cloud of particles. The motion vectors emanate from one point, the focus of expansion (FOE; large white dot), that indicates the direction of movement. (b) Performance of a harbor seal during optic flow testing (modified from Gläser et al. 2014). The seal's performance is depicted in detection of deviations (delta phi) between a cross superimposed on the simulation and the actual FOE (in %) as a function of the angular distance between the cross and the FOE (delta phi; in deg). The seal was still able to detect deviations of 0.6° between cross and FOE (50% threshold). Inset shows the optic flow projection used for testing optic flow perception in a harbor seal. The gray ring was shading the region in which the FOE (white dot, not visible during experiments) could be programmed. A green cross was superimposed that could either match or deviate by an angular distance (delta phi) from the FOE.

from the projection in order to receive a reward. By varying the angle of deviation, the 50% threshold (see the "Psychophysics" section) was estimated. The seal quickly found access to the complex optic flow stimulus and was directly going with the flow indicating that the flow pattern created an illusion of self-motion. The seal could still detect a deviation of 0.6° between cross and FOE (**Figure 7.8b**) which is comparable to the best performances published so far for monkeys (see Britten and Van Wezel 2002) and humans (see Warren, Morris, and Kalish 1988). With this sensitivity to optic flow, harbor seals can most likely rely on optic flow information for goal-directed locomotion or assessing parameters important for **path integration** such as traveled distance. In conclusion, with the documented sensitivity to optic flow fields, harbor seals and perhaps other marine mammals can probably also benefit from particle load in the water contrary to the notion that particles only impede underwater vision (see the "Vision in Turbid Waters" and "How to Find Prey Without Vision" sections).

OPTIC AND HYDRODYNAMIC FLOW PERCEPTION COMBINED

When a marine mammal is in its habitat, it will often see, feel, hear, taste, and smell at the same time. To better understand this, we need to analyze the integration of information from these senses, such as hydrodynamics and vision combined. In detail, one challenge is that harbor seals have to cope with optic flow induced by external water movements such as ocean currents. The question becomes whether harbor seals are able to analyze optic flow fields generated by their own movements even if the dissolved particles are displaced by water movements also, in the presence of drift for example? If harbor seals were not able to solve the task using vision alone, they might be able to sense the external water movement with the help of their vibrissae. By integrating optic and hydrodynamic flow information they might then be able to subtract the

optic flow induced by external water movements from the overall flow pattern.

The finding of visual and hydrodynamic flow perception, both mostly in harbor seals, will influence research in the field of foraging, underwater locomotion as well as orientation and navigation.

TOOLBOX

Sensory research of the past decades involving marine mammals has shown that there are still many discoveries to be made. In our opinion, the basis for these discoveries is one central and important method which is "seeing/observing-perceiving-thinking." We would like to encourage everybody to use this central method in order to develop and finally work on ideas that even seem to be "crazy" at first thought.

Operant conditioning

Sensory abilities are often analyzed in behavioral experiments for which the experimental subjects need to learn specific tasks. When learning tasks, the subjects are trained with the help of **operant conditioning** (Skinner 1938). The basis for operant conditioning is that the experimental subject is behaviorally active and that it learns from the consequences of its behavior. That is, upon showing a specific behavior, which initially might only happen incidentally, the specific behavior is reinforced whereas others have no consequences or are negatively reinforced or even punished. Over time, the frequency with which a subject shows the desired behavior will increase, the animal learns, meaning it alters its behavior from the consequences of its behaviors.

When training mammals such as seals, **positive reinforcement** is sufficient for successful learning. Positive reinforcement usually consists of a **primary reinforcer**, a piece of food, and a **secondary reinforcer**, such as a whistle. The application of a secondary reinforcer is essential for the precise timing of the reinforcement which might be difficult with a primary reinforcer under some conditions, and furthermore it bridges the gap until the primary reinforcer can be delivered. A secondary reinforcer needs to be established using **classical conditioning** (Pavlov 1902).

Psychophysics

Sensory abilities of an organism are mainly investigated in psychophysical experiments. The aim of **psychophysics** is to link sensation and perception with physical stimuli.

The physical stimuli need to be measured precisely to know the physical properties of the stimulus presented to an experimental subject. Furthermore, when presenting stimuli underwater, it is crucial to regularly measure and calibrate the stimuli as well as noise to ensure that the stimulus does not get attenuated on its way from the emitter to the receiver or gets masked by noise. Good access to physics and measurement techniques is a prerequisite for a good psychophysicist.

The well-controllable and measurable stimuli when systematically changed allow a researcher to assess sensory thresholds determined as **absolute thresholds**, which is defined as the smallest amount of stimulus energy necessary to produce a sensation, or as **difference thresholds**, which is defined as the amount of change in a stimulus required to produce a just noticeable difference in sensation (Gescheider 1997).

In order to obtain a sensory threshold, two main experimental methods are usually applied in marine mammal research, the **staircase method/method of limits** or the **method of constant stimuli** (Gescheider 1997). Using the staircase method, the intensity of the stimulus is decreased step by step by one unit until the experimental subject ceases to respond correctly. Whenever the subject answers incorrectly – defined as a transition point (or reversal) – the stimulus intensity is again increased by one unit. Thresholds are then calculated by averaging over approximately six transition points. A threshold can also be approached starting with subthreshold intensity values. However, with respect to the test subject's motivation, a descending series

offers the advantage that the session starts with suprathreshold intensity values that are easily detectable. The staircase method allows determining a threshold within one session which is very effective and sometimes even necessary if, for example, changes in the environment need to be considered (see Weiffen et al. 2006). However, the thresholds obtained are often not the lowest as the experimental subjects rapidly learn that if they respond incorrectly, the task will be easier afterward.

Using the method of constant stimuli, a set of stimuli is selected before data collection with stimulus intensities well above and below the putative threshold. This set of stimuli is repeatedly presented to the experimental subject and its response is noted. Finally, the threshold can be calculated from the **psychometric function** that depicts the performance (in %) of the subject averaged over all presentations as a function of the stimulus intensity. Threshold estimations with the method of constant stimuli are more time consuming than the estimation with a staircase method as several sessions are needed before data collection can be completed. However, the thresholds are usually lower in comparison to those obtained with a staircase method as the subject cannot develop expectancies.

In psychophysical experiments, two experimental paradigms are generally used to present the stimuli, a **go/no go paradigm** or a **two-alternative-forced-choice (2AFC) paradigm**. In a go/no go-experiment, two stimulus categories are defined, the go stimulus and the no go stimulus. When confronted with the go stimulus, the experimental subject has to give an answer by moving its head, for example, to a response target; when confronted with the no go stimulus, the subject needs to stay stationary for a predefined time interval. Altogether four responses are possible: a hit (go stimulus and go response), a miss (go stimulus but no go response), a correct rejection (no go stimulus and no go response), and a false alarm (no go stimulus but go response). As a high hit rate can be achieved by always answering with a go response, during analysis, hits and false alarms need to be considered, and it is required to document a low false alarm rate to make the data included in the hit rate valid. Using a go/no go paradigm additionally requires a perfect control of the time interval as some subjects tend to give a go response in the final milliseconds of stimulus presentation of the no go time interval when stimulus intensity is close to threshold.

In a 2AFC paradigm, the experimental subject always has to discriminate between two stimuli in the same trial which can be presented either simultaneously or successively. In each trial, the subject is forced to give a predefined response such as indicating the position of the positive stimulus (S+), which is followed by reinforcement, whereas choosing the negative stimulus (S−) is not reinforced. In a 2AFC experiment, side preferences need to be documented and counteracted by an equal number of trials with the S+ on either side (Gellermann 1933). Generally, the 2AFC procedure leads to a threshold faster in comparison to the go/no go paradigm as in the latter only the performance in respect to one stimulus category, usually the go stimuli, is analyzed to determine a threshold. However, double the number of trials is needed including the no go stimuli as go and no go stimuli need to be balanced in order to be able to approach the threshold.

Particle image velocimetry

PIV measurements are fundamental to all hydrodynamic experiments in order to control qualitative and quantitative aspects of the water movements used as stimuli. With the help of PIV, water movements can be visualized and water velocities can be measured. During PIV measurements, tracer particles are added to the water. These particles are neutrally buoyant and keep position unless a water movement is generated. A fanned-out laser illuminates the particles in one layer. A top view video camera records the light reflected from particles in the region of interest (ROI) first without (background flow) and then with hydrodynamic event such as induced by, e.g., a paddle that is moved from left

to right through the ROI. The video raw material can then be analyzed offline in the PIV software DaVis 7.2 (LaVision GmbH, Göttingen, Germany). With this software, velocity vectors are calculated by correlating the displacement of tracer particles in subsequent images.

CONCLUSIONS

Marine mammals have senses that are specifically adapted to their lifestyle and habitat. This chapter highlighted that the visual system of most marine mammals shows numerous adaptations that allow vision in dim light conditions that are encountered by the animals at depth or at night. Even in turbid waters, vision can provide information during foraging when in close vicinity of the prey such as during benthic foraging or in the final stages of pursuit of pelagic prey. However, when the use of vision reaches its limits, the marine mammals are not lost. Among the other senses, they have a sophisticated vibrissal system. The vibrissae are perfectly adapted to function underwater, but also in air in the semiaquatic marine mammals and allow the animal to judge object parameters when in direct contact with objects. Vibrissae can also be used to detect hydrodynamic information such as faint breathing currents or hydrodynamic trails left behind by prey, conspecifics, or predators. Besides hydrodynamic flow perception, marine mammals, at least harbor seals, are also able to perceive and interpret optic flow. Optic flow is elicited when swimming through turbid waters. We conclude that even, or especially in turbid waters, marine mammals gain optic information to be used for numerous tasks in the context of orientation and navigation.

REVIEW

1 Why do marine mammals often encounter dim light conditions?

2 How does light get attenuated under water?

3 How can you increase the sensitivity of an eye?

4 Can you give an example of how eye characteristics reflect the specific lifestyle or habitat of a species?

5 Which sensory information could marine mammals use when foraging on benthic prey?

6 What is a hydrodynamic trail, and what is the most important parameter of a hydrodynamic trail that seals utilize to judge in which direction a trail generator has swum?

7 What are the differences between the anatomy of follicle sinus complexes of marine mammals and terrestrial animals?

8 Which vibrissal types can be found in pinnipeds and how do they differ in performance?

9 How is optic flow defined?

10 What is psychophysics and what is it mainly aiming at?

CRITICAL THINKING

1 What specific adaptations would you expect to find in the eyes of marine mammals living in regions with pack ice or perpetual ice?

2 Can you think of any function that the vibrissae on different body regions in, for example, sirenians or baleen whales could have?

3 Is there more "(sensory) flow information" you can think of that a marine mammal would want to perceive besides optic or hydrodynamic flow?

FURTHER READING

Bauer G.B., J.C. Gaspard, D.E. Colbert, et al. 2012. Tactile discrimination of textures by Florida manatees (*Trichechus manatus latirostris*). *Marine Mammal Science* 28(4): E456–E471.

Czech-Damal N., A. Liebschner, L. Miersch, et al. 2012. Electroreception in the Guiana dolphin (*Sotalia guianensis*). *Proceedings of the Royal Society of London B* 279: 663–668.

Geurten B., B. Niesterok, G. Dehnhardt, et al. 2017. Saccadic movement strategy in an aquatic species: the harbor seal (*Phoca vitulina*). *Journal of Experimental Biology* 220: 1503–1508.

Krüger Y., W. Hanke, L. Miersch, et al. 2018. Detection and direction discrimination of single vortex rings by harbour seals (*Phoca vitulina*). *Journal of Experimental Biology* 221(8): jeb170753.

Reichmuth C., C. Casey, and A. Friedlaender. 2022. In-situ observations of the sensory hairs of Antarctic minke whales (*Balaenoptera bonaerensis*). *The Anatomical Record* 305(3): 568–576.

Walls G.L. 1942. *The vertebrate eye and its adaptive radiation*. New York: Hafner Press.

GLOSSARY

Absolute threshold: The smallest amount of stimulus energy to produce a sensation.

Absorption: A process during which matter takes up energy of photons. As a consequence, the intensity of light is attenuated. Please note that some wavelengths of light might get absorbed quicker than others depending on the medium.

Active touch: Active tactile exploration of objects. Tactile perception is mediated by pressure, skin stretch, vibration, and temperature sensors.

Benthic prey: Prey that lives close to, on, or in the sea bed.

Bioluminescence: Some organisms are able to produce and emit light either themselves or with the help of bioluminescent bacteria.

Breathing current: A breathing current is generated by water that is expelled from the gill opening of a fish.

Burst-and-glide-swimming: A type of swimming in which undulatory bursts are alternating with periods in which the body is held straight and the organism is gliding forward.

Classical conditioning: As a result of classical conditions a previously neutral stimulus elicits the response (salivation) associated with an unconditioned stimulus (food). This is achieved by repeatedly pairing the neutral stimulus with the unconditioned stimulus.

Coherent motion: When objects move coherently, they move into the same direction. Coherence can vary between 100% meaning that all objects in a display move into the same direction and 0% meaning that all objects move randomly.

Dark adaptation: Dark adaptation is the ability of the eye to adjust its sensitivity to dim light conditions.

Difference threshold: The amount of change in a stimulus required to produce a just noticeable difference in stimuli.

Electroreception: The ability to sense electric fields. In passive electroreception, an electric field produced by an organism is perceived. In active electroreception, an electric field is generated and distorted by objects differing in conductivity and capacity in the environment.

Focus of expansion: The singular point from which all visual motion seems to emanate during movement of an observer. In pure translations, the heading corresponds to the focus of expansion.

Follicle sinus complex: The unit of the vibrissa and its follicle is called follicle sinus complex.

Go/no go experiment: In a go/no go experiment, the animal has to wait for a specific time interval when confronted with the no go stimulus, but it has to give a response within the respective time interval when confronted with the go stimulus.

Guard hairs: Guard hairs form one layer of the fur. These are long hairs that stick out of the undercoat, the bottom layer of the fur.

Hydrodynamics: The flow and movement properties of water.

Hydrodynamic flow perception: It is the ability of an organism to detect water movements.

Hydrodynamic trail: The water disturbances caused by every object that is pulled or actively swimming through the water.

Lateral line: A sensory organ of aquatic vertebrates that is able to detect water movements and vibrations.

Method of constant stimuli: A predefined set of stimuli spanning the threshold that is repeatedly presented to the experimental subject.

Numerical simulation: Numerical simulation is a technique of computational fluid dynamics.

Operant conditioning: Operant conditioning describes a method of learning during which an organism learns through the consequences of its own behavior.

Optic flow: The pattern of visual motion elicited on the retina of a moving observer.

Optokinetic eye movements: The optokinetic nystagmus is a basic motion stabilizing reflex. During optokinetic stimulation, the eye makes a pursuit eye movement in direction of the motion stimulus during which the external motion is ideally nullified. The pursuit eye movement is followed by a fast saccade against stimulus direction in order to recenter the eye in the orbit.

Particle image velocimetry: It is a technique to visualize water movements and measure water velocities.

Path integration: Path integration describes a process during which an organism is continuously documenting distances travelled and directions taken during a trip which provides an estimate of its current position relative to a start position and allows integrating a homing vector which will lead the organism back to its starting point.

Pelagic prey: Prey that lives in the free water column.

Positive reinforcement: An event that increases the probability of a response to reoccur in the future.

Primary reinforcer: A stimulus that is inherently reinforcing as it satisfies biological needs such as hunger.

Psychometric function: A psychometric function is plotting the correct choices (mostly in percent) as a function of the stimulus parameter that is varied during testing.

Psychophysics: It is the discipline that links sensation and perception to physical stimuli.

Pupil/pupillary opening: The pupil is the aperture in the iris that allows the light to enter the eye.

Pursuit eye movements: Pursuit eye movements function to closely follow a moving object.

Random dot display: In a random dot display, a specific number of dots is displayed that can either move completely randomly or with a certain percentage of coherence.

Retina: Photosensitive layer in the backside of the eye which transforms light into nerve impulses.

Retinal ganglion cells: These are neurons that give rise to the optic nerve which sends visual information via action potentials to the brain. Ganglion cells receive their input from amacrine and bipolar cells within the retina.

Rhodopsin: Rhodopsin is one of the visual pigments within the photoreceptor cells of the retina.

Rods, cones: Rods and cones are photoreceptors found within the retina that absorb photons. Rods are very sensitive and function in dim light (scotopic vision), whereas cones are less sensitive and function in bright light (photobic vision). Usually, color vision is based on the presence of different cone types.

Saccade: A fast (up to 1000°/s) eye movement.

Scattering: A process by which the paths of photons deviate from a straight path as the photons encounter particles.

Secondary reinforcer: A stimulus that is associated with the primary reinforcer and thus acquires reinforcing properties.

Staircase method/method of limits: In a staircase method, the intensity of the stimulus is decreased step by step by one unit until the experimental subject ceases to respond correctly. At this point, the transition point, stimulus intensity is again increased by one unit until the experimental subject starts to answer correctly.

Tapetum: Reflective layer behind the retina that reflects photons that were not absorbed by the retina on the first passage back enabling absorption during the second passage of the retina. The tapetum thus increases the sensitivity of the eye.

Two-alternative-forced choice experiment: In a two-alternative-forced choice experiment, the experimental subject has to discriminate between two stimuli. Its task is to always choose the positive stimulus and reject/ignore the negative stimulus.

Vision: Vision is mediated by the eyes and describes an organism's ability to process the information contained in light.

REFERENCES

Bauer, G.B., R.L. Reep, and C.D. Marshall. 2018. The tactile senses of marine mammals. *International Journal for Comparative Psychology* 31: 1–28.

Blake, R.W. 1983. Functional design and burst-and-coast swimming in fishes. *Canadian Journal of Zoology* 61: 2491–2494.

Bleckmann, H. 1994. *Reception of hydrodynamic stimuli in aquatic and semiaquatic animals*, ed. W. Rathmayer. Stuttgart/Jena/New York: Gustav Fischer Verlag.

Britten, K.H., and R.J.A. Van Wezel. 2002. Area MST and heading perception in macaque monkeys. *Cerebral Cortex* 12: 692–701.

Dehnhardt, G. 1994. Tactile size discrimination by a California sea lion (*Zalophus californianus*) using its mystacial vibrissae. *Journal of Comparative Physiology A* 175: 791–800.

Dehnhardt, G. 2002. Sensory Systems. In *Marine mammal biology: An evolutionary approach*, ed. A.R. Hoelzel, 116–141. Oxford: Blackwell Publishing.

Dehnhardt, G., W. Hanke, S. Wieskotten, et al. 2014. Hydrodynamic Perception in Seals and Sea Lions. In *Flow sensing in air and water*, ed. H. Bleckmann, 147–167. Berlin/Heidelberg: Springer.

Dehnhardt, G., and B. Mauck. 2008. Mechanoreception in Secondarily Aquatic Vertebrates. In *Sensory evolution on the threshold: Adaptations in secondarily aquatic vertebrates*, ed. J.G.M. Thewissen and S. Nummela, 295–314. Berkeley, Los Angeles: University of California Press.

Dehnhardt, G., B. Mauck, and H. Bleckmann. 1998. Seal whiskers detect water movements. *Nature* 394: 235–236.

Dehnhardt, G., B. Mauck, W. Hanke, et al. 2001. Hydrodynamic trail-following in harbor seals *(Phoca vitulina)*. *Science* 293: 102–104.

Dehnhardt, G., B. Mauck, and H. Hyvärinen. 1998. Ambient temperature does not affect tactile sensitivity of mystacial vibrissae in harbour seals. *The Journal of Experimental Biology* 201: 3023–3029.

Dehnhardt, G., L. Miersch, C.D. Marshall, et al. 2020. Passive Electroreception in Mammals. In *The senses: A comprehensive reference*, ed. B. Fritzsch, 385–392. Oxford: Elsevier.

Gaspard, J.C.I., G.B. Bauer, R.L. Reep, et al. 2013. Detection of hydrodynamic stimuli by the Florida manatee *(Trichechus manatus latirostris)*. *Journal of Comparative Physiology A* 199(6): 441–450.

Gellermann, L.W. 1933. Chance orders of alternating stimuli in visual discrimination experiments. *Journal of Genetic Psychology* 42: 206–208.

Gescheider, G.A. 1997. *Psychophysics: The fundamentals*. 3rd ed. New York/Toronto/London/Sydney: Lawrence Erlbaum Associates.

Gibson, J.J. 1950. *Perception of the visual world*. Boston: Houghton Mifflin.

Gläser, N., B. Mauck, F. Kandil, et al. 2014. Harbour seals can perceive optic flow underwater. *PloS One* 9(7): e103555.

Gläser, N., S. Wieskotten, C. Otter, et al. 2011. Hydrodynamic trail following in a California sea lion *(Zalophus californianus)*. *Journal of Comparative Physiology A* 197: 141–151.

Hanke, F.D., W. Hanke, K.-P. Hoffmann, et al. 2008. Optokinetic nystagmus in harbor seals *(Phoca vitulina)*. *Vision Research* 48(2): 304–315.

Hanke, W., and H. Bleckmann. 2004. The hydrodynamic trails of *Lepomis gibbosus* (Centrarchidae), *Colomesus psittacus* (Tetraodontidae) and *Thysochromis ansorgii* (Cichlidae) investigated with scanning particle image velocimetry. *The Journal of Experimental Biology* 207: 1585–1596.

Hanke, W., M. Witte, L. Miersch, et al. 2010. Harbor seal vibrissa morphology suppresses vortex-induced vibrations. *Journal of Experimental Biology* 213: 2665–2672.

Hyvärinen, H., A. Palviainen, U. Strandberg, et al. 2009. Aquatic environment and differentiation of vibrissae: Comparison of sinus hair systems of ringed seal, otter and pole cat. *Brain, Behavior and Evolution* 74: 268–279.

Johnson, G.L. 1901. Contributions to the comparative anatomy of the mammalian eye, chiefly based on ophthalmoscopic examination. *Philosophical Transactions of the Royal Society of Biological Characters* 194: 1–82.

Jonas, J.B., U. Schneider, and G.O. Naumann. 1992. Count and density of human retinal photoreceptors. *Graefe's Archive for Clinical and Experimental Ophthalmology* 230(6): 505–510.

Levenson, D.H., and R.J. Schusterman. 1997. Pupillometry in seals and sea lions: Ecological implications. *Canadian Journal of Zoology* 75: 2050–2057.

Levenson, D.H., and R.J. Schusterman. 1999. Dark adaptation and visual sensitivity in shallow and deep-diving pinnipeds. *Marine Mammal Science* 15(4): 1303–1313.

Lythgoe, J.N., and H.J.A. Dartnall. 1970. A 'deep sea rhodopsin' in a marine mammal. *Nature* 227: 995–996.

Mass, A.M., and A.Y. Supin. 1989. Distribution of ganglion cells in the retina of an Amazon river dolphin *Inia geoffrensis*. *Aquatic Mammals* 15(2): 49–56.

Mass, A.M., and A.Y. Supin. 2018. Eye optics in semiaquatic mammals for aerial and aquatic vision. *Brain Behavior and Evolution* 92: 117–124.

McKay Strobel, S., J.M. Sills, M.T. Tinker, et al. 2018. Active touch in sea otters: In-air and underwater texture discrimination thresholds and behavioral strategies for paws and vibrissae. *Journal of Experimental Biology* 221(18): jeb181347.

Meredith, R.W., J. Gatesby, C.A. Emerling, et al. 2013. Rod monochromacy and the coevolution of cetacean retinal opsins. *PLoS Genetics* 9(4): e1003432.

Miersch, L., W. Hanke, S. Wieskotten, et al. 2011. Flow sensing in pinniped whiskers. *Philosophical Transactions of the Royal Society of London B Biological Sciences* 366: 3077–3084.

Murphy, C.T., M. Marx, W.N. Martin, et al. 2022. Feeling for food: Can rostro-mental hair arrays sense hydrodynamic cues for foraging North Atlantic right whales? *The Anatomical Record* 305(3): 577–591.

Niesterok, B., Y. Krüger, S. Wieskotten, et al. 2017. Hydrodynamic detection and localization of artificial flatfish breathing currents by harbour seals *(Phoca vitulina)*. *Journal of Experimental Biology* 220: 174–185.

Pavlov, I.P. 1902. *The work of the digestive glands*, London: Griffin.

Peichl, L., and K. Moutairou. 1998. Absence of short-wavelength sensitive cones in the retinae of seals *(Carnivora)* and African giant rats *(Rodentia)*. *European Journal of Neuroscience* 10: 2586–2594.

Pilleri, G. 1974. Side-swimming, vision and sense of touch in *Platanista indi* (Cetacea, Platanistidae). *Experientia* 30(1): 100–104.

Reep, R.L., M.L. Stoll, C.D. Marshall, et al. 2001. Microanatomy of facial vibrissae in the Florida manatee: The basis of specialized sensory function and oripulation. *Brain Behavior and Evolution* 58: 1–14.

Reichmuth, C., C. Casey, and A. Friedlaender. 2022. In-situ observations of the sensory hairs of Antarctic minke whales *(Balaenoptera bonaerensis)*. *The Anatomical Record* 305(3): 568–576.

Skinner, B.F. 1938. *The behavior of organisms: An experimental analysis*. New York: Appleton Century.

Tautz, J. 1989. *Medienbewegung in der Sinneswelt der Arthropoden - Fallstudien zu einer Sinnesökologie*. Stuttgart/New York: Gutsav Fischer Verlag.

Warrant, E.J., and N.A. Locker. 2004. Vision in the deep sea. *Biological Review* 79: 671–712.

Warren, W.H., M.W. Morris, and M. Kalish. 1988. Perception of translational heading from optical flow. *Journal of Experimental Psychology* 14(4): 646–660.

Weiffen, M., B. Mauck, G. Dehnhardt, et al. 2014. Sensitivity of a harbor seal *(Phoca vitulina)* to coherent visual motion in random dot displays. *SpringerPlus* 3: 688.

Weiffen, M., B. Möller, B. Mauck, et al. 2006. Effect of water turbidity on the visual acuity of harbor seals *(Phoca vitulina)*. *Vision Research* 46: 1777–1783.

Wieskotten, S., G. Dehnhardt, B. Mauck, et al. 2010a. Hydrodynamic determination of the moving direction of an artificial fin by a harbour seal *(Phoca vitulina)*. *Journal of Experimental Biology* 213: 2194–2200.

Wieskotten, S., G. Dehnhardt, B. Mauck, L. et al. 2010b. The impact of glide phases on the trackability of hydrodynamic trails in harbour seals *(Phoca vitulina)*. *Journal of Experimental Biology* 213: 3734–3740.

Wieskotten, S., B. Mauck, L. Miersch, et al. 2011. Hydrodynamic discrimination of wakes caused by objects of different size or shape in a harbour seal *(Phoca vitulina)*. *Journal of Experimental Biology* 214: 1922–1930.

CHAPTER 8

HYDRATION

Miwa Suzuki
Nihon University, Kanagawa, Japan

Rudy M. Ortiz
University of California, Merced, CA

Miwa Suzuki

Miwa explored what cetaceans perceive as stress from an endocrinological approach in her student days and earned her Ph.D. in 2001 in agricultural science from the University of Tokyo. Since joining the research profession, she has been working mainly on the regulatory mechanisms of osmotic pressure as a type of metabolic stress. She works on these questions both in her laboratory and in the field in Japan, where a wide variety of cetaceans live and many aquaria are located. She has also worked with fasting northern elephant seals at University of California (UC), Merced.

Rudy M. Ortiz

Rudy earned his Ph.D. in 2001 from UC Santa Cruz, studying the endocrinology of prolonged fasting in northern elephant seal pups with a focus on renal physiology and substrate metabolism. He is a full professor of endocrinology and physiology with research experience on many marine mammals including dolphins, seals, and manatees. Expanding from marine mammal work, Rudy's current research extends into the study of altered nutrition and metabolic dysfunction on cardiovascular and renal diseases and lipid metabolism in humans and rodent mode.

Driving Question: How do marine mammals hydrate their body in seawater?

INTRODUCTION

Living organisms are filled with water; for example, 60–70% of the content of the human body is water. Cells are full of water along with various components, and water is involved in various ways in almost every step of cellular activity (Jéquier and Constant 2010). In addition, gases, nutrients, ions, and waste products are carried throughout the body via blood and lymphatic fluids, and this flow would not be possible without water. Therefore, how to secure and retain water in the body is perhaps the most important issue for an organism.

Because all living organisms on earth evolved from an aquatic precursor, we can say life essentially was born in the sea. Our body fluids contain ions of various minerals present in seawater, as minerals are essential for cellular activity. It is also important for cellular

DOI: 10.1201/9781003297468-8

activity to maintain proper ion concentrations in body fluids.

During vertebrate evolution, some fish species invaded land from the sea. Their descendants developed water retention mechanisms to withstand desiccation and successfully adapted the land environment and dissipated on land, giving rise to amphibians, reptiles, birds, and mammals. Ultimately, mammals were designed to live in a dry environment. Interestingly, the ancestors of many marine mammals, such as cetaceans and pinnipeds, emerged from terrestrial mammals and returned to an aquatic environment. Through the processes of evolution, marine mammals succeeded in fully adapting to life in water. Most pinnipeds spend much of their time in the sea, including feeding, although their reproductive activities take place on land. Aside from the fact that some species live in fresh or brackish water, for those species that are strictly marine by definition the ability to maintain the hydration state of their bodies is perplexing. Similarly, the mechanisms evolved to maintain constant body water content are equally robust as mechanisms to maintain salt and mineral balance (electrolytes). Thus, the mechanisms required to maintain body water complement the mechanisms required to maintain electrolytes as these must go hand in hand to achieve homeostasis (metabolic balance). This chapter introduces the mechanisms by which marine mammals maintain water and ionic balance, or osmotic regulation and homeostasis (osmoregulation).

THE IMPORTANCE OF WATER FOR LIFE

Water molecules and hydration

Before diving into the mechanisms of osmoregulation in marine mammals, it is important to understand why water is involved in almost every aspect of sustaining life. Water molecules consist of one oxygen atom and two hydrogen atoms bonded together, but folded rather than lined up (**Figure 8.1**). Due to this folding, the

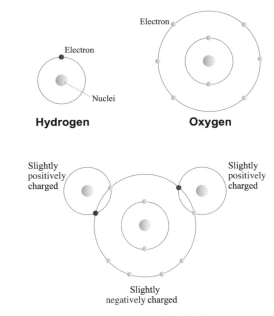

Figure 8.1 **Molecular structure of water.**

oxygen atom side carries a slightly negative charge and the hydrogen atom sides a slightly positive charge. This polarity allows water molecules to electrically attract and "hydrate" various substances. In other words, water electrically attracts and binds ions and other substances or attaches itself to a substance and creates a layer of water around it. Hydration allows those substances to exist, move, and function in body and cellular fluids. Only in the hydrated state can proteins function as they should (with very few exceptions). For all organisms then, water is required to maintain proper cellular activity to sustain life.

The body is filled with water

Because components that form a body such as proteins cannot function without water, this means that an organism must maintain adequate amounts of water in its body. The total pool of body water is compartmentalized with most of the water in cells (67%) and the remainder (33%) outside of the cells (extracellular volume), and these compartments are surrounded and

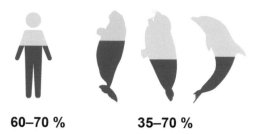

60–70 %   **35–70 %**

Figure 8.2 Total body water contents in human and marine mammals.

protected from dehydration by the skin which forms the barrier between the living organism and the environment. This total body water (TBW) pool serves as the media by which organic and nonorganic substances are dissolved and provides the requisite pool from which complex metabolic processes are supported to maintain the life. Furthermore, the vascular system, which transports blood throughout the body to distribute oxygen and other nutrients and exchange information among organs, contains most of the extracellular volume of water. In humans, 60–70% of body mass is water. In cetaceans and pinnipeds, the TBW to body mass ratio is smaller than in human, with the range of 35–50% due to the large proportion of body fat, whereas TBW is about 70% in dugongs (**Figure 8.2**) (Hui 1981; Nordøy et al. 1992; Smoll, Beard, and Lanyon 2020).

WHERE TO FIND WATER?

For most vertebrates and especially mammals, obtaining fresh, solute-free water to drink is relatively easy. Wild animals drink water from puddles, rivers, or morning dew on plants. However, for marine mammals living in the open ocean, fresh water for drinking is unavailable. Seawater is about 3.5% salt (mainly sodium and chloride) and is too salty for most mammals to drink because their kidney system cannot remove that much salt. Therefore, robust physiological mechanisms to maintain body water and minimize losses had to evolve for marine mammals. Cetaceans, pinnipeds,

and sirenians have developed unique renal (kidney) anatomies that may facilitate efficient Na^+ (sodium) excretion and water retention, and their bodies contain relatively large TBW pools. This section describes how marine mammals procure water.

Drink seawater as water source?

Marine fish and reptiles drink seawater and discard excess salt through different types of extra-renal organs leaving solute-free freshwater in their bodies, but at an energetic cost. Seabirds have salt glands that function like cellular pumps to remove chloride and use a similar strategy to secure water. On the other hand, voluntary and intentional ingestion of seawater (mariposia) is not considered common among marine mammals, even though most of them can produce urine that is hyperosmotic to seawater.

Measuring saltwater drinking in marine mammals is not easy and is based on water flux measurements using chemically labeled atoms of water and on blood and urine composition analyses. In cetaceans and pinnipeds, the amount of seawater ingested has been calculated to be a relatively low percentage of their total water intake through water flux measurements (Depocas, Hart, and Fisher 1971; Ridgway 1972), suggesting that they ingest seawater incidentally rather than actively drinking it. However, in some cases seawater drinking has been more substantial. In tropical pinnipeds, seawater drinking is greater than in colder-water species where mariposia does not appear to be common if at all (Costa and Trillmich 1988), suggesting that seawater intake may be used to reduce body temperature. It has also been suggested that mariposia is induced by fasting or dehydration to secure water (Nordøy 2020). Dugongs have relatively high water turnover rate and have the renal anatomy to obtain free water from seawater (Lanyon, Newgrain, and Alli, 2006), but they do not appear to actively ingest seawater (Smoll, Beard, and Lanyon, 2020). Manatees can excrete hyperosmotic urine (high salt content) (Ortiz, Worthy, and MacKenzie, 1998), but their water turnover rates do not indicate active seawater consumption (Ortiz, Worthy, and Byers,1999). Their urine-concentrating ability may be necessary to facilitate the excretion of excess salt due to the consumption of marine plants (Reich and Worthy 2006). On the other hand, the sea otter has a high metabolic rate (see **Chapter 2**), consumes large amounts of salty food (clams etc.), and actively consumes seawater, which is exception among marine mammals. Active mariposia in the sea otter is not a problem because their kidneys can excrete much higher concentrations of Na^+ and Cl^- than seawater, and water flux measurements suggested that they may drink seawater to increase the urinary osmotic space for dumping excess urea (Costa 1982).

Like water, Na^+ (or salt) is also indispensable and required to sustain life because it contributes to osmotic balance, transmits nerve-cell stimulation, stimulates appetite, regulates pH, and assists digestion and absorption. While excessive salt intake is detrimental to cardiovascular health (Kotchen, Cowley, and Frohlich, 2013), living in a marine environment apparently does not make marine mammals prone to cardiovascular consequences or induce inappropriately elevated Na^+ levels in circulation. Recently, it was reported in humans that high salt intake led to the synthesis of urea, a key molecule for increasing urine concentration, by catabolizing muscle proteins to conserve water while expelling salt (Kitada et al. 2017). Thus, in a situation where animals live in salty water and are at risk of salt overload and dehydration, avoiding the consequences of mariposia may be behaviorally important to maintain homeostasis and minimize the energetic cost to excrete a hyperosmotic urine and other metabolic burdens.

Water from food: free and metabolic water
Preformed water
Unless they are fasting, marine mammals apparently meet their water requirements from their diet. Marine fish have developed an excellent water acquisition technique by actively drinking seawater and desalinizing it using chloride excretion cells. Fish have about 70% TBW ratios. When eating fish, marine mammals are essentially parasitic (Willmer, Stone, and Johnstson,2009) because they steal this free water that was preformed by the fish. Marine mammals meet a significant portion of their water requirements from preformed water in their diets (**Figure 8.3**).

To obtain as much of the water as possible from the fish they consume, the intestines of marine mammals are evolved to efficiently absorb water. Cetaceans and pinnipeds have relatively longer intestines, especially the small intestine, than land carnivores (Stevens and Hume 2004). Water in the food is absorbed in the small and large intestines, so a longer small intestine may be advantageous for efficient water absorption. Generally, in terrestrial mammals, water is absorbed via the cellular junction of absorptive intestinal cells (paracellular pathway), or sodium–glucose cotransporters may also transport water in absorptive cells (Loo, Wright, and Zeuthen, 2002). In the small intestine of the common bottlenose dolphin

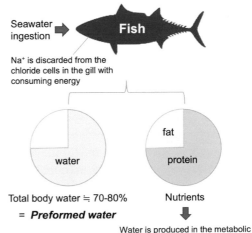

Figure 8.3 **Water utilized by carnivorous marine mammals.**

(*Tursiops truncatus*), the cellular water transport channel (aquaporin-1; AQP1) resides in the apical membrane of absorptive cells (Suzuki 2010). Therefore, AQP1 may contribute to the rapid and effective absorption of water (transcellular pathway). A similar arrangement of AQP at the apical membrane is observed in the large intestine of desert rodents or rats with water deprivation (Gallardo et al. 2001) suggesting that AQPs are upregulated and distributed to facilitate water absorption when necessary. These anatomical and histological features of the intestine may facilitate efficient absorption of water.

Metabolic water

In cetaceans and pinnipeds, the respiratory quotient (RQ) is around 0.71–0.75 (see **Chapter 2**). RQ is the ratio of the amount of carbon dioxide emitted to the amount of oxygen consumed in aerobic respiration, which breaks down nutrients for energy production. An RQ = 1 represents strict glucose metabolism, a range of 0.8–0.9 represents a mix of glucose and protein metabolism, and 0.7 represents lipid-dependent metabolism. Therefore, a range of 0.71–0.75 reported in marine mammals indicates that cetaceans and pinnipeds use primarily fat as the source of energy. Mono- and polyunsaturated fats are abundant in fish including palmitic acid, oleic acid, eicosapentaenoic acid, and docosahexaenoic acid. For example, palmitic acid ($C_{16}H_{32}O_2$), when completely metabolized, uses 23 molecules of oxygen (O_2) to generate 16 CO_2 and 16 H_2O ($C_{16}H_{32}O_2 + 23O_2 \rightarrow$ 16 CO_2 + 16 H_2O), giving a RQ of 23 ÷ 16 = 0.696 or 0.70. Because most fats have a long carbon chain backbone, each of which has hydrogen atoms attached, a large amount of water is produced after oxidative phosphorylation as a byproduct of metabolism. This water, called metabolic water, is derived from the metabolism of all organic substrates and is a critical source of water, especially during fasting conditions where body stores are metabolized for water (**Figure 8.3**). On balance, the metabolism of 1 g of fat can produce 1.1 g of water.

Water during fasting

We have discussed in the previous section that food is the primary source of water for marine mammals; however, prolonged food deprivation or fasting is a common event in the life history of all marine mammals and it poses a unique challenge. During fasting, obtaining preformed water in the diet is not possible, putting a strict reliance on metabolic water. Surprisingly, for the vast majority of studies while marine mammals are fasting, indications of dehydration have not been observed. They balance the internal water by catabolizing fat, muscle, and glycogen stores, and sometimes by mariposia.

Fasting marine mammals depend primarily on metabolic water from oxidation of their large fat stores (see **Chapter 5** and **11**). Some advantages of preferentially metabolizing fat are the production of relatively larger volumes of water compared to the metabolism of other substrates (glucose or protein) and relatively more energy despite needing more oxygen to burn that fat (Worthy and Lavigne 1987; Ortiz et al. 2010; Ortiz, Worthy, and Byers 1999). Marine mammals store large amounts of subcutaneous fat, which is available for metabolism during prolonged periods (weeks to months) of fasting. While lipid is preferentially metabolized during prolonged fasting, some protein (from muscle) may also contribute to the production of metabolic water. In northern elephant seal pups, estimations from changes in tissue volumes during fasting suggest that 434 g/day of water was derived from fat and 161 g/day from muscle breakdown (Lester and Costa 2006) As mentioned above, muscle catabolism may also confer a physiological benefit in terms of water retention by generating urea, the final metabolite of protein catabolism, and increasing urine osmolality. However, the contribution of protein catabolism, especially from muscle, to the generation of metabolic water is unusual and, when present, is minimized to help maintain protein and reduce the nitrogen load on the kidneys. Therefore, lipid metabolism is the preferred source of energy and subsequent metabolic water for prolonged fasted marine mammals. Marine mammals rely on the conservation of respiratory water to maintain water balance especially during fasting (Worthy and Lavigne 1987).

REDUCING WATER LOSS

There are multiple pathways for water to leave an animal's body including excreta (feces, sweat, or urine) and to a lesser extent evaporation (breathing and skin). The bodies of marine mammals are designed to prevent too much water from leaving the body for each route. The following sections describe these evolved adaptions.

Keeping water out of the skin

The skin (epidermis) of marine mammals forms the boundary between the internal and external environments. Terrestrial mammals must prevent water from evaporating through their skin in dry atmospheres. Marine mammals do not have to worry about such water loss because of humidity gradients, but they must prevent water deprivation through differences in osmotic pressure gradients. The osmotic pressure of seawater (1,030 mOsm/L) is more than three times higher than that of plasma in humans (285 ± 5 mOsm/L) and marine mammals (300–350 mOsm/L). If a human remains immersed in salt water, water in the body is slowly pulled out of the body through the skin driven by the difference in osmotic pressure between seawater and body fluid. Marine mammals have evolved adaptations to resist such water loss.

The epidermis of cetaceans is thick with proliferative pool and composed of lipokeratinocytes. These skin cells are unique to cetacean and are quite rich in lamellar bodies and non-membrane-bound lipid droplets as well as keratin filaments. The lipokeratinocytes carry out a waterproofing function and serve as natural barrier against hyperosmotic environments in addition to their contribution to buoyancy, insulation and so on (Menon et al. 1986). Although there are some exceptions, marine mammals usually do not have sweat glands (Ridgway 1972), thus eliminating the potential for evaporative loss from sweat (see **Chapter 5**). Taken together, cetacean skin is characterized by low to no water permeability and high lipid content, which contribute to the conservation of body water.

Nonetheless, there is no consensus on the potential of water to move across the skin in marine mammals, especially cetaceans. Telfer, Cornell, and Prescott, (1970) did not detect the flux of water across the skin in bottlenose dolphins; however, Hui (1981) measured significant flux of water across the skin in common dolphins. Using radioisotopes of water and sodium, Hui calculated that water influx through the skin accounted for about 70% of total water turnover. In addition, dolphins in freshwater may obtain water through the skin depending on the osmotic gradient of the environment (Andersen and Nielsen 1983). The possibility of water entering and exiting through the skin of marine mammals continues to be debated even now, and further study of water fluxes in the skin is needed.

Reducing respiratory water loss

When you exhale in cold weather, your breath appears white because the water in your warm exhalation is cooled and condensed. This a simple demonstration of water vapor in exhaled air that clearly illustrates respiratory water loss with every breath. In seals, body water is conserved by a countercurrent heat exchange system in the nasal cavity which rapidly cools the exhaled air and allows the water vapor to condense to the epithelial lining of the nasal membranes with this water being absorbed, greatly minimizing respiratory water loss (Huntley et al. 1984). This system, coupled with low breathing frequency, rapid thermal changes associated with nasal pressure potential, and high ambient humidity on the sea surface, allows delphinids such as bottlenose dolphins and Pacific white-sided dolphins to lose only 30% to 77% of the water that a similarly sized terrestrial mammal would lose through respiration (Coulombe, Ridgway, and Evans, 1965).

Reducing urinary water loss

The largest route of water loss from the mammalian body is urinary excretion. For example, a healthy adult human drinks about 35 mL/kg/day of body mass. This rate means a 50-kg person drinks about 1.75 L of water/day, consumes about the same amount of water from food, and excretes about 1 to 1.5 L/day of urine, accounting for a majority of the 2.4 L/day of total water lost by various routes. The following sections describe how the kidneys of marine mammals regulate water and electrolytes by altering their concentrations in urine as well as their contributions to regulating other organic and nonorganic compounds to maintain biochemical homeostasis.

Urine-concentrating ability of the kidneys

The kidneys of marine mammals possess adaptive mechanisms to concentrate their urine more than seawater, making these animals excellent concentrators of urine. Maximum urine osmolality represents the concentrating ability of the animal, with most urine samples measured reaching values greater than 2,000 mOsm/L in dolphins, pinnipeds, and sea otters. This is about two times the osmotic value of seawater. The ratio of urine to plasma osmolality is another indicator of urine-concentrating ability and is

higher in dolphins and seals than in humans. The maximum urine osmolality of the bottlenose dolphin is around 2,700 mOsm/L, but comparatively lower than some exceptional cases such as a domestic cat (5,000 mOsm/L) and desert rodent (9,000 mOsm/L). Interestingly, an inverse relationship exists between body size and urine-concentrating ability (Beuchat 1990). Cetaceans and pinnipeds have the above-average urine-concentrating ability among animals of their size and are able to concentrate their urine to reduce urinary water loss (**Figure 8.4**). In addition, the ability to concentrate their urine greater than sea water allows these animals, in theory, to drink seawater and have a positive gain in free water; however, mariposia is not a common behavior in most marine mammals (but the potential to extract solute-free water from salt water exists). The characteristics that allow the kidneys to concentrate urine are discussed in next paragraph.

It has been shown that common bottlenose dolphins do not change their plasma osmolality when drinking seawater and continue to produce hypertonic urine with high sodium ion concentrations, and that feeding seawater together with protein enhances their urine-concentrating ability (Ridgway and Venn-Watson 2010). These observations demonstrate the excellent urine-concentrating ability of dolphin's kidney

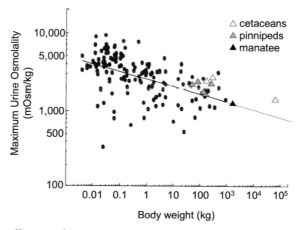

Figure 8.4 Interspecific allometry between maximum urine osmolality and body weight in 146 species of mammals including animals inhabiting arid regions (modified from Beuchat 1990). Maximum urine osmolalities recorded in marine mammals shown in triangle symbols are plotted above the regression line.

and also indicates that urea, the end product of protein metabolism, contributes significantly to this ability.

Importance of urea for water retention and urine-concentrating ability

Urea is a simple organic compound that is made from the ammonia molecules that are released during the breakdown of protein, and thus, it represents an end product of metabolism. It functions as an osmotic regulator to retain water in the body through several mechanisms of action. For example, the presence of urea in high amounts in body fluids increases the osmotic pressure, preventing water loss due to osmotic gradients. Because urea traps water, its relatively high concentration in the horny layer of the epidermis helps to moisturize the skin. In addition, urea is an important molecule for the production of concentrated urine. Urea is reabsorbed from urine in the collecting ducts of the kidney and reused between the circulation and the renal tubules. Urea accumulation in the renal interstitium increases the osmotic pressure in the tissue, promoting the reabsorption of water from the filtrate in the renal tubule. In this way, urea is an important substance for water retention, and the expression of the urea transporters (UT-A and UT-B) in the lining of the renal tubule is closely related to the urine-concentrating ability of the animal.

In cetaceans, interesting studies have been reported on urea. Wang et al. (2015) analyzed the genomic information of several species of cetaceans and compared these sequences to homologous genes in terrestrial mammals. They found that the gene coding for the urea transporter 2 (UT-A2) was positively selected for in the cetacean lineage. In addition, detailed analysis of the positive selection site of UT-A2 demonstrated that urea uptake and permeability may be enhanced by the molecular structure of UT-A2. While not directly evaluated, the evolution of UT-A2 in the cetacean kidney likely contributes significantly to enhanced urea uptake and accumulation in the renal medullary interstitium to increase osmotic pressure.

Plasma urea concentrations in cetaceans are relatively higher than in closely related terrestrial mammals Artiodactyla (ungulates), which may reflect their carnivorous nature with a higher protein diet (Birukawa et al. 2005), but even when compared to carnivorous terrestrial mammals (dog), their plasma urea level is reported to be significantly high (Suzuki et al. 2018). Compared to mice, plasma urea concentrations in cetaceans are significantly greater due to their lower urea turnover, suggesting that urea regulation may be an important adaptation for the conservation of body water (Miyaji et al. 2010). Such high urea concentrations in body fluids may help not only to increase the urine-concentrating ability of the kidneys, but also to increase osmotic pressure in the tissue interstitium to reduce the osmotic gradient with seawater, reducing the risk of solute-free water from being pulled out of the body.

Water immersion does not induce increased urination

Many of us have experienced the sensation of needing to urinate immediately after jumping into a swimming pool or the ocean (or any other body of water). The immersion of water up to your neck (where baroreceptors are located) induces diuresis (excretion of urine) by increasing cardiac output and arterial pressure, triggered by increased hydrostatic pressure that compresses peripheral arteries or by sensation of cold stimuli. These responses to water immersion ultimately cause the acute distention of arteries inducing a baro-reflex that suppresses the antidiuretic hormone, vasopressin, resulting in a diuresis (Epstein 1992). This reflex arc is known as the Henry–Gauer reflex (Gauer and Henry 1976).

In marine mammals which are chronically or frequently immersed in water, activation of this reflex arc would in theory result in exacerbated urinary water loss. This reflex may be permanently suppressed and/or nonexistent in aquatic mammals as an adaptation to conserve water. In harbor seals (*Phoca vitulina*), urine flow ceased during diving (Murdaugh et al. 1961). This rapid reduction in urine flow is probably due to a significant decrease in glomerular filtration rate (GFR), which is the rate at which a volume of blood passes through the glomerulus. It was reported that GFR decreased in freely diving seals probably because of the diving response, which is associated with a reduction in cardiac output (Davis et al. 1983). The decrease in GFR induced by immersion or diving coincides well with the decrease in blood flow to the kidneys (Zapol et al. 1979). In seals, heart rate and blood flow decrease when diving, but blood pressure is maintained (Blix et al. 1976; see **Chapter 3**). This is also true in cetaceans where it is suspected that blood pressure remains constant during diving due to a specialized vascular system developed to prevent rapid reductions in blood pressure, but heart rate decreases (Nagel et al. 1968; Mompeó et al 2020). The local shunting of blood flow to the kidneys associated with the diving response is likely a major hemodynamic change evolved by marine mammals to conserve water during immersion or diving.

THE KIDNEY AS KEY OSMOTIC REGULATOR

The primary function of the kidneys is to selectively filter electrolytes, water, and certain metabolic byproducts including blood glucose and urea from the plasma and excrete excess substances. Plasma osmolality of animals is determined by the sum of all dissolved molecules in solution (water), including electrolytes (mainly sodium), glucose, and urea (blood urea nitrogen, BUN), and can be approximated by the formula:

$$plasma\ osmolality = Na\,(mEq/L)*2$$
$$+ glucose\ (mg/dL)/18$$
$$+ BUN\ (mg/dL)/2.8.$$

In this section, the anatomy of the kidney in marine mammals is introduced in relation to its osmotic regulatory capacity.

Anatomical and histological features of the kidney

The kidneys of marine mammals are highly lobulated. In cetaceans, the lobulation is particularly prominent, with the right and left kidneys each divided into hundreds of reniculi (small kidneys: plural form of reniculus or renicule) in small cetaceans and into thousands in larger cetaceans (**Figure 8.5**). In sirenians, the lobulation is less pronounced and superficial. Each small kidney has cortical tissue and a single medullary pyramid, which is inserted into a single calyx, and the unit is bundled together within the renal capsule (**Figure 8.6**).

All mammalian kidneys consist of the cortex and medulla. Plasma is filtered through the glomeruli located in the cortex. As plasma is filtered through the multiple layers of glomerulus, the new fluid that is formed in the renal tubule (nephron) is known as the ultrafiltrate. As the

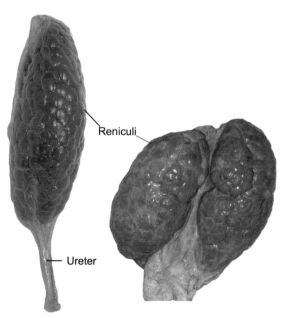

Figure 8.5 A dissected one-side kidney of spotted dolphin (*Stenella attenuate*, left) and both-side kidneys of spotted seal (*Phoca largha*, right). The renal capsule of the dolphin kidney was removed. A number of reniculi are assembled to form a single kidney. (Photo by Keiichi Ueda and Chika Shirakata)

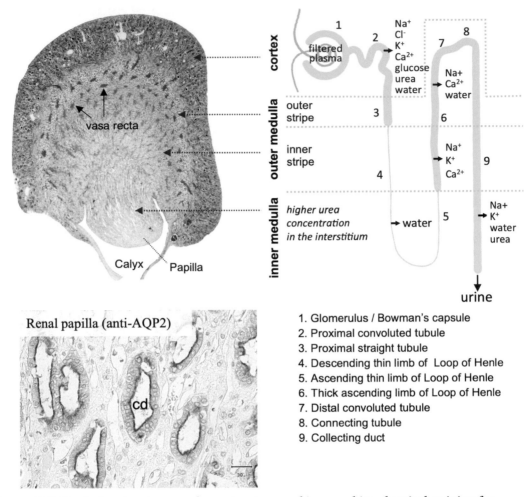

Figure 8.6 **Renicular structure, nephron structure, and immunohistochemical staining for aquaporin 2 (AQP2) in common bottlenose dolphin. The basic structure of the reniculus is identical to that of terrestrial mammal kidneys. Vasa recta, a bundle of the straight arterioles and venules, is prominent (upper left). The nephron structure is also the same as that of terrestrial mammals (upper right). Intense positive reactions for AQP2 antibody are observed at the apical membrane of the principal cells in the collecting duct (lower). cd: collecting duct.**

1. Glomerulus / Bowman's capsule
2. Proximal convoluted tubule
3. Proximal straight tubule
4. Descending thin limb of Loop of Henle
5. Ascending thin limb of Loop of Henle
6. Thick ascending limb of Loop of Henle
7. Distal convoluted tubule
8. Connecting tubule
9. Collecting duct

filtrate flows through the proximal tubule of the nephron, electrolytes and organic compounds, including urea, are passively reabsorbed. The filtrate then enters the loop of Henle, the segment of tubule that is descended into the medulla, where water is primarily absorbed by the presence of osmotic gradients. The filtrate passes through the loop of Henle and ascends up into the distal tubule in the cortical section of the kidney, where the ligand-dependent sodium and potassium channels actively transport electrolytes. The distal tubule leads to the cortical collecting ducts, where ligand-sensitive water channels are stimulated to actively reabsorb water (or inhibited if excess water needs to be excreted). Once the filtrate passes through the final segments of the collecting ducts and no further exchange of water, electrolytes, or other organic compounds occurs, the final urine is formed and flows from the renal pelvis to the bladder, from which it is ultimately excreted (**Figure 8.6**).

Renal structures related to urine-concentrating ability

The regulation of water and salt balance in the internal milieu is dependent on the physiological responses to the changes in external input and losses from the kidney. In situations where the salt load is excessive and water availability is limited, such as those in which marine mammals are placed, the urine-concentrating ability of the kidneys is important to excrete excess salt and increase water retention to maintain osmotic homeostasis. As indicated in sections above, marine mammals have excellent urine-concentrating ability. The histological factors of the kidney involved in urine concentration are not simple, and many factors are relevant such as the length of the renal papilla, number of nephrons, percentage of nephrons with longer loops of Henle, the structure of the collecting ducts, the vascular bundles in the inner medulla, and the degree of development of each medullary layer among other features (Bankir and De Rouffignac 1985). For lobulated kidneys, the loop of Henle is shorter than in unlobulated kidneys with same size, because the small kidney is the functional unit. If loops of Henle are longer (the greater the descent into the inner medulla), then greater is the urine-concentrating ability. Then why do marine mammals with reniculate kidneys possess excellent urine-concentrating ability?

The physiological significance of renal lobulation

The kidneys of most terrestrial mammals are not segmented. However, humans, which some advocate evolved at the seashore (aquatic ape hypothesis), and terrestrial mammals that consume high-salt products also have multilobed kidneys with multiple renal cones. Although not as extreme as in marine mammals, it has been speculated that the divided lobe is associated with the ability to expel salt (Williams 2006).

Compared to the unlobulated kidney, the lobulation relatively increases the surface area, i.e., the cortex, as well as the corticomedullary borders and renal papillae. The lobulation of kidneys must have been advantageous for the adaptation of animals to their environment.

One conceivable reason for increasing the renal cortical area is to increase the total mass of the renal glomeruli to increase the filtration rate of the plasma, facilitating the removal of excess salt efficiently from the body. However, Maluf and Gassman (1998) examined the morphology of glomeruli in the kidneys of the killer whale and concluded that although their number was greater than those of land mammals, the mass was smaller and the total glomerular capacity might not be different from those of land mammals. They simultaneously showed that the kidney-to-body mass ratio, excluding the renal capsule, is comparable to that of terrestrial mammals. This suggests that the kidneys are not likely lobulated for the strategy of increasing filtration to excrete excess salt (and other substances). In common bottlenose dolphins, for example, GFR measured by the inulin clearance method ranged from 131 to 465 mL/min (Malvin and Rayner 1968) and the median estimated GFR was calculated to be 188 mL/min (Venn-Watson et al. 2008). If we correct for their minimum average body mass (200 kg), the relative GFR (per 100 g body mass) is 0.07–0.23 mL/min and the median is 0.09 mL/min. These values are comparable to those of animals such as humans and pigs (0.1–0.2 mL/min per 100 g BM) (Benjamin et al. 2015), suggesting that dolphins filter equivalent volumes of blood through their lobulated kidneys as other animals.

On the other hand, the increased mass of the renal papillae may contribute to urine-concentrating ability. In general, water reabsorption from the filtrate is mainly mediated by the cell wall transport protein aquaporin 2 (AQP2) in the chief cells of the renal collecting ducts. In the kidneys of common bottlenose dolphins, the distribution of AQP2 is found to be particularly abundant in the apical membrane of the collecting ducts of the well-developed renal papilla (Suzuki et al. 2008; **Figure 8.6**). This suggests that the renal papilla is the primary site of urine concentration through active water reabsorption. As described above, cetaceans have excellent urine-concentrating ability considering their body size. Lobulation increases the volume of the renal papilla, which in turn accommodates more AQP2-expressing renal collecting ducts and increases the space for the accumulation of urea, which facilitates the reabsorption of water. From an osmoregulatory perspective, it is reasonable to conclude that marine mammal kidneys are highly lobulated to concentrate urine at a high level by increasing the thickness of the medulla, especially the renal papillae, while maintaining a GFR that is equivalent to that of terrestrial mammals (Cozzi, Huggenberger, and Oelschläger 2016).

Relative medullary thickness

The excellent urine-concentrating ability in marine mammals can be inferred, at least partially, from the thickness of the renal medulla. In mammals in general, the ratio of the mean thickness of the medulla to the cube root of the kidney volume (relative medullary thickness, RMT = 10*medullary thickness/cube root of the product of kidney length, width and thickness) increases in proportion to urine-concentrating capacity (Beuchat 1990). In other words, the thicker the medulla, the greater the urine-concentrating ability. In marine mammals, there is considerable variation in RMT. In cetaceans, RMT is relatively large, but in pinnipeds, it ranges widely from 1.1 to 7.5 (Bester 1975; Vardy and Bryden 1981; Suzuki et al. 2008). The medulla of the kidneys of West Indian manatee is reported to be six times thicker than the cortex (Hill and Reynolds III 1989). The RMT values in marine mammals suggest that they have a high urine-concentrating capacity, although the variability among pinnipeds needs further investigation to better understand the significance of the RMT to their urine-concentrating ability.

Prominent and complicated vasa recta

Surrounding the loop of Henle is a convoluted capillary network, the vasa recta. As blood flows into this capillary rete from arteries and out through veins, it forms a countercurrent exchange system in which blood flows slowly and exchanges substances with the loop of Henle, which is believed to maintain a concentration gradient of substances in the renal medulla and contribute to urine concentration (Pallone et al. 2003). In cetaceans, vasa recta are large and conspicuous (**Figure 8.7**). Although the efficiency of countercurrent exchange is thought to be reduced in reticulated straight arterioles such as in cetaceans (Pfeiffer 1997), interestingly, similar complex and prominent vasa recta development is observed in desert rat,

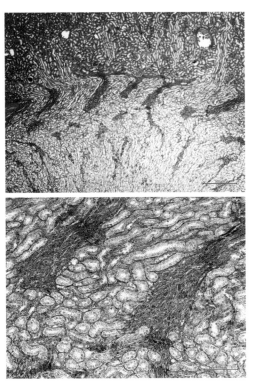

Figure 8.7 Well-developed *vasa recta* in the kidney of common bottlenose dolphin. It is prominently observed in the outer medulla (top). The *rete*-like structure can be seen especially in the outer medulla (bottom).

which has a greater urine-concentrating ability (Bankir et al. 1979). This suggests that the remarkable development of the vascular bundles may be related to the improvement of urine concentration, but further studies are needed.

Kidney function during fasting

As mentioned previously, many species of pinnipeds and cetaceans fast as part of their life history s, relying on metabolic water obtained by through energy production using primarily their vast stores of fat and to some extent muscle, respectively. During fasting, renal hemodynamics and the reabsorption of electrolytes and free water from urine are also altered to maintain fluid and electrolyte homeostasis. See **Chapters 8** and **11**.

The northern elephant seal, which has a prolonged fasting period after weaning and during molting, mating, and lactation, is an excellent model to study fasting physiology. Fasting for more than 2.5 months in weaned northern elephant seal pups reduces renal nitrogen (primarily urea) load and GFR, resulting in reduced urine output and enhanced water conservation (Adams and Costa 1993). Lactating/fasting female elephant seals increase protein catabolism to produce calorically dense milk, which in turn increases GFR to discard the unnecessary products, but also increases urea reabsorption to concentrate urine and prevent urinary water loss (Crocker et al. 1998). Coupled with reduced evaporative water loss from the respiratory system (Lester and Costa 2006), these changes in kidney function contribute to the conservation of body water during fasting. In addition, they can maintain plasma electrolyte concentrations even after nearly 3 months of fasting, which is likely achieved by increasing renal reabsorption of Na^+ and K^+ in exchange for H^+ excretion (Ortiz et al. 2000).

For logistical reasons, controlled studies of renal function in larger cetaceans are impossible; however, valuable observations of urinary concentrations have provided reasonable assumptions regarding their renal function. Unique measurements made in fasted humpback and Bryde's whales suggested that fat and/or protein catabolism and urine concentrations were increased without any signs of dehydration (Bentley 1963; Priddel and Wheeler 1998). More robust and controlled studies have been performed in bottlenose dolphins. During fasting, GFR decreased, leading to decreased urinary excretion of electrolytes and water, which likely contributed to the maintenance of osmotic homeostasis (Venn-Watson et al. 2008; Ortiz et al. 2010). Further elaborate renal studies measured the renal capabilities of bottlenose dolphins during varying and well-controlled acute fasting and osmotic manipulations (Ridgway and Venn-Watson 2010). These unique studies showed that the dolphin's kidneys appropriately respond to exogenous manipulations to maintain electrolyte homeostasis. The results provided insights to the renal mechanisms induced to support the animal's needs as they transition in and out of waters of varying salinities and/or during periods of reduced food availability.

ENDOCRINE CONTROL OF PLASMA OSMOLALITY

Hormones are intrinsic factors that regulate signaling pathways at the cellular level to help maintain homeostasis of water and electrolyte balance. There are many hormones involved in osmoregulation. Perhaps because water and ion balance are closely related to blood pressure, a suite of vasoactive hormones, such as arginine vasopressin (AVP), angiotensin II (Ang II), atrial natriuretic peptide (ANP), vasoactive intestinal peptide, and urotensin II, contribute to the acute (minutes-hours) perturbations in water and electrolyte balance, as presented in **Figure 8.8**. On the other hand, steroids (aldosterone and cortisol), prolactin, and growth hormone are more involved in longer-term water and electrolyte responses. In marine mammals, studies have focused more on the acute responses of the hormones to hyperosmotic water ingestion or fasting than on the chronic effects.

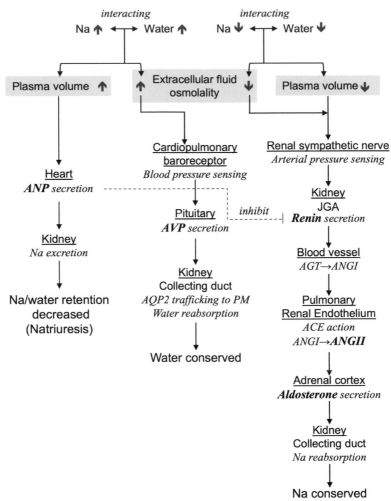

Figure 8.8 General schematics of the functions of osmoregulatory hormones. ACE: angiotensin-I-converting enzyme, AGT: angiotensinogen, ANG: angiotensin, AVP: arginine vasopressin, JGA: juxta-glomerular apparatus in the kidney, ANP: atrial natriuretic peptide.

Natriuretic peptides

Natriuretic peptides, as the name implies, are hormones that stimulate the urinary excretion of Na^+ to protect against excessive salt and water retention. The peptides induce diuresis (increased urination) and natriuresis (sodium excretion in the urine) through regulation of renal hemodynamic and direct tubular actions (Levin, Gardner, and Samson 1998). The peptide family consists of atrial natriuretic peptide (ANP), brain natriuretic peptide (BNP), and C-type natriuretic peptide (CNP). ANP is secreted mainly from the atrial wall in response

to increased cardiac pressure and distention, and it promotes natriuresis by increasing renal filtration, inhibiting Na^+ absorption by suppressing renin and aldosterone release, and antagonizing the actions of Ang II (**Figure 8.8**).

While ANP may be a highly functional and relevant hormone in marine mammals because of the abundance of Na^+ in their environment, only a few reports of changes in natriuretic peptides and their potential osmoregulatory function exist. Precursors of ANP and BNP were cloned in three species of dolphins and the peptides were also detected from the plasma of bottlenose dolphins (Naka et al. 2007). Plasma ANP concentrations were also determined in six pinniped species (Zenteno-Savin and Castellini 1998). In fasting northern elephant seal pups, urinary ANP excretion more than doubled in response to a hyperosmotic saline infusion, but levels were not significantly increased with an iso-osmotic infusion. In addition, the hyperosmotic infusion was associated with a sustained elevation in plasma Na$^+$ despite the natriuresis suggesting that ANP may function to promote natriuresis in pinnipeds during conditions of increased Na$^+$ intake, but its potential to restore osmotic homeostasis may take a little than the duration of the study (Ortiz et al. 2002). Further studies to ascertain the functional relevance of ANP and the other natriuretic peptides in marine mammals are warranted.

Arginine vasopressin

Primarily in response to lack of fresh water, thirst stimulates a series of neural and hemodynamic responses that ultimately concentrate urine to prevent water loss. In mammals in general, the volume of water excreted is primarily regulated by the octapeptide, arginine vasopressin (AVP) (formerly antidiuretic hormone, ADH) (**Figure 8.9**). AVP is secreted from the posterior pituitary gland by stimulation of the osmoreceptor or baroreceptors primarily in response to increases in plasma osmolality or plasma volume, respectively (Robertson, Shelton, and Athar 1976). When AVP binds to its receptor (V2R) on the principal cell of the collecting duct in the kidney, the water channel, AQP2, is transferred and inserted into the apical membrane of chief cells that line the renal

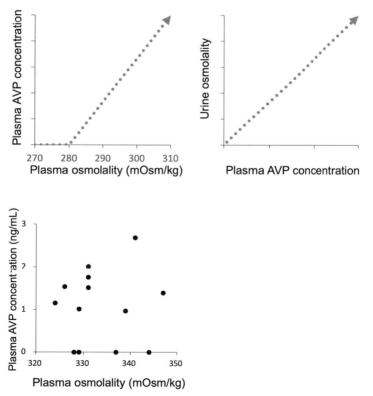

Figure 8.9 Relationship between plasma arginine vasopressin (AVP) concentration and plasma/urine osmolality. Generally, plasma AVP concentration is correlated with plasma and urine osmolality in proportional (top), but the relationship is not observed in common bottlenose dolphin (bottom; Suzuki Unpublished data).

tubule and contact with filtrate. AQP distribution on the membrane facilitates the reabsorption of water from the filtrate along an osmotic gradient formed by urea and sodium accumulated in the renal interstitium. AVP is presumably important for marine mammals who cannot access freshwater; however, the traditional function of AVP in these animals remains controversial and not well-defined.

Administration of synthetic AVP in harbor seals induced typical responses leading to water conservation including decreased urine flow and increased urinary electrolyte concentrations, urine osmolality, and osmotic clearance (Bradley, Mudge and Blake 1954). In studies on several species of seals, fasting increased plasma AVP concentrations and urine osmolality, which are positively correlated with each other, and decreased urine flow (Hong et al. 1982). In West Indian manatees, a significant correlation between plasma vasopressin and osmolality was detected further suggesting that AVP may function as an antidiuretic, as in terrestrial mammals, in these marine mammals (Ortiz, Worthy, and MacKenzie 1998). However, the dynamic relationships among AVP, plasma osmolality, and urine osmolality are not readily observed and less consistent in some species of marine mammals. For example, in cetaceans, circulating concentrations of AVP were relatively low and often fail to be directly related to plasma osmolality (Malvin, Bonjour, and Ridgway 1971; see **Figure 8.9**). While a postmeal increase in plasma NaCl levels were associated with an increase in urinary AVP levels suggesting that AVP secretion at least is regulated in a typical manner (Ballarin et al. 2011), the antidiuretic function of AVP is still inconclusive in cetaceans. Studies on fasting elephant seal pups suggest that static, chronic measures of plasma AVP and osmolality are not sufficient indicators of AVP function and that acute challenges like iso- and hyperosmotic saline administration may better elucidate the functionality of AVP in marine mammals (Ortiz et al. 2002).

The association between water retention and AVP in marine mammals exposed to a constant salt load warrants more thorough study.

Renin–angiotensin–aldosterone system

The renin–angiotensin–aldosterone system (RAAS) is important for the control of sodium excretion and blood pressure. Renin, the rate-limiting step in this system, is an enzyme secreted by specialized cells in the kidney in response to decreasing plasma Na^+ concentration and blood pressure. Renin is secreted from renal juxtaglomerular apparatus (JGA) when changes in renal arterial pressure and reduced tubular Na^+ stimulate renal sympathetic nerves (**Figure 8.8**). Conversely, renin secretion is suppressed by an increase in tubular Na^+ delivery to the JGA. Renin converts circulating angiotensinogen (secreted by the liver) to Ang I, which is subsequently converted to Ang II by angiotensin-converting enzyme (ACE), which is secreted by the lungs. In turn, Ang II stimulates the secretion of aldosterone from adrenal gland and aldosterone then stimulates the renal reabsorption of Na^+ from the distal tubule resulting in a decrease in urinary Na^+ excretion (Eaton and Pooler 2009). In marine mammals including cetaceans, pinnipeds, and sirenians, RAAS likely possesses a typical mammalian function as described above for terrestrial mammals. Because manatees that inhabit freshwater have reduced availability to Na^+, manatees are more sensitive to RAAS than pinnipeds and cetaceans (Ortiz 2001).

Positive selection of osmoregulation-related genes in marine mammals

Recent remarkable innovations in the field of genomics have allowed us to model what osmoregulatory genes may have evolved in marine mammals. As mentioned above, four osmoregulation-related genes including AQP2, SLC14A2 (urea transporter 2), ACE, and AGT were positively selected during evolution in cetaceans, suggesting that cetaceans may have adapted to enhance urine-concentrating ability and RAAS to balance water and electrolytes (Xu et al. 2013). Wang et al. (2015) analyzed genomic information of several species of cetaceans and land mammals and also came to the same conclusion, in addition to enhancements of UT-A2 function in cetaceans as noted in the previous section. The proteins translated from these genes may complement the actions of UT-A2 to facilitate the urine-concentrating ability in these animals. These findings suggest that improved osmotic regulation ability through endocrine systems may have undergirded their adaptive radiation to the ocean.

TOOLBOX

As described above, the osmolality of body fluid is determined by the balance between water and osmotic substances like ions and urea. The amount of water and osmotic substances is regulated primarily by the kidneys but other organs including the brain, liver, heart, intestines, skin, and lungs also contribute to osmoregulation.

Blood and urine analyses

A thorough analysis of plasma and urine is essential for the appropriate study of body fluid regulation. Blood and urine samples are collected from animals to determine the osmotic pressure of body fluids, their constituents, and the amount of water or osmotic substances excreted. Generally, osmotic pressure can be measured using the freezing point depression method, which is based on the principle that the more substances dissolved in a liquid, the lower the temperature at which it becomes a solid. Ions can be measured by ion-selective electrode method, enzyme method, or atomic absorption method. Blood and urine urea concentrations can be measured by enzyme-based methods, which detect the ammonia nitrogen that makes up urea. Systemic regulation is often mediated by

hormones such as Ang II, aldosterone, ANP, and vasopressin, all of which can be easily measured by antibody-based reactions. Recent advances in measurement technology have been remarkable and provide untargeted, large platform analyses of primary carbon metabolites in an approach known as metabolomics. Metabolomic approaches allow for the simultaneous analysis of hundreds to thousands of metabolites, making it possible to measure many molecules at once, and have recently been used extensively in research on marine mammals. These "-omic" approaches can be used to extensively study genes (transcriptomics) and proteins (proteomics) as well. In any case, efficient and rapid collection, processing, and storage of samples in an appropriate manner is important for generating robust data. No matter how excellent and advanced the measurement techniques are, if the samples have deteriorated or compromised, it is very challenging to obtain appropriate data. It is not easy to properly process samples from marine mammals as collecting them at the poolside of an aquarium or in the field can be very challenging, but every effort should be made to do so.

Tissues and cells

To understand the characteristics and functions of each organ involved in osmotic regulation, it is necessary to obtain organs from healthy individuals and examine their tissue structures and the expression and distribution of molecules that function in them. However, there are little opportunities to obtain various organs from marine mammals because of various regulations to access marine mammals. While there is a fair amount of basic information on the tissue structure of each organ of marine mammals, data on molecules that function in each tissue in relation to osmotic pressure are quite limited. In model organisms such as rodents, it is common practice to elucidate the function of target molecules by demonstrating in which cells those are distributed, under what conditions their abundance and behavior change, and what happens when they are knocked out by genetic modifications. In marine mammals, however, only a small fraction of tissues, such as skin, can be collected in vivo while maintaining a low level of invasiveness to the individual. It is desirable to establish a variety of cell lines to enable genetic manipulation, but currently only a limited number of cell lines exist.

Isotopes

Isotopic dilution technique with isotopes of water ($2H^{216}O$) and sodium ($^{24}Na^+$), for example, can be used to determine how much water and sodium an animal stores in its body and how much those are ingested, utilized, and disposed of each day. These labeled molecules differ slightly in mass from normal molecules, and some are radioactive so their concentration can be determined. If a known amount of heavy water is injected directly into a blood vessel, and once that known amount of isotope equilibrates with the body water pool, the isotopes concentration measured from a blood sample can be used to calculate the TBW pool of the animal using already validated and established equations. If the rate of depletion of heavy water from the blood within a certain time can be determined, the amount of water replaced in the body fluid (turnover rate) can be calculated from that rate. It should be noted that the amount of injected isotope must be measured very accurately. Isotopes are present in small amounts in the environment and must be subtracted as a background level in a blood or urine sample prior to administering isotope. In the case of heavy water, its level can be measured by infrared spectrometry and in the case of radio-labeled water (^3H_2O), it is measured by scintillation. In addition, several atoms can be measured simultaneously by nuclear magnetic resonance (NMR) spectral analysis.

Genes

As described in the previous sections, genomic studies have advanced our understanding of the evolution of physiological mechanisms that regulate water and electrolytes within the cetacean lineage. Genome-wide analyses provide the opportunities to assess if genes are positively or negatively selected, and it has the advantage of using only a small amount of sample for analysis. Comprehensive and panoramic investigations by genome-wide analysis provide a wealth of information. It should be noted, however, that inferences about physiological mechanisms based on genomic findings need to be confirmed by functional studies with appropriate caution.

FUTURE RESEARCH

Despite their relatively larger size across taxa, marine mammals remain very visible subjects for the study of osmoregulation. Every marine mammal taxon includes species that inhabit not only seawater but freshwater also or transition between bodies of water of different salinities, making this group of mammals a unique and highly intriguing subject for the study of osmoregulation. Furthermore, various groups of researchers have been successful in using marine mammals for the study of issues that impact human health such as cardiovascular complications, diabetes, and obesity, each of which are conditions that impair kidney function. From the perspective of the kidney, most marine mammals inhabit strictly marine (or hyperosmotic) environments, and while active seawater consumption (mariposia) is not common in most marine mammals, those that practice this behavior are not known to suffer the consequences induced by frequent bouts of high salt intake such as hypertension. Elucidating the mechanisms of water and sodium retention in marine mammals should provide many clues to ameliorating diseases caused by excessive salinity. Relatively long breath-hold periods (apnea) associated with marine mammal exceptional diving abilities are associated with periods of ischemia followed by reperfusion-reoxygenation at rebreathing. These intermittent bouts of ischemia-reperfusion-reoxygenation are similar to events in humans with cardiovascular, renal, and sleep disorders as well as transplantation of organs and have their own consequences that do not appear to be prevalent in prolong-diving marine mammals. In addition, most species naturally experience prolonged bouts of absolute food deprivation, which could increase muscle wasting (cachexia) and subsequently increase the nitrogen load on the kidneys, which has further consequences. However, marine mammals appear to have evolved robust physiological mechanisms that allow them to tolerate extreme conditions with little or no consequences making them very unique and intriguing models for many conditions. Independent of studies with applications toward human health, many questions remain on the renal mechanisms evolved to tolerate a strictly marine environments as well as the mechanisms that provide the flexibility to accommodate transitions to habitats of different salinities.

Their primarily aquatic habitats, size, and protection by federal and international laws pose challenges and limitations to the research using marine mammals. For example, researchers are often limited to the type (blood, urine, and superficial biopsies) and amount of sample that can be legally obtained for research. Regardless of the research questions (basic biological or modeling human health conditions), with some creativity and the advances in research technologies, many of the limitations inherent with working with marine mammals can be managed to provide profound discoveries especially in the area of kidney function.

The physiological responses observed in marine mammals seem to be quite consistent with reactions or phenomena that are ubiquitous in animals. There remain many unresolved and unexamined basic biological questions around the exceptional renal capabilities of this group of animals that warrant further investigation.

EQUATIONS

$$\text{plasma osmolality} = Na\,(mEq/L)*2$$
$$+ glucose\,(mg/dL)/18$$
$$+ BUN\,(mg/dL)/2.8$$

$$RMT = 10 * \text{medullary thickness/cube root}$$
$$\text{of the product of kidney length,}$$
$$\text{width and thickness}$$

GLOSSARY

Aldosterone: A steroid hormone secreted by the adrenal cortex that primarily regulates the reabsorption of Na^+ in the kidney and colon.

Angiotensin II (Ang II): An eight-amino-acid peptide cleaved from angiotensin I by angiotensin-converting enzyme that primarily regulates blood pressure and the adrenal secretion of aldosterone.

Angiotensinogen: An alpha-2 globulin produced by the liver and serves as the precursor protein for the formation of angiotensin I.

Arginine vasopressin: A nine-amino-acid peptide secreted by neurohypophysis (posterior pituitary) that primarily regulates the reabsorption of free water in the collecting duct of the renal tubule in mammals.

Atrial natriuretic peptide (ANP): A 28-amino-acid peptide produced by the atrium that functions primarily to induce urinary Na^+ excretion (natriuresis).

Baroreceptor: A type of mechanoreceptor that can relay information from blood pressure to the autonomic nervous system.

Countercurrent heat exchange system: A design in which fluids at sufficiently different temperatures flow closely in opposite directions to facilitate heat exchange down the temperature gradient.

Concentration gradient: A gradient generated by differences in solute concentrations across a membrane that separates two compartments. Depending on the permeability of the separating membrane, water moves from the compartment with the lower concentration of dissolved particles to the compartment with the greater concentration.

Diuresis: Increased urine output.

Electrolytes: Elements that are decomposed into ions, which carry a charge (+ or −) when dissolved in water or body fluids.

Glomerular filtration rate (GFR): The rate at which a volume of blood passes through the glomeruli.

Homeostasis: A tendency toward equilibrium between interdependent variables, especially those maintained by physiological processes.

Horny layer: The outermost layer of the epidermis formed by the shedding of dead cells.

Hyperosmotic: Increased osmotic content of a solution.

Hypertonic: Having a greater osmotic pressure or tonicity in a fluid compared to another fluid.

Isotopic dilution: A technique to study the kinetics of an isotopically labeled molecule that emulates the endogenous, unlabeled molecule. The rate at which an isotope is diluted in a compartment estimates the turnover of the endogenous, unlabeled molecule.

Ligand: A molecule that binds to another molecule, usually larger than itself.

Mariposia: The voluntary and deliberate consumption of seawater.

Metabolic water: The water produced during substrate-level metabolism.

Natriuresis: Urinary sodium excretion.

Nephron: The functional unit of the kidney. A convoluted, tube-like structure that produces urine in the process of removing waste and excess substances from the blood.

Osmoregulation: The potential to maintain a constant osmotic pressure in the body fluids of an organism by controlling the concentration of water and salt.

Osmosis: The movement of water across a membrane.

Osmotic pressure: The pressure produced by the concentration of dissolved particles in solution.

Preformed water: The water that occurs naturally in food, for example.

Relative medullary thickness: The ratio of the mean thickness of the medulla to the cube root of the kidney volume.

Reniculus: Refers to each lobe of the multilobed kidney, which acts as an independent renal functional unit.

Total body water (TBW): Total amount of water present in the body of an animal, accounting for all compartments of the body including intracellular space and extracellular space (blood and intercellular space).

REFERENCES

Adams, S.H., and D.P. Costa. 1993. Water conservation and protein metabolism in northern elephant seal pups during the postweaning fast. *Journal of Comparative Physiology B* 163(5): 367–373.

Andersen, S.H., and E. Nielsen. 1983. Exchange of water between the harbor porpoise, *Phocoena phocoena*, and the environment. *Experientia* 39(1): 52–53.

Ballarin, C., L. Corain, A. Peruffo, et al. 2011. Correlation between urinary vasopressin and water content of food in the bottlenose dolphin (*Tursiops truncatus*). *The Open Neuroendocrinology Journal* 4(1): 9–14.

Bankir, L., and C. De Rouffignac. 1985. Urinary concentrating ability: Insights from comparative anatomy. *American Journal of Physiology-Regulatory, Integrative and Comparative Physiology* 249(6): R643–R666.

Bankir, L., B. Kaissling, C. de Rouffignac, et al. 1979. The vascular organization of the kidney of *Psammomys obesus*. *Anatomy and Embryology* 155(2): 149–160.

Benjamin, A., D.J. Gallacher, A. Greiter-Wilke, et al. 2015. Renal studies in safety pharmacology and toxicology: A survey conducted in the top 15 pharmaceutical companies. *Journal of*

Pharmacological and Toxicological Methods 75: 101–110.

Bentley, P.J. 1963. Composition of the urine of the fasting humpback whale (*Megaptera nodosa*). *Comparative Biochemistry and Physiology* 10(3): 257–259.

Bester, M.N. 1975. The functional morphology of the kidney of the Cape fur seal, *Arctocephalus pusillus* (Schreber). *Madoqua* 1975(4): 69–92.

Beuchat, C.A. 1990. Body size, medullary thickness, and urine concentrating ability in mammals. *American Journal of Physiology-Regulatory, Integrative and Comparative Physiology* 258(2): R298–R308.

Birukawa, N., H. Ando, M. Goto, et al. 2005. Plasma and urine levels of electrolytes, urea and steroid hormones involved in osmoregulation of cetaceans. *Zoological Science* 22(11): 1245–1257.

Blix, A.S., J.K. Kjekshus, I. Enge, et al. 1976. Myocardial blood flow in the diving seal. *Acta Physiologica Scandinavica* 96(2): 277–280.

Bradley, S.E., G.H. Mudge, and W.D. Blake. 1954. The renal excretion of sodium, potassium, and water by the harbor seal (*Phoca vitulina* L.): Effect of apnea; sodium, potassium, and water loading; pitressin; and mercurial diuresis. *Journal of Cellular and Comparative Physiology* 43(1): 1–22.

Costa, D.P. 1982. Energy, nitrogen, and electrolyte flux and sea water drinking in the sea otter *Enhydra lutris*. *Physiological Zoology* 55(1): 35–44.

Costa, D.P., and F. Trillmich. 1988. Mass changes and metabolism during the perinatal fast: A comparison between Antarctic (*Arctocephalus gazella*) and Galapagos fur seals (*Arctocephalus galapagoensis*). *Physiological Zoology* 61(2): 160–169.

Coulombe, H.N., S.H. Ridgway, and W.E. Evans. 1965. Respiratory water exchange in two species of porpoise. *Science* 149(3679): 86–88.

Cozzi, B., S. Huggenberger, and H. Oelschläger. 2016. Natural history and evolution of dolphins: short history of dolphin anatomy. *Anatomy of dolphins: Insights into body structure and function.* 1st ed. Amsterdam: Elsevier, pp. 1–20.

Crocker, D.E., P.M. Webb, D.P. Costa, et al. 1998. Protein catabolism and renal function in lactating northern elephant seals. *Physiological Zoology* 71(5): 485–491.

Davis, R.W., M.A. Castellini, G.L. Kooyman, et al. 1983. Renal glomerular filtration rate and hepatic blood flow during voluntary diving in Weddell seals. *American Journal of Physiology-Regulatory, Integrative and Comparative Physiology* 245(5): R743–R748.

Depocas, F., J. Hart, and H.D. Fisher. 1971. Seawater drinking and water flux in starved and Fed Harbor seals, *Phoca vitulina. Canadian Journal of Physiology and Pharmacology* 49(1): 53–62.

Eaton, D.C. and J. P. Pooler. 2009. *Vander's renal physiology*. 7th Edition. New York: McGraw-Hill Medical.

Epstein, M. 1992. Renal effects of head-out water immersion in humans: A 15-year update. *Physiological Reviews* 72(3): 563–621.

Gallardo, P., L.P. Cid, C.P. Vio, et al. 2001. Aquaporin-2, a regulated water channel, is expressed in apical membranes of rat distal colon epithelium. *American Journal of Physiology-Gastrointestinal and Liver Physiology* 281(3): G856–G863.

Gauer, O.H., and J.P. Henry. 1976. Neurohormonal control of plasma volume. *Cardiovascular Physiology II* 9:145–190.

Hill, A., and J.E. Reynolds III. 1989. Gross and microscopic anatomy of the kidney of the West Indian manatee, Trichechus manatus (*Mammalia: Sirenia*). *Cells Tissues Organs* 135(1): 53–56.

Hong, S.K., R. Elsner, J.R. Claybaugh, et al. 1982. Renal functions of the Baikal seal *Pusa sibirica* and ringed seal *Pusa hispida. Physiological Zoology* 55(3): 289–299.

Hui, C.A. 1981. Seawater consumption and water flux in the common dolphin *Delphinus delphis. Physiological Zoology* 54(4): 430–440.

Huntley, A.C., D.P. Costa, and R.D. Rubin. 1984. The contribution of nasal countercurrent heat exchange to water balance in the northern elephant seal, *Mirounga angustirostris. Journal of Experimental Biology* 113(1): 447–454.

Jéquier, E., and F. Constant. 2010. Water as an essential nutrient: The physiological basis of hydration. *European Journal of Clinical Nutrition* 64(2): 115–123.

Kitada, K., S. Daub, Y. Zhang, et al. 2017. High salt intake reprioritizes osmolyte and energy metabolism for body fluid conservation. *The Journal of Clinical Investigation* 127(5): 1944–1959.

Kotchen, T.A., A.W. Cowley Jr, and E.D. Frohlich. 2013. Salt in health and disease: A delicate balance. *New England Journal of Medicine* 368(13): 1229–1237.

Lanyon, J.M., K. Newgrain, and T.S.S Alli. 2006. Estimation of water turnover rate in captive dugongs (*Dugong dugon*). *Aquatic Mammals* 32(1): 103–108.

Lester, C.W., and D.P. Costa. 2006. Water conservation in fasting northern elephant seals (*Mirounga angustirostris*). *Journal of Experimental Biology* 209(21): 4283–4294.

Levin, E.R., D.G. Gardner, and W.K. Samson. 1998. Natriuretic peptides. *New England Journal of Medicine* 339(5): 321–328.

Loo, D.D., E.M. Wright, and T. Zeuthen. 2002. Water pumps. *The Journal of Physiology* 542(1): 53–60.

Maluf, N.S.R., and J.J. Gassman. 1998. Kidneys of the killer whale and significance of reniculism. *Anatomical Record* 250(1): 34–44.

Malvin, R.L., J.P. Bonjour, and S.H. Ridgway. 1971. Antidiuretic hormone levels in some cetaceans. *Proceedings of the Society for Experimental Biology and Medicine* 136(4): 1203–1205.

Malvin, R.L., and M. Rayner. 1968. Renal function and blood chemistry in Cetacea. *American Journal of Physiology-Legacy Content* 214(1): 187–191.

Menon, G.K., S. Grayson, B.E. Brown, and P.M. Elias. 1986. Lipokeratinocytes of the epidermis of a cetacean (*Phocena phocena*). *Cell and Tissue Research* 244(2): 385–394.

Miyaji, K., K. Nagao, M. Bannai, et al. 2010. Characteristic metabolism of free amino acids in cetacean plasma: Cluster analysis and comparison with mice. *PloS One* 5(11): e13808.

Mompeó, B., L. Pérez, A. Fernández, et al. 2020. Morphological structure of the aortic wall in three Delphinid species with shallow or intermediate diving habits: Evidence for diving adaptation. *Journal of Morphology* 281(3): 377–387.

Murdaugh Jr, H.V., B. Schmidt-Nielsen, J.W. Wood, et al. 1961. Cessation of renal function during diving in trained seal (*Phoca vitulina*). *Journal of Cellular and Comparative Physiology* 58(3): 261–265.

Nagel, E.L., P.J. Morgane, W.L. McFarland, et al. 1968. Rete mirabile of dolphin: Its pressure damping effect on cerebral circulation. *Science* 161(3844): 898–900.

Naka, T., E. Katsumata, K. Sasaki, et al. 2007. Natriuretic peptides in cetaceans: Identification, molecular characterization and changes in plasma concentration after landing. *Zoological Science*, 24(6): 577–587.

Nordøy, E.S., D.E. Stijfhoorn, A. Råheim, et al. 1992. Water flux and early signs of entrance into phase 111 of fasting in grey seal pups. *Acta Physiologica Scandinavica* 144(4): 477–482.

Nordøy, F.A.I.E.S. 2020. Evidence of seawater drinking in fasting subadult hooded seals (*Cystophora cristata*). *Journal of Animal Behaviour and Biometeorology* 7(2): 52–59.

Ortiz, R.M., G.A.J. Worthy, and D.S. MacKenzie. 1998. Osmoregulation in wild and captive West Indian manatees (*Trichechus manatus*). *Physiological Zoology* 71(4): 449–457.

Ortiz, R.M., G.A. Worthy, and F.M. Byers. 1999. Estimation of water turnover rates of captive West Indian manatees (*Trichechus manatus*) held in fresh and salt water. *Journal of Experimental Biology* 202(1): 33–38.

Ortiz, R.M., C.E. Wade, and C.L. Ortiz. 2000. Prolonged fasting increases the response of the renin–angiotensin–aldosterone system, but not vasopressin levels, in postweaned northern elephant seal pups. *General and Comparative Endocrinology* 119(2): 217–223.

Ortiz, R.M. 2001. Osmoregulation in marine mammals. *Journal of Experimental Biology* 204(11): 1831–1844.

Ortiz, R., C. Wade, D. Costa, et al. 2002. Renal effects of fresh water-induced hypo-osmolality in a marine adapted seal. *Journal of Comparative Physiology B* 172(4): 297–307.

Ortiz, R.M., B. Long, D. Casper, et al. 2010. Biochemical and hormonal changes during acute fasting and re-feeding in bottlenose dolphins (*Tursiops truncatus*). *Marine Mammal Science* 26(2): 409–419.

Pallone, T.L., M.R. Turner, A. Edwards, et al. 2003. Countercurrent exchange in the renal medulla. *American Journal of Physiology-Regulatory, Integrative and Comparative Physiology* 284(5): R1153–R1175.

Pfeiffer, C.J. 1997. Renal cellular and tissue specializations in the bottlenose dolphin (*Tursiops truncatus*) and beluga whale (*Delphinapterus leucas*). *Aquatic Mammals* 23: 75–84.

Priddel, D., and R. Wheeler. 1998. Hematology and blood chemistry of a Bryde's whale, *Balaenoptera edeni*, entrapped in the Manning River, New South Wales, Australia. *Marine Mammal Science* 14(1): 72–81.

Reich, K. J., and G.A.J. Worthy. 2006. An isotopic assessment of the feeding habits of free-ranging manatees. *Marine Ecology Progress Series* 322: 303–309.

Ridgway, S., and S. Venn-Watson. 2010. Effects of fresh and seawater ingestion on osmoregulation in Atlantic bottlenose dolphins (*Tursiops truncatus*). *Journal of Comparative Physiology B* 180(4): 563–576.

Ridgway, S.H. 1972. Homeostasis in the aquatic environment. *Mammals of the sea*. Springfield. IL: Charles C. Thoma. pp. 590–747.

Robertson, G.L, R.L. Shelton, and S. Athar. 1976. The osmoregulation of vasopressin. *Kidney International* 10(1): 25–37.

Smoll, L.I., L.A. Beard, and J.M. Lanyon. 2020. Osmoregulation and electrolyte balance in a fully marine mammal, the dugong (*Dugong dugon*). *Journal of Comparative Physiology B* 190(1): 139–148.

Stevens, C.E, and I.D. Hume. 2004. *Comparative physiology of the vertebrate digestive system.* Cambridge, UK: Cambridge University Press

Suzuki, M. 2010. Expression and localization of aquaporin-1 on the apical membrane of enterocytes in the small intestine of bottlenose dolphins. *Journal of Comparative Physiology B* 180(2): 229–238.

Suzuki, M., N. Endo, Y. Nakano, et al. 2008. Localization of aquaporin-2, renal morphology and urine composition in the bottlenose dolphin and the Baird's beaked whale. *Journal of Comparative Physiology B* 178(2): 149–156.

Suzuki, M., M. Yoshioka, Y. Ohno, et al. 2018. Plasma metabolomic analysis in mature female common bottlenose dolphins: Profiling the characteristics of metabolites after overnight fasting by comparison with data in beagle dogs. *Scientific Reports* 8(1): 1–11.

Telfer, N., L.H. Cornell, and J.H. Prescott. 1970. Do dolphins drink water? *Journal of the American Veterinary Medical Association* 157(5): 555–558.

Vardy, P.H., and M.M. Bryden. 1981. The kidney of *Leptonychotes weddellii* (Pinnipedia: Phocidae) with some observations on the kidneys of two other southern phocid seals. *Journal of Morphology* 167(1): 13–34.

Venn-Watson, S., C.R. Smith, C. Dold, et al. 2008. Use of a serum-based glomerular filtration rate prediction equation to assess renal function by age, sex, fasting, and health status in bottlenose dolphins (*Tursiops truncatus*). *Marine Mammal Science* 24(1): 71–80.

Wang, J., X. Yu, B. Hu, et al. 2015. Physicochemical evolution and molecular adaptation of the cetacean osmoregulation-related gene UT-A2 and implications for functional studies. *Scientific Reports* 5(1): 1–17.

Williams, M.F. 2006. Morphological evidence of marine adaptations in human kidneys. *Medical Hypotheses* 66(2): 247–257.

Willmer, P., G. Stone, and I. Johnston. 2009. *Environmental physiology of animals.* Malden: Wiley-Blackwell..

Worthy, G.A.J., and D.M. Lavigne. 1987. Mass loss, metabolic rate, and energy utilization by harp and gray seal pups during the postweaning fast. *Physiological Zoology* 60(3): 352–364.

Xu, S., Y. Yang, X. Zhou, et al. 2013. Adaptive evolution of the osmoregulation-related genes in cetaceans during secondary aquatic adaptation. *BMC Evolutionary Biology* 13(1): 1–9.

Zapol, W.M., G.C. Liggins, R.C. Schneider, et al. 1979. Regional blood flow during simulated diving in the conscious Weddell seal. *Journal of Applied Physiology* 47(5): 968–973.

Zenteno-Savin, T., and M.A. Castellini. 1998. Plasma angiotensin II, arginine vasopressin and atrial natriuretic peptide in free ranging and captive seals and sea lions. *Comparative Biochemistry and Physiology Part C: Pharmacology, Toxicology and Endocrinology* 119(1): 1–6.

David Rosen
University of British Columbia, Vancouver, BC

Lisa Hoopes
Georgia Aquarium, Atlanta, GA

David Rosen

David Rosen is an Assistant Professor at the Institute for the Oceans and Fisheries at the University of British Columbia, Canada. He has been studying the physiology, energetics, nutrition, and behavior of marine mammals for more than 30 years. While he first chose to study marine mammals because "they are so cool," his main research focus was to contribute to science-based conservation strategies for threatened marine mammal populations. As Director of the Marine Mammal Energetics and Nutrition Laboratory, one of his main tools is to conduct studies with marine mammals held temporarily or permanently under human care. As an offshoot, he has also become involved in legislative and practical efforts to ensure the welfare of marine mammals in zoos and aquariums.

Lisa Hoopes

Dr. Lisa Hoopes is the Director of Research & Conservation at Georgia Aquarium (Atlanta, GA, USA) where she leads a team of scientists who conduct research globally with the aim to better understand our oceans and the animals that inhabit them. Lisa completed her B.S. in Zoology from Michigan State University, and M.S. and Ph.D. in Wildlife and Fisheries Science from Texas A&M University. She is a physiological ecologist by training, and her research over the past 20 years has focused on understanding feeding ecology, comparative and applied nutrition, and bioenergetics of aquatic species. She has worked on projects spanning a wide variety of marine taxa including sharks and rays, bottlenose dolphins, African penguins, horseshoe crabs, and weedy sea dragons. Lisa is particularly interested in how discoveries about foraging ecology, diet, and nutritional requirements of wild populations can inform and improve the care of animals managed in zoos and aquariums.

Driving Question: What kind of food do marine mammals need to stay healthy?

BASICS OF ANIMAL NUTRITION

What is nutrition and why is it important?

Nutrition is the process of converting food into living tissues through the intake of nutrients. Nutrients are chemical compounds present in food that are essential for an organism's growth and the maintenance of life. Nutrients support

DOI: 10.1201/9781003297468-9

the building and repairing of tissues, provide energy, and regulate and maintain body processes. A nutritionally appropriate diet provides fuel for cellular work, supplies raw materials to construct organic molecules, and furnishes essential nutrients that animals are unable to synthesize on their own. Nutrition has a tremendous impact on an animal being able to perform the essential functions of life. Proper nutrition is the cornerstone of good health and nutrients play a role in disease prevention and immune system function, often tipping the balance between healthy and diseased states. Many nutrients have an optimal level of intake, with either too much or too little resulting in negative effects on the health of an animal.

What nutrients do animals need from their diet?

Despite great diversity in food preferences, all marine mammals require the same types of nutrients at the cellular level: proteins, fats, carbohydrates, vitamins, minerals, and water. Nutrients that are required by the body in large amounts are called **macronutrients** and these consist of proteins, fats, and carbohydrates. **Micronutrients**, such as vitamins and minerals, are required by the body in small amounts.

Energy

Energy is not a nutrient, but rather the end product from the metabolism of proteins, fats, and carbohydrates and it serves as the fuel to power biochemical processes. The ideal energy source can vary depending on the natural diet of the animal. Carnivores are generally better equipped to use proteins and fats as energy sources, while herbivores are typically better adapted to use carbohydrates as an energy source. Energy intake requirements vary among the marine mammal groups and can vary seasonally in marine mammals due to life history events like migration and reproduction that require periods of **hypophagia** (controlled reduction in food intake) and associated preparatory periods of **hyperphagia** (increased food consumption).

Protein

Proteins are found in all living cells and they are connected with all phases of activity that constitute the life of a cell. Protein is primarily used by the body for building and repairing anatomical structures and for doing work (**Figure 9.1**). Proteins are the major structural components of tissues like skin, muscle, hair, and tendons. Proteins can also provide energy when needed and the synthesis of protein from the diet is essential for animal growth. Proteins are also involved in a number of other biochemical, immunological, and regulatory activities.

Ingested protein must be broken down into smaller usable units known as **amino acids**. These amino acids become part of the body's "amino acid pool" which form the crucial building blocks for the maintenance and growth of muscles and other tissues during normal use and repair. Amino acids continually move into the pool from tissue degradation (**catabolism**) and move out of the pool during periods of protein synthesis (**anabolism**). Net protein turnover is the balance between protein synthesis and degradation, and it is influenced by the level and quality of dietary protein intake, as well as physiological and metabolic requirements. Not all protein sources are equal – protein quality can be evaluated in terms of the amount of nitrogen absorbed from the diet and the proportion of dietary amino acids obtained. Proteins are not stored for later use, so excess protein must be converted into glucose or triglycerides and either used to supply energy or stored in the body, usually as fat. If enough energy is not obtained from dietary sources (e.g., during periods of fasting), the body will catabolize both lipid reserves and muscle tissue to release energy (**see Chapter 11**).

Most required amino acids can be synthesized by the body, but a few can only be obtained from the diet. These are referred to as **essential amino acids**. While there is some variation of what constitutes essential amino acids among species, all animals require amino acids like arginine, histidine, isoleucine, leucine, lysine, methionine, phenylalanine, threonine, tryptophan, and valine.

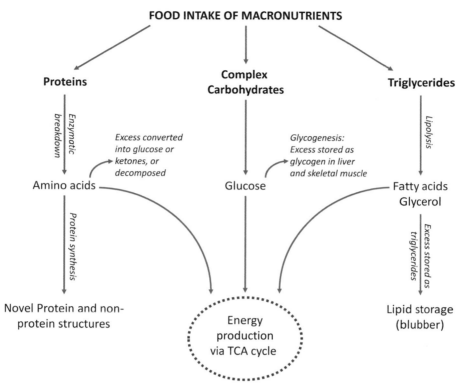

Figure 9.1 Macronutrients and their major roles in animal health. This schematic illustrates how the body uses protein, carbohydrates, and lipids from ingested food, including what happens to these compounds when ingested amounts are greater than current requirements. See Chapter 11 to see how animals use compounds during periods of nutritional deficits.

Nonessential amino acids are those that can be synthesized by the animal, assuming there is enough dietary nitrogen. Nitrogen from protein is also needed to produce essential compounds like heme (a component of hemoglobin), nucleic acids (components of RNA and DNA), and creatinine.

An animal's requirement for dietary protein is not constant, varying with both seasonal and developmental physiological requirements. For example, more protein is needed during phases of rapid growth in young animals, during pregnancy, and to support lactation. Protein metabolism plays a particularly important role in pinnipeds during **catastrophic molting** that occurs during periods of fasting (**see Chapter 11**).

Fat

Fat is an energy-dense nutrient and foods high in fat provide marine mammals with a readily available fuel source (**Figure 9.1**). Fats belong to a larger class of molecules called lipids. Lipids are essential to many aspects of marine mammal biology. Lipids are primarily deposited and stored directly under the skin in marine mammals as a layer of fat known as blubber which provides insulation from cold ocean temperatures (see **Chapter 5**). The blubber layer also contributes to buoyancy and streamlining in marine mammals. Particular lipids within cell membranes may ensure normal cellular function under the high pressure of deep diving. Lipids play an important role as a fuel source, particularly during long periods of fasting (see **Chapter 11**). In pinnipeds and cetaceans, life history events like migration and reproduction are tied to the seasonal management of lipids. In these species, lipids are rapidly assimilated and deposited into the blubber layer during periods of excess energy intake. Stored lipids can be later mobilized to support metabolism, energy requirements, and high-fat milk production for young ones during extended periods without feeding. Lastly, ingestion of lipids is necessary for many fat-soluble vitamins (A, E, D) to be effectively absorbed by the body. Major types of lipids include fatty acids, waxes, steroids, and phospholipids.

Fatty acids are the building blocks of fats in much of the prey marine mammals eat. Fatty acids are composed of a straight carbon atom backbone with hydrogen atoms along the length of the chain and at one end of the chain and a carboxyl group (–COOH) at the other end. **Essential fatty acids** cannot be synthesized by the animals and must come from the diet. Fatty acids are an important dietary source of fuel for animals, serve as key structural components for cells, and are vital to the functioning of most organ systems. These play important roles in signal transduction pathways, cellular fuel sources, composition of hormones and lipids, modification of proteins, and in energy storage.

Cholesterol is another form of lipid which is the principal **sterol** produced by most animals. Cholesterol is mainly synthesized in the liver but a certain level is required in the diet. While cholesterol has gotten a bad reputation in human health, cholesterol is essential for many important functions, such as making hormones and building cells, and is the precursor to many reproductive hormones (testosterone and estradiol), vitamin D, and bile salts. Cholesterol is also a component of the outermost layer of animal cells (plasma membrane) which is responsible for the transport of materials and cellular recognition and it is also involved in cell-to-cell communication.

Phospholipid is another type of lipid that is needed in the diet. Phospholipids are a major structural component of cell membranes. Phospholipids are also needed to support the absorption of fats and fat-soluble nutrients, such as omega-3 fatty acids ("good" fat). In the intestine, phospholipids form a "shell" around these fats, allowing them to be absorbed and distributed through the body. Phospholipids also help in preventing fat from accumulating in the liver and play a crucial role in communicating with the immune system to mount an effective immune response.

Carbohydrates

Carbohydrates are molecules that provide energy for cells. These include sugars, starches, and fibers. Carbohydrates are broken down in the digestive system rapidly into glucose, which is transported in the blood after eating (**Figure 9.1**). Excess energy is stored as glycogen in the liver and muscle and can be mobilized to satisfy subsequent energy demands. Unlike humans, most marine mammals do not consume carbohydrates in their diet; rather protein and fat are converted to glucose by the body through the processes of gluconeogenesis and ketosis, respectively (**Chapter 11**).

Carbohydrates do play a large role in the nutrition of herbivorous marine mammals like manatees (*Trichechus* spp.) and dugongs (*Dugong dugong*) which feed on aquatic vegetation like seagrass, water lettuce, and marine algae. These herbivorous species have lower dietary protein requirements compared to other marine mammals and have evolved to consume diets containing a variety of carbohydrates, including highly fibrous material like cellulose. Mammals lack the enzymes to digest these insoluble carbohydrates. Like terrestrial herbivores (e.g., horses and elephants), manatees and dugongs have developed a complex digestive system to process the large quantities of plant material. These animals possess a fermentation vat located in a pouch (called a **cecum**) off the large intestine/colon to assist in the breakdown of this plant fiber. The cecum is filled with bacteria, protozoa, and fungi who break down the plant material into fatty acids that can be absorbed across the wall of the colon directly into the bloodstream. These fatty acids provide a significant portion of a hindgut fermenter's dietary energy.

Vitamins

Vitamins are organic compounds that are generally required in trace amounts in the diet to meet needs for growth, reproduction, and health. Fat-soluble vitamins (A, D, E, and K) are absorbed with fats in the diet, transported in the blood, and stored in the liver and fatty tissues, such as blubber, until needed. Since the body can store these vitamins for months at a time, excess consumption can lead to toxicity. Water-soluble vitamins are not stored in the body; they are used immediately, and any excess amount is eliminated in urine. Water-soluble vitamins include vitamin C and the B vitamins (thiamin [B1], riboflavin [B2], niacin [B3], pantothenic acid [B4], pyridoxine [B6], biotin [B7], folic acid or folate [B9], and cobalamin [B12]). Vitamin functions are detailed in *Table 9.1*. Requirements for these vitamins change as per size, age, growth rates, environmental factors, and interrelationships with other nutrients.

Trace minerals

Minerals are inorganic elements consumed by animals that are considered essential for health.

Calcium (Ca), phosphorus (P), magnesium (Mg), sulfur (S), sodium (Na), potassium (K) and chloride (Cl) are required by the body in larger quantities (macrominerals), while iron (Fe), manganese (Mn), copper (Cu), iodine (I), zinc (Zn), cobalt (Co), molybdenum (Mo), and selenium (Se) are all considered microminerals since they are required in very small quantities. Mineral functions are described in *Table 9.2*.

Water

Water is required for digestion, absorption, elimination of waste products, and thermoregulation. The majority of marine mammals' water needs are met through their diet. The metabolic breakdown of fat and carbohydrates (the main fuel sources in most prey eaten by marine mammals) produces water as a byproduct. While marine mammals generally do not drink seawater, some species are known to take in seawater during feeding events (large baleen whales lunge feeding) or deliberate drinking (sea otters)

Table 9.1 **Function of fat-soluble and water-soluble vitamins**	
FAT-SOLUBLE	**ROLE OR PHYSIOLOGICAL FUNCTION**
Vitamin A	Required for vision, tendon strength, bone development, and skin and mucous membrane health. Integral in disease resistance and healing through its effect on the immune system
Vitamin D	Regulates calcium to strengthen bones, supports muscle strength, and controls gene activity of the immune system
Vitamin E	Acts as a biological antioxidant to protect cells from damage, maintains integrity of central nervous system
Vitamin K	Regulates blood clotting and supports bone health
Water-Soluble	
Thiamin, B1	Growth, appetite, digestion, nerve activity, and energy production
Riboflavin, B2	Growth and development of fetus, energy production, maintenance of eye and mucous membranes
Niacin, B3	Acts as co-enzymes in the metabolism of carbohydrates, fats, and amino acids
Pantothenic acid, B5	Lipid and protein metabolism
Pyridoxine, B6	Growth, amino acid and carbohydrate metabolism, cholesterol and steroid synthesis
Biotin, B7	Fatty acid metabolism, growth, maintenance of skin, hair, and bone marrow
Folic acid or folate, B9	Fetal nervous system development, blood cell production, and synthesis of DNA
Cobalamin, B12	Acts as co-enzymes in protein and lipid synthesis
Vitamin C	Important antioxidant that protects cells from damage, enhances bone and collagen formation for wound healing, enhances absorption of iron, supports immune system

Table 9.2 Functions of macro- and trace minerals

MACROMINERALS	ROLE OR PHYSIOLOGICAL FUNCTION
Ca	Integral to growth and integrity of bone and teeth, nerve function, muscle contraction, blood coagulation, and cell permeability
P	Involved in bone and teeth formation, protein synthesis, and energy metabolism
Mg	Essential for normal skeletal development, nerve impulse conduction, muscle contraction and relaxation
S	Component of amino acids necessary to synthesize and repair DNA
Na	Major cation involved in osmotic pressure and acid-base balance of body fluids, nutrient transfer to cells, waste removal, and water balance
K	Major cation of intracellular fluid, plays a role in osmotic pressure, acid-base balance, muscle activity, and normal heart function
Cl	Major anion involved in osmotic pressure and acid-base balance of body fluids
Trace Minerals	
Fe	Major constituent of hemoglobin in red blood cells and myoglobin in muscle cells
Mn	Essential for normal bone formation, amino acid metabolism, fatty acid synthesis, growth, and sex hormone production
Cu	Role in hemoglobin formation, enzyme systems, hair development and pigmentation, bone development, reproduction, and lactation
I	Essential component of thyroid hormones, regulation of metabolic rate
Zn	Needed for bone development, protein synthesis and metabolism
Co	Component of vitamin B12
Se	Antioxidant that works with vitamin E against cellular damage

on occasion. Marine mammals have specialized kidneys with multiple lobes ("reniculate") which increase urine concentration to handle the high levels of salt in seawater and prevent dehydration (**Chapter 8**).

DIVERSITY AND SPECIALIZATION IN MARINE MAMMAL DIETS

Classifying marine mammal dietary strategies

As a group, marine mammals have a wide range of diets and there are innumerable ways of describing them; even then, many species do not fall neatly into one category. Marine mammals collectively eat prey from all trophic food levels in the ocean. Their diet can be described by their ecological niche: from those which are exclusively **herbivores** (the sirenians), grazers

feeding on krill and other small organisms at lower trophic levels (mysticete whales), **piscivores** that exploit fish and squid (most of the odontocete whales and pinnipeds), up to top **carnivores** – predators that take other marine mammals or birds (e.g., killer whales, *Orcinus orca*; polar bears, *Ursus maritimus*; and leopard seals, *Hydrurga leptonyx*) (**Figure 9.2**).

Alternatively, marine mammals can be classified by their mode of foraging: suction feeders (walrus, *Odobenus rosmarus*), filter feeders (baleen whales), pursuit and pierce feeders (most of the odontocete whales and pinnipeds), and grazers (the sirenians), etc. These divisions provide information on the range of potential prey items and behavioral aspects of foraging strategies, but little insight into specific nutritional needs.

Marine mammal prey preferences can also be classified according to the relationship

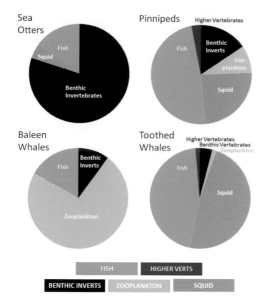

Figure 9.2 Summary of diets for pinnipeds, sea otters, baleen whales, and toothed whales showing proportions of prey biomass consumed from five prey categories: benthic invertebrates, large zooplankton, squids, fish, and higher vertebrates (marine mammals, seabirds, and turtles). (Figure reproduced from Trites and Spitz 2018)

between their diet and their overall lifestyles. For example, one such study (Spitz, Ridoux, and Brind'Amour 2014) found that smaller marine mammals that lead high-energy lifestyles (high swimming speed supported by underlying muscular physiology) and travel in large groups tend to target high energy density fish or other vertebrates (**Figure 9.3**). These marine mammals might be viewed as "elite athletes" that require plenty of high-energy food to satisfy their internal engines. Large marine mammals, such as baleen whales, that have moderate diving abilities and lead a moderate-energy lifestyle, primarily target small prey that appear in large aggregations near the surface. Finally, marine mammals with a medium body size that might be described as having a low-energy lifestyle (low swimming speed but high diving capacity) tend to forage on low-quality prey (low energy and protein content) that live in the deep-ocean environment.

Choosing specific prey: diversity vs. specialization

Animals can only consume prey that are sufficiently abundant in their environment and which they can realistically catch and consume – which are often two very different considerations. But beyond those logistical considerations, how do individuals choose among all the potentially available prey items? The short answer is that animals must consume items that satisfy their nutritional requirements, that is, provide the compounds they need to function optimally. Unfortunately, this does not really answer the question of what prey they should select and how they do so.

Humans are taught that "good nutrition" results from a well-balanced diet which is achieved by consuming items from several key food groups. Some marine mammals are **dietary specialists** who manage to fulfill their nutritional requirements from a single, or a limited number, of food sources. However, their numbers are relatively low, as it is rare that a single prey item can satisfy the full suite of nutritional requirements. Modern nutritional theory proposes that individuals select prey using a Geometric Framework for determining what combination of prey most readily fulfills their overall needs (Simpson and Raubenheimer 1993). This theory is used to explain why animals will switch from feeding on an abundant prey source to selectively target a less abundant one that may contain required nutrients.

On the other end of the continuum from specialists are the **dietary generalists**. These species of marine mammals consume a large variety of prey items. They are sometimes referred to as "opportunist foragers," which implies that the food they consume is largely determined by what they happen to come across. This is largely erroneous: while animals may have significant flexibility in what they will eat, they would not survive long if they did not practice some degree of selection in what they consumed. While most species of marine mammals generally limit their prey preferences to one general "type" (e.g., exclusively

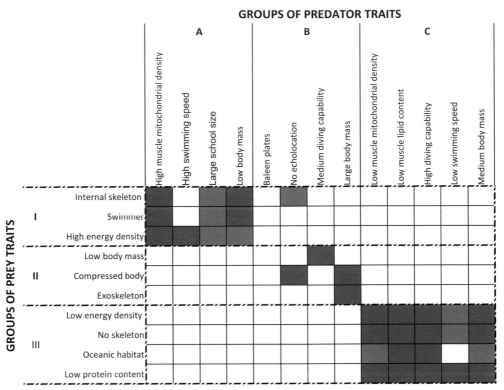

Figure 9.3 Schematic representing a statistical analysis of the relationship among three groups of traits in predators with three groups of traits of their prey. Dark boxes indicate strong positive correlations, gray boxes indicate positive correlations, and white boxes indicate no significant relationship. (Modified from Spitz, Ridoux, and Brind'Amour 2014)

fish, crustaceans, zooplankton, or plants, etc.) there are some notable exceptions. For example, while leopard seals in the Antarctic consume krill (*Euphausia superba*) most of the year, fish, cephalopods, sea birds, penguins, and seals can make up a significant portion of their diet in the summer (Hall-Aspland and Rogers 2004).

Dietary specialization is partly a function of anatomy. As animals evolve, their anatomy changes to become more efficient at catching and consuming their prey. As a result, we tend to think that diet specialization and physical modifications go hand in hand (**Chapter 10**). An obvious example is seen in baleen whales that are anatomically adapted to consume very small prey (plankton and small fishes) in very large numbers. In a case of **convergent evolution**, the teeth of crabeater seals (*Lobodon carcinophaga*) have evolved with extra projections

to function much like the baleen of cetaceans to strain out the krill they eat from mouthfuls of seawater. Walruses have developed muscular oral cavities and a piston-like tongue that can produce tremendous suction power that allows them to suck out the meat from the clams and other bivalve mollusks that form a large portion of their diet. However, it is always important not to oversimplify the evolutionary relationship between physical form and diet.

All of these aforementioned specializations describe external anatomical features. However, as with all groups of animals, there can also be adaptations in the digestive system related to diet specialization (although sometimes the link between diet and digestive anatomy is not clear). The digestive systems of pinnipeds and sea otters (*Enhydra lutris*) very closely resemble that of terrestrial carnivores. They possess a relatively simple **monogastric stomach**, with a longer small intestine (where macronutrients are primarily broken down and absorbed), and a shorter large intestine (where water and electrolytes are reabsorbed prior to fecal expulsion) compared to terrestrial mammals (Stevens and Hume 2004). The longer small intestine may assist in high levels of nutrient and water retention (**Chapter 8**). It has even been proposed that the large intestinal mass and its capacity to process large meals is a mechanism that allowed for the evolution of gigantism in cetaceans (Stevens and Hume 2004). However, this adaptation comes at a cost since intestines are energetically expensive to maintain. Interspecific comparisons demonstrate that resting energy expenditure of carnivorous mammals scales directly with small intestinal length. Intestinal length has been proposed to explain why marine mammals might appear to have higher energy requirements than similarly sized terrestrial mammals, a difference that disappears when compared to other carnivores (Williams et al. 2001).

The digestive apparatus of cetaceans is in some ways more complex than that of pinnipeds, particularly in the variety of different arrangements of multichambered stomachs they exhibit. In this respect, they are very similar to ruminant Artiodactyla (even-toed ungulates or hooved mammals), despite a lack of evolutionary connection and the fact that many are piscivores rather than herbivores. Odontocetes generally have three clearly distinguishable stomachs: a forestomach (which functions akin to a bird's crop), a second glandular stomach (akin to the typical single mammalian stomach), and a pyloric stomach. For unknown reasons, beaked whales have a plethora of variations on the basic cetacean model, including a lack of forestomach and multiple main and/or pyloric stomach compartments (Mead 2007).

Some food items require more complex digestive apparatus to break down and extract their nutritional content. Zooplankton are difficult to digest due to the high level of **wax esters** that they accumulate as energy reserves.

Crabeater seals are much less efficient at digesting krill than minke whales (*Balaenoptera acutorostrata*), due to the presence of the specialized digestive enzymes and various bacteria in the stomachs of the latter (Mårtensson, Nordøy, and Blix 1994b). As previously noted, sirenians are herbivores that use hindgut fermentation to break down their cellulose diet similar to elephants and horses, perhaps a reflection of their evolutionary lineage from grass-eating land mammals. They possess a single-chambered stomach and most of the digestive fermentation occurs in the large intestine (Reynolds and Rommel 1996).

Killer whales are an interesting example of a social, rather than physical, evolution of dietary preferences. Many different populations are found around the globe, and they have surprisingly distinct diets even in the same geographic range. The population in the North Pacific can be distinguished by three "ecotypes," which are largely similar in physical appearance, but differ dramatically in their diets (Ford 2019). Resident killer whales (which are more commonly found near the shore) are not just salmon specialists, but also have a particular preference for a single species (Chinook salmon; *Oncorhynchus tshawytscha*). Transient killer whales hunt strictly for other marine mammals and seabirds, despite the large geographic overlap with resident killer whales. And offshore killer whales specialize in preying on sharks, resulting in teeth which are dramatically worn down by the rough skin of their prey (Ford 2019). There are suggestions that some orcas that feed on sharks both in northern and southern oceans are specifically targeting the high-fat livers (Engelbrecht, Kock, and O'Riain 2019). A similar demonstration of dietary specialization is seen in the North Atlantic where some ecotypes ("type 2") target cetaceans while others ("type 1") consume a variety of prey, from pinniped to numerous fish species. The five types of killer whale in the Antarctic all have different types and levels of dietary specializations, variously concentrating on a combination of whales, seals, penguins, and/or fish. The hypothesis is that these strong dietary specializations seen among some – but not all – populations of killer whales are maintained by the strong social structure of this species (Ford 2019).

Adapting to changes in prey availability

Every animal's dietary strategy has presumably been selected for in light of a species' physiological needs, environment, and even social structure. Their chosen diet can be viewed as an "optimal strategy" under a given set of circumstances. However, because these dietary strategies are the product of multiple competing costs and benefits, reliance on these strategies may have different repercussions when environmental circumstances change.

Most marine mammals exhibit natural plasticity in their diets and unexpected changes in their prey base can simply result in adjustments in the proportion of usual items in their diet. However, these new diets may represent a necessary compromise to an altered environment rather than an alternate optimal strategy, thereby imposing some cost. For example, beluga whales in the Eastern Bering Sea primarily consume Arctic cod (*Boreogadus saida*), a species highly sensitive to climate change. They also normally consume capelin (*Mallotus villosus*), particularly the female adult whales. However, between 2011 and 2014 when there was a decrease in cod abundance due to changing ocean conditions, there was a shift within the population to a greater reliance on capelin in the diet (Choy et al. 2020). This diet shift also coincided with a significant decrease in the whales' body condition.

In cases when an animal cannot simply alter the proportion of their usual prey in their diet, animals must shift from their "preferred" prey items to a greater reliance on alternate prey to meet their nutritional requirements. In many cases, these nutritional challenges associated with changes in the prey base are reflected as an increase in **diet diversity**. For example, ringed seals (*Pusa hispida*) in Hudson Bay preferentially select sand lance (*Ammodytes hexapterus*) as a primary prey item. In years when sand lance is less available (due to environmental changes), the proportion of this fish in their diet is relatively low, their dietary diversity is much greater, and their body condition is poorer (Chambellant, Stirling, and Ferguson 2013). Similarly, fin whales (*Balaenoptera physalus*) have been observed to alter their diet in response to a decline in their primary prey, Arctic krill (*Thysanoessa* spp.) (Jory et al. 2021). The majority (60%) of fin whales demonstrated a clear shift from their normal dietary specialization of either krill or lipid-rich pelagic fish to a much more diverse, generalist diet composed of various zooplanktons and fish prey. This shift also coincided with a period of decreased body condition within the population.

However, the opposite trend may have occurred with the endangered Steller sea lion (*Eumetopias jubatus*). The population of Steller sea lions in western Alaska declined ~85% from 1980s to 2000s, while the population in eastern Alaska slowly increased. A strong statistical relationship was found between the rate of population decline and diet diversity among the sea lion rookeries, such that animals from the healthy eastern population had a relatively diverse diet that included many different fish species, while the diet of those from the western population was dominated by only one or two fish species (Merrick, Chumbley, and Byrd 1997). However, a subsequent analysis indicated that rates of population declines were also strongly related to diet quality. Populations experiencing the fastest declines were dominated by two relatively low-fat fish species, Atka mackerel (*Pleurogrammus monopterygius*) and walleye pollock (*Gadus chalcogrammus*) (Winship and Trites 2003).

A more unique example is seen in the previously mentioned southern resident killer whales in the North Pacific who are specialists on Chinook salmon. When summer Chinook salmon abundance is lower (or environmental conditions make it more difficult to acquire) there is almost no prey shift to other abundant salmon species (Hanson et al. 2021). This dietary obstinance may be causing nutritional stress and contributing to their continued threatened status.

While many cases of changes in the prey base are brought about by anthropogenic causes, such as fishing pressure and global warming, some species have had to alter their diet based on the ecological changes caused by their own prey preferences. For example, when prey is plentiful, sea otters target energy-rich species (e.g., urchins, abalone, clams) that are easy to capture and eat, largely bypassing less nutritious items regardless of abundance (Costa 2009). However, they are also very efficient feeders and can quickly deplete local invertebrate prey populations and be forced to move onto other less-energy-rich food items. In the longer term, the lack of urchins can alter the local community through promotion of kelp growth, further altering the prey available to otters.

In a similar vein, NW Pacific transient killer whales are an interesting example of a "serial specialist." It has been hypothesized, although not without critics, that transient killer whales along the western coast of North America have sequentially consumed their way through a succession of marine mammal species, leaving a series of threatened or endangered species in their wake (Springer et al. 2003). According to the Sequential Megafaunal Collapse theory, after the cessation of modern whaling cut off a valuable source of cetacean corpses, killer whales first switched to harbor seals as a food source, followed by sea lions, then fur seals, and finally onto nutritionally questionable sea otters to meet their dietary requirements.

DIFFERENCES BETWEEN INTAKE AND NUTRITION

It may not make sense at first, but the amount of food that an animal needs to take in is substantially greater than its actual energy or nutritional requirements. This is because the assimilation of ingested food energy or nutrients into usable biological components is not 100% efficient. The processes of breaking down and absorbing nutrients in the food as well as the removal of waste products means that food energy and nutrition intake must be higher than physiological demands.

Energetic efficiency: an overview

Most of the research on digestive efficiencies in marine mammals has been focused on the absorption of energy, although this is intimately related to the absorption of different macronutrients. This emphasis is partly because energetic efficiency is much easier to experimentally determine than the efficiency of nutrient absorption. **Gross energy intake** (GEI, or ingested energy, IE) is the "raw" biochemical energy contained in the food consumed. GEI is calculated by multiplying the mass of the ingested food item by its energy density. Depending on its ultimate use, the energy content of food can be expressed on either a wet weight (including normal water content) or dry mass basis (by chemically removing or mathematically discounting the water).

Energy loss during digestion comes through three avenues: fecal energy loss (FEL), urinary energy loss (UEL), and the **heat increment of feeding** (HIF; **Figure 9.4**). These three losses are somewhat dependent on each other, as they occur in a distinct sequence during the digestive process. As a result, intermediate calculations of "available" energy can be made at each step of the process.

A certain amount of GEI is expelled via FEL, and the subtraction of FEL from ingested energy gives the "apparent" **digestible energy** (DE). DE represents the energy that is absorbed and enters the bloodstream of the animal, and it can be expressed as a proportion of GEI (DE%). The "apparent" designation is because not all of the energy that is lost into the feces originates from food. Digestive enzymes, sloughed intestinal lining, and bacteria add to the energy value of the feces and therefore lead to underestimates of the true DE.

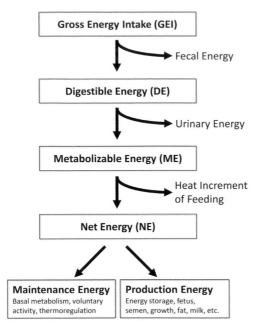

Figure 9.4 A typical bioenergetic scheme illustrating the transformation from the gross energy ingested (GEI) in food to the net energy (NE) that is biologically available to the animal to perform various biological functions (i.e., maintenance and production). The decrease from GEI to NE is due to unavoidable digestive losses from fecal energy, urinary energy, and the heat increment of feeding. The names for the energy remaining during intermediate steps are also included.

Additional energy is lost as various products in the urine, partly due to deamination of proteins. Technically, UEL represents the difference between digestible energy and **metabolizable energy** (ME) and should therefore be expressed as a percentage of DE because the losses are proportional to absorbed nitrogen and not the ingested nitrogen (such that fecal loss is irrelevant). However, UEL is commonly reported as a proportion of GEI (%UEL) for ease of calculations.

ME is the energy that remains available after accounting for the energy lost through the excreta (feces and urine) and can be expressed as a percentage of GEI (ME%). However, not all of this energy is "available" to the animal for use.

The mechanical and biochemical processes of digestion cause an increase in metabolic rate, which is known as the HIF (National Research Council (U.S.) 1981), also referred to as specific dynamic action (SDA) (Secor 2009). The heat generated through HIF is generally considered a waste product.

The end product after all of these digestive processes is termed **net energy** (NE), such that:

$$NE = GEI - FEL - UEL - HIF$$

These three processes of digestive efficiency must be taken into account when calculating the level of GEI required to satisfy energy requirements. These costs can be substantial. For example, in many carnivores, HIF accounts for 20% of GEI, while 8% of GEI is lost as FEL, and 12% passes through as UEL, such that food intake must be about 40% greater than physiological requirements to achieve the desired NE intake.

Fecal energy loss

FEL is the amount of energy defecated. Conversely, the efficiency of energy retention in this process – termed either digestive efficiency (DE%) or assimilation efficiency (AE%) – is the value often reported. FEL can be determined experimentally by collecting all of the feces excreted over a sufficiently long period of time. However, this is logistically challenging, not least because holding an animal for extended periods in a collection chamber can alter its normal digestive processes. Alternately, FEL can be estimated by measuring the energy content of a sample of feces, and then determining what proportion of the total ingested food mass the sample represents using either an added marker or a naturally occurring biomarker.

The efficiency with which animals absorb different components of a meal varies as a function of the length and morphology of the digestive tract, food type, seasonality, and nutritional state. Measurements performed on a variety of pinnipeds indicate that DE% is consistently high (~93–97%), but predictably varies with prey composition and mass of food ingested.

In general, DE% is higher for higher-energy/lipid-content prey and lower for prey with higher protein content (Rosen and Worthy 2018). The significant decrease in DE% with increasing protein content of the diet may be explained by the fact that among all of the components in food, the breakdown and assimilation of proteins to obtain energy takes the most time and effort (Blaxter 1989). However, the dynamics of protein and lipid absorption are not independent. For example, in northern fur seals (*Callorhinus ursinus*) digestibility of both protein and lipid were high across different diets, although macronutrient retention of lipids (96.0–98.4%) was significantly higher than that of proteins (95.7–96.7%). However, while higher protein diets increased protein retention, they also resulted in decreased lipid digestibility (Diaz-Gomez et al. 2020). To complicate matters further, it is known that lipid absorption generally varies with both the amount and the specific type of fatty acids in the diet (Mu and Høy 2004).

Studies with harp seals (*Pagophilus groenlandicus*) have reported high DE% when consuming fish, but low DE% (73–82%) on a shrimp (*Pandalus borealis*) diet, likely due to the large component of indigestible chitin (Mårtensson, Nordøy, and Blix 1994a). Crabeater seals demonstrate a similarly low DE% on a krill diet (Mårtensson, Nordøy, and Blix 1994b).

There have been attempts to better understand digestion in cetaceans through *in vitro* simulations of digestibility in their multicompartmental stomach. These studies suggest that baleen whales have very high DE% of 92% for herring (*Clupea harengus*) that is comparable to pinnipeds, and a surprising efficiency of 83% for krill (Nordøy, Sørmo, and Blix 1993). Olsen, Aagnes, and Mathiesen (1994) suggested that their ability to effectively digest krill was enhanced by the presence of symbiotic chitonolytic bacteria in the forestomach that help to break down the chitin.

Similar to seals, sea otters have been shown to have DE% ranging from 77% to 88% for a variety of invertebrate prey (Fausett 1976).

However, variation between individuals and seasons has been shown to result in a range of DE% values, from a low of 66.3% for a female eating crab in the winter, to a high of 96.5% for a female eating abalone in the winter.

For herbivores, DE% is inversely correlated with the level of crude fiber content of the food. DE% of West Indian manatees (*Trichechus manatus*) has been determined for both lettuce (90%; 10–12% crude fiber) and water hyacinth (80%; 12–17% crude fiber) diets (Lomolino and Ewel 1984). These measured DE% values for manatees are higher than for most herbivores, especially nonruminants, and have been attributed to the extremely slow food passage time (5–6 days) and extensive digestive tract (18 m) in this species (Lomolino and Ewel 1984). Manatees also have one of the highest digestibility coefficients for cellulose (80%) of any mammalian herbivore (Burn 1986).

Quantifying DE% in marine mammals can be difficult, as digestive physiology is complex and can change through time. Periods of low food intake can result in physical shortening and decreased total mass of the intestine. This is part of an energy-saving strategy as intestines are energetically expensive to maintain, but it also reduces immediate gut transport capacity and digestive efficiency. Transport cells within the intestinal wall also change to optimize absorption of specific diets; as a result, sudden shifts in diet composition may result in a "mismatch" with current intestinal transport function. In hindgut fermenters, such as sirenians, the composition of gut microflora will have significant impacts on DE%. For example, wild manatees eating seagrasses had significantly higher DE% than long-term managed animals consuming seagrass for short periods of time (46.9% vs. 36.2%, respectively), suggesting potential modification of gut flora over time (Worthy and Worthy 2014).

Urinary energy loss

Urinary energy losses represent the energy lost through urinary excretion mainly due to the deamination of proteins. Hence, UEL is proportional to nitrogen intake, such that animals fed higher nitrogen diets experience higher UE losses. Unfortunately, UEL can only be directly quantified from complete urine collection. Few studies have used this method as it is logistically difficult in marine mammals, particularly for any species aside from pinnipeds. Mean energy excreted in the urine of harp, ringed, and gray seals (*Halichoerus grypus*) was relatively similar at 6.9% to 9.5% of DE intake when all were provided a herring diet (Parsons 1977; Keiver, Ronald, and Beamish 1984; Ronald et al. 1984). The effects of diet composition were seen as the 1.5X higher UEL% (7.3% vs. 4.8% GE) of harbor seals (*Phoca vitulina*) fed a high (91%) protein diet compared to those on a lower (56%) protein diet (Ashwell-Erickson and Elsner 1981), similar to results reported for northern fur seals (Miller 1978).

Given the logistical difficulties of total urine collection, alternate methods of estimating UEL% have been employed. As the major component of UEL is urea, UEL can be estimated from food protein intake levels (Keiver, Ronald, and Beamish 1984). For example, calculations suggest that UEL% of northern fur seals fed a capelin diet (81.6% protein) would be 2.7 times greater (26.7% vs. 9.9%) than that of herring diet (51.4% protein) (Diaz-Gomez, Rosen, and Trites 2016), which is very similar to what has been found empirically (Miller 1978).

Determinations can also be performed on combined urine and feces samples, although this does not allow partitioning of energy losses between urinary and fecal routes, but only estimates of ME. This method has been used in a few studies to confirm the effect of diet type on ME available to California sea lions (*Zalophus californianus*) and bottlenose dolphins (*Tursiops truncatus*) (Shapunov 1971; Costa 1986).

Heat increment of feeding

The HIF is the increase in energy expenditure associated with the breakdown and assimilation of food. It is distinct from the cost of physically catching and ingesting the food. HIF is generally considered to be waste energy, and is usually calculated as a proportion of GEI of a meal.

The magnitude and duration of HIF varies with the size and composition of the diet. In general, carbohydrates increase metabolism by 4% to 30% for 2–5 hours after ingestion, lipids by 4% to 15% for 7–9 hours, and proteins by 30% to 70% for as long as 12 hours after ingestion (Hoch 1971). Unfortunately, for food with a combination of nutrients, it is impossible to accurately predict HIF from meals simply from knowledge of its **proximate composition**, but rather it must be empirically determined.

HIF has been measured in only a few species of marine mammals, including sea otters and various pinnipeds. There are currently no data available for cetaceans. It has been suggested that sirenians, due to their almost continuous hindgut fermentation and prolonged food passage times, do not experience a distinct HIF, as is also true for other nonruminant herbivores (Blaxter 1989), although the actual cost of grazing in sirenians has been estimated between 3.4% and 5.4% of GEI (Gallivan and Best 1986).

Studies report values between 4.3% and 16.8% of GEI expended as HIF for pinnipeds consuming a variety of fish species. As with other mammals, there is a relatively consistent trend for HIF to increase with higher protein densities (also lower energy/lipid density) among pinnipeds (see Table 3 in Rosen and Trites 1997). In mammals, HIF is affected by the mass of food ingested (Secor 2009), and studies with pinnipeds similarly confirm that larger meal sizes increase both the extent and duration of HIF. In Steller sea lions, HIF lasted 6–8 hours for a 2-kg meal of herring and 8–10 hours for a 4-kg meal (Rosen and Trites 1997), consistent with studies in northern elephant seals (*Mirounga angustirostris*) (Barbour 1993).

A theoretical exception to the concept of HIF as a waste product occurs when the animal is experiencing environmental temperatures which are below its **thermoneutral zone** (the point where it has to expend additional energy to maintain body temperature; **Chapter 5**). Under these conditions, the heat generated via digestion could be conserved and used to offset some of the necessary increase in metabolism needed to stay warm. Costa and Kooyman (1984) provided evidence that sea otters may use the short and intense increase in metabolism from HIF to offset heat loss while resting, and to maintain body temperature. However, this strategy of sea otters, with their high energy requirements and decreased thermal tolerance, may not be typical of marine mammals in general. For example, no evidence of using HIF for thermal substitution was found for the much larger Steller sea lion (Rosen and Trites 2003).

Nutritional content of prey: why does it matter?

Why does the energetic or nutritional content of individual prey items matter? After all, if food availability is not a limiting factor, can animals not fulfill their nutritional requirements by simply eating more of a less nutritious prey type?

There are several reasons why increasing the quantity of ingested food to make up for deficiencies in quality is not an optimal strategy. First, as we have previously discussed, the efficiency of nutrient extraction during the digestive process – particularly DE% and the HIF – is strongly influenced by the mass of ingested food. As more food mass must be ingested to achieve the same GEI, a greater proportion of that energy is lost through feces and HIF. This results in an animal having to consume even more prey, further degrading the net benefit.

In addition, the amount of prey that an animal can consume and process is anatomically limited. The effect of this finite digestive capacity has been demonstrated in studies with Steller sea lions and northern fur seals where the quality and availability of prey was altered (Rosen and Trites 2004; Rosen, Young, and Trites 2012). In these studies, the animals increased their food intake when it was only provided every other day in order to maintain a relatively constant average GEI. They also increased their intake to maintain a set GEI when provided food with

a lower energy content. The question of how they could determine their energy intake particularly between different prey items is largely unknown, but still quite impressive! However, this compensation ability reached a ceiling when they were offered lower quality prey every other day. They simply were not able to sustain a high enough level of food mass intake to meet their usual energy intake levels.

SEASONAL AND DEVELOPMENTAL CHANGES IN NUTRITION

Changes in food intake to meet changing nutritional needs

The primary function of food intake is to obtain sufficient energy and nutrients to power all aspects of life: maintenance, reproduction, growth, activity, thermoregulation, etc. As the emphasis on these different life history-specific costs change during an animal's lifetime, so do its nutritional requirements.

Like other mammals, marine mammals exhibit higher growth rates early in life. High levels of nutritional (particularly protein) and energy intake are required to support this somatic growth. As a result, marine mammals typically have the greatest levels of energy intake (relative to body mass) during this early developmental period that usually reaches an asymptote as physical maturity is reached (**Figure 9.5**). There is a broad rule of thumb among zookeepers that an adult marine mammal eats the equivalent of ~5% of its body mass daily. Innes et al. (1987) derived a predictive mathematical relationship (ingested food mass = 0.123 ∗ body mass$^{0.8}$) that reflects the fact that larger mammals eat proportionally less than smaller ones. As useful as such estimates may be, they are oversimplifications that fail to account for differences between consumer species and energy density of diets.

Even as adults, life history events result in changes in nutritional requirements. Some of these events result in an increase or change in food intake to satisfy immediate needs and keep

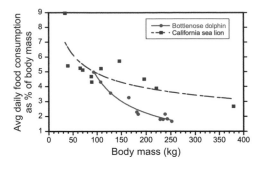

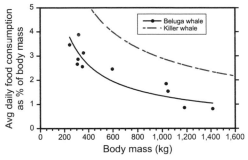

Figure 9.5 Changes with age in average food mass intake (kg d⁻¹) as a proportion of body mass for four species of marine mammals under human care. Data are presented from previously published studies for (top panel) California sea lions and bottlenose dolphins and (bottom panel) beluga and killer whales. Note the different scales on both axes. (Adapted from Rosen and Worthy (2018) and original sources therein)

the animal in a proper nutritional state. For example, the direct cost of pregnancy is not, in itself, generally considered energetically expensive, although it is also very difficult to measure experimentally as there are a myriad of concurrent changes in physiology and energy allocation. However, the successful completion of the cycle from fertilization to birth can be severely impacted by inadequate nutrition. Although no studies have specifically identified special nutritional needs during pregnancy in marine mammals, most terrestrial mammals have specific nutritional requirements during this period. Pregnant mammals typically require more protein and energy to support the growth of a normal, healthy fetus. While these demands may be met by simply increasing the level of food

intake, a switch to a more nutrient rich prey may be required if they are already consuming a high biomass of food. Poor nutrition and its direct effect on body condition can impact reproduction in terms of low conception rates, increased likelihood of spontaneous abortions, and unhealthy or undersized offspring (which ultimately impact the female's lifetime reproductive success).

The curious relationship between nutritional needs and intake

It is not unreasonable to expect a very tight relationship between nutritional requirements and food intake. Surprisingly, this is not always the case. **Daily energy expenditure** (DEE; variously referred to as average daily metabolic rate or field metabolic rate) is the sum total of energy used by an individual in a 24-hour period. For healthy animals, DEE will correspond to average energy obtained from foods over an extended time period. However, DEE does not necessarily equate to energy intake on a daily basis. It is not unusual for the intake levels of marine mammals to not be at maintenance levels, even when the demands for developmental growth are taken into account.

In other words, it is not surprising to see significant asynchrony between energy expenditure and energy requirements in marine mammals, particularly on a seasonal basis. For most marine mammals, surviving natural, predictable periods of negative energy balance (either due to increased expenditures or decreased intake) means building up an adequate lipid layer, which is the prime source for compensating for energy deficits due to insufficient food intake.

The most extreme example of this nutritional imbalance occurs when animals are fasting. As will be discussed in detail in **Chapter 11**, fasting is a normal part of the life cycle of many pinnipeds and cetaceans. Fasting occurs whenever an animal has a more important activity to perform, even in the presence of available food. For marine mammals, fasting is often associated with reproduction, breeding, migration, and/or the molt. Marine mammals are physiologically adapted to fasting. It is a natural, hormone-mediated response to predictable periods of either low food availability or a lack of opportunities to forage. In order to prepare for such periods, marine mammals must increase their food intake above normal levels – **hyperphagia** – in anticipation of these periods when food intake is lower – **hypophagia**. As they must rely heavily or completely on internal stores to meet their immediate nutritional requirements during fasting, they must consume prey in excess of immediate energetic expenditures, resulting in the buildup of lipid stores (**Figure 9.1**). These stores are commonly packed away in the blubber layer just below the skin, but significant abdominal lipid deposits also occur. These lipid energy reserves are then utilized at other times of year when energy expenditures outstrip food intake levels.

Although there are exceptions, these seasonal changes that result in animals being in either positive (gaining lipids) or negative (losing stores) energy balance are often tied to natural cycles in prey availability in the environment. For example, spring increases in food intake and resultant blubber energy stores may be keyed to the reproductive cycles of their prey, such as the availability of numerous lipid-rich pre-spawning fish. Conversely, it has been suggested that some Amazonian manatees fast for extended periods during the dry season when regular food sources are not available. Their extensive blubber reserves and low metabolic rate enable them to fast for up to 200 days (Best 1983).

These seasonal cycles in food intake are so physiologically ingrained that appetite is affected even when food access, either availability or foraging opportunity, is not restricted. Studies with marine mammals under human care have demonstrated that they retain natural seasonal cycles of hyper- and hypophagia, as well as cycles of body mass loss and gain, even under unnatural conditions of constant food availability (Rosen, Thometz, and Reichmuth 2021) (**Figure 9.6**). These seasonal cycles are controlled by hormones that affect metabolism, appetite, and activity, and are mediated by

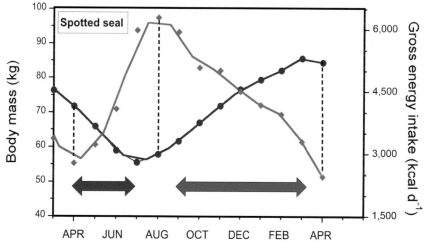

Figure 9.6 The complex relationship between seasonal changes in mean monthly body mass (in blue) and gross energy intake (in green) of spotted seals under human care. As denoted by the dashed vertical lines, minimum food intake was observed in April and maximum in August. The period from April to August was characterized by high energy expenditures (red arrow), where body mass decreased despite an increase in ingested energy. Conversely, the fall and winter was a period of energy conservation (blue arrow) when body mass steadily increased despite a significant drop in food consumption. (Data originally published in Rosen, Thometz, and Reichmuth 2021)

environmental cues like daylight and temperature. In most species where these cycles have been documented, the regularity and magnitude of these seasonal changes intensify as the animals reach sexual maturity.

In contrast to gestation, lactation is a very expensive event. Among many other physiological changes that occur, female marine mammals increase their body stores in anticipation of these costs. The degree to which they do so depends on which of the two general lactational strategies they employ to manage these costs. **Capitol breeders** generally rely almost completely on body reserves to manufacture milk for their offspring while the females largely or completely fast over the investment period. In phocid seals, lactation periods tend to be relatively short, and the females produce lipid-rich milk in order to transfer as much energy to the pups as possible in the shortest period of time (thereby limiting the time females have to fast). Hooded seals (*Cystophora cristata*) have the shortest lactation period of any marine mammal (as little as 3.5 days), and they produce milk that is up to 61% lipids (the highest of any mammal) (Oftedal, Boness, and Bowen 1988). Several studies have demonstrated that female seals in better body condition invest more into their pups, and those pups are then larger at weaning. Polar bears have an unusual strategy where they actually give birth and begin to nurse their cubs while still hibernating (and therefore fasting) in their dens. However, upon emergence, the mothers must quickly regain their body condition through intense foraging.

Most marine mammals, however, are **income breeders**. While initial lactation costs during the perinatal period may be satisfied from internal reserves, over the longer term they rely on external foraging to offset the costs of lactation. For some species, such as sirenians, otters, and many cetaceans, maternal foraging and nursing can occur almost simultaneously; while for others, such as otariid seals, the female alternates foraging trips at sea with nursing episodes on land. The pups, of course, are on the opposite schedule – fasting on land when their mothers are at sea and nursing when they return (**see Chapter 11**).

Many baleen whale species, such as gray whales (*Eschrichtius robustus*), blue whales (*Balaenoptera musculus*), and humpback whales (*Megaptera novaeangliae*), have even more extreme geographic separation of foraging and reproductive costs. For example, humpback whales undertake one of the longest recorded mammalian migrations of ˜9,000 km between high-latitude feeding grounds in the summer/autumn and low-latitude breeding grounds where they mate and give birth in the winter/spring. As will be discussed in **Chapter 11**, it is assumed that most species fast during the migration itself (based primarily upon travelling speeds), and – due to reduced feeding opportunities in the tropics – most of these whale species also fast during their residency on the breeding grounds. While there is growing evidence that some food intake occurs during both the migration and reproduction phases of the annual cycle, food intake is certainly severely curtailed at a time with high reproductive expenditures. Regardless, the energetic burden of rearing a calf over such a long period of diminished food intake requires accumulating tremendous energy reserves beforehand.

Just as females have to prepare for these reproductive events by increasing their intake and energy reserves in anticipation of these high levels of energy expenditures, males in many species also exhibit large seasonal fluctuations in food intake and energy expenditure. For species where there is intense male–male competition – whether associated with guarding and pursuit of females or holding on-land or aquatic territories – foraging opportunities may be severely curtailed or nonexistent. Hence, males in many species also "bulk up" in anticipation of the reproductive season. In northern elephant seals, adult territorial males must prepare to not feed for 5–6 months each year and lose one-third of their body mass. Males that do not start the breeding season with sufficient body energy reserves will not be large enough to compete for a territory, nor will they be able to hold that territory before starving to death.

Failure to adequately prepare or compensate for these changes in nutritional requirements can have severe consequences. If animals find themselves unexpectedly nutritionally taxed, they will have to realign their energy expenditures among different requirements. In general, animals will display innate priorities for allotting limited energy to specific functions, and these priorities may be dependent upon age, season, and reproductive status. For example, females with offspring will often buffer environmental perturbations by using up more of their internal energy reserves during lactation in order to maintain a constant milk supply to her offspring. However, there is a limit: females may have to curtail milk supplies or even abandon their offspring prematurely, particularly if females are themselves young and if the risk of further investment in the current offspring will impact their own lifetime reproductive success. Failure to compensate for more extreme deficits may have lifelong (e.g., impacting long-term growth) or population-level impacts (e.g., reproductive success or survival).

SPECIAL NUTRITIONAL CONSIDERATIONS FOR CAPTIVE MARINE MAMMALS

Marine mammals maintained under human care do not have the luxury of choosing their own food. Rather, it is the responsibility of professional staff to ensure the nutritional requirements of their charges are being met by offering them appropriate types and quantities of prey items. While there is still much work to be done to understand the peculiarities of marine mammal nutritional requirements, great progress has been made in recent years. The following section details some of the basic knowledge that professional husbandry staff employ in order to guarantee the health of marine mammals under their care.

What to feed?

A nutritionally appropriate diet for marine mammals in human care consists of the essential nutrients (i.e., proteins, lipids, carbohydrates including fiber, minerals, vitamins, fatty acids, and amino acids) found in the wild diet, and balanced for the current life history stage and the unique conditions under which the animal is managed. While this sounds easy, the challenge for zoos and aquariums is to provide these essential nutrients in a diet that is readily and commercially available, palatable, and obtained in a sustainable manner. For example, while polar bears in the high Arctic feed predominantly on a diet of seal and whale blubber, this is impractical for the feeding of captive polar bears. Instead, these facilities take advantage of the plasticity in their native diet to provide other more readily sourced diet options, including high-fat marine fishes, whole terrestrial prey and carcasses, fruits, vegetables, and browse.

In nature, marine mammals feed on a wide variety of food items which can vary seasonally. However, marine mammals in captivity are generally provided with a narrower range of food items with which to meet their nutritional requirements (which themselves often also vary seasonally). Diets consisting of a single species of fish are unlikely to provide balanced nutrition. Similarly, one diet will not serve the nutritional needs of all carnivores or piscivores. Compounding this challenge is the distinct lack of information on the dietary and nutrient requirements of most marine mammal species. In the absence of this information, understanding wild diet and foraging ecology, metabolism and energetics, and digestive anatomy and passage time of the target species can be helpful in choosing nutritionally appropriate food items to feed.

How much to feed?

Energy requirements can be considerably lower for marine mammals maintained in zoos and aquariums due to the absence of energetic costs associated with searching and hunting for food. For this reason, diets are carefully crafted and often adjusted to prevent overfeeding and obesity based on body weight and activity level of the individual. All fish are not of equal nutritional value. A sound nutritional program for marine mammals at zoos and aquariums includes an understanding of the composition and quality of food being offered. Diet items are routinely sent for nutritional testing of macro- and micronutrients and energy density so that a complete nutrient profile of the entire diet can be evaluated for nutritional health.

Food intake in marine mammals varies considerably, depending on fat content of fish (or other prey), water temperature, and activity level. Under human care, adult bottlenose dolphins generally eat the equivalent of 4% to 6% of their own body weight per day (BW/day). Adult seals and sea lions consume 5% to 8% BW/day in fish. Captive manatees can be maintained on a diet of lettuce, cabbage, aquatic plants (e.g., water hyacinth), and vegetables at a daily ration of 5% to 10% BW/day, although these food items often provide less fiber than wild diets (Siegal-Willott et al. 2010). In contrast, sea otters, which have a higher metabolism than most other marine mammals, can consume 25% to 30% BW/day in very expensive seafood!

Nutritional disorders

Generally, marine mammals in aquariums are fed a diet that is primarily seafood-based and which has been frozen and then thawed prior to feeding. The logistics and difficulty in providing this diet can lead to some special nutritional concerns. Historically, marine mammals under human care have developed nutritional disorders due to the loss or alteration of certain nutrients in seafood during long-term frozen food storage and subsequent improper thawing and/or handling. These disorders are rare in modern zoos and aquariums largely due to frequent nutritional testing and quality controls, improved thawing and food handling techniques, diet formulation, and vitamin and mineral supplementation.

While many of these ailments can be treated once detected, prevention is key. Many diet-related health issues can be avoided by paying close attention to fish quality, storage times, and handling practices.

Thiamin (vitamin B1) deficiency

Under long-term frozen storage and subsequent thawing, the nutrient composition of seafood can be altered, including overall water loss. Thiamin and other B vitamins are water-soluble, and can therefore be leached from seafood during the thawing process, leading to deficiencies. Proper thawing of all frozen seafood should be done slowly under refrigeration to minimize these water-based losses. Additionally, some species of fish (e.g., herring, smelt) and shellfish (e.g., mussels, clams, shrimp) contain high levels of thiaminase, a naturally occurring enzyme that breaks down thiamin. Thiamin is a cofactor in numerous life-sustaining enzymes that are required for basic cellular metabolism and thiamin deficiency can result in neurological symptoms and ultimately death. Deficiency can be prevented, and in most cases reversed, by monitoring frozen seafood carefully and by supplementing thiamin orally or via injection. Thiamin deficiency is now relatively rare in well-managed aquarium marine mammal populations; however, instances of thiamin deficiency within the food web affecting wild populations of marine mammals and seabird populations is increasing.

Vitamin E deficiency

During long-term frozen storage, amino acids and unsaturated lipids in seafood can undergo oxidation. These oxidative processes can destroy vitamin E levels and other important antioxidants in fish. Vitamin E plays an important role in maintaining the integrity of cellular membranes and deficiency can result in steatitis, an inflammatory condition of adipose tissue from excessive intake of rancid fats. Proper storing and thawing of seafood and vitamin supplementation can prevent vitamin E deficiency.

Scrombroid poisoning

Some species of Scrombroid fishes (e.g., tuna and mackerel) can develop large amounts of histimine in their flesh when "spoiled" due to improper storage and handling, which can induce an acute histamine or allergic reaction. Symptoms include anorexia, lethargy, a red inflamed mouth or throat, and conjunctivitis and increased tears. Affected animals can be treated with antihistamines.

Iron storage disease

Bottlenose dolphins under human care are reported to develop iron storage disease, which is rarely reported in the wild. Iron storage disease, or hemochromatosis, causes the body to absorb and store too much iron from the diet, and it occurs in humans and other animal taxa managed under human care (e.g., rhinos, fruit bats, lemurs, and toucans). The cause of iron storage disease in bottlenose dolphins is unknown but may relate to an underlying metabolic imbalance, disease state, or dietary management. Treatment for iron storage disease has traditionally been limited to extracting large volumes of blood which contain bound iron, but more modern treatments include feeding a substance that binds to iron (a natural or pharmacological chelator) which is then excreted from the body. While the role of diet in managing iron storage disease is unclear, emerging research on the diets of managed dolphin collections suggests that a more native diet consisting of warm-water fish species (e.g., mullet, croaker, pinfish) versus a traditionally fed cold-water fish diet (e.g., capelin, herring) may improve numerous nutritional and metabolic blood markers (Venn-Watson et al. 2015).

Proper vitamin supplementation

Vitamin levels in fish and other seafood can be highly variable in lean versus fattier species, with vitamins A, E, and D occurring in higher levels in species like herring, mackerel, and silversides and in lower (or even undetectable) levels in leaner fishes like capelin and most invertebrates. While feeding fattier fishes certainly conveys an advantage of making sure all fat-soluble vitamins are included in the diet, these fishes are calorically dense and therefore can adversely affect weight and body condition. Conversely, lean fish and invertebrates are desirable from a weight management perspective, but are often lacking in essential vitamins. To solve some of these tradeoffs, most zoos and aquariums feed a varied diet which includes vitamin and mineral supplementation. While there is agreement that thiamin and vitamin E supplementation are generally always required in managed marine mammal diets, the need for vitamin A, vitamin C, and vitamin D supplementation is unresolved and highly variable based on taxa, diet composition, and institution. Routine vitamin testing in seafood, regular diet formulation, and consultation with an animal nutritionist can be helpful to understanding vitamin needs in managed marine mammals.

FUTURE CONCERNS/RESEARCH

Our current knowledge of the nutritional requirements of marine mammals comes from two main approaches: detailing what specific types of food they eat in the wild and observing the health effects of different diets fed to animals in zoos and aquariums. While knowledge of marine mammal nutrition has always been important for understanding the ecology of wild marine mammals and ensuring the well-being of those under human care, the need is becoming even more important given the rapid changes that are occurring in our oceans.

Human-induced climate change and fishing practices are rapidly combining to alter the abundance and distribution of marine mammal prey in the oceans. As humans proceed to "fish down the food chain" (Pauly, Froese, and Palomares 2000), critical fish stocks around the globe are being overexploited. The effects of industrial fishing are increasingly magnified by the effects of human-induced climate change. Warming sea surface temperatures will have an increasingly dramatic impact on the composition, size, and distribution of prey species. These changes in food availability and distribution are increasingly forcing marine mammal populations to switch to alternate prey sources. They are also impacting the selection of diet items available to feed marine mammals in aquariums.

Understanding the impacts of these forecast changes in prey availability on marine mammals is complex. However, it is nearly impossible to make accurate predictions without a better understanding of the nutritional requirements of this diverse group of animals. Improving our knowledge of marine mammal nutrition – a serious task given the range of foraging and digestive specializations across this diverse group – is essential if we are to formulate science-based management strategies for ensuring their conservation in the wild and protecting their health while under human care.

TOOLBOX

Determining the nutritional value of prey/requirements of animals

An animal's optimal diet should satisfy their nutritional and energetic requirements. Whether a specific diet fully does so is based on three variables: the animals' compound-specific requirements, the chemical composition of prey, and the prey availability. To make things interesting, all of these factors can change for both the consumer and the prey with developmental stage, reproductive status, and season.

Scientists are not sure how animals "choose" their specific diet. However, a basic assumption is that the diet they choose to consume under normal circumstances meets their nutritional requirements. To understand what those requirements might be, scientists can work backwards by performing detailed chemical analyses of the prey they ingest.

The most common way that all food can be classified is by the amount of moisture, protein, fat, and carbohydrate it contains. The proportion of these macronutrients determine the proximate composition of the food item. From a nutritional perspective, the exact nature of the lipids is important. The laboratory techniques for quantifying these specific elements are very well established. In addition to these macronutrients, there are additional important micronutrient compounds that must be supplied in the diet, such as vitamins, minerals, and amino acids. Some of the consequences for failing to satisfy these latter requirements are dealt with earlier in this chapter (see *Tables 9.1* and *9.2*). Identifying micronutrients is more difficult. While the laboratory techniques for quantifying micronutrients are also straightforward, there is a vastly greater list of potential compounds to evaluate. A complete analysis is both time-consuming and expensive.

Energy intake is another aspect of nutritional requirements. Laboratory analysis of the energy content of prey can be undertaken using bomb calorimetry, which involves the complete combustion of a sample in oxygen within a special apparatus. The heat released during this process is a direct measure of the chemical energy contained in the food.

While bomb calorimetry can provide a measure of the total energy density of prey items, the proximate composition of prey can also be used to estimate the gross energy content. Quite simply, the amount of each macronutrient can be multiplied by its energy density. Although exact conversions depend on details of the chemical structure of the specific element, commonly used values are 39.3 kJ g^{-1} and 18.0 kJ g^{-1} for the energetic density of lipids and protein, respectively. In general, the fat content is the major determinant of the energy value of the food, while the protein content will influence the level of digestive losses.

Proximate composition and energy data is often presented in terms of dry mass, on the assumption that water content is not relevant to energy content. However, knowledge of energy density per wet mass is relevant for calculations of energy density per fed mass, which is required when calculating food intake requirements.

Further, the water content gives you information regarding the amount of preformed water that is available to the animal (**see Chapter 8**).

The basic drawback of this type of composition analysis to determine the nutritional requirements of animals is that this tells us what they are consuming, but not what is absolutely needed. For example, an animal can safely consume a higher level of water-soluble vitamins as the excess can be excreted in the urine. Another problem is that not all of the ingested food is absorbed by the animal (some is excreted as urine and feces).

Another issue when analyzing the nutritional requirements or intake of marine mammals is the difficulty in finding relevant data on proximate composition of food in the literature or obtaining appropriate samples from the wild. There is a growing body of literature detailing the proximate composition of prey commonly consumed by wild marine mammals. However, it is important to note that many prey species show geographic, age-related, and seasonal changes in proximate composition and consequently energy content. Therefore, previously published (or analyzed) values should be used cautiously. Changes in proximate composition are often related to reproductive condition, with gravid females being exceptionally high in lipid content. In comparison, "spent" females may have very low lipid content; a change that can occur very rapidly. For example, the fat content of herring can drop from 15% to 20% in the winter to just 2% to 4% during early spring (Røjbek et al. 2014).

How do we know what marine mammals eat?

The most direct way of determining what kind of prey a marine mammal consumes is through direct observation. This is traditionally only possible in select circumstances. For example, it is relatively straightforward to record what an otter is consuming, as it only ingests its food at the surface and can often be observed from the shoreline. Cetaceans feeding on dense concentrations of forage fish near the surface (e.g., herring balls) are also readily observable.

In most cases, direct observation is not possible, and so other techniques must be applied to identify diet. Traditionally, the remains of prey in the stomach or digestive tracts of individuals killed as part of a commercial or scientific hunt were examined. Aside from ethical considerations for this type of terminal sampling, this method can only identify items from the last meal, and the results may be biased due to differential digestion of different prey in the stomach.

Fecal sample analysis relies on the fact that certain hard parts of the prey – such as fish otoliths, eyeballs, and vertebrae, as well as beaks and pens of cephalopods – are not readily digested and can be found in the animal's feces. The remaining hard parts can often be identified to genera or species, and even size of prey. The technique has been refined for many marine mammals to account for biases due to differential degree of digestion, passage rate, and even animal activity level. Typically, prey species are classified simply either as "present or absent" in a single scat sample, a measure known as "frequency of occurrence". More detailed analyses of hard parts that accounts for the number of individual prey items represented and incorporating estimates of prey mass provide a better estimate of the biomass and nutritional contribution of different prey types to the overall diet. However, diet analysis from hard part remains is intrinsically limited to identifying the last meal or two that the animal consumed, and certain prey items (such as cephalopods, invertebrates, marine mammals, or prey not consumed whole) may not predictably leave hard parts to identify. In addition, this technique is largely restricted to pinnipeds that produce discrete fecal samples on land compared to most other marine mammals that defecate in the water.

Technological advances, such as the development of DNA metabarcoding, have resulted in genetic analysis being increasingly used to identify prey items from fecal remains (Ando et al. 2020). Ingested prey items will leave a genetic fingerprint in the predator's scat which can be amplified and identified in the lab. While this technique can be used in conjunction with traditional hard part analyses, one of the main benefits is that it can be applied to fecal material with minimal identifiable hard parts (such as with cetacean feces). As a bonus, the predator also sheds DNA into the scat, allowing identification of the sex of the consumer. Genetic analysis can also be applied in cases where prey remains can be found in the water, often the result of prey sharing or ripping apart large prey items. One current limitation to this technique is the availability of known prey DNA in the sequence library needed to identify prey, although this limitation will be reduced as more studies are undertaken. The next step in genetic scat analysis is based on the theory that the relative amount of different prey items consumed should result in a similar ratio of genetic material in the feces. Development of this technique may provide a novel tool to quantify the biomass of different prey consumed.

Two other techniques – fatty acid analysis and stable isotope analysis – are designed to give a longer-term picture of what a predator is consuming. They are both based on the adage "you are what you eat," in that the chemical structure of the prey will be integrated into the chemical profile of the predator. Fatty acid signature analysis (FASA) is based on the fact that the lipid of prey is made up of distinct proportions of identifiable fatty acids (either as standalone compounds or as triglycerides, phospholipids, and cholesteryl esters), which differ from each other in the number of carbon atoms and nature of the carbon bonds (Iverson 2009). These fatty acids are integrated (with various levels of transformation) into the various tissues of the consumer. Examining the profile of different tissues of the predator can allow you to reconstruct the combination of potential prey that contributed to that tissue. In addition, as different tissues have unique turnover rates, they can each be used to look at changes in diet over varying time frames. For example, red blood cells and milk provide diet information over days, while blubber can integrate signals over weeks. In addition, studies have worked at combining the fatty acid profiles (using complex mathematical modeling) to estimate the relative quantities of different prey items (quantitative fatty acid analysis; QFASA).

In much the same way, the ratio of stable isotopes in the prey and in consumer's tissues can be used to identify broader aspects of diet (vs. distinct prey items). Stable isotopes are naturally occurring, nonradioactive forms of atoms that differ in their atomic weight due to differences in their number of neutrons. Most elements can be found in two or more forms, with the lighter version usually the more common. Since they do not decay, their ratios do not alter with time but do vary predictably due to environmental and biological processes. In dietary research, carbon and nitrogen isotopes are most often used. The ratio of nitrogen isotopes (specifically, $^{15}N/^{14}N$) can be used to indicate average trophic level of the diet, with enrichment of ^{15}N increasing ~3–4‰ every tropic dietary level due to a process known as fractionation. In the literature, you will commonly see isotope ratios expressed as delta (∂) values, which are differences in the ratio compared to a recognized international standard. Carbon stable isotopes generally indicate location of foraging. The ratio of carbon isotopes ^{13}C to ^{12}C can provide information on the relative contribution of prey from different geographic areas, with higher ^{13}C values associated with areas of high primary productivity, allowing for differentiation between different marine habitat types (e.g., open ocean, nearshore, sea grass, kelp forests). As with fatty acids, analysis of different tissues (e.g., hair vs. red blood cells) can provide insights into diet over different time frames. In addition, stable isotopes can be integrated into a variety of tissues that grow in a predictable pattern, such as baleen, teeth, claws, and vibrissae. This provides a record of foraging patterns over time by analyzing isotopic ratios along the length of the tissue.

Technological developments have also been used to advance the most direct (and traditional) measure of prey intake – direct observation. Scientists can now mount forward facing video cameras on individuals to identify prey. Advancements in digital recording and battery life have made this tool smaller and longer-lasting, and some systems incorporate infrared imaging to observe foraging at depth. Autonomous underwater vehicles (UAVs) can also be used for making detailed observations of underwater foraging behavior. Similarly, overhead drones can make detailed observations of the foraging behavior of animals with minimal disturbance.

Data tags affixed to the animal or sonar devices can be used to follow individual animals and identify when they forage on a distinct prey patch, which can then be sampled via nets (such as herring balls or plankton patches) or characterized by hydro-acoustics (Sato, Trites, and Gauthier 2021). Some tags record changes in the orientation and acceleration of predators to allow 3D reconstruction of a foraging dive, and can even record sound to identify the use of echolocation when hunting or the "crunch" of a large kill (Holt et al. 2019).

CONCLUSIONS

Marine mammals encompass a diverse phylogenetic, ecological, and physiological group of animals. As such, they have equally disparate nutritional requirements. In many regards, their nutritional requirements are like many terrestrial mammals, although scientists are starting to uncover some important differences. The more we learn about their specific needs, the more accurately we can predict the effects of changes on their food supply in the wild on their health and population status, and the better we can support animals under human care.

REVIEW

1 What does it mean for an animal to have "proper nutrition"?
2 What are macronutrients and what is the main function of each?

3 What are micronutrients? Give three examples.
4 How does the process of digestion affect the energy derived from prey?
5 What are some of the ways you can classify different diets of marine mammals?
6 What are some of the reasons that marine mammals generally eat a variety of prey items?
7 How does anatomy relate to diet in marine mammals?
8 Why do food requirements of marine mammals change over their lifetime?
9 Why do food requirements of marine mammals change during the year?
10 In what different ways do marine mammals respond to changes in the availability of their usual diet items?

CRITICAL THINKING

1 How will climate change affect the nutrition of marine mammals?
2 How can marine mammals under human care be used to inform us about the nutrition of wild marine mammals and vice versa?
3 How would you design a study to look at the diet of an arctic cetacean, such as a narwhal or bowhead whale?

FURTHER READING

Bluhm, B.A., and R. Gradinger. 2008. Regional variability in food availability for Arctic marine mammals. *Ecological Applications* 18(sp2): S77–S96.

Evans, P.G., and A. Bjørge. 2013. Impacts of climate change on marine mammals. Marine Climate Change Impacts Partnership (*MCCIP*) *Science Review* 2013: 134–148.

Simpson, S., and D. Raubenheimer. 1993. A multi-level analysis of feeding behaviour: The geometry of nutritional decisions. *Philosophical Transactions of the Royal Society B: Biological Sciences* 342(1302): 381–402.

Spitz, J., V. Ridoux, and A. Brind'Amour. 2014. Let's go beyond taxonomy in diet description:

Testing a trait-based approach to prey–predator relationships. *Journal of Animal Ecology* 83(5): 1137–1148.

Williams, T.M., J. Haun, R.W. Davis, et al. 2001. A killer appetite: Metabolic consequences of carnivory in marine mammals. *Comparative Biochemistry and Physiology A* 129: 785–796.

GLOSSARY

Amino acids: Small molecules that are the building blocks of proteins. While each of the 20 amino acids have a similar core chemical structure, they each have a specific unique side chain that gives it a different physiological function.

Anabolism and catabolism: Anabolism refers to the process of building biological structures, while catabolism refers to the breakdown of structures into smaller components. In relation to tissues, growth is an anabolic process that requires energy and the appropriate component parts, whereas tissue catabolism breaks down tissue (e.g., lipid reserves or muscles) usually to supply energy to be used for other physiological functions. The process of digesting foods is also a catabolic process.

Capitol breeders: Animals (such as phocid seals) that use internal stored energy for reproduction.

Catastrophic molting: Molting is the annual process of replacing the top layer of skin and/or fur and catastrophic molting refers to when this occurs over a short period of time with the tissues being shed in large patches rather than gradually.

Carnivore: An organism that eats mostly meat or the flesh of animals. Carnivores can specialize on consuming different types of animals. For example, **piscivores** are carnivorous animals that eat primarily fish.

Cecum: A pouch-like structure at the start of the large intestine. In most mammals it is the site of fluid and salt absorption. In herbivores, the cecum is enlarged and filled with bacteria that aid in the digestion of plant matter and absorption of nutrients.

Cholesterol: A type of lipid that is an essential structural component of animal cell membranes. Cholesterol is the major **sterol** of vertebrates.

Convergent evolution: The process where distantly related organisms independently evolve similar traits (such as a feature of anatomy) to adapt to similar necessities or challenges.

Daily energy expenditure: The amount of energy (often number of calories) you expend throughout a 24-hour period by doing any type of physiological activity: breathing, moving, thermoregulation, growth, etc.

Diet diversity: A measure of how many different prey species an animal consumes, which can sometimes also incorporate how evenly they consume multiple prey items. There are many mathematical ways to quantify diet diversity in addition to counting the number of different items consumed, such as the Shannon Index and the Simpson's Index.

Dietary specialists: An animal that consumes only a few particular prey items.

Dietary generalists: An animal that has a varied diet.

Digestible energy: The amount of food energy absorbed by the body after initial digestion, after accounting for the loss of energy after removal of feces.

Essential amino acids and essential fatty acids: Compounds that animals are unable to synthesize on their own and must obtain from their diet as they are required for normal physiological function.

Fatty acids: A class of lipid biomolecules that are present in all organisms and serve as both an important dietary source of fuel for animals and a key structural component for cells.

Gross energy intake: The raw chemical energy contained in the food that an animal ingests. Not all of this energy is available to the animal for use (see **Net Energy**).

Heat increment of feeding: The increase in an animal's rate of energy expenditure (seen as an increase in heat production or oxygen consumption) due to the mechanical and biochemical processes of digestion.

Herbivore: An animal anatomically and physiologically adapted to eating plant material as the main component of its diet.

Hyperphagia and hypophagia: Respectively referring to the consumption of an above average or below average amount of food. They are considered regulated states, in that the level of food consumption is driven by internal hormonal controls and not food availability per se.

Income breeders: Animals (such as sea lions, fur seals, and walrus) that rely on feeding during the reproductive period to offset energetic costs.

Macronutrients: Nutrients consumed at relatively high levels. Fats, proteins, and carbohydrates are classified as macronutrients.

Micronutrients: Nutrients consumed at relatively low levels, specifically vitamins and minerals.

Minerals: Inorganic elements consumed by animals that are considered essential for health.

Metabolizable energy: The amount of food energy absorbed by the body after accounting for the loss of energy from production of feces and urine.

Monogastric stomach: A simple, single-chambered stomach.

Net energy: The amount of energy from food that is physiologically available to an animal, after accounting for digestive losses through feces, urine, and the heat increment of feeding.

Phospholipids: A type of lipid containing a hydrophilic phosphate group "head". It is an important component of membranes, and it also affects immune health, gut health, liver health, and other functions.

Piscivores: An animal anatomically and physiologically adapted to eating fish for the main component of its diet, a type of carnnivore.

Proximate composition: With reference to food, it is the proportion (%) of different macromolecules – proteins, fats, carbohydrates, and water that the item contains.

Sterol: A type of lipid that regulates biological processes and is important for maintaining the structure of cell membranes. Cholesterol is the major sterol of vertebrates.

Thermoneutral zone: The range of ambient temperatures in which homeotherms maintain internal body temperatures with minimal metabolic regulation. In temperatures outside the thermoneutral zone, animals must increase their metabolic rate to generate heat (temperatures below their thermoneutral zone) or dump excess heat (temperatures above their thermoneutral zone) to maintain their internal body temperature.

Vitamins: A set of organic molecules with diverse biochemical functions that an organism needs in small quantities for proper physiological functioning.

Wax esters: Naturally occurring molecules that are composed of a single fatty acid molecule and a long-chain (fatty) alcohol. They are present in animals, plants, and microbial tissues, and play a variety of functions ranging from energy storage to waterproofing.

REFERENCES

Ando, H., H. Mukai, T. Komura, et al. 2020. Methodological trends and perspectives of animal dietary studies by noninvasive fecal DNA metabarcoding. *Environmental DNA* 2(4): 391–406.

Ashwell-Erickson, S., and R. Elsner. 1981. The energy cost of free existence for Bering sea harbor and spotted seals. In *The Eastern Bering Sea Shelf: Oceanography and Resources*, ed. D.W. Hood, and J.A. Calder, 869–899. Seattle: University of Washington Press.

Barbour, A.S. 1993. Heat increment of feeding in juvenile northern elephant seals. M.Sc. thesis, Marine Sciences, University of California, Santa Cruz.

Best, R.C. 1983. Apparent dry-season fasting in Amazonian manatees (*Mammalia: Sirenia*). *Biotropica* 15(1): 61–64.

Blaxter, K. 1989. *Energy Metabolism in Animals and Man*. Cambridge: Cambridge University Press.

Burn, D.M. 1986. The digestive strategy and efficiency of the West Indian manatee (*Trichechus manatus*). *Comparative Biochemistry and Physiology A* 85(1): 139–142.

Chambellant, M., I. Stirling, and S.H. Ferguson. 2013. Temporal variation in western Hudson Bay ringed seal (*Phoca hispida*) diet in relation to environment. *Marine Ecology Progress Series* 481: 269–287.

Choy, E.S., C. Giraldo, B. Rosenberg, et al. 2020. Variation in the diet of beluga whales in response to changes in prey availability: Insights on changes in the Beaufort Sea ecosystem. *Marine Ecology Progress Series* 647: 195–210.

Costa, D.P. 1986. *Assessment of the Impact of the California Sea Lion and the Northern Elephant Seal on Commercial Fisheries*. San Diego: Institute of Marine Sciences, University of California.

Costa, D.P. 2009. Energetics. In *Encyclopedia of Marine Mammals*, ed. W.F. Perrin, B. Wursig and J.G.M. Thewissen, 387–394. San Diego: Academic Press.

Costa, D.P., and G.L. Kooyman. 1984. Contribution of specific dynamic action to heat balance and thermoregulation in the sea otter *Enhydra lutris*. *Physiological Zoology* 57(2): 199–203.

Diaz-Gomez, M., D.A.S. Rosen, I.P. Forster, et al. 2020. Prey composition impacts lipid and protein digestibility in northern fur seals, *Callorhinus ursinus*. *Canadian Journal of Zoology* 98(10): 681–689.

Diaz-Gomez, M., D.A.S. Rosen, and A.W. Trites. 2016. Net energy gained by northern fur seals (*Callorhinus ursinus*) is impacted more by diet quality than by diet diversity. *Canadian Journal of Zoology* 94(2): 123–135.

Engelbrecht, T.M., A.A. Kock, and M.J. O'Riain. 2019. Running scared: When predators become prey. *Ecosphere Naturalist* 10(1): e02531.

Fausett, J. 1976. Assimilation efficiency of captive sea otters, *Enhydra lutris*. M.A. thesis, California State University, Long Beach.

Ford, J.K.B. 2019. Killer whales: Behavior, social organization, and ecology of the oceans' apex predators. In *Ethology and Behavioral Ecology of Odontocetes*, ed. B. Wursig, 239–259. Switzerland: Springer.

Gallivan, G.J., and R.C. Best. 1986. The influence of feeding and fasting on the metabolic rate and ventilation of the Amazonian manatee (*Trichechus inunguis*). *Physiological Zoology* 59(5): 552–557.

Hall-Aspland, S., and T. Rogers. 2004. Summer diet of leopard seals (*Hydrurga leptonyx*) in Prydz Bay, Eastern Antarctica. *Polar Biology* 27(12): 729–734.

Hanson, M.B., C.K. Emmons, M.J. Ford, et al. 2021. Endangered predators and endangered prey: Seasonal diet of Southern resident killer whales. *PLoS One* 16(3): e0247031.

Hoch, F. 1971. *Energy Transformations in Mammals: Regulatory Mechanisms*. Philadelphia: W. B. Saunders.

Holt, M.M., M.B. Hanson, C.K. Emmons, et al. 2019. Sounds associated with foraging and prey capture in individual fish-eating killer whales, *Orcinus orca*. *The Journal of the Acoustical Society of America* 146(5): 3475–3486.

Innes, S., D.M. Lavigne, W.M. Earle, et al. 1987. Feeding rates of seals and whales. *Journal of Animal Ecology* 56(1): 115–130.

Iverson, S.J. 2009. Tracing aquatic food webs using fatty acids: From qualitative indicators to quantitative determination. In *Lipids in Aquatic Ecosystems*, ed. M.T. Arts, M.T. Brett and M.J. Kainz, 281–308. New York: Springer.

Jory, C., V. Lesage, A. Leclerc, et al. 2021. Individual and population dietary specialization decline in fin whales during a period of ecosystem shift. *Scientific reports* 11(1): 1–14.

Keiver, K.M., K. Ronald, and F.W.H. Beamish. 1984. Metabolizable energy requirements for maintenance and faecal and urinary losses of juvenile harp seals (*Phoca groenlandica*). *Canadian Journal of Zoology* 62: 769–776.

Lomolino, M., and K. Ewel. 1984. Digestive efficiencies of the West Indian manatee (*Trichechus manatus*). *Florida Scientist* 47(3): 176–179.

Mårtensson, P.-E., E.S. Nordøy, and A.S. Blix. 1994a. Digestibility of crustaceans and capelin in harp seals (*Phoca groenlandica*). *Marine Mammal Science* 10(3): 325–331.

Mårtensson, P.-E., E.S. Nordøy, and A.S. Blix. 1994b. Digestibility of krill (*Euphasia superba* and *Thysanoessa* sp.) in minke whales (*Balaenoptera acutorostrata*) and crabeater seals (*Lobodon carcinophagus*). *British Journal of Nutrition* 72(5): 713–716.

Mead, J.G. 2007. Stomach anatomy and use in defining systemic relationships of the cetacean family Ziphiidae (beaked whales). *The Anatomical Record: Advances in Integrative Anatomy and Evolutionary Biology* 290(6): 581–595.

Merrick, R.L., M.K. Chumbley, and G.V. Byrd. 1997. Diet diversity of Steller sea lions (*Eumetopias jubatus*) and their population decline in Alaska: A potential relationship. *Canadian Journal of Fisheries and Aquatic Sciences* 54(6): 1342–1348.

Miller, L.K. 1978. *Energetics of the Northern fur Seal in Relation to Climate and Food Resources of the Bering Sea*. Washington DC: U.S. Marine Mammal Commission.

Mu, H., and C.-E. Høy. 2004. The digestion of dietary triacylglycerols. *Progress in Lipid Research* 43(2): 105–133.

National Research Council (U.S.). 1981. *Nutritional Energetics of Domestic Animals & Glossary of Energy Terms*, 2nd edition. Washington, D.C.: National Academy Press.

Nordøy, E.S., W. Sørmo, and A.S. Blix. 1993. In vitro digestibility of different prey species of minke whales (*Balenoptera acutorostrata*). *British Journal of Nutrition* 70: 4485–4489.

Oftedal, O.T., D.J. Boness, and W.D. Bowen. 1988. The composition of hooded seal (*Cystophora*

cristata) milk: An adaptation for postnatal fattening. *Canadian Journal of Zoology* 66(2): 318–322.

Olsen, M.A., T.H. Aagnes, and S.D. Mathiesen. 1994. Digestion of herring by indigenous bacteria in the minke whale forestomach. *Applied and Environmental Microbiology* 60(12): 4445–4455.

Parsons, J.L. 1977. Metabolic studies on ringed seals (*Phoca hispida*). M.Sc., University of Guelph.

Pauly, D., R. Froese, and M.L. Palomares. 2000. Fishing down aquatic food webs: Industrial fishing over the past half-century has noticeably depleted the topmost links in aquatic food chains. *American Scientist* 88(1): 46–51.

Reynolds, J.E., and S.A. Rommel. 1996. Structure and function of the gastrointestinal tract of the Florida manatee (*Trichechus manatus latirostris*). *Anatomical Record* 245(3): 539–558.

Røjbek, M.C., J. Tomkiewicz, C. Jacobsen, et al. 2014. Forage fish quality: Seasonal lipid dynamics of herring (*Clupea harengus* L.) and sprat (*Sprattus sprattus* L.) in the Baltic sea. *ICES Journal of Marine Science* 71(1): 56–71.

Ronald, K., K.M. Keiver, F.W.H. Beamish, et al. 1984. Energy requirements for maintenance and faecal and urinary losses of the grey seal (*Halichoerus grypus*). *Canadian Journal of Zoology* 62: 1101–1105.

Rosen, D.A.S., N.M. Thometz, and C. Reichmuth. 2021. Seasonal and developmental patterns of energy intake and growth in Alaskan ice seals. *Aquatic Mammals* 47(6): 559–573.

Rosen, D.A.S., and A.W. Trites. 1997. Heat increment of feeding in Steller sea lions (*Eumetopias jubatus*). *Comparative Biochemistry and Physiology A* 118(3): 877–881.

Rosen, D.A.S., and A.W. Trites. 2003. No evidence for bioenergetic interaction between digestion and thermoregulation in Steller sea lions (*Eumetopias jubatus*). *Physiological and Biochemical Zoology* 76(6): 899–906.

Rosen, D.A.S., and A.W. Trites. 2004. Satiation and compensation for short-term changes in food quality and availability in young Steller sea lions (*Eumetopias jubatus*). *Canadian Journal of Zoology* 82(7): 1061–1069.

Rosen, D.A.S., and G.A.J. Worthy. 2018. Nutrition and energetics. In *CRC Handbook of Marine Mammal Medicine*, 3rd edition, ed. F.M. Gulland,

L.A. Dierauf, and K.L. Whitman, 695–738. Boca Raton: CRC Press.

Rosen, D.A.S., B.L. Young, and A.W. Trites. 2012. Rates of maximum food intake in young northern fur seals (*Callorhinus ursinus*) and the seasonal effects of food intake on body growth. *Canadian Journal of Zoology* 90(1): 61–69.

Sato, M., A.W. Trites, and S. Gauthier. 2021. Southern resident killer whales encounter higher prey densities than northern resident killer whales during summer. *Canadian Journal of Fisheries and Aquatic Sciences* 78(11): 1732–1743.

Secor, S. 2009. Specific dynamic action: A review of the postprandial metabolic response. *Journal of Comparative Physiology B* 179(1): 1–56.

Shapunov, V. 1971. Food requirements and energy balance in the Black sea bottlenose dolphin (*Tursiops truncatus ponticus* Barabasch). *Morphology and Ecology of Marine Mammals Israel Program for Scientific Translations, Jerusalem, Israel*: 207–212.

Siegal-Willott, J.L., K. Harr, L.-A.C. Hayek, et al. 2010. Proximate nutrient analyses of four species of submerged aquatic vegetation consumed by Florida manatee (*Trichechus manatus latirostris*) compared to romaine lettuce (*Lactuca sativa var. longifolia*). *Journal of Zoo and Wildlife Medicine* 41(4): 594–602.

Simpson, S., and D. Raubenheimer. 1993. A multi-level analysis of feeding behaviour: The geometry of nutritional decisions. *Philosophical Transactions of the Royal Society B: Biological Sciences* 342(1302): 381–402.

Spitz, J., V. Ridoux, and A. Brind'Amour. 2014. Let's go beyond taxonomy in diet description: Testing a trait-based approach to prey–predator relationships. *Journal of Animal Ecology* 83(5): 1137–1148.

Springer, A.M., J.A. Estes, G.B. van Vliet, et al. 2003. Sequential megafaunal collapse in the North Pacific Ocean: An ongoing legacy of industrial whaling? *Proceedings, National Academy of Sciences, USA* 100(21): 12223–12228.

Stevens, C.E., and I.D. Hume. 2004. *Comparative Physiology of the Vertebrate Digestive System*. Cambridge: Cambridge University Press.

Trites, A.W., and J. Spitz. 2018. Diet. In *Encyclopedia of Marine Mammals*, ed. B. Wursig, J.G.M. Thewissen, and K.M. Kovacs, 255–259. Philadelphia: Elsevier.

Venn-Watson, S.K., C. Parry, M. Baird, et al. 2015. Increased dietary intake of saturated fatty acid heptadecanoic acid (C17: 0) associated with decreasing ferritin and alleviated metabolic syndrome in dolphins. *PLoS One* 10(7): e0132117.

Williams, T.M., J. Haun, R.W. Davis, et al. 2001. A killer appetite: Metabolic consequences of carnivory in marine mammals. *Comparative Biochemistry and Physiology A* 129(4): 785–796.

Winship, A.J., and A.W. Trites. 2003. Prey consumption of Steller sea lions (*Eumetopias jubatus*) off Alaska: How much prey do they require? *Fisheries Bulletin, United States* 101(1): 147–163.

Worthy, G.A.J., and T.A.M. Worthy. 2014. Digestive efficiencies of ex situ and in situ West Indian manatees (*Trichechus manatus latirostris*). *Physiological and Biochemical Zoology* 87(1): 77–91.

Christopher D. Marshall
Texas A&M University, Galveston, TX

Jeremy A. Goldbogen
Stanford University, Pacific Grove, CA

Christopher D. Marshall

Christopher Marshall earned his Ph.D. in physiological sciences and functional morphology in 1997 from the University of Florida. He is currently a professor of marine biology at Texas A&M University, Galveston Campus (Galveston, TX, USA) and holds an Appointment in the Department of Ecology and Conservation Biology, Texas A&M University (College Station, TX, USA). His work is driven by questions of how animals work and how they hunt. Christopher is particularly interested in connections between morphology, behavioral performance, ecology, a field termed "ecomorphology." His work is comparative and evolutionary in nature. He has worked on every major group of marine mammals, as well as sea turtles, fishes, and invertebrates, focusing on both extant and extinct species. Since much of his work is with threatened and endangered species, his interests also include conservation biology and providing data-driven solutions for management of these species.

Jeremy A. Goldbogen

Jeremy Goldbogen is a comparative physiologist who studies the integrative biology of marine organisms. He started his research career studying the biomechanics of locomotion in hummingbirds and Antarctic sea butterflies (pteropods) as an undergraduate student at the University of Texas at Austin. Jeremy then completed his M.Sc. in marine biology from the Scripps Institution of Oceanography at the University of California, San Diego. He later moved on to earn his Ph.D. from the University of British Columbia in Vancouver, Canada, with a thesis titled "Mechanics and energetics of rorqual lunge feeding." He returned to Scripps as a postdoctoral researcher for one year before joining the Cascadia Research Collective in Olympia, WA for two years. He is now professor of oceans at Stanford University, located at the Hopkins Marine Station in Pacific Grove, CA.

Driving Question: How do marine mammals catch their prey?

DOI: 10.1201/9781003297468-10

INTRODUCTION

Feeding is an essential behavior that is required to obtain the energy needed to fuel a wide range of life functions including metabolism, locomotion, and reproduction. Therefore, feeding adaptations are directly tied to fitness and are subject to strong **selection pressures**. The physical forces experienced in an aquatic environment impart strong selection pressures upon all vertebrates and have influenced the evolution of foraging mechanics and strategies. This basic conceptual framework is fundamental in understanding marine mammal feeding mechanics of this diverse group of predators and grazers that evolved from terrestrial ancestors. Marine mammals have an extensive evolutionary history of anatomical and biomechanical innovations for prey capture in water (reviewed by Marshall and Goldbogen, 2016 and Marshall and Pyenson, 2019). As a group, these functional innovations reveal numerous evolutionary experiments in feeding mechanics that include aquatic herbivory, several forms of biting, suction, and filter feeding.

The group "marine mammals" collectively represents several independent mammal lineages that made the transition back to aquatic environments at different evolutionary timescales. Mapping traits associated with different feeding mechanics onto their evolutionary history (**phylogeny**) demonstrates a diversity of independent feeding methods and foraging strategies among marine mammals. The ancestors of cetaceans (whales and dolphins) and sirenians (sea cows or manatees and dugongs) were among the first mammals to make the transition back to the sea, followed by pinnipeds (walruses, otariid, and phocids), sea otters, and polar bears. These evolutionary innovations for aquatic feeding in each of these lineages are a driving factor in marine mammal evolution and their success.

The investigation of feeding mechanics among any vertebrate group is an integrative and comparative endeavor that involves morphological, physiological, developmental, behavioral performance, and ecological studies. In addition to understanding the evolution of mammalian feeding in aquatic environments, understanding prey capture tactics and feeding performance are important considerations for **trophic** ecological questions since these behaviors can determine prey choice due energetic needs and constraints. Feeding has a direct bearing on organismal fitness by determining the behavioral capacity, or **performance,** of an animal to exploit environmental resources. Feeding morphology and physiology (e.g., head, jaw, teeth shape and associated musculature, muscle and bone physiology) influences an organism's ecology through constraints of behavioral performance, prey choice, and ecological niche (i.e., diving capacity and foraging depth). Therefore, feeding and foraging performance is an important link between morphology and ecology in marine mammal biology.

Marine mammals represent the most recent vertebrate group that underwent a major evolutionary transition from land to sea (Pyenson, Kelley, and Parham, 2014). This major shift from terrestrial to a primarily aquatic lifestyle required a complex suite of feeding adaptations, which are exhibited among several diverse lineages of cetaceans, sirenians, and pinnipeds. Collectively, these species feed on a wide variety of resources that span all trophic levels including marine algae, aquatic angiosperms (flowering plants), zooplanktons (such as krill and copepods), fish, squid, and even marine birds such as penguins, and other marine mammals. Therefore, marine mammals have become important consumers in diverse ocean ecosystems worldwide by utilizing multiple innovations that enable highly successful feeding.

ORDER CETACEA: WHALES AND DOLPHINS

Cetaceans represent a radiation of carnivorous marine mammals descended from terrestrial even-toed ungulates (hoofstock; artiodactyls) that diversified approximately 50 million years ago (**Ma**) during the establishment of the cir-cum-Antarctic current system in the Southern Ocean (Fordyce and Barnes, 1994). Given our current understanding of the Southern Ocean as a whale feeding hotspot, this restructuring of the oceans presumably changed the abundance and diversity of oceanic resources in a dramatic way (Steeman et al., 2009; Pyenson, Kelley, and Parham, 2014). Current cetaceans are represented by two major groups that exhibit very divergent feeding strategies: toothed whales (Odontoceti) and baleen whales (Mysticeti). Odontocetes possess adaptations of the skull (and other physiological systems) for the integrated dual functions of prey capture and echolocation to target single prey items, whereas baleen whales feed in bulk on aggregations of small prey (like krill) using baleen as a filter. Both echolocation and bulk filter feeding represent major evolutionary innovations that underlie the ecological success of this adaptive expansion of marine mammals.

Cetacea: Odontoceti

It is thought that toothed whales evolved **echolocation** to feed at night in shallow waters on daily vertically migrating cephalopods (e.g., squid) and later this physiological adaptation developed in many odontocete lineages to exploit deep prey that were also available during the day. The deepest diving toothed whale lineages are largely squid hunters (**teuthophagous**), as exemplified by beaked and sperm whales, but their diets can also be supplemented with fish in many geographic regions (Gaskin, 1982). Several other odontocete lineages (e.g., oceanic dolphins and porpoises) also feed on both fish and cephalopods, although narwhals and belugas may also feed on crustaceans and **benthic** (i.e., river bottom, lake bottom, or sea floor) invertebrates. These generalized and perhaps opportunistic feeding preferences reflect the flexibility of feeding strategies afforded by echolocation, a key evolutionary innovation that is a hallmark of toothed whale functional ecology and evolution.

Odontocete (and all cetacean) skulls are among the most derived among mammals. The morphology of the entire skull (snout, nose, braincase, ear bones, and jaw) and the **biomechanics** of jaw opening and closing are drastically modified. The unusual morphology of cetacean skulls is due to overlapping and telescoping of bones that shorten the cranium, and the elongation of the facial region by lengthening of the bones associated with the jaw (maxilla, premaxilla, and mandible; **Figure 10.1**). Such modifications of the skull are linked to feeding, respiration, and the generation and reception of sound used for echolocation (**see Chapter 6**). The ratio of facial length to cranial length is variable among cetaceans and results in species with very short blunt skulls (e.g., pygmy and dwarf sperm whales, and pilot whales) or long, narrow rostra (snouts, e.g., river dolphins). The dentition of odontocete jaws varies from virtually none (e.g., beaked whales) to several hundred

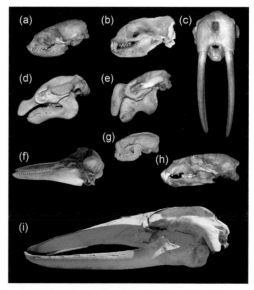

Figure 10.1 Representative skulls of major marine mammal clades. (a) True seals (harbor seal), (b) sea lion and fur seals (Steller sea lion), (c) walrus, (d) manatees (West Indian manatee), (e) dugongs (dugong), (f) Odontoceti (bottlenose dolphin), (g) otters (sea otter), (h) polar bear, and (i) Mysticeti (gray whale). Skull images are not to scale.

simple and almost identical homodont teeth (e.g., river dolphins). These traits are related to diet. It is thought that genetic constraints in early dental development have been released in cetaceans. This is further supported by the possible loss of complexity in the enamel of modern odontocetes (Werth, Loch, and Fordyce, 2020). Changes in the development of teeth can drive such morphological evolution as observed in odontocetes. The feeding mechanics of cetaceans are also among the most specialized and varied among mammals. They range from several forms of **feeding** involving biting to suction feeding, while mysticetes are generally categorized as filter feeders. In odontocetes, and many aquatic **tetrapods** (four-limbed animals including frogs, mammals, turtles, and others), there is a dichotomy of cranial morphology associated with their feeding mode. Feeding primarily on fish (**piscivory**) tends to be associated with long narrow rostra and mandibles, and jaws filled with numerous teeth. This "**ecomorph**" of a long narrow rostra with numerous teeth has evolved independently several times (McCurry et al., 2017) among different aquatic vertebrate groups (e.g., extinct marine reptiles, crocodiles, and some odontocetes) and it is an adaptation for high velocity of the jaw tips to capture elusive prey, at the expense of bite force (think fast, but weak jaws; **Figure 10.1**). On the other extreme, teuthophagous toothed whales are associated with short, blunt rostra and jaws, a reduction in tooth number (or function), an expanded basihyoid bone (**hyoid**), and the use of suction as the primary feeding mode. However, there are exceptions to this pattern, as observed in sperm whales and beaked whales. Both of these groups feed on squid, yet show the anatomical head features more commonly seen in piscivorous groups. It is thought that the muscles associated with the enlarged hyoid apparatus (Reidenberg and Laitman, 1994) result in a greater force of tongue depression that presumably increases negative pressure. However, the shape of the mouth and head, as well as and tongue shape may be just as important in directing the negative pressure toward the front of the mouth (Marshall and Pyenson, 2019). Such traits and correlated performance measures are well known for bony fish.

To acquire prey, toothed whales use two primary mechanisms: **ram feeding** that involves biting and **suction**. Ram feeding occurs when the whale's attack speed and agility outperform their prey (Weihs and Webb, 1984). In contrast, suction is generated from negative intraoral pressures through the rapid depression of the hyoid and tongue. These two capture mechanisms are not necessarily exclusive, and some toothed whale species can use these strategies either together or in sequence. Suction appears to be an important mechanism for prey capture in odontocetes, as demonstrated by the well-developed tongue musculature and increased surface area of the hyoid for attachment of these muscles in some species such as pygmy and dwarf sperm whales (Bloodworth and Marshall, 2005, 2007). The hyolingual apparatus (hyoid and tongue) exhibits a greater proportion of extrinsic muscle fibers (connections of the tongue to other structures) and a lower proportion of intrinsic muscle fibers (connections within the tongue itself), as compared to both terrestrial mammals and other aquatic mammals (Marshall and Pyenson, 2019).

Suction in marine mammals is generated by the rapid depression and retraction of the **hyolingual** apparatus (but in some cases by the addition of fast jaw opening), which results in a rapid increase in mouth volume and concomitant decrease in pressure (Marshall and Pyenson, 2019). This mechanism to generate suction during feeding has been experimentally demonstrated (Werth, 2006a) in odontocetes of varying head shape and bluntness of the rostrum (common dolphin, white-sided dolphin, and harbor porpoise). The greatest suction capability in this study of odontocetes was found in harbor porpoises, which also possessed the bluntest rostrum in the study (Werth, 2006b). In addition, this mechanism to produce suction has been documented in several toothed whale species including pygmy sperm whales, pilot whales, and belugas, as well as harbor porpoises (Marshall and Pyenson, 2019).

The morphologies of the skull and hyoid bones alone do not determine, nor limit, feeding mode and performance. The function of soft tissues must also be considered to fully understand the integrative biology of feeding in cetaceans. In addition to blunt rostra and wide jaws, odontocetes that can produce a more circular mouth aperture (pursing) with lip muscles can generate higher negative pressures and greater suction performance (i.e., higher negative pressure = greater performance; Bloodworth and Marshall, 2005; Kane and Marshall, 2009). For example, beluga whales (*Delphinapterus leucas*) often exhibit discrete ram and suction components during feeding, but the latter is greatly enhanced by pursing of the lips to occlude the lateral **gape** (angle of corner of mouth) or sides of the mouth (Kane and Marshall, 2009). Lip pursing behaviors act to form a small circular aperture to magnify negative intraoral pressures, a mechanism that is convergent with more basal vertebrates (Marshall and Pyenson, 2019) like numerous fishes. Toothed whales that exhibit this pursing ability tend to be larger, deep divers that have reduced dentition. Although this circular mouth adaptation is represented in several toothed whale families (Werth, 2006b), many odontocetes, such as bottlenose and pacific white-sided dolphins, have limited lip pursing abilities and instead use ram feeding primarily to capture prey (Marshall and Pyenson, 2019). In these species, ram and suction may be more synchronized, as demonstrated in long-finned pilot whales, where submaximal **gape** angles or soft tissue adaptations effectively occlude lateral gape to enhance suction performance during capture events (Kane and Marshall, 2009).

Cetacea: Mysticeti

Baleen whales (Mysticeti) evolved from toothed whale ancestors presumably to exploit aggregations of prey, rather than single prey items. Baleen whales do not have teeth as adults, but instead have calcified protein keratin baleen racks that act as the primary feeding structure. These racks hang from the top of the rostrum and there are no baleen on the bottom jaw. All baleen whales are obligate bulk filter feeders that engulf a volume of prey-laden water that is subsequently filtered out of the mouth using the two racks of vertically oriented baleen plates. Each baleen rack consists of an array of plates that are obliquely angled relative to the whale's long body axis. There is a tremendous amount of morphological diversity among mysticetes with respect to plate and fringe morphology, and these variants correlate generally with prey preference and ecological niche.

There are different modes of filter feeding represented among different baleen whale families. These modes include suction and ram feeding, the latter of which can be either intermittent or continuous. However, these functional elements show up in very different ways in baleen whales compared to toothed whales. Mysticetes have very large skulls (**Figure 10.1**) and most lack soft tissues that completely occlude the lateral gape, so enhanced suction performance may not be possible in most species. However, gray whales are known to use suction to feed on benthic invertebrates along the sea floor (Nerini, 1984). Gray whales (*Eschrichtius robustus*) have well-developed **hyolingual** (hyoid and tongue) musculature and ventral groove blubber that facilitate the depression of the hyoid and tongue to generate suction (Werth, 2007). However, it is unclear how adequate suction is produced without lateral occlusion of the gape or if baleen can function to occlude the gape and facilitate flow past the anterior tips of the jaws. Gray whales have been observed both rolling onto their lateral sides during feeding. Therefore, the flow of water and prey may enter and exit through the side of the mouth, and their relatively short baleen plates may enable this unique suction mechanism. In addition, gray whales may use the substrate and the morphology of the sea floor to passively increase suction distance. Interestingly, other observations suggest that gray whales are generalist filter feeders and may also use ram in combination with suction, and filtering, to feed throughout the water column (Pyenson and Lindberg, 2011). Despite several observational and anatomical studies, the filter feeding mechanisms in gray whales remain poorly understood.

The two mysticete families Balaenidae and Balaenopteridae (such as bowhead and blue whales *Balaenoptera musculus*, respectively) are extremely different from each other in both the anatomy of the feeding apparatus and the **hydrodynamic** (motion of fluid) mechanics employed to capture and filter prey from seawater. Broadly, balaenids have large and stiff tongues that direct continuous flow of prey-laden water past long baleen plates, whereas balaenopterids have largely flaccid tongues for inversion and expansion of the large ventral pouch. Balaenids swim at slow, steady speeds (<1 m/s) to continuously drive water into the mouth (ram feeding; Simon, Johnson, and Madsen, 2009). Such a strategy may be required to efficiently filter water with an enlarged mouth aperture that incurs significant drag and large energy costs (Potvin and Werth, 2017). Despite this large anterior opening, there are two much smaller posterior openings of the mouth where water exits the mouth after being filtered through the baleen (Werth and Potvin, 2016). The difference in area between the anterior and posterior openings creates a **Venturi effect**, or suction, in front of the mouth that should reduce a bow wave from forming that would deflect prey away from the mouth as the individual swims forward. In addition, structures around the mouth, such as a mobile tongue influences how the steady flow of water moves through the baleen to improve the efficiency of filtration.

In contrast, Balaenopterids (rorquals, such as blue and minke *Balaenoptera sp.* whales) exhibit a dynamic process that involves a lunge, or a rapid acceleration to high speed (Goldbogen et al., 2006; Simon et al., 2012), and the subsequent engulfment of a large volume of prey-laden water. After the target volume of prey and water is engulfed, the mouth closes just enough to leave a small area of the baleen rack exposed for filtration (Goldbogen et al., 2017). The engulfed water is then driven past the baleen plates through the contraction of the expanded ventral pouch. This so-called lunge filter feeding mechanism is enabled by an integrated suite of morphological and mechanical adaptations that facilitates the lunge feeding process: a tongue that can invert to become the lining of the expanded oropharyngeal cavity (Werth et al., 2019), hyper-expandable ventral groove blubber, and elongated and curved jaws (Goldbogen, Potvin, and Shadwick,, 2010; Pyenson, Goldbogen, and Shadwick, 2013). In many large-bodied rorqual species, the size of the engulfed volume is commensurate with the whale's body size (Goldbogen, Pyenson, and Shadwick, 2007; Kahane-Rapport, and Goldbogen, 2018; Kahane-Rapport et al., 2020). This is due in large part to the positive relationship of the skull and ventral pouch; whereby large whales exhibit relatively larger oropharyngeal cavities. As a result, larger rorquals have big heads and short tails, which may reflect the investment of growth in the anterior region at the expense of the posterior region (Goldbogen, Potvin, and Shadwick, 2010). The relationship of this ventral pouch relative to body size appears to have important consequences for rorqual diving physiology, feeding performance, and their ecological niche (Goldbogen et al., 2012; Kahane-Rapport et al., 2020).

ORDER SIRENIA: MANATEES AND DUGONGS

Sirenians were among the first mammal lineages to return to aquatic habitats. The earliest sirenians appear in the fossil record ~50 Ma in the middle to late Eocene and our modern dugongs and manatees first appeared in the middle Eocene and Oligocene, respectively (Marshall and Pyenson, 2019). Our modern assemblage includes three manatee species: West Indian manatees (*Trichechus manatus*), West African manatees (*T. senegalensis*), and Amazonian manatees (*T. inunguis*) and two species within the family Dugongidae, dugongs and the recently extinct Steller's sea cow (**see Chapter 5** for more about the Steller's sea cow). Sirenians are distinct among marine mammals in that they are **herbivorous** and possess many adaptations for grasping, excavating, and chewing aquatic plants. Although sirenians may be superficially similar in both morphology and diet, they are quite different in that manatees are generalists, whereas dugongs are benthic specialists. They also differ considerably in their trophic ecology and feeding mechanics. There are several overarching functional themes in sirenian feeding mechanics that include the degree of deflection of the rostrum (**Figure 10.1**), tooth replacement, and the vibrissal (whisker)–muscular complex. Among all sirenians, the skull is dominated by a broad narial basin (associated with the nose) and a rostrum (snout) composed of enlarged premaxillary bones. The variation of the degree of rostral deflection of these premaxillary bones relative to the angle of the upper jaw plane reflects the location within the water column where feeding occurs most efficiently (Velez-Juarbe, Domming, and Pyenson, 2012; Marshall and Pyenson, 2019). On one extreme of rostral deflection are Amazonian manatees. They feed primarily at the surface upon "floating meadows" of the Amazonian rivers, lakes and floodplains, and inundated vegetation of grasses. West African manatees also primarily consume floating vegetation – although recent evidence does show consumption of fish and freshwater mollusks. Both species inhabit turbid aquatic habitats where submerged aquatic plants are not widely supported. Correspondingly, these manatees possess the least deflected snouts (~25–42° and 15–40°, respectively; Marshall and Pyenson, 2019). On the other extreme of rostral deflection are dugongs. Dugongs are benthic-feeding specialists that consume primarily sea grasses. They consume both the above- and below-ground biomass. Below-ground biomass contains **rhizomes** which dugongs often target,

and are a rich source of carbohydrate "packets." Dugongs possess the greatest rostral deflection (70°). West Indian manatees are the ultimate generalists among sirenians. They inhabit a variety of habitats that span coastal marine, estuarine, and freshwater ecosystems. They consume more than 60 species of aquatic vegetation that are distributed throughout these habitats. Much of their diet is comprises sea grasses, brackish and freshwater submerged aquatic vegetation, but many terrestrial grasses and other types of vegetation are also consumed. The locations of these food resources span the water column from benthic, midwater, floating, semi-immersed, to terrestrial. Their intermediate snout deflection (29–52°) allows them to consume vegetation throughout these various locations. More recently, it has been demonstrated that such correlations are much more complex and interesting. For example, tusk size and body size in addition to degree of snout deflection are among several suites of important morphological traits that determined resource partitioning in extinct dugong species (**Figure 10.2**; Velez-Juarbe et al., 2012).

Unlike other marine mammals, sirenians chew their food. Plants are processed using teeth and/or upper and lower palatal pads that are covered in keratin (cornified, thickened skin). This processing reduces particle size, increases food surface area, and ruptures the tough plant cell walls. Herbivores digest plants either using their forestomach (such as ox and other hoofstock) or their **hindgut** (small and large intestines), which consists of expanded sections of the large intestine and a cecum (such as horses). Sirenians use hindgut fermentation to digest the cellulose and cell contents of the plants they consume (Marshall and Pyenson, 2019). Plant matter, particularly grasses (which contain high levels of silica) can be highly abrasive. The abrasive nature of grasses has driven **selection pressures** (factors that cause selection of favorable traits) that have resulted in numerous adaptations by herbivores, both terrestrial and aquatic, to resist tooth wear, since the life of an herbivore only lasts as long

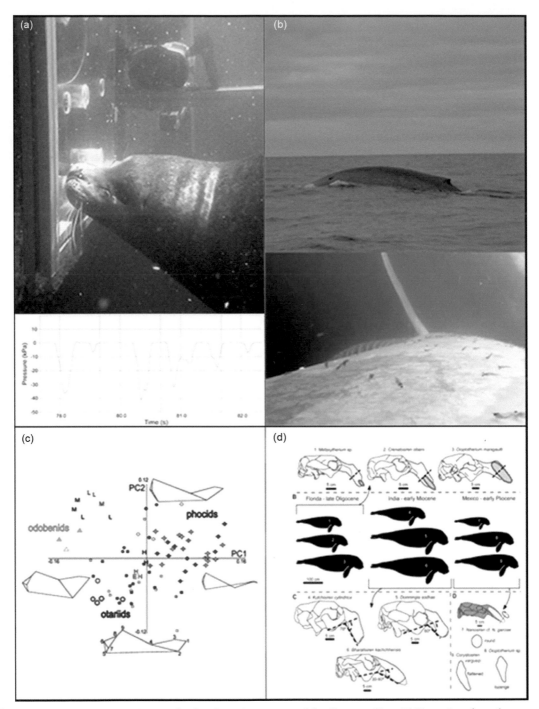

Figure 10.2 Representative methods of marine mammal feeding studies. (a) Functional performance, kinematics, and biomechanics (Marshall et al., 2014). (b) Animal-borne tags on free-ranging animals (Photo by Jeremy Goldbogen. Unpublished video tag from: Jeremy Goldbogen, Dave Cade, Ari Friedlaender, John Calambokidis. NMFS Permits: #14534–2 and 14809). (c) Comparative and integrative morphology, biomechanics, and geometric morphometrics (Jones, Ruff, and Goswani, 2013). (d) Integration of morphology and paleoecology (Velez-Juarbe, Domming, and Pyenson, 2012).

as their teeth last. Classic examples include the **hypsodont teeth** (tall tooth crowns) of grazers such as horses, the open-rooted ever-growing incisors of rodents (that result in continued tooth growth), and the serial replacement of multi-rooted molars of elephants (Unger, 2010). In this latter example, molars erupt slowly over the life of the animal. All manatees exhibit a novel mechanism of cheek-tooth replacement, which differs significantly from dugongs (and other herbivores). At any one time six to eight cheek-teeth are erupted and functional. As they wear, these teeth migrate horizontally from the posterior region of the tooth row to the anterior region (Marshall and Pyenson, 2019). Teeth at the anterior locations (near the mouth) have worn considerably and have little to no crown remaining. Once the tooth has become nonfunctional, the roots are reabsorbed and the tooth falls out. New molars erupt at the posterior tooth row (back of the mouth) and migrate anteriorly as replacements. Manatees apparently have an unlimited number of molars that can be replaced in this way. This is an evolutionary novel solution to coping with an abrasive diet. Dugongs do not possess such a conveyor belt mechanism to resist an abrasive diet; they are thought to be at a disadvantage compared to manatees when consuming grasses. Instead, dugong teeth are open-rooted, simple peg-like molars that consist of dentin covered by cementum (a hard substance that differs from enamel). These molars erupt slowly over their life time, through anterior drift, also known as molar progression (Unger, 2010). Unlike manatees, however, dugongs have a finite number of molars (six). Interestingly, there is evidence that suggests dugong cheek-teeth may not be functional at all. Instead, the enlarged, heavily cornified keratin palatal pads may function to process and transport sea grasses into the mouth. Although manatees also possess similar cornified palatal pads, those of dugongs are much more robust and cover a larger surface area (Marshall and Pyenson, 2019).

The facial muscles of all sirenians form a short muscular snout that is capable of highly complex and varied movements. Other examples include tongues, elephant trunks, and squid tentacles. These are "muscular organs" with no bones. In sirenians, the complex facial muscles surround a series of six fields of modified whiskers or **vibrissae**, which are located on both the broad and expanded upper and lower lips (Marshall and Pyenson, 2019; Marshall et al., 2022). Vibrissae are specialized hairs that transmit tactile (touch sensation) information from the environment to the brain (Bauer, Reep, and Marshall, 2018; **see Chapter 7**). Although, these vibrissae are similar to vibrissae of other mammals, manatee vibrissae are called bristles since they are short and thick. Manatee bristles are also unusual in that they function in both manipulation and grasping as well as sensory roles (Marshall, Sarko, and Reep, 2022). When contracted, facial muscles protrude the largest pair of bristles on the upper and lower lips, which are then used to handle vegetation (Marshall and Pyenson, 2019). How this is accomplished differs between manatees and dugongs (Marshall et al., 2003). In manatees, the paired upper bristle fields are protruded anteriorly and then toward the center of the mouth in a grasping motion, pushing vegetation in the mouth. This is a prehensile-like pinching action. This movement is alternated with a sweeping motion of the lower bristle fields to further push vegetation into the mouth. In dugongs, the upper bristle fields have a more horizontal distribution across the lip margin but can also be protruded, as much as 6 cm. However, rather than moving toward the midline, the upper bristles are moved to each side in a breast-stroke-like motion. As dugongs graze along the sea floor, this functions to part the sea grass in front of the animal and introduce plant material into the side of the mouth. As in manatees, the bristle fields on the lower jaw alternate and sweep vegetation further into the mouth. When feeding upon small species of sea grasses, this mechanism can be used to literally excavate the root system, including rhizomes, from the sea floor. This action is manifested as a signature "feeding trail" on the sea floor. During feeding events, this behavior creates substantial disruption of the sea floor. Feeding trails span the width of the dugong rostrum and can be as deep as 5 cm and as long as 10 m. Up to 90% of the vegetation can be removed from these feeding trails.

ORDER CARNIVORA: PINNIPEDIA

Pinnipeds are a lineage of carnivores that include sea lions and fur seals, true seals, and walruses. Each family has successfully transitioned back to the aquatic environment, albeit at different timescales. Living pinnipeds consume a diversity of prey that include fish, cephalopods, bivalves, crustaceans, invertebrates, and large "warm blooded" prey (penguins, seabirds, other marine mammals; King, 1983; Pauly et al., 1998). Our knowledge of how pinnipeds feed is still largely descriptive and lacks the detailed functional analyses found for fishes. However, new studies are beginning to change this perspective. Three themes in the feeding function of pinnipeds are: (1) loss of **mastication** (chewing), (2) a reduced and simplified dentition, and (3) importance of suction feeding. Based on head and tooth morphology, as well as functional studies, pinniped feeding had been generally characterized as biting, suction, grip-and-tear, or filter feeding, but more recent performance work has demonstrated that pinnipeds are multimodal in their feeding mechanics and typical use several of these feeding modes (e.g., Marshall, Kovaks, and Lydersen, 2008; Hocking, Evan, and Fitzgerald, 2013; Hocking et al., 2014; Marshall et al., 2014, 2015; Kienle et al., 2018).

Functional studies of terrestrial carnivores demonstrate that head, tooth, and jaw morphology are good predictors of feeding performance and diet (**Figure 10.2**). It is also well known that certain biomechanical skull elements of mammals are associated with specific types of prey (Radinsky, 1981a, 1981b). Therefore, working backward, capture of specific prey suggests that head and tooth morphology are also associated with specific prey capture modes in pinnipeds (Jones and Goswami, 2010; Jones, Ruff and Goswani, 2013; Kienle and Berta, 2016).

The skulls and lower jaws of pinnipeds are the least specialized among marine mammals (with a few notable exceptions) and are more like those of canids (i.e., members of the dog family, **Figure 10.1**). Modifications of head and tooth morphology, particularly shape, have long been thought to be associated with feeding adaptations (Marshall and Pyenson, 2019). In general, pinniped skulls are rounded and sharply delineated from the facial region (rostrum or snout) of the skull; they do not possess sinuses with air chambers. The skulls of sea lions fur seals are less variable than those of true seals, and many are **sexually dimorphic**. Compared to closely related terrestrial carnivores, pinniped dentition is reduced. Their premolars and molars tend to be uniform in cusp number, size and often shape, resulting in similar shapes. Premolars and molars are often collectively referred to as cheek-teeth or postcanine teeth. As a consequence, pinnipeds have lost the shearing cusps seen in terrestrial carnivores (Unger, 2010). However, among true seals, and perhaps some sea lions and fur seals, there are some interesting cheek-teeth that are associated with specific feeding modes and trophic ecology (Peredo, Ingle, and Marshall, 2022).

The skulls of sea lions and fur seals tend to be longer than true seals and show less variation in their postcanine teeth. Their canines are large relative to true seals and are thought to be used for grip-and-tear feeding (biting large prey and shaking the head to remove flesh). The cheek-teeth are all similar, but possess the addition of a large shelf-like cusps. Although distinctive tooth types such as those found in crabe-ater (*Lobodon carcinophaga*) and leopard seals (*Hydrurga leptonyx*) are not found among sea lions and fur seals, feeding specializations still exist (Peredo, Ingle, and Marshall, 2022). Many sea lions and fur seals possess interesting feeding mechanics for capturing prey. For example, Antarctic fur seals feed heavily upon zooplankton including krill. Their postcanine teeth are among the smallest of any sea lion or fur seal, and are thought to be modified for filtering krill. The skulls of South American sea lions are perhaps the most different in this group. Their skulls are large and strong, the upper palate is arched and long, the jaws are short and broad, and the lip and face muscles are well developed to broadening the snout. This suite of characteristics is commonly associated with suction feeding specialization (King, 1983) and indirect data strongly suggest that South American sea lions are capable of generating powerful suction. These traits are exemplified in the skulls of walruses – well-known suction specialists.

The number of teeth of true seals is reduced relative to sea lions and fur seals. Among true seals, there is a diversity of dental types which are likely adaptations for catching fish and squid, but outstanding exceptions for filter feeding exists. At least two species, leopard and crabeater seals, possess intricate lobes of the cheek-teeth that are thought to assist in filter feeding. The distinctive postcanine teeth of crabeater seals possess three long shearing cusps. Observations of captive crabeater seals suggest they ingest krill using suction, then push water out while filtering and keeping the krill inside their mouths using these teeth like a sieve (Ross et al., 1976; Klages and Cockcroft, 1990). Leopard seals are known to feed upon penguins and other seals using a grip and tear feeding mode. The enlarged canines and strong jaw and neck muscles are useful for biting and grip-and-tear feeding modes. However, leopard seals also possess pronounced lobes on their postcanine teeth that are effective for filtering as in crabeater seals. Consumption of krill by leopard seals was once thought to be minimal but is now know to be important seasonally. Functional feeding data demonstrates they are capable of suction feeding when consuming krill and then using their postcanine teeth as a sieve to filter feed on krill (Klages and Cockcroft, 1990; Hocking, Evan, and Fitzgerald, 2013). Biting and suction feeding modes are likely still the most commonly used feeding mode. Notable among true seals is

the suction feeding capability of bearded seals (*Erignathus barbatus*). Their dental and orofacial morphology, feeding performance (suction and hydraulic jetting), and trophic ecology are similar to walruses, a known suction specialist. Bearded seals consume **infaunal** invertebrates such as bivalves and tubeworms, as well as fish. The upper palate of bearded seals is also vaulted, although not to the degree observed in South American sea lions (*Otaria flavescens*) and walruses (*Odobenus rosmarus*). They can generate negative pressures that match the capability of walruses (Marshall, Kovaks, and Lydersen,, 2008). Face and lip muscles close the lateral lips to prevent negative pressure loss, while the broad lips at the anterior are pursed to create a circular aperture. These two features create a pipette like structure that enhances suction generation and direct those negative pressures in front of the animal (Marshall, 2016). Such behavior has also been reported for walruses (Fay, 1982), but also for harbor seals and Steller sea lions (*Eumetopias jubatus*; Marshall, Rosen, and Trites, 2015).

The skulls of walruses are distinctive in its size, fusion, and generally unusual morphology (**Figure 10.1**) compared to other mammals. The tusks dominate its morphology. The cheekbones in the walrus are enlarged to accommodate and anchor the tusks to the skull (Fay, 1982; King, 1983). The frontal orientation of the cheekbones broadens and shortens the rostrum, which is advantageous for benthic feeding. The tusks of walruses are not used for feeding. Instead, tusks are used for male–male interactions and for hauling out of the water onto ice; walruses commonly use their tusks to pull and lift their bodies from the water. In fact, the history of their Latin name (*Odobenus*) means tooth walker. The tusks are ever-growing upper canines that can grow up to a meter in length in males (Fay, 1982; King, 1983; Unger, 2010). The upper and lower incisors of walruses (typically present between the right and left canines (tusks)) are lost, leaving a wide space between the two tusks on the upper jaw and canines on the lower jaw.

This loss creates a circular space that is part of the formation of a pipette-like shape that assists in generating negative pressure. Walruses can generate suction higher than 1 atmosphere of pressure (Marshall and Pyenson, 2019). The up-and-down movement of a piston-like tongue is thought to be responsible for generating the substantial negative pressures. The vaulted palate enhances suction generation since there is a greater volume within the mouth to act on. Walruses are infaunal benthic specialists consuming mostly **bivalves** (e.g., clams, mussels, and oysters) and other invertebrates, but with occasional exceptions such as marine birds and other marine mammals. Functional studies have demonstrated that in addition to suction, walruses can perform the opposite behavior – hydraulic jetting or squirting water out of the mouth with high pressure. Large bivalves are excavated from the sea floor by alternating suction with hydraulic jetting to remove the sediment around their prey.

SUCTION AND BITING FEEDING MODES

Unlike some odontocetes, it is noteworthy that extreme rostral elongation has not evolved among pinnipeds. This suggests that biting and suction feeding modes are not biomechanical tradeoffs as in odontocetes, but are perhaps synergistic in pinnipeds. Short, wide jaws with high mechanical advantage for biting feeding modes should also be advantageous for suction feeding if other morphological adaptations are also present. Indeed, evidence from captive performance studies shows that suction is widespread across most pinniped groups. Pinnipeds that are presumed, or known, to employ suction feeding tend to have short, wide rostra with jaws that have scoop-like anterior ends. There is also a trend toward simplification of tooth morphology and reduced functionality, lost functionality, or loss of certain classes of teeth. The postcanine teeth of walruses, bearded seals, elephant seals, are simple oval peg-like teeth

with simple cusps that are quickly worn flat with age, having with little-to-no crowns (Fay, 1982; King, 1983; Unger, 2010). These teeth may not be functional, or limited in function, in adults (Peredo, Ingle, and Marshall, 2022). Monk seals depart from this pattern. Although no specific functional feeding studies of monk seals have been conducted, animal-borne camera studies of feeding (Parrish et al., 2002), as well as anecdotal evidence, strongly support a suction capability in these species. Their broad, large teeth suggest these shallow water foragers crush many prey items.

The biomechanics and shape of the lower jaw differ among pinnipeds (Jones, Ruff, and Goswani, 2013). The shape of pinniped jaws appears to separate species based on their level of relatedness and each family displays distinct morphologies. In addition, there are fundamental differences in jaw closing muscles of sea lions and fur seals. Their jaws are relatively longer than those of true seals. However, male–male combat and the influence of mating strategies appears to also act as a strong evolutionary driver in pinniped mandibular function (Jones, Ruff, and Goswani, 2013).

Harbor seals (*Phoca vitulina*) are the ultimate generalists among pinnipeds. A feeding performance study of harbor seals (Marshall et al., 2014) suggests that suction may be widespread among all true seals. Harbor seals are the most widely distributed true seal, and despite some disparity, there are up to five subspecies recognized. Harbor seals are opportunistic foragers that exhibit a generalized feeding ecology. For example, they feed upon a wide diversity of small- to medium-size fishes that include herring, anchovy, cod, hake, trout, smelt, shad, scorpionfish, rockfish, prickleback, greenling, sculpin, capelin, sandlance, salmon, flatfish, a variety of cephalopods, and invertebrates that are mostly crab and shrimp species, but also include mollusks. Therefore, harbor seals are ideal candidates to test questions regarding feeding performance and the presence of suction feeding among true seals and all pinnipeds

(**Figure 10.2**). In fact, animal-borne cameras attached to harbor seals provide convincing evidence that they likely use several feeding modes (suction and biting; Bowen et al., 2002). Harbor seals displayed a wide repertoire of behaviorally flexible feeding strategies, which likely forms the basis of their opportunistic, generalized feeding ecology and concomitant wide breadth of diet. It is this multimodal feeding mode that has likely allowed this seal to be so successful.

ORDER CARNIVORA: SEA OTTERS

Sea otters (*Enhydra lutris*) are the smallest and most recent group of marine mammals to return to marine habitats. They are thought to have arisen in the North Pacific during the Pleistocene and have only become fully aquatic in the last 1–3 million years. They are considered carnivores and two subfamilies exist. These are mainly terrestrial predators (e.g., fishers, martens, and wolverines) and all otters, including sea otters.

Although feeding adaptations have been examined to some extent in terrestrial species, fewer data exist for otters. Morphological and behavioral diversity among otters is reflected in their diet and foraging behaviors (Radinsky, 1981a, 1981b; Goswami, 2006;). For example, in river otters, some jaw-opening muscles are larger compared to terrestrial carnivores; this is thought to allow the rapid jaw movement necessary for catching elusive fish with their mouths underwater (Goswami, 2006). This is also reflected in their skull morphology (**Figure 10.1**); river otters possess broad protuberances where these enlarged muscles attach (Goswami, 2006). River otters also possess sharp **carnassials** (a pair of upper and lower shearing cheek-teeth common in carnivores) necessary for piercing and processing fish. In contrast, sea otters possess short, blunt skulls with blunt, wide cheek-teeth used for crushing hard, benthic prey.

Recent functional data (Timm-Davis, DeWitt, and Marshall, 2015; Law, Young, and Mehta, 2016) suggests that sea otters exhibit further specialized feeding methods. Sea otters forage on the bottom, in waters as deep as 40 m. Their diet is varied but shellfish and urchins comprise a large portion of what they eat. Otters use their forepaws to dig out clams, and to pry shellfish and urchins from the rocky substrate, sometimes using tools. Food is usually consumed at the surface; behavioral observations suggest that otters do not use their teeth underwater, even when feeding on fish. Upon surfacing, fish are killed by a bite to the head. A rock or some other tool is usually carried in a flap of skin under the arm and is used to pound open shellfish. The spines of urchins are simply bitten off and the shell of the urchin is crushed with the cheek-teeth.

It is commonly understood that sea otters consume hard-shelled prey and forage primarily upon infaunal as well as **epibenthic** invertebrates. However, Northern and Russian sea otters also prey on bottom fish and occasionally on tetrapods. Sea otters differ from other otters in that the morphological traits are correlated with consuming hard-shelled prey. These traits include their extremely blunt and wide skull and jaw, and a wide cheek-teeth relatively to their body size. Sea otters also possess a relatively high mechanical advantage of the masseter muscle (one of the main jaw-closing muscles) compared to other otters, which is thought to increase force at the most posterior molars as an adaptation for crushing hard-shelled animals. This feature in other mammals is known to provide additional control over mastication. An analysis of sea otter skulls show that they have taller and wider lower jaws, and shorter and blunter skulls than other otters (Timm-Davis, Dewitt, and Marshall, 2015), which is consistent with increased bite force at the carnassials of other mammals. Furthermore, the blunt crushing carnassials, and all molars in general, are known to possess relatively thicker enamel that is resilient in its structure and is 2.5 times stronger than human enamel (Marshall and Pyenson, 2019). Both morphological and behavioral investigations of sea otter biting demonstrate that they can display wide gapes (Timm-Davis, DeWitt, and Marshall, 2015) while maintaining powerful bite forces – useful adaptations for feeding upon bivalves. This ability to consume hard-shelled prey is referred to as **durophagy.** Thus, sea otters demonstrate a newly described mode of feeding termed as durophagous biting (Timm-Davis, DeWitt, and Marshall 2015).

ORDER CARNIVORA: POLAR BEARS

Polar bears (*Ursus maritimus*) are considered to be marine mammals since they spend a significant time in the marine environment and rely upon the consumption of other marine mammals to survive. This species arose only less than 1 million years ago from brown bears, which are omnivorous (Slater et al., 2010). Although they tend to be completely carnivorous (Slater et al., 2010), when required, polar bears will resort to **omnivory** and supplement their diet with fruits, vegetation, and various other food material. However, recent data demonstrates that the skull morphology of polar bears (**Figure 10.1**) does differ from brown bears, and other bears, reflecting the increased biomechanical selection pressures for consuming a primarily marine mammal diet (Slater et al., 2010). Polar bears tend to grasp their prey with their mouths and break the neck or skull of their prey with their large jaw muscles and strong teeth. Interestingly, their skull morphology and feeding biomechanics demonstrates a potential tradeoff leaving polar bears less effective at processing tough food matter associated with omnivory (Slater et al., 2010). A study of the head and teeth morphology of all bears in relation to diet demonstrates significant morphological separation among bears that exhibit omnivory, herbivory, **carnivory**, and **insectivory**. Only polar and brown bears exhibit significant carnivory. Feeding adaptations include a reduction in molar size, flexible lower jaws, and relatively small flesh-shearing blades (**carnassials**). Polar bear feeding adaptions have more in common with omnivorous canids. The less than robust

feeding apparatus that might be expected from polar bears can be explained by the fact that polar bears target ringed and bearded seal pups and seals that are under 2 years of age (King, 1983). These seals are much smaller in body size and much more vulnerable than larger prey. However, polar bears can take large prey such as walruses, beluga whales also (King, 1983; Sacco and Van Valkenburgh, 2004).

TOOLS AND METHODS FOR STUDYING FEEDING IN MARINE MAMMALS

As discussed throughout this book, marine mammals are notoriously difficult to study both in the wild and in captivity. Consequently, much work on feeding mechanics focused on morphology. Morphology can be used to predict function, but with caution, and these types of studies should be considered functional hypotheses until function can be verified using experimental methods. Regardless, head and tooth morphology can be very instructive for investigating feeding mechanisms. Indeed, new methods such as computed tomography, 3D reconstruction, 3D printing, finite element analyses (a computer mesh model of a skeleton that predicts how forces move through the structure), physical and computational biomechanical modeling, materials science, and geometric morphometrics (shape analyses) are providing new and exciting frontiers in morphological research. Advances in camera, video, computing, electrophysiology capability, and the reduction in size and increased portability of equipment are now allowing pool side performance experiments that could only have been conducted in the laboratory in the past. This translates into additional focused and experimental captive studies, which quantitatively measures movement, performance, and physiology. Such experiments interface well with field studies in which a cadre of animal-borne tags and instruments on free-ranging animals (**Figure 10.2**) collect data from foraging bouts and events in the open ocean. Such tools include integrated systems that incorporate time-depth recorders with video systems, oceanographic data collection systems, and 3D movement sensors (accelerometers, magnetometers, and gyroscopes) that can record behavior, body movement, visual perspective, acoustics and physiology from free-swimming marine mammals. New unmanned aerial vehicles (UAVs) are providing new access and perspective to animal behavior and beyond.

CONCLUSIONS

It's an exciting time to be a marine mammalogist. The future of functional studies of marine mammal feeding is bright. Advances in new technologies as described above will allow researchers to close the gap between experimental captive studies and field studies. As new fossils are discovered, new tools will allow researchers to investigate the evolution of feeding and elucidate functional transitions from land to sea and the evolution of a wide range of feeding modes and foraging behaviors. In total, the impact of synthesis and integration will advance our understanding of mammalian evolution and the functional pathways that aquatic mammals took during their transition to land as well as the broader evolutionary and ecological patterns of vertebrate natural history.

REVIEW QUESTIONS

1 What is the primary difference in how baleen whales and toothed whales catch their prey?
2 Are there true seals or sea lions and fur seals that eat krill?
3 How do sirenians (manatees and dugong) differ in their feeding mechanics compared to other marine mammals?
4 What are major feeding differences in feeding mechanics between manatees and dugongs?
5 Which are more diverse in their feeding mechanics, true seals or sea lions and fur seals?

6 Which marine mammals are suction specialists and how to they achieved maximum suction capability?
7 How do the teeth of sea otters differ from other otters?
8 Do polar bears feeding differently than other bears?

CRITICAL THINKING

1 Why do you think some whales evolved baleen, and others teeth for catching prey? Is there an advantage for one over the other?
2 How can blunt, wide skulls be effective for both suction feeding and biting?
3 What are the underlying biological reasons that have driven the evolution of diverse feeding mechanisms observed among marine mammals?

FURTHER READING

Chakrabarty P. *Explaining Life Through Evolution.* New Delhi: Penguin Books India PVT, Limited. 2022. ISBN: 9780670095100, 0670095109

Davis RW. *Marine Mammals: Adaptations for an Aquatic Life.* Cham: Springer International Publishing. 2019. ISBN: 9783319982809, 331998280X

Irschick DJ and Higham TE. *Animal Athletes: An Ecological and Evolutionary Approach.* Oxford: Oxford University Press. 2016. ISBN: 9780199296552, 0199296553

Jarrett B and Shirihai H; Illustrated by Brett Jarrett. *Whales, Dolphins and Seals: A Field Guide to the Marine Mammals of the World.* New York: MacMillan. 2021. ISBN: 1472969669, 9781472969668

Kardong K. *Vertebrates: Comparative Anatomy, Function, Evolution.* New York: McGraw Hill Education. 2019. ISBN10: 1259700917 ISBN13: 9781259700910

Leatherwood S, Randall RR, and Foster L. *The Sierra Club Handbook of Whales and Dolphins.* San Francisco: Sierra Club Books. 1983. ISBN: 9780871563408, 0871563401

Marsh H. *Ethology and Behavioral Ecology of Sirenia*. Cham : Springer. 2022. ISBN: 9783030907426, 3030907422

Stewart BS, Reeves R, and Leatherwood S. *The Sierra Club Handbook of Seals and Sirenians*. San Francisco: Sierra Club Books. 1992. ISBN: 9780871566560, 0871566567

GLOSSARY

Benthic: Sea floor.

Biomechanics: Application of mechanical engineering to organisms.

Bivalves: Invertebrates with two shells such as clams, mussels, and oysters.

Carnassials: An upper and lower pair of cheek-teeth common in carnivores used to shear prey.

Carnivory/carnivore: Consumption of meat; member of Carnivora family, a distinct mammalian lineage that consume meat.

Durophagy/durophagous: The ability to crush and consume hard or hard-shelled prey.

Echolocation: Ability to navigate or locate objects using sound.

Ecomorph: Nonrelated organisms that have similar morphology.

Epibenthic: On top of the sea floor.

Feeding: Consumption of matter that provides energy for development, sustenance, maintenance, and survival.

Gape: Angle of the corner of the mouth.

Herbivory/herbivore: Consumption of plants; organisms that consume plants.

Hindgut: Combined small and large intestines.

Hydrodynamic: Motion of fluid.

Hyoid: A bone to which the tongue is attached.

Hyolingual: Referring to the hyoid and tongue.

Hypsodont teeth: Tall tooth crown.

Infaunal: Within the sea floor.

Insectivory/insectivore: Consumption of insects, a subset of carnivory; organisms that consume insects.

Ma: Millions of years ago.

Mastication: Processing of food items using teeth, jaw, and associated muscles; chewing.

Phylogeny: Evolutionary history of a group of organisms.

Performance: Also known as behavioral performance; a behavior that can be directly measured such as bite force or suction.

Omnivory/omnivorous: Consumption of both plant and animal matter; organisms that eat both plant and animal matter.

Piscivory/piscivorous: Consumption of fish, a subset of carnivory; organisms that consume fish.

Ram feeding: Increasing the swimming speed to overtake or outperform the agility of prey.

Rhizome: Carbohydrate "packets" associated with seagrasses and other plants.

Selection pressure: Factors that cause selection of favorable traits in evolution.

Sexually dimorphic/sexual dimorphism: Male and females of the species have different morphologies and often different body sizes.

Tetrapods: four-limbed vertebrates including frogs, mammals, turtles, and others.

Teuthophagy/teuthophagous: Consumption of squid, a subset of carnivory.

Trophic: Referring to food or an ecological food web.

Venturi effect: Flow of fluid through a constricted section of a pipe-like structure; suction.

Vibrissae: Specialized hairs (whiskers) that transmit touch information from the environment to the brain.

REFERENCES

Bauer, G.B., R.L. Reep,, and C.D. Marshall, 2018. The tactile senses of marine mammals. *International Journal of Comparative Psychology* 31. DOI:10.46867/ijcp.2018.31.02.01

Bloodworth, B., and C.D. Marshall, 2005. Feeding kinematics of *Kogia* and *Tursiops* (Odontoceti: Cetacea): Characterization of suction and ram feeding. *Journal of Experimental Biology* 208: 3721–3730.

Bloodworth, B., and C.D. Marshall, 2007. A functional comparison of the hyolingual complex in pygmy and dwarf sperm whales (*Kogia breviceps* & *K. sima*), and bottlenose dolphins (*Tursiops truncatus*). *Journal of Anatomy* 211: 78–91.

Bowen, W.D., D. Tully, D.J. Boness, et al. 2002. Prey dependent foraging tactics and prey profitability in a marine mammal. *Marine Ecology Progress Series* 244: 235–245.

Fay, F.H., 1982. Ecology and biology of the Pacific walrus, *Odobenus rosmarus divergens Illiger*. *North American Fauna* 74: 1–279.

Fordyce, R.E., and L.G. Barnes. 1994. The evolutionary history of whales and dolphins. *Annual Review of Earth and Planetary Sciences* 22: 419–455.

Gaskin, D.E. 1982. *The Ecology of Whales and Dolphins*. New York: Heinemann.

Goldbogen, J.A., D.E. Cade, J. Calambokidis, et al. 2017. How baleen whales feed: The biomechanics of engulfment and filtration. *Annual Review of Marine Science* 9: 367–386.

Goldbogen, J.A., J. Calambokidis, D.A. Croll, et al. 2012. Scaling of lunge feeding performance in rorqual whales: Mass-specific energy expenditure increases with body size and progressively limits diving capacity. *Functional Ecology* 26: 216–226.

Goldbogen, J.A., J. Calambokidis, R.E. Shadwick, et al. 2006. Kinematics of diving and lunge-feeding in fin whales. *Journal of Experimental Biology* 209: 1231–1244.

Goldbogen, J.A., J. Potvin, and R.E. Shadwick. 2010. Skull and buccal cavity allometry increase mass-specific engulfment capacity in fin whales. *Proceedings of the Royal Society of London B* 277: 861–868.

Goldbogen, J.A., N.D. Pyenson, and R.E. Shadwick. 2007. Big gulps require high drag for fin whale lunge feeding. *Marine Ecology Progress Series* 349: 289–301.

Goswami, A. 2006. Morphological integration in the carnivoran skull. *Evolution* 60: 122–136.

Hocking, D.P., A.R. Evan, and E.M.G. Fitzgerald. 2013. Leopard seals (*Hydrurga leptonyx*) use suction and filter feeding when hunting small prey underwater. *Polar Biology* 36: 211–222.

Hocking, D.P., M. Salverson, E.M.G. Fitzgerald, et al. 2014. Australian fur seals (*Arctocephalus pusillus doriferus*) use raptorial biting and suction feeding when targeting prey in different foraging scenarios. *PLoS One* 9 p. e112521.

Jones, K.E. and A. Goswami. 2010. Quantitative analysis of the influences of phylogeny and ecology on phocid and otariid pinniped (Mammalia; Carnivora) cranial morphology. *Journal of Zoology (London)* 280: 297–308.

Jones, K.E., C.B. Ruff, and A. Goswami. 2013. Morphology and biomechanics of the pinniped jaw mandibular evolution without mastication. *Anatomical Record* 296: 1049–1063.

Kahane-Rapport, S.R., and J.A. Goldbogen. 2018. Allometric scaling of morphology and engulfment capacity in rorqual whales. *Journal of Morphology* 279(9): 1256–1268.

Kahane-Rapport, S.R., M.S. Savoca, D.E. Cade, et al. 2020. Lunge filter feeding biomechanics constrain rorqual foraging ecology across scale. *Journal of Experimental Biology* 223(20) jeb224196.

Kane, E.A., and C.D. Marshall,2009. Comparative feeding kinematics and performance of odontocetes: Belugas, Pacific white-sided dolphins, and long-finned pilot whales. *Journal of Experimental Biology* 212: 3939–3950.

Kienle, S.S., and A. Berta. 2016. The better to eat you with: The comparative feeding morphology of phocid seals (Pinnipedia, Phocidae). *Journal of Anatomy* 228(3): 396–413.

Kienle, S.S., H. Hermann-Sorensen, D.P. Costa, et al. 2018. Comparative feeding strategies and kinematics in phocid seals: Suction without specialized skull morphology. *Journal of Experimental Biology* 221(15): jeb179424.

King, J.E. 1983. *Seals of the World*. 3rd ed. Ithaca: Cornell University Press.

Klages, N.T.W., and V.G. Cockcroft. 1990. Feeding behavior of a captive crabeater seal. *Polar Biology* 10: 403–404.

Law, C.J., C. Young, and R.S. Mehta. 2016. Ontogenetic scaling of theoretical bite force in southern sea otters (Enhydra *lutris* nereis). *Physiological and Biochemical Zoology*, 89(5): 347–363.

Marshall, C.D., 2016. Morphology of the bearded seal (*Erignathus barbatus*) muscular-vibrissal complex: A functional model for phocid subambient pressure generation. *Anatomical Record* 299: 1043–1053.

Marshall, C.D., and J. Goldbogen. 2016. Marine mammal feeding mechanisms. In: *Marine Mammal Physiology: Requisites for Ocean Living*.

M.A. Castellini and J.A. Mellish, eds. Boca Raton: CRC Press, pp. 95–117.

Marshall, C.D., and N.D. Pyenson. 2019. Feeding in aquatic mammals: An evolutionary and functional approach. In: *Feeding in Vertebrates: Anatomy, Biomechanics, Evolution.* Cham: Springer, pp. 743–785.

Marshall, C.D., K. Kovacs, and C. Lydersen. 2008. Feeding kinematics, suction, and hydraulic jetting capabilities in bearded seals (*Erignathus barbatus*). *Journal of Experimental Biology* 211: 699–708.

Marshall, C.D., D. Rosen, and A.W. Trites. 2015. Feeding kinematics and performance of basal otariid pinnipeds, Steller sea lions (*Eumetopias jubatus*), and northern fur seals (*Callorhinus ursinus*): Implications for the evolution of feeding modes. *Journal of Experimental Biology* 218: 3229–3240..

Marshall, C.D., D. Sarko, and R.L. Reep. 2022. Morphological and sensory innovations for an aquatic lifestyle. In: *Sirenian Ethology.* H. Marsh, ed. Cham: Springer, pp. 19–65.

Marshall, C.D., H. Maeda, M. Iwata, et al. 2003. Orofacial morphology and feeding behaviour of the dugong, Amazonian, West African and Antillean manatees (Mammalia: Sirenia): Functional morphology of the muscular – vibrissal complex. *Journal of Zoology (London)* 259: 1–16.

Marshall, C.D., S. Wieskotten, W. Hanke, et al. 2014. Feeding kinematics, suction, and hydraulic jetting performance of harbor seals (*Phoca vitulina*). *PLoS One* 9: e86710.

McCurry, M.R., C.W. Walmsley, E.M.G Fitzgerald, et al. 2017. The biomechanical consequences of longirostry in crocodilians and odontocetes. *Journal of Biomechanics* 56: 61–70.

Nerini, M. 1984. A review of gray whale feeding ecology. In: *The Gray Whale Eschrichtius robustus.* M.L. Jones, S.L. Swartz and S. Leatherwood, eds. Orlando: Academic Press, Inc., pp. 423–449.

Parrish, F.A., K. Abernathy, G.J. Marshall, et al. 2002. Hawaiian monk seals (*Monachus schauinslandi*) foraging in deep-water coral beds. *Marine Mammal Science* 18: 244–258.

Pauly, D., A.W. Trites, E. Capuli, et al. 1998. Diet composition and trophic levels of marine mammals. *ICES Journal of Marine Science.* 55: 467–481.

Peredo, C.M., D.N. Ingle, and C.D. Marshall. 2022. Puncture performance tests reveal distinct feeding ecomorphs in pinniped teeth. *Journal of Experimental Biology* 225: jeb244296.

Potvin, J., and A.J. Werth. 2017. Oral cavity hydrodynamics and drag production in balaenid whale suspension feeding. *PLoS One* 12(4): e0175220.

Pyenson, N.D., and D.R. Lindberg. 2011. What happened to gray whales during the Pleistocene? The ecological impact of sea-level change on benthic feeding areas in the North Pacific Ocean. *PLoS One* 6: e21295.

Pyenson, N.D., J.A. Goldbogen, and R.E. Shadwick. 2013. Mandible allometry in extant and fossil Balaenopteridae: The largest vertebrate skeletal element and its role in rorqual lunge-feeding. *Biological Journal of the Linnean Society* 108: 586–599.

Pyenson, N.D., N.P. Kelley, and J.F. Parham. 2014. Marine tetrapod macroevolution: Physical and biological drivers on 250 Ma of invasions and evolution in ocean ecosystems. *Palaeogeography, Palaeoclimatology, Palaeoecology* 400: 1–8.

Radinsky, L.B., 1981a. Evolution of skull shape in carnivores. 1. Representative modern carnivores. *Biological Journal of the Linnean Society* 15: 369–388.

Radinsky, L.B., 1981b. Evolution of skull shape in carnivores: 2. Additional modern carnivores. *Biological Journal of the Linnean Society* 16: 337–355.

Reidenberg, J.S., and J.T. Laitman.1994. Anatomy of the hyoid apparatus in Odontoceti (toothed whales): Specializations of their skeleton and musculature compared with those of terrestrial mammals. *Anatomical Record* 240: 598–624.

Ross, G.J.B., F. Ryan, G.S. Saayman, et al. 1976. Observations on two captive crabeater seals at the Port Elizabeth Oceanarium. *International Zoo Yearbook* 16: 160–164.

Sacco, T., and B. Van Valkenburgh. 2004. Ecomorphological indicators of feeding behaviour in bears (Carnivora: Ursidae). *Journal of Zoology (London)* 263: 41–54.

Simon, M., M. Johnson, and P.T. Madsen. 2012. Keeping momentum with a mouthful of water: Behavior and kinematics of humpback whale lunge feeding. *Journal of Experimental Biology*, 215: 3786–3798.

Simon, M., M. Johnson, P. Tyack, et al. 2009. Behaviour and kinematics of continuous ram filtration in bowhead whales (*Balaena mysticetus*).

Proceedings of the Royal Society of London Series B 276: 3819–3828.

Slater, G.J., B. Figueirido, L. Louis, et al. 2010. Biomechanical consequences of rapid evolution in the polar bear lineage. *PLoS One* 5: e13870.

Steeman, M.E., M.B. Hebsgaard, R.E Fordyce, et al. 2009. Radiation of extant cetaceans driven by restructuring of the oceans. *Systematic Biology* 58: 573–585.

Timm-Davis, L.L., T. DeWitt, and C.D. Marshall. 2015. Divergent skull morphology supports two trophic specializations in otters (Lutrinae). *PLoS One*. 10.1371/journal.pone.0143236.

Unger, P.S. 2010. *Mammal Teeth: Origin, Evolution and Diversity*. Baltimore: Johns Hopkins Press. p. 304.

Velez-Juarbe, J., D.P. Domning, and N.D. Pyenson. 2012. Iterative evolution of sympatric seacow (Dugongidae, Sirenia) assemblages during the past ~26 million years. *PLoS One* 7: e31294.

Weihs, D., and P.W. Webb. 1984. Optimal avoidance and evasion tactics in predator-prey interactions. *Journal of Theoretical Biology* 106: 189–206.

Werth, A.J. 2006a. Odontocete suction feeding: Experimental analysis of water flow and head shape. *Journal of Morphology* 267: 1415–1428.

Werth, A.J. 2006b. Mandibular and dental variation and the evolution of suction feeding in Odontoceti. *Journal of Mammalogy* 87: 79–588.

Werth, A.J. 2007. Adaptations of the cetacean hyolingual apparatus for aquatic feeding and thermoregulation. *Anatomical Record*. 290: 546–568.

Werth, A.J. and J. Potvin. 2016. Baleen hydrodynamics and morphology of cross-flow filtration in balaenid whale suspension feeding. *PLoS One*, 11(2): e0150106.

Werth, A.J., C. Loch, and R.E. Fordyce. 2020. Enamel microstructure in Cetacea: A case study in evolutionary loss of complexity. *Journal of Mammalian Evolution* 27(4): 789–805.

Werth, A.J., M.A. Lillie, M.A. Piscitelli, et al. 2019. Slick, stretchy fascia underlies the sliding tongue of rorquals. *The Anatomical Record* 302(5): 735–744.

Allyson Hindle

University of Nevada, Las Vegas (UNLV), Las Vegas, NV

David Rosen

University of British Columbia, Vancouver, Canada

Allyson Hindle

Allyson Hindle has been a faculty member in the School of Life Sciences at the University of Nevada, Las Vegas since 2019. She has been studying metabolic physiology in mammals for 20 years. She works with marine mammals as well as hibernators; both types of animals experience physiological extremes in their natural environments. She uses physiological telemetry to study freely behaving animals in the field, combined with molecular and genomic approaches in the lab, leading the team that analyzed the genome of the Weddell seal. Her research has taken her from the North Pacific to Antarctica.

David Rosen

David Rosen is an assistant professor at the Institute for the Oceans and Fisheries at the University of British Columbia, Canada. He has been studying the physiology, energetics, nutrition, and behavior of marine mammals for more than 30 years. While he first chose to study marine mammals because "they are so cool," his main research focus is to contribute to science-based conservation strategies for threatened marine mammal populations. As director of the Marine Mammal Energetics and Nutrition Laboratory, one of his main tools is to conduct studies with marine mammals held temporarily or permanently under human care. As an offshoot, he has also become involved in legislative and practical efforts to ensure the welfare of these marine mammals in zoos and aquariums.

Driving Question: How do marine mammals survive long periods without feeding?

INTRODUCTION

Marine mammals forage in the oceans, but many species also spend long periods of time away from their feeding grounds. Some species even spend long periods of time on land. In fact, most marine mammals experience some predictable periods of fasting throughout their lives. But surprisingly, these fasting periods often occur during life stages when additional energy is needed for important tasks such as long-distance migrations, growth in young animals, lactation in females, and territory defense in males. How can these critical activities happen during fasting without animals starving to death? By combining changes in behavior and

DOI: 10.1201/9781003297468-11

physiology to coincide with fasting, animals can survive these periods with minimal long-term consequences.

Although a general model can describe fasting physiology in most vertebrates, the physiology of fasting in marine mammals is constantly surprising scientists. This is partially because not all marine mammals can fast to the same degree. This chapter will discuss the marine mammals that are champions of fasting, as well as those with poor fasting abilities. Among the many different marine mammals that withstand extended fasts, even closely related species seem to cope with fasting in very different ways. As a further complication, many physiological mechanisms shift as the fast goes on, allowing the animal to adjust to ongoing challenges. As is typical in other fasting-adapted vertebrates, lipid stores are the most important fuel that support long periods without food in marine mammals. Most marine mammals have large amounts of lipid deposited in the blubber layer that lies beneath their skin (see also **Chapter 5**). This blubber layer is a critical component of their fasting tolerance and sets them apart from other vertebrates. While fasting often occurs during energetically demanding life history stages, such as the breeding season and growth, marine mammals can manage these activities and extend their fast through hormonal control and by reducing other energetic costs.

WHAT IS FASTING AND WHEN DOES IT OCCUR?

Fasting refers to the physiological state of living with a complete lack of food intake. In the strictest sense, short-term fasting occurs constantly in most mammals, since individuals are not continuously eating – or more accurately they are digesting previously acquired food or prey. In this chapter, we will not be discussing such short-term events (what is referred to as **dynamic equilibrium**). We will address "fasting" as a complete lack of food intake over longer intervals of days, weeks, or even months.

It might be surprising to learn that scientists do not wholly agree upon clear definitions for the terms "fasting state" and "starvation." Both imply a lack of food intake, and while they are sometimes used interchangeably, they actually define two different physiological states. Broadly, the terms fasting and starvation differ in the severity of the effects of lack of food intake. Where fasting can be sustained over a period of time, a "starvation state" is characterized by life-threatening or potentially irreversible physiological and anatomical changes.

Fasting and starvation can also be distinguished by cause or motivation. A fasting animal is unwilling to eat (or forage) even though food might be available. For marine mammals, this happens when another life requirement occurs which is not compatible with foraging in the ocean. Fasting could result during periods of time on land (e.g., in seals and sea lions), when these animals nurse their pups, molt, or in some species defend breeding territories. Fasting might also result when foraging must happen in specific ocean locations that they cannot access, for example, during long migrations or while defending underwater breeding territories. While fasting occurs during hibernation in terrestrial animals also, no marine mammals exhibit these strategies. The sole exception is the pregnant female polar bear, *Ursus maritimus*. In contrast, starvation refers to the condition when an animal is willing to consume food but extrinsic factors, such as lack of prey, inclement weather, or predator avoidance, limit food consumption.

This difference between fasting and starvation is also related to predictability; fasting is a natural part of life history and may therefore be anticipated. Since fasting is "natural" and timed to life history, an animal will be more physiologically prepared to withstand this lack of food intake at certain times of the year or life stage. We expect that the impacts of a natural fasting episode be less detrimental than those incurred during "unpredicted" episodes of starvation.

Pinnipeds

Almost all pinniped species fast, and their fasting episodes also coincide with periods when they are available on land for intensive study. As a result, most of our knowledge about marine mammal fasting physiology comes from this group.

Importantly, most pinnipeds experience their first fasts during their initial year of life, a time when physical growth is critical to survival, and their ability to forage is physically and behaviorally limited. After a period of near-constant energy intake, thanks to maternal milk during nursing, many phocid pups are abruptly weaned. This leads to a postweaning fast, which often corresponds to their first annual molt. In northern elephant seals (*Mirounga angustirostris*), this postweaning fast lasts for 9–12 weeks. For the most part, pups remain on land during this period assumedly to minimize the costs and

duration of the molt. This means that fasting is a consequence of the environmental boundary imposed by the molt. While some physiological adaptations are designed to facilitate their survival during this fasting period, the extended postweaning fasts may also give them time to develop diving capacity.

Otariid strategies of weaning often differ from those of phocid pups, with many otariids weaning later in life compared to phocids (**Figure 11.1**). This corresponds to a longer nursing period while the pups are developing, and some otariid species may avoid complete fasting in the postweaning period if they can gain some experience of swimming and foraging while the female is still in attendance (Fowler et al. 2007). The more fundamental difference is that otariid pups characteristically contend with regular, intermittent fasts throughout their nursing period when their mothers periodically return

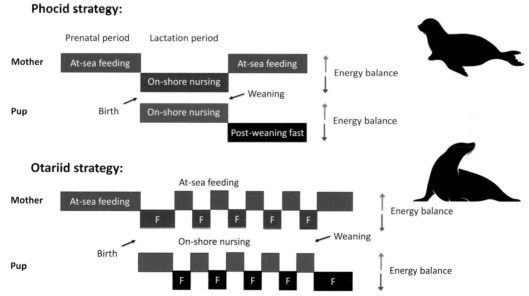

Figure 11.1 Different lactation strategies of phocid seals and otariids necessitated by pups being on land and food being offshore. In many cases, phocids nurse their pups from birth to weaning while continuously fasting on shore. This can encompass a period of time ranging from 4 days (hooded seals) to several weeks. Otariids nurse their pups for several days during the perinatal period and then go to sea to feed. They subsequently alternate nursing their pups on shore (when mothers must fast) with feeding trips to sea (while their pups fast ashore) for the duration of the lactation period, which may last for more than a year.

to sea to feed. The complex way females balance time and energy devoted to lactation while managing their own fast is covered in **Chapter 12**. The nursing/foraging cycle of otariid females can be viewed as a balancing act between maternal and offspring fasting episodes. The length of maternal foraging trips and the resulting fasting episodes of pups can last from days to weeks, depending on species and stage of lactation. These episodes can make up a significant portion of the total nursing period for pups. For example, episodes of 4–5 day foraging bouts by Antarctic fur seal mothers (*Arctocephalus gazella*) interspersed with 1–2 days on shore mean that their pups are potentially fasting for 71% of the total nursing period (Boyd 1991; Lunn et al. 1993). On the other hand, the time that the mother is attending her pup on land necessitates that she herself fasts while expending tremendous energy on milk production.

Pinniped males also fast during the breeding season. For many species, breeding strategies rely on continuously defending territories on beaches or breathing holes underwater, often for months at a time. This requires them to expend large amounts of energy fighting and mating, thus eliminating the opportunity to feed for the entire reproductive season. To put it in perspective, northern elephant seal males lose about one-third of their starting body mass over the 3-month breeding period (Crocker, Houser, and Webb 2012). As one might expect, males are often in very poor physical shape by the end of the breeding season, and many are often not observed in subsequent years (LeBoeuf 1974). Individuals in poor condition that are not observed to return to the rookeries in later years are presumed to have died due to detrimental rates of protein loss over the extended, expensive fast (Sharick et al. 2015).

Cetaceans

Many large baleen whales travel great distances across oceans, alternating between summer feeding grounds in high-latitude cold waters and winter reproductive areas in low-latitude warm waters (Corkeron and Connor 1999). The trip itself – which can be thousands of kilometers - requires considerable energetic output, and it is not clear whether these whales feed during migrations based on the speeds that they travel (Horton et al. 2011). Once on the calving grounds, many whale species likely fast due to limited food availability in these low-productivity waters. The warmer water may provide a more hospitable thermal environment and potentially fewer predators for their calves (Corkeron and Connor 1999), but it does coincide with high maternal energetic costs of reproduction and lactation. Similar patterns of extensive migrations are not generally observed in smaller-bodied odontocetes, but there is evidence of fasting during periodic "maintenance" migrations undertaken by Orcas (*Orca orcinus*; (Durban and Pitman 2012)).

ENERGY SUBSTRATES

Energy is constantly required to maintain cellular and biochemical **homeostasis** (see **Chapter 5**). As we shall discuss later, although there are strategies to limit and streamline energy expenditure requirements during fasting, energy use will always exceed external energy intake, which is zero during fasting. Energy requirements during fasting episodes are therefore met solely by substantial contributions from endogenous (internal) body reserves. **Catabolism** is the process by which energy-containing molecules, or substrates, are broken down to release energy that can be used by the body to do work. Fuel stores are unequally accumulated within body tissues in three forms: fat (lipid), protein, and carbohydrate.

The most common energy source for a feeding organism is glucose – a carbohydrate that is packaged and stored as glycogen in the liver and muscles. Despite the importance of glucose in meeting the energy requirements of body tissues, including the brain and red blood cells, glycogen stores are relatively small in most marine mammals, and these animals also do not acquire large amounts of glucose from their prey. By far, lipids and proteins provide the largest energy reserves. Lipids, stored as fat, are the preferred fuel to sustain during fasting for several reasons. They have the highest caloric density, meaning that more energy is liberated per gram of fat catabolized (approx. 40 kJ g^{-1} depending on the composition of lipid stores) compared to proteins (approx. 18 kJ g^{-1}). Lipids also provide more metabolic water on breakdown than proteins (107 g vs. 40 g per 100 g tissue), an important resource that is also limited or completely lacking from external sources during fasting (**Figure 11.2**).

Unlike many terrestrial mammals that store abdominal fat, most lipids used as an energy reserve for marine mammals are situated in the hypodermal (subcutaneous) blubber layer. This blubber layer can account for 50% of the total body mass, and while providing critical insulation, its extent is generally far greater than required for thermoregulation (see **Chapter 5**). During **lipolysis**, enzymes known as lipases

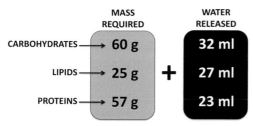

Figure 11.2 **Lipids are the most energy-dense substrate; less lipid mass is required to generate 1000 kJ of energy compared to carbohydrates or proteins. These estimates are respectively based on glucose as the carbohydrate source, the complete oxidation of an "average" fat, and the catabolism of a protein to urea. Different sources or configurations of carbohydrate, lipid and protein energy stores will yield slightly different values. Although catabolism of carbohydrates to generate 1000 kJ of energy produces the most metabolic water, lipids yield the most water per gram. (Data from Schmidt-Nielson 1997; Edeny 1977).**

break down **triacylglycerols** present in the blubber layer into **glycerol** and **free fatty acids** (FFA). Glycerol is available as an immediate energy source. FFAs must be activated for transport into the mitochondria, where they are broken down into acetyl-CoA subunits (via beta oxidation), which enter the citric acid cycle (to eventually produce energy). Fatty acids are also partially oxidized in the liver into four-carbon **ketone bodies**, which are later oxidized as fuel by other tissues (**Figure 11.3**).

In addition to their lower energy density, it is least desirable to catabolize proteins because they are energetically expensive to make and are a critical element of tissues. All enzymes, molecular chaperones, receptors, and many building blocks of the cell itself are proteins. Protein catabolism diminishes the integrity of lean body tissues. In fact, death by starvation is linked to depletion of lean body mass which is why obese animals can still starve to death. Limiting protein catabolism during the fast has a direct effect on survival (Øristland and Markussen 1990; Oritsland 1990). Protein catabolism also generates nitrogen end products that must be recycled or removed in urine. This is at odds with the strategy to minimize urinary water loss observed in many animals without access to external water stores (see **Chapter 8**).

However, protein stores do play a critical role in providing glucose during a fast. A constant supply of glucose is required by neurons and the central nervous system (which lack enzymes to oxidize FFAs, although they can use ketone bodies for a portion of their energy requirements) and red blood cells (which lack mitochondria and therefore the enzymes for both FFA and ketone oxidation). Glycogen stores in the liver are usually depleted within a few days of fasting. Fortunately, the liver and kidneys can then produce glucose through the process of **gluconeogenesis** which helps to maintain a reduced but relatively constant blood glucose level throughout extended fasts. Gluconeogenesis forms glucose from the noncarbohydrate precursors **glycerol** (from **lipolysis** of adipose

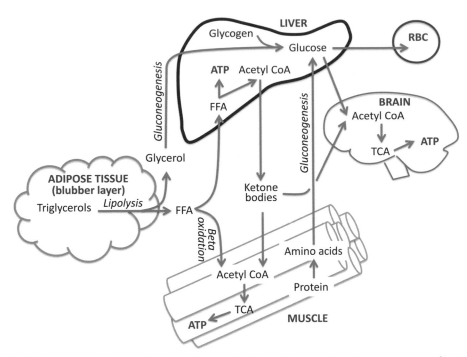

Figure 11.3 Summary of some of the major metabolic pathways related to energy production during a fast. Adapted from Lieberman and Marks (2009). Triacylglycerols catabolized via lipolysis from adipose tissues form the major source of energy during fasting. The resulting free fatty acids (FFA) are transformed via beta oxidation into acetyl coenzyme A (acetyl CoA), which produce energy (ATP) at target tissues via the tricarboxylic acid (TCA) cycle. Some of the FFAs that enter the liver are oxidized directly, but most are converted to ketone bodies (acetoacetate and β-hydroxybutyrate), which are released into the blood (the liver cannot use ketone bodies itself). The ketone bodies are oxidized in the mitochondria of target cells into acetyl-CoA, which enters the TCA cycle. Certain tissues require glucose to function. This can be originally met by liver glycogen stores, which are rapidly depleted. New glucose can be formed in the liver (gluconeogenesis) from the smaller glycerol segment of triacylglycerols and from amino acids freed via protein catabolism (proteolysis). The central nervous system can derive some of its energy from oxidation of ketone bodies, but red blood cells are obligate glucose consumers. In the later stages of fasting, protein becomes a more dominant metabolic substrate.

triacylglycerols) and amino acids (via **prote-olysis**, the breakdown of proteins to individual amino acids). Since most fatty acids cannot provide carbon for gluconeogenesis, only the small glycerol portion of the vast store of food energy contained in adipose tissue, i.e., triacylglycerols, can enter the gluconeogenic pathway. Therefore, some level of protein catabolism is constantly required during fasting to provide the amino acid components to manufacture glucose for these glucose-obligate systems.

These differences in the abundance, benefits, and consequences of fueling fasting metabolism through different substrates results in many vertebrates, including marine mammals, undergoing a predictable progression in fasting physiology. This series of adaptive changes was formalized in the late 1980s and 1990s by scientists such as George Cahill and Yves Cherel into three "classical phases of fasting" (**Figure 11.4**).

Phase I: Readily accessible but limited carbohydrate stores (such as glycogen) are used to

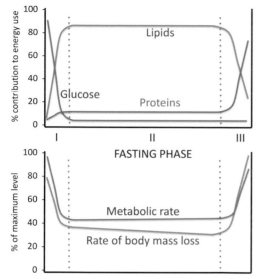

Figure 11.4 Predicted changes in physiological parameters according to the classical three phases of fasting as detailed by Cherel, Robin, and Maho (1988). Significant changes include a shift in metabolic fuel use (top panel) and rates of body mass loss and mass-specific metabolic rates (bottom panel).

fuel the initial stages of fasting (Cherel, Robin, and Maho 1988). Glycogen stores are primarily found in the liver and muscles. Animals also transition to a protein-sparing metabolism and increase mobilization of lipid resources. Despite energetic contributions from other tissue sources, carbohydrates are depleted in hours to days. The shift may be shorter in carnivorous/piscivorous species whose diets contain little carbohydrate.

Phase II: This period represents the most physiologically stable fasting state that can last for weeks to months, where fasting marine mammals rely almost entirely on lipid stores. Unfortunately, not all biochemical processes can be continued by lipid catabolism alone. As mentioned, protein sources play a limited but vital role by supplying amino acids to replenish circulating glucose via gluconeogenesis. Animals able to accumulate greater body fat reserves prior to fasting will sustain this physiological state longer; however, lipid stores are finite.

Phase III: The blubber layer cannot be completely depleted without infringing on thermoregulatory capabilities, causing further energetic imbalances due to increased heat loss (Rosen, Winship, and Hoopes 2007). At this point, marine mammals must end their protein-sparing strategy; however, increased protein catabolism represents the "end-game" physiological state associated with impending exhaustion of fuel reserves. Although Phase III physiology can be reversible upon refeeding, at this point animals must quickly end their fast or they might suffer severe physical consequences including death. Phase III is sometimes classified as starvation rather than fasting. The critical impacts to the body resulting from this phase are the reason that reliance on protein catabolism only lasts days to weeks. Fasting in most marine mammals does not typically extend into Phase III.

Body mass dynamics often reflect these changes in fasting physiology in combination with changes in energy expenditure. Rates of mass loss tend to decline early in the fasting period, reaching stable and relatively low rates for most of the episode. A terminal switch to protein catabolism is often accompanied by rapidly increasing rates of mass loss.

Shifts in substrate metabolism are also apparent as recognizable changes in biochemical signatures. These include changes in levels of enzymes necessary to breakdown different fuels, as well as metabolite by-products and end products of those reactions. Circulating FFAs derived from lipid (triacylglycerol) breakdown are metabolized into fuel that can be used directly by tissues or they can be partially oxidized in the liver to produce the ketone bodies β-hydroxybutyrate (β-OHB) and acetoacetate (**Figure 11.2**). Water-soluble ketones can then be used as energy by cells instead of glucose. Ketone bodies can pass the blood–brain barrier, providing a portion of the vital fuel for the brain and central nervous system. As a result of these shifts in metabolic substrate, plasma levels of FFA and β-OHB increase rapidly during the initial phases of fasting and remain high as indicators of an almost complete reliance on lipid catabolism (**Figure 11.3**). Later increases in protein breakdown are mirrored by elevated rates of nitrogen excretion and blood urea nitrogen (BUN) levels.

Phocids

One of the best-studied scenarios is the postweaning fast of phocid seals that can last up to 10 weeks. In postweaning gray seal pups (*Halichoerus grypus*), the energetic contribution from lipids can reach up to 94–97% of total energy use (Oritsland et al. 1985; Reilly 1991). Similarly, northern elephant seal pups may derive up to 98% of their energetic requirements from lipids over their 10-week fast (Pernia, Hill, and Ortiz 1980; Houser and Costa 2001). For harp seal pups (*Phoca groenlandica*), 86–97% of energy is derived from lipids in the first 8 weeks of their fast (Worthy and Lavigne 1983; Nordøy, Ingebretsen, and Blix 1990; Nordøy, Aakvaag, and Larsen 1993), and 72–80% of this energy comes from blubber lipids (Worthy and Lavigne 1983; Kovacs and Lavigne 1985). While extended fasting episodes (>8 week) are associated with increased protein catabolism, it still only accounts for up to 15.7% of energy use (Worthy and Lavigne 1983).

While phocid seals rely extensively on lipid metabolism during the bulk of their postweaning fast, the underlying metabolic shift to this substrate is not instantaneous. Many seals appear to rely initially on protein stores, for up to 1–3 weeks (Worthy and Lavigne 1983, 1987). For example, during the initial 3 days of the very brief hooded seal (*Cystophora cristata*) pup fast, 16% of the total mass loss was from fat while 28% was from protein (the remainder was water) (Lydersen, Kovacs, and Hammill 1997). The pups also have high initial rates of mass loss, which rapidly decrease to lower, stable rates by day 5 (Bowen, Boness, and Oftedal 1987). Harp seal pups similarly use greater levels of protein in the initial 2 weeks of their fast before switching almost entirely to lipid catabolism

(Worthy and Lavigne 1983, 1987). This protein to lipid switch is partly reflected in changes their blood chemistry (Nordøy, Aakvaag, and Larsen 1993). While gray seal pups exhibit uniformly low levels of protein oxidation starting at day 4 of their postweaning fast (Nordøy, Ingebretsen, and Blix 1990) their rates of mass loss, which continue to drop over the first 10 days of fasting, appear to be partly disconnected from changes in metabolic substrate use.

Plasma FFAs increase continuously during the fasts of harp and gray seal pups (Nordøy and Blix 1991), reaching levels even higher than seen in northern elephant seals (Castellini, Costa, and Huntley 1987). These increased FFAs are accompanied by linearly increasing β-OHB in harp and gray seal pups after day 8 (Nordøy and Blix 1991). On the other hand, northern elephant seal pups produce ketone bodies almost immediately upon the onset of fasting (Castellini and Costa 1990). These levels increase until ~day 55, before declining sharply, soon followed by the pup's departure to sea. It is interesting that northern elephant seal pups likely have the longest postweaning fast and seem able to switch to a total reliance on lipid metabolism almost immediately (Castellini and Costa 1990; Adams and Costa 1993). Protein catabolism indicative of "endstage" fasting is scarce among phocid seals, as most begin foraging before lipid reserves are overly depleted (Castellini and Rea 1992; Houser and Costa 2003). Scientists *have* observed marine mammals entering Phase III fasting, but it is not common. For example, "starveling" northern elephant seal weanlings have 2–4 times the protein catabolism rates as "healthy" animals of the same age (Houser and Costa 2003). Not surprisingly, these "starveling" pups do not survive if they enter Phase III fasting before they molt, and they appear more likely to strand if they go to sea after completing their molt (Houser and Costa 2003).

In theory, the ability of pups to remain in Phase II fasting should depend upon the extent

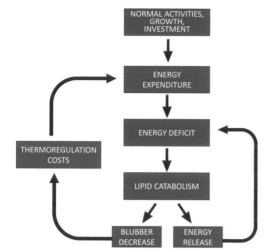

Figure 11.5 Schematic representing spiraling thermoregulatory consequences of depletion of the blubber layer. During Phase II fasting, marine mammals primarily rely on the catabolism of lipids stored in the blubber to fuel the normal costs of existence. However, at some point, the blubber layer becomes so thin that its effectiveness as insulation decreases. The animal then has to expend more energy to thermoregulation, increasing the amount of lipids that need to be catabolized, thereby further decreasing the blubber layer and increasing the thermoregulatory costs.

of their individual lipid reserves. As previously noted, depletion of lipid reserves will not only lead to greater reliance on protein reserves (eventually leading to Phase III) but this can also have thermoregulatory consequences (**Figure 11.5**). Many studies have examined the thermal effects of changes in blubber reserves during fasting in phocid seal pups, while others have documented the effect of thermal challenges, body composition, and substrate use. Northern elephant seal pups exhibit differing energy usage patterns depending on body mass and body composition at the end of nursing. Fatter pups catabolize more lipids and spare proportionally more protein than leaner pups. As a consequence, leaner pups, catabolizing relatively more protein versus lipid, conclude their postweaning fast with sufficient remaining lipid stores for

thermoregulation during their first at-sea foraging trip (Noren and Mangel 2004). There is evidence that species which normally fast in water are more adapted to preserving their lipid layer for thermoregulation than species which fast on land (Worthy and Lavigne 1987). Experimentally, gray seal pups (which normally fast on land) had thinner blubber layers after a 10-week fast in water compared those fasting on land, whereas harp seals (which normally fast partly in water) showed no differences in body composition whether fasting on land or in water (Worthy and Lavigne 1987).

Maintaining stable circulating glucose levels is critical for proper nervous system functioning and this can be particularly challenging for pups. Initial drops in circulating glucose are sometimes observed early in the fast, before production catches up to utilization. Harp seal pups maintain high, constant plasma glucose levels over 32-day weaning fasts, except for an early, insignificant decline (Worthy and Lavigne 1982; Nordøy, Aakvaag, and Larsen 1993). In contrast, gray seal pups exhibit a constant 20% decline in glucose levels that reflects decreasing replacement rates over their 52-day fast (Nordøy, Ingebretsen, and Blix 1990). Plasma glucose concentrations in northern elephant seal pups are quite variable. While they may sometimes decline slightly during the fast, they generally remain at consistent high levels, with no obvious changes related to either the start of weaning or the decrease in circulating ketones that is thought to signal the end of weaning (Castellini and Costa 1990; Ortiz, Wade, and Ortiz 2001; Crocker et al. 2012).

Breeding male phocids also rely on lipid metabolism during their extensive fasts. Male northern elephant seals meet only 7% of energy expenditure through protein breakdown (Crocker, Houser, and Webb 2012). Protein use declines with initial proportion of body fat (so thinner males use more protein catabolism), emphasizing the importance of adequate prefast lipid stores. Circulating metabolites in these fasting males differ from expectations for fasting mammals relying on lipolysis, and also differ from fasting females and pups (Castellini and Costa 1990; Houser, Champagne, and Crocker 2007). β-OHB levels are very low and only increase slightly during fasting in contrast to females and pups, suggesting ketone regulation differs with life history. Furthermore, BUN levels (and serum FFA) are consistent over the fasting and are unrelated to protein catabolism (Crocker, Houser, and Webb 2012).

Food and water restriction may also occur outside of the breeding season and some experiments have attempted to mimic these "unpredicted" but natural episodes. Juvenile harbor seals (*Phoca vitulina*) experimentally fasted for 2 weeks derived most (~75%) of their energy from fat, showed a marked (20%) decrease in metabolism (Markussen, Ryg, and Øritsland 1992; Markussen 1995), and a linear decline in body mass typical of lipid-based metabolism.

Otariids

In comparison to the postweaning fasts experienced by phocid pups, otariid pups must endure repeated fasts during the nursing period due to regular maternal foraging trips, which can last from days to weeks. Antarctic fur seal pups transition to a lipid-based metabolism within 2–3 days of their 5-day nursing fast (Arnould, Green, and Rawlins 2001). Protein turnover accounts for only 5.4% of total energy expenditure over the course of their fast, during which they exhibit decreases in plasma BUN, triglyceride concentrations, and circulating glucose levels, and increases in β-OHB.

Subantarctic fur seal pups (*Arctocephalus tropicalis*) undergo relatively prolonged nursing fasts compared to most other otariids which initially last for 10 days but extend up to 3–4 weeks just prior to weaning (Georges and Guinet 2000). On average, mass loss during these fasts comprised 56.3% lipids and 9.6% protein (the remainder was water), suggesting that 93% of their energy requirements are derived from lipids (Beauplet, Guinet, and Arnould 2003). Curiously, protein

catabolism was twice as high in female pups, despite their greater lipid reserves.

Based on blood chemistry, 6-week-old Steller sea lion pups (*Eumetopias jubatus*) make a swift metabolic transition to fasting within 16 hours, with rapidly decreased plasma BUN concentrations and increased β-OHB (Rea, Rosen, and Trites 2000). Subsequent increases in plasma BUN implies that the pups reverted to protein catabolism after only 2.5 days of fasting. Older Steller sea lions undergoing 7–14-day experimental fasts (Rea et al. 2009) also demonstrated both age- and season-specific responses. Similar to fasting pups, BUN decreased rapidly in both juveniles and subadults. This decline was more rapid for juveniles during the nonbreeding season, when animals had slightly higher prefasting lipid reserves) compared to during the breeding season. Significantly increased BUN concentrations observed at the end of the nonbreeding season fasts suggest that subadult Steller sea lions are unable to maintain a protein-sparing metabolism for a full 14-day fast during this season. Subadult and juvenile sea lions also exhibited lower circulating ketone body concentrations compared to pups, suggesting age-related differences in substrate use. Breeding male otariids also undergo substantial fasts, but almost nothing is known about their physiology during this period.

Other marine mammals

We know virtually nothing about the metabolic basis of fasting physiology in any other marine mammal species. Sea otters are very intolerant to fasting, becoming rapidly hypoglycemic after even overnight fasts. Presumably, this is due to their relatively small lipid stores coupled with a very high metabolic rate. Cetaceans likely support fasting from massive lipid reserves contained in their blubber, but emerging studies suggest that small cetaceans, such as dolphins, have limited tolerance for fasting (Houser et al. 2021). On the other hand, large baleen whales appear to be the champions of fasting. It is understandably difficult to study metabolic fuel use over periods of fasting in large whales in the open ocean which is why less is known about these species, compared to smaller marine mammals. One of the few available studies analyzed urine samples collected from migrating humpback whales. Scientists used the sodium content in the urine to account for saltwater ingestion and urea concentrations to predict protein catabolism during the fast. By subtracting out these two sources of water from the urine sample, the metabolic water produced from the breakdown of lipids was estimated to be 96.7%, which meant that only 3% came from proteins (Bentley 1963). Polar bears are also extremely tolerant to fasting; summer land-based bears have highly elevated FFA levels in their blood (Nelson et al. 1983). As Arctic seal predators, they rely heavily on seasonal sea ice to access productive feeding grounds and can experience extended periods where marine prey is inaccessible. They are also known to recycle nitrogen from urea during fasting to limit protein loss.

CONSERVING ENERGY DURING FASTING

The impact of fasting depends on the accrued energetic deficit, which is a direct consequence of an animal's energy expenditure. It is not surprising that animals may make behavioral and physiological adjustments to minimize their energy requirements during a fast. Potential energetic savings can come from decreases in activity levels, thermoregulatory costs, and **resting metabolic rate** (**RMR**; Cherel, Robin, and Maho 1988). For example, many fasting pups are minimally active. Subantarctic fur seal pups spend more time sleeping than other otariid pups, contributing to lower daily rates of energy expenditure and lower rates of mass loss while their mothers were foraging (Beauplet, Guinet, and Arnould 2003). What is surprising is that energy conservation efforts often happen during critical energy-demanding life stages.

For adult females, these conflicts between energy expenditure and energy conservation are during lactation and mating, while for adult males these are during the defense of breeding territories. There are very few studies that have explored what options reproductive marine mammals employ to minimize costs during these expensive reproductive periods. Reduced physical activity will obviously lower energy use; however, the option to do so is highly dependent on coinciding behavioral requirements. Female pinnipeds spend a lot of time resting when hauled out, likely reducing activity costs while nursing their pups. Males holding a reproductive territory may also rest between frequent bouts of high-energy behavior (for both defense and mating activities). However, as demonstrated in elephant seals, more successful males expend more energy over the mating period (Deutsch, Haley, and Le Boeuf 1990).

The molting period is another inflexible life history stage that imposes additional energy requirements and also happens during a fast for most marine mammals. Molting has both direct costs (skin and hair replacement) and potential indirect thermoregulatory costs associated with an increase in skin temperature and metabolism that facilitates pelage replacement. Thermoregulation costs during the molt are reduced in pinnipeds by spending more time hauling out on land, given the higher thermal conductivity (potential rate of heat loss) of water versus air (see also **Chapter 5**). Some species, such as elephant seals, have a more condensed and drastic molt than others; the shorter molting period allows them to haul out and fast over the entire process. Cetaceans also experience an annual molt, when they shed the outer layer of their skin. Obviously, cetaceans do not have the option to exit the ocean water and limit their thermoregulatory costs during the molt, as pinnipeds are able to do. Instead, killer whales, for example, migrate to warmer waters to molt, and there is evidence that they do not forage during these trips (Durban and Pitman 2012). The assumption is that regenerating skin in warmer waters provides net energy savings and shortens the total molt time despite the additional locomotory costs and extended fast that it requires.

The seasonal migrations of baleen whales between calving and foraging grounds were originally thought to occur for thermoregulatory benefits to calves. These migrations can be unbelievably extensive; humpback whales (including mothers and calves), migrating from Antarctic feeding grounds to Pacific wintering areas off the coast of Central America, travel approximately 8,300 km and largely fast during the trip as well as on the calving grounds. While adult whales have sufficient blubber to withstand the colder waters in the feeding grounds, it was suggested that there was a thermal benefit to the smaller calves by relocating to warmer waters. While this may be partly true, it is now hypothesized that the main benefit of this migration is a reduction in predation on the inexperienced calves.

RESTING METABOLIC RATE

Metabolic depression is a term for the decrease in resting metabolic rate (RMR) – the total energy expenditure of an inactive animal within its thermoneutral zone. Metabolic depression is often cited as a common physiological adaptation to periodic food shortages, among many species (Keys et al. 1950), that serves to limit rates of mass loss despite insufficient energy intake. This strategy may be key to maximizing protein sparing during extended fasts (Henry, Rivers, and Payne 1988), by limiting glucose consumption rates and therefore catabolism of protein for gluconeogenesis. Among vertebrates, rapid depression of RMR is often observed early in the fast, leading to a new steady state fasting level (i.e., the amount of energy expended per kg body mass). As mass-specific RMR remains constant, decreases in overall (or absolute) metabolic rate throughout the fast are due to decreasing body mass. There are also indications of increasing metabolic rate toward the end of the natural fasting period, as cells and tissues are prepared for the resumption of foraging and digestion.

Depressed RMR can occur by at least three avenues: selective loss of metabolically active tissue, downregulation of energetically expensive processes, and reduction of cellular metabolism. While protein catabolism and loss of lean tissue will reduce total metabolism (vs. loss of metabolically inert lipids), mass-specific decreases in metabolism can most effectively be produced by targeted catabolism of an animal's most energy-demanding tissues. One common strategy is to shrink portions of the digestive tract, which represents a large portion of total body mass and requires continuous turnover of the cell lining, which is relatively expensive to maintain. Fortunately, during periods of fasting, the gut is not required, and can be "restructured" when food becomes available again. The kidney is another metabolically expensive organ. However, unlike the gut, there is no evidence that the kidney becomes physically smaller during fasts. This is partly due to its complexity and partly due to the fact that it is still required to process metabolic waste and help maintain fluid and electrolyte homeostasis through resorption of endogenous electrolytes and body water (see **Chapter 8**). However, energy is saved if the level of processing through the kidney is significantly reduced during fasts (usually measured as changes in glomerular filtration rate), due to decreased protein turnover and increased water conservation. Metabolic depression can also occur through downregulation of cellular processes. This hypometabolic state is not the result of major biochemical reorganization, but of molecular controls operating at a level "above" that of allosteric regulation of enzymes and "below" that of gene expression.

Phocids

Several studies have examined changes in RMR during the postweaning fast of phocid pups. Collectively, these illustrate apparent species-specific differences in patterns of changes in metabolism during the fast. Fasting gray seal pups demonstrate the expected rapid decrease mass-specific RMR, which reaches a reached stable level 45% below initial levels by day 10, and continues largely unchanged through day 47 of the fast (E. S. Nordøy, Ingebretsen, and Blix 1990). A subsequent study confirmed the constancy of postweaning mass-specific metabolism in both gray seals and harp seals fasting for 8 weeks in air, although they also found an increase in RMR over the fast for individuals fasting (but not tested) in water, potentially due to differences in blubber and mass loss incurred due to higher thermal costs (Worthy and Lavigne 1987). Such changes in RMR seem to conform to the standard fasting model.

However, a study on northern elephant seals revealed a different pattern. During the course of their 10-week molting fast, metabolism of pups gradually declined by 35% (L. D. Rea and Costa 1992). This may explain why fasting pups lose mass at a faster rate during the first 4 weeks of the fast than during the following 4-week period. Metabolism also notably increased during the final week of the fast, just prior to when they set off to sea to forage.

A few studies have been conducted on fasting in older phocids. Both experimentally fasted juvenile and adult harbor seals and gray seals display rapid (within 24 hour) decreases in mass-specific metabolism (Markussen, Ryg, and Øritsland 1992; Boily and Lavigne 1995; Markussen 1995) indicative of rapid entrance into an adaptive "fasting state."

Otariids

Several studies have examined changes in the RMR of otariid pups during the nursing period. Mass-specific RMR decreased significantly while Antarctic fur seal pups were fasting over 5 days (normal for maternal foraging trips) (Arnould, Green, and Rawlins 2001). However, it is unclear whether part of the large drop might have been due to the large increase in metabolism attributable to digestion of the milk (i.e., the heat increment of feeding) in their stomachs at the start of the fasting period, highlighting a difficulty in measuring and interpreting resting metabolism in intermittent feeders. Subantarctic fur seal pups show increasing fasting abilities, including a decrease in mass-specific metabolism during their on-shore fasts, which themselves increase from 5 to 30 days over the course of their extended period of maternal investment (Verrier et al. 2011).

However, the metabolic response of mammals during life history stages or seasons when food shortages normally occur may be different from the response during periods when food shortages are unexpected. Young Steller sea lion pups (6–14 weeks old) significantly decrease metabolism in response to 48-hour fasts. However, 6 to 24-month-old northern fur seals (*Callorhinus ursinus*) showed no changes in RMR when subject to a similar fast (Rosen, Volpov, and Trites 2014). The difference might be attributable to the fact that, at this age, Steller sea lions would still be undergoing natural fasts during the nursing stage, while northern fur seals would already be weaned. By contrast, juvenile Steller sea lions outside their natural fasting period showed a 31% decrease in RMR over a 14-day experimental fast (Rosen and Trites 2002).

Other marine mammals

Polar bears likely exhibit episodes of metabolic depression throughout the year and perhaps episodes of hibernation during denning (Nelson et al. 1983). Curiously, there are no published studies of energy-saving strategies among any other types of marine mammals. Hence, our vision of what makes up a "marine mammal" response is likely highly skewed. This is obviously an important area for future comparative research into the bioenergetic strategies of fasting among marine mammals.

HORMONE CONTROL

Changes in an animal's physiology during a fast – including metabolic rate, appetite, growth, and choice of metabolic substrate – are all part of a controlled shift that is mediated by a suite of hormones. In this section, we will focus on those hormones thought to be most important in fasting physiology of marine mammals, but the list is by no means exhaustive.

Glucocorticoids are steroid hormones known most commonly as a biochemical marker of physiological stress ("stress hormone"). However, they have multiple specific purposes (such as within the immune system) including serving an important regulatory role during fasting. In marine mammals, the primary glucocorticoid is cortisol. Plasma cortisol levels increase during prolonged fasting. Among its functions, cortisol helps to provide energy by increasing lipolysis and the associated mobilization of FFAs. It also assists in maintaining circulating glucose concentrations via increased gluconeogenesis from protein stores. It is believed that during fasts the impacts of increased cortisol on lipid mobilization predominates over protein wasting effects (Ortiz, Wade, and Ortiz 2001).

The increase in lipolysis and lipid oxidation during extended fasting is thought to be regulated by increased levels of **growth hormone (GH)** and decreased levels of **leptin**. GH is a peptide hormone that, true to its name, stimulates cellular growth. However, it also functions as a "stress hormone," increasing in response to fasting in most species. It is important in the conservation of protein during fasting by raising the concentration of FFAs (via increased lipolysis). In non-fasting animals, it also stimulates production of the hormone **IGF-1 (insulin-like growth factor 1)** which is important in protein production (anabolism). However, during fasting there is both an elevation of GH and suppression of IGF-1 (possibly due to reduced hepatic GH receptors) (Crocker et al. 2012). Inhibition of IGF-1 secretion allows hypersecretion of GH during fasting without diverting energy to tissue growth (Crocker et al. 2012).

Leptin is a relatively recently discovered hormone that is often referred to as the "satiety hormone." It is produced in fat cells and in feeding animals it regulates the amount of fat stored in the body through adjustment of the hunger response. In fed animals, as fat deposition surpasses a critical point, the fat cells release increasing levels of leptin, thus decreasing the sensation of hunger (increasing satiety) and increasing energy expenditures, promoting lipid oxidation, and reducing triacylglycerol synthesis. Episodes of fasting are usually associated with lowered leptin concentrations as part of an animal's strategy for limiting energy expenditures (Crocker et al. 2012).

This is partly because decreases in leptin are also associated with decreases in the **thyroid hormones** T3 (triiodothyronine) and T4 (thyroxine) in fasting animals. These hormones, produced by the thyroid gland, are primarily responsible for regulation of metabolism, and therefore decreased levels facilitate the decreases in metabolic expenditures associated with metabolic depression. This is despite the fact that the energy-demanding process of fat mobilization might require higher thyroid hormone levels (Ortiz, Wade, and Ortiz 2001). During extended fasting, reductions occur in levels of both T4 and T3, the former assisted by the conversion of T4 to the biologically inactive reverse T3 (rT3).

Ghrelin antagonizes the action of leptin and is known as the "hunger hormone." It is produced by specialized cells that line the stomach and the pancreas when the stomach is empty, while secretion stops when the stomach is stretched. It acts on hypothalamic brain cells, and its neural receptors are found on the same cells in the brain as the receptors for leptin. It serves to both increase hunger and increase gastric acid secretion and gastrointestinal motility to prepare the body for food intake. It increases both appetite and fat mass by triggering receptors that stimulate production of neuropeptide Y. Ghrelin also functions as a growth-hormone-releasing peptide. While leptin and ghrelin usually work antagonistically in feeding animals, ghrelin levels should be lower in naturally fasting animals as a means of suppressing appetite (including suppression of neuropeptide Y).

As will become apparent, the results of experimental studies so far indicate that we have only begun to understand the hormonal control of physiological processes during fasting in marine mammals and, in fact, a new theoretical framework may be required given how poorly the experimental data to date match our current expectations.

Pinnipeds

As with most aspects of fasting physiology, most studies of hormonal changes have been conducted on northern elephant seal pups. In this species, many of the hormonal changes that support fasting occur earlier than expected, indicating that preparatory actions in the theory occur prior to weaning. Cortisol levels have been observed to increase between early and late nursing, and then again during early fasting, with the largest increase (more than doubling) seen between early and late fasting (Ortiz et al. 2003). All thyroid hormones decrease from early to late nursing, and then do not increase during fasting (except for a small increase in total T4). No significant changes have been observed in leptin levels during the fast (Ortiz, Wade, and Ortiz 2001), suggesting that leptin does not have an expected role in regulating body fat in fasting (or nursing) northern elephant seal pups (Ortiz et al. 2003).

Studies on the pups of other phocid species also yield unexpected results. Plasma cortisol levels in harp seal pups remained stable throughout their fast and at levels similar in older, feeding pups (Nordøy, Aakvaag, and Larsen 1993). Neither was there any observed decrease in thyroid hormones. Similarly, plasma cortisol levels remained relatively low and stable in gray seals, except for a rise in some animals toward the end of their 52-day fast (Nordøy, Ingebretsen, and Blix 1990). In juvenile elephant seals, cortisol concentrations do not change during their seasonal fast and neither do T3. However, both T4 and GH concentrations decrease dramatically during the fast (Kelso et al. 2012).

Some of the hormone changes observed in male northern elephant seals during the breeding fast are closer to those predicted by fasting theory (Crocker et al. 2012). For example, there was a 43% decrease in GH, with no matching decrease in IGF-1. This reduction in GH should function to reduce lipolysis and increase hepatic glucose production, which would seem maladaptive in a fasting animal. However, it has been proposed that reductions in GH may be required to suppress more serious anabolic actions given that some level of protein catabolism is required for gluconeogenesis. While leptin concentrations decreased by 11% (although they did not follow changes in fat mass), ghrelin concentrations did not change. This suggests that the high levels of ghrelin and dropping levels of leptin may function to suppress the drive to forage while optimizing rates of lipid oxidation (Crocker et al. 2012). High levels of ghrelin also run counter to the observed reductions in GH, as increased ghrelin is usually associated with increased GH production in fasting animals. This suggests a loss of the ability of ghrelin to stimulate GH secretion in fasting adult seals. Furthermore, the observed decreases in leptin were not associated with any significant changes in thyroid hormones. An exception was that changes in total T3 were directly related to changes in daily energy expenditure. Cortisol levels also did not change during the fast.

Cetaceans

Only a single study has examined multiple hormone changes in a fasting cetacean. This study was conducted on two fasting adult bottlenose dolphins, a species that likely rarely experiences prolonged fasts (Ortiz et al. 2010). The results confirmed the expected switch to lipid metabolism, but the swiftness of the response was more typical of mammalian species not adapted to regular fasting episodes. Plasma fatty acids doubled by 24 hours and increased 2.5-fold by 38 hours of fasting. Conversely, BUN decreased 17% by 24 hours of fasting and 22% by 38 hours. Plasma glucose decreased 25% between 14 and 24 hours and levels returned to baseline by 38 hours of fasting.

Neither plasma total T3 nor free T4 were changed. Mean total T4 increased 19% by 38 hours of fasting, while mean rT3 showed an initial 30% decrease by 24 hours of fasting, but returned to baseline levels by 38 hours. The increase in total T4 might be due to decreased clearance rates (vs. increased production) while the eventual recovery of rT3 by 38 hours might reflect preferential deiodination of T4 to decrease cellular metabolism. Measured plasma cortisol levels were undetectable in these animals during fasting.

TOOLBOX

Methods for measuring mass loss and body condition

Most pinniped studies use serial measures of body composition and body mass to calculate changes in the mass of specific tissues over time (usually differentiated into lipid and fat-free or lean mass). While a variety of methods can be used to determine body composition (whole-body dissections, direct imaging, ultrasonic measurements of lipid depth, bioelectrical impedance), the most common method currently employed uses dilution of a chemical marker to indirectly estimate body composition through measures of whole-body water content.

This method is based on the knowledge that different tissue types have different water content. This differential water content of tissues is the basis for how your home scale takes body fat measurements. Briefly, a small dose of isotopically distinct water (either deuterium oxide or tritiated water) is injected into the animal and allowed time (usually several hours) to equilibrate with the rest of the animal's body water. A blood sample is analyzed for the resulting concentration of the chemical marker in the serum which, combined with the known amount injected, yields an estimate of the total body water. This value is then converted to estimates of lean and lipid mass through published mathematical equations previously generated from empirical studies (often involving carcass

analysis). The accuracy of the technique is dependent upon the applicability of the mathematical models converting body water to body composition (these equations are often species- and even age-specific) and the hydration state of the animal (which may be an issue in fasting animals).

For many marine mammals, including most cetaceans, it is impractical to consider any techniques that require repeated capture and handling, and so new, innovative measures have to be developed. Photogrammetry – using photographs to make morphological measurements – have been investigated as a possible tool (de Bruyn et al. 2009) but is usually limited to detecting large changes in body condition. The animal's diving behavior, which can be more readily monitored via attached dive recorders, may also provide an indirect measure of its relative body stores. For many marine mammals, a portion of the natural dive sequence is made up of an unpowered glide. The rate at which a marine mammal ascends or descends through the water during a glide is dependent upon the hydrodynamic drag (a factor of body shape), water depth, and its buoyancy. An individual's buoyancy is altered by changes in the relative proportion of lipid mass in their body (as well as factors as the amount of air they dive with). Hence, rates of ascent/descent while gliding and changes in stroke rate have been used to detect differences in body condition in a number of species of marine mammals (Aoki et al. 2011). There have even been attempts to determine the metabolic status of cetaceans by chemically analyzing their captured exhalations (Aksenov et al. 2014).

Measuring metabolism and calculating metabolic depression

Metabolic rate is technically the amount of energy liberated or expended in a given unit of time by an animal. While the earliest studies of metabolism measured the amount of heat an animal produced (often by measuring the change in temperature of a surrounding water bath or ice mixture), later studies realized that aerobic metabolism consumed a set amount of oxygen (and produced a set amount of carbon dioxide) depending on the exact fuel source. Respirometry, the science of measuring the rate of oxygen consumption, has become the standard method of measuring rates of metabolism. The method is quite simple; usually ambient air is drawn at a known rate through a sealed chamber containing the organism (or at least within which it must breathe). The excurrent airflow is sampled to determine the concentration of oxygen (and often carbon dioxide). Knowing the rate of airflow and the difference in gas concentrations between the sampled and the ambient air (which is essentially constant) allows one to calculate the rate of oxygen consumption. Often, metabolic rate is presented as a rate of oxygen consumption; while this can be converted to a rate of energy use, this step also involves several assumptions. One of these assumptions is the nature of the metabolic substrate, which can be elucidated by examining the ratio of carbon dioxide produced to oxygen consumed (known as the respiratory quotient).

Unfortunately, respirometry is of limited application in the field. Scientists have developed proxies to indirectly estimate rates of energy expenditure from wild marine mammals, including heart rate, flipper strokes, and body acceleration (Iverson et al. 2010). By far, the most common method involves the differential turnover of two isotopically labeled waters, known as the doubly labeled water method. Still, most of these methods cannot provide estimates of RMR, but only estimate an animal's average metabolic rate over time, known as its field metabolic rate (FMR), which may be affected by parallel changes in activity, thermoregulation, and other factors.

Determining changes in resting metabolism over the course of a fast is complicated by the fact that the animal is simultaneously losing body mass, which will inherently decrease total energy expenditures. In addition, body mass loss during different phases of the fast may be due to loss of tissues that are metabolically active (protein) or relatively metabolically inert (lipid stores). For metabolic depression to occur, the decrease in metabolic rate during the fast must not be a mere consequence of body mass loss. Some authors have suggested that true metabolic depression can be said to occur if there is a clear decrease in this rate when it is expressed per unit body mass (mass-specific metabolism) (Cherel, Robin, and Maho 1988). While some might argue that this incorrectly assumes a specific relationship between body mass and metabolic rate, it is certainly a conservative way of calculating changes in metabolism.

Plasma metabolites as biomarkers of fasting phase

We can analyze the metabolites present in a blood sample for clues about the fasting state of an animal. Metabolites are the end products and the intermediaries of all chemical reactions in the body. They can therefore provide evidence about the rate and amount of specific substrate types that are broken down across time during fasting. Examples of these metabolites are mentioned earlier in this chapter and include glucose, FFAs, and BUN.

Metabolites can be analyzed in two ways to study fasting. Specific known biomarkers associated with metabolic processes can be analyzed in a plasma or serum sample. Glucose and BUN, for example, are routinely measured in human and veterinary clinical applications and are available on standard blood chemistry panels. FFAs vary by chain length and number of double bonds, meaning that researchers must decide whether to simply measure all circulating lipid, or to measure the specific composition of the lipid pool. The latter requires more sophisticated analytical methods, including gas chromatography. Metabolites can also be analyzed by large-scale screens of the metabolome (the entire metabolite population), taking advantage of rapidly advancing technology and analytical methods to find patterns and signals in large data sets. The benefit of evaluating the entire metabolome is that more data is retrieved from the single sample, which allows further investigation into the details of substrate turnover and the interaction between fasting and other physiological processes (e.g., lactation, stress, molt).

Metabolites are low-molecular-weight chemicals that are identified by metrics such as their mass, charge, pH, and hydrophobicity. Metabolomics platforms at research facilities and in industry combine liquid or gas chromatography to separate metabolites (by charge, pH, hydrophobicity, etc.) from a biosample, and then the mass of separated metabolites is determined by mass spectrometry. Knowing the specifics of mass along with charge, etc., identifies metabolites from catalogues of hundreds of known (and synthetic) chemicals; these platforms are also able to quantify each metabolite within a biosample. Using metabolites as the target of large-scale screening is especially useful for studying marine mammals because they can be consistently identified by their composition and does not rely on comprehensive genomic information (needed to identify proteins and transcripts based on their sequences), which is not yet available for many marine mammal species.

A limitation to using metabolite signatures to study fasting is that they require tissue or blood sample collection. Samples must also be quickly processed (e.g., to separate the plasma or serum from red blood cells) and stabilized by freezing to limit the degradation of metabolites. We know the most about the biochemistry of fasting in elephant seals, and it is not hard to imagine that this is partly because they fast on accessible beaches. Collecting and carefully preserving a blood sample from a migrating, fasting whale in pelagic waters is more daunting.

Tracking metabolic tracers

Similar to the doubly labeled water method, the dilution and breakdown of tracer chemicals can be used to track substrate use during fasting. This is even a more complicated field sampling strategy than a single blood draw. An initial blood sample is collected for baseline information, and subsequent samples are collected to determine the rate of tracer chemical turnover. Potentially, several samples must be collected to confirm this rate. Often, the animal's blood volume (typically measured by dilution of IV-injected Evans blue dye) is also determined. Depending on the timeframe and the study species, it is necessary to sedate the animals to collect blood or give injections. The length of these procedures can complicate sampling strategy. However, this type of experiment provides very useful, specific data about the metabolism of substances, and has been successfully accomplished in fasting elephant seals for labeled urea, fatty acids and glucose, to name a few (Pernia, Hill, and Ortiz 1980; Castellini, Costa, and Huntley 1987; Houser and Costa 2001).

CONCLUSIONS

There are many specific details of fasting physiology that we have only begun to explore. While our depth of understanding of basic fasting physiology is rapidly increasing, its breadth is still very narrow. The bulk of our knowledge on hormonal control mechanisms, for example, comes from a single species, the northern elephant seal. Studies with other pinnipeds have supported a basic conclusion that marine mammals are well-adapted to their natural fasts, but how they accomplish this feat differs significantly between species and from terrestrial mammals. But these scientific glimpses are exceedingly limited. Perhaps the most important focus for future research is to explore fasting physiology in a wider range of species, with particular emphasis on cetaceans. This increased knowledge will also allow us to determine what evolutionary processes have honed species-specific fasting strategies, even among closely related animals.

The question of the fasting capacity of marine mammals is of more than mere academic interest, but also has important conservation implications. For example, in 2014–2015, unprecedented numbers of stranded, starving

California sea lions (*Zalophus californianus*) began arriving on beaches of the US west coast. Scientists believe that anomalously warm coastal waters shifted the food base, requiring lactating mothers to prolong their trips to sea. This resulted in an extension beyond natural fasting durations that the pups cannot endure, driving them into the water to begin their own foraging too early.

With concern increasing over the impacts to wild populations of both natural and anthropogenic disturbance, it is timely that we start addressing how predictable periods of fasting interact with unpredicted perturbations. How much of an additional physiological burden is imposed by threats such as disease, pollution, and human harassment? How might survival and reproductive capacity be impacted by environmental changes resulting in slightly longer or more expensive fasts, slightly smaller energy reserves, or altered seasonal timing of fasts? Future research will help us understand how changing oceans could affect marine mammal populations.

REVIEW QUESTIONS

1 Why do marine mammals fast? What is the difference between fasting and starvation?
2 How does an animal's physiology change over the course of the fast (between stages I, II and III)?
3 What types of on-board energy reserves provide fuel for the fasting animal, and how does this change during a fast?
4 How can an animal minimize energy requirements during a fast?
5 What hormones control fasting physiology in marine mammals?
6 Why do marine mammals fast?
7 What are some examples of "high-energy" behaviors that marine mammals still must perform when they are fasting?
8 How is fasting in marine mammals different than in terrestrial animals? What aspects are the same?
9 What organs and tissues are the most important in marine mammals to support fasting?
10 Identify differences between fasting strategies in different types of marine mammals.

CRITICAL THINKING

1 In your opinion, what other activities in the life of a marine mammal conflict the most with fasting?
2 Which type of marine mammal do you think demonstrates the most impressive fasting capabilities and why?
3 What changes in the environment could impact how long marine mammals are able to fast?

FURTHER READING

Guppy, M., and P. Withers. 1999. Metabolic depression in animals: Physiological perspectives and biochemical generalizations. *Biological Review* 74: 1–40.

Kleiber, M. 1975. *The fire of life: An introduction to animal energetics*, New York: Robert E. Krieger Publ. Co.

Lieberman, M., and A.D. Marks. 2009. *Marks' basic medical biochemistry: A clinical approach*, Philadelphia: Lippincott Williams & Wilkins.

Storey, K.B., and J.M. Storey. 2004. Metabolic rate depression in animals: Transcriptional and translational controls. *Biological Reviews* 79(1): 207–233.

GLOSSARY

Catabolism: Process by which energy-containing molecules, or substrates, are broken down to release energy that can be used by the body to do work.

Dynamic equilibrium: A state in which forward reactions are occurring at the same rate as corresponding reverse reactions.

Fasting: The physiological state of living with a complete lack of food.

Free Fatty Acids (FFAs): Fatty acids that are not bound into triglycerides.

Ghrelin: Hormone produced by specialized cells present in the lining of the stomach and pancreas that antagonizes the action of leptin. Commonly known as the "hunger hormone."

Glucocorticoids: Steroid hormones, known most commonly as a biochemical marker of physiological stress. These also serve essential roles in fasting.

Gluconeogenesis: Formation of glucose from glycerol and amino acid precursors.

Glycerol: Forms the backbone of triacylglycerol molecules.

Growth hormone (GH): Hormone that stimulates cellular growth.

Homeostasis: Tendency in cells and organisms toward a stable equilibrium of physiological processes.

IGF-1: Insulin-like growth factor I, which is important in protein production.

Leptin: Hormone made by fat cells that is often referred to as the "satiety hormone."

Lipolysis: Process of breakdown of lipid (fat) substrates.

Ketone bodies: Fatty acids modified in the liver that have a ketone group.

Proteolysis: Breakdown of proteins into individual amino acids.

Resting metabolic rate (RMR): Metabolism of a resting, inactive animal.

Thyroid hormones: T3 (triiodothyronine) and T4 (thyroxine), produced by the thyroid gland, are primarily responsible for regulation of metabolism.

Triacylglycerols: Dietary lipids containing three fatty acids bound to a glycerol backbone.

REFERENCES

Adams, S., and D. Costa. 1993. Water conservation and protein metabolism in northern elephant seal pups during the postweaning fast. *Journal of Comparative Physiology B* 163(5): 367–373.

Aksenov, A.A., L. Yeates, A. Pasamontes, et al. 2014. Metabolite content profiling of bottlenose dolphin exhaled breath. *Analytical Chemistry* 86(21): 10616–10624.

Aoki, K., Y.Y. Watanabe, D.E. Crocker, et al. 2011. Northern elephant seals adjust gliding and stroking patterns with changes in buoyancy: Validation of at-sea metrics of body density. *Journal of Experimental Biology* 214(17): 2973–2987.

Arnould, J.P., J. Green, and D. Rawlins. 2001. Fasting metabolism in Antarctic fur seal (*Arctocephalus gazella*) pups. *Comparative Biochemistry and Physiology Part A: Molecular & Integrative Physiology* 129(4): 829–841.

Beauplet, G., C. Guinet, and J.P. Arnould. 2003. Body composition changes, metabolic fuel use, and energy expenditure during extended fasting in subantarctic fur seal (*Arctocephalus tropicalis*) pups at Amsterdam Island. *Physiological and Biochemical Zoology* 76(2): 262–270.

Bentley, P. 1963. Composition of the urine of the fasting humpback whale (*Megaptera nodosa*). *Comparative Biochemistry and Physiology* 10(3): 257–259.

Boily, P., and D.M. Lavigne. 1995. Resting metabolic rates and respiratory quotients of gray seals (*Halichoerus grypus*) in relation to time of day and duration of food deprivation. *Physiological Zoology* 68(6): 1181–1193.

Bowen, W., D.J. Boness, and O.T. Oftedal. 1987. Mass transfer from mother to pup and subsequent mass loss by the weaned pup in the hooded seal, *Cystophora cristata. Canadian Journal of Zoology* 65(1): 1–8.

Boyd, I. 1991. Environmental and physiological factors controlling the reproductive cycles of pinnipeds. *Canadian Journal of Zoology* 69(5): 1135–1148.

Castellini, M.A., D.P. Costa, and A.C. Huntley. 1987. Fatty acid metabolism in fasting elephant seal pups. *Journal of Comparative Physiology B* 157(4): 445–449.

Castellini, M.A., and L.D. Rea. 1992. The biochemistry of natural fasting at its limits. *Experientia* 48(6): 575–582.

Castellini, M.A., and D.P. Costa. 1990. Relationships between plasma ketones and fasting duration in neonatal elephant seals. *American Journal of*

Physiology-Regulatory, Integrative and Comparative Physiology 259(5): R1086–R1089.

Cherel, Y., J.-P. Robin, and Y.L. Maho. 1988. Physiology and biochemistry of long-term fasting in birds. *Canadian Journal of Zoology* 66(1): 159–166.

Corkeron, P.J., and R.C. Connor. 1999. Why do baleen whales migrate? *Marine Mammal Science* 15(4): 1228–1245.

Crocker, D.E., D.S. Houser, and P.M. Webb. 2012. Impact of body reserves on energy expenditure, water flux, and mating success in breeding male northern elephant seals. *Physiological and Biochemical Zoology* 85(1): 11–20.

Crocker, D.E., R.M. Ortiz, D.S. Houser, et al. 2012. Hormone and metabolite changes associated with extended breeding fasts in male northern elephant seals (*Mirounga angustirostris*). *Comparative Biochemistry and Physiology Part A: Molecular & Integrative Physiology* 161(4): 388–394.

de Bruyn, P.N., M.N. Bester, A.R. Carlini, et al. 2009. How to weigh an elephant seal with one finger: A simple three-dimensional photogrammetric application. *Aquatic Biology* 5(1): 31–39.

Deutsch, C.J., M.P. Haley, and B.J. Le Boeuf. 1990. Reproductive effort of male northern elephant seals: Estimates from mass loss. *Canadian Journal of Zoology* 68(12): 2580–2593.

Durban, J., and R. Pitman. 2012. Antarctic killer whales make rapid, round-trip movements to subtropical waters: Evidence for physiological maintenance migrations? *Biology Letters* 8(2): 274–277.

Edney, E.B. 1977. Metabolic Water. In *Water balance in land arthropods*, 189–195. Berlin: Springer.

Fowler, S., D. Costa, J. Arnould, et al. 2007. Ontogeny of oxygen stores and physiological diving capability in Australian sea lions. *Functional Ecology* 21(5): 922–935.

Georges, J.-Y., and C. Guinet. 2000. Maternal care in the subantarctic fur seals on Amsterdam Island. *Ecology* 81(2): 295–308.

Henry, C., J. Rivers, and P. Payne. 1988. Protein and energy metabolism in starvation reconsidered. *European Journal of Clinical Nutrition* 42(7): 543–549.

Horton, T.W., R.N. Holdaway, A.N. Zerbini, et al. 2011. Straight as an arrow: Humpback whales swim constant course tracks during long-distance migration. *Biology Letters* 7(5): 674–679.

Houser, D.S., and D.P. Costa. 2001. Protein catabolism in suckling and fasting northern elephant seal pups (*Mirounga angustirostris*). *Journal of Comparative Physiology B* 171(8): 635–642.

Houser, D.S., C.D. Champagne, and D.E. Crocker. 2007. Lipolysis and glycerol gluconeogenesis in simultaneously fasting and lactating northern elephant seals. *American Journal of Physiology-Regulatory, Integrative and Comparative Physiology* 293(6): R2376–R2381.

Houser, D.S., and D.P. Costa. 2003. Entrance into stage III fasting by starveling northern elephant seal pups. *Marine Mammal Science* 19(1): 186–197.

Houser, D.S., D. Derous, A. Douglas, et al. 2021. Metabolic response of dolphins to short-term fasting reveals physiological changes that differ from the traditional fasting model. *Journal of Experimental Biology* 224(9): jeb238915.

Iverson, S.J., C.E. Sparling, T.M. Williams, et al. 2010. Measurement of Individual and Population Energetics of Marine Mammals. In *Marine mammal ecology and conservation: A handbook of techniques*, eds. I.L. Boyd, W.D. Bowen, and S.J. Iverson, 165–189. New York: Oxford University Press.

Kelso, E.J., C.D. Champagne, M.S. Tift, et al. 2012. Sex differences in fuel use and metabolism during development in fasting juvenile northern elephant seals. *Journal of Experimental Biology* 215(15): 2637–2645.

Keys, A., J. Brozek, A. Henschel, et al. 1950. *The biology of human starvation*, Volume II. Minneapolis: University of Minnesota Press.

Kovacs, K.M., and D. Lavigne. 1985. Neonatal growth and organ allometry of Northwest Atlantic harp seals (*Phoca groenlandica*). *Canadian Journal of Zoology* 63(12): 2793–2799.

LeBoeuf, B. 1974. Hectic life of alpha bull-elephant seal as fighter and lover. *Psychology Today* 8(5): 104.

Lunn, N., I. Boyd, T. Barton, et al. 1993. Factors affecting the growth rate and mass at weaning of Antarctic fur seals at Bird Island, South Georgia. *Journal of Mammalogy* 74(4): 908–919.

Lydersen, C., K. Kovacs, and M. Hammill. 1997. Energetics during nursing and early postweaning fasting in hooded seal (*Cystophora cristata*) pups from the Gulf of St Lawrence, Canada. *Journal of Comparative Physiology B* 167(2): 81–88.

Markussen, N.H. 1995. Changes in metabolic rate and body composition during starvation and semistarvation in harbour seals. *Developments in Marine Biology* 4: 383–391. Elsevier.

Markussen, N.H., M. Ryg, and N.A. Øritsland. 1992. Metabolic rate and body composition of harbour seals, *Phoca vitulina*, during starvation and refeeding. *Canadian Journal of Zoology* 70(2): 220–224.

Nelson, R.A., G.E. Folk Jr., E.W. Pfeiffer, et al. 1983. Behavior, biochemistry, and hibernation in black, grizzly, and polar bears. *Bears: Their Biology and Management* 5: 284–290.

Nordøy, E.S., and A.S. Blix. 1991. Glucose and ketone body turnover in fasting grey seal pups. *Acta Physiologica Scandinavica* 141(4): 565–571.

Nordøy, E.S., A. Aakvaag, and T.S. Larsen. 1993. Metabolic adaptations to fasting in harp seal pups. *Physiological Zoology* 66(6): 926–945.

Nordøy, E.S., O. Ingebretsen, and A. Blix. 1990. Depressed metabolism and low protein catabolism in fasting grey seal pups. *Acta Physiologica Scandinavica* 139(1–2): 361–369.

Noren, D., and M. Mangel. 2004. Energy reserve allocation in fasting northern elephant seal pups: Inter-relationships between body condition and fasting duration. *Functional Ecology* 18: 233–242.

Øristland, N., and N. Markussen. 1990. Outline of a physiologically based model for population energetics. *Ecological Modelling* 52: 267–288.

Oritsland, N. 1990. Starvation survival and body composition in mammals with particular reference to *Homo sapiens*. *Bulletin of Mathematical Biology* 52(5): 643–655.

Oritsland, N., A. Påsche, N. Markussen, et al. 1985. Weight loss and catabolic adaptations to starvation in grey seal pups. *Comparative Biochemistry and Physiology A, Comparative Physiology* 82(4): 931–933.

Ortiz, R.M., D.S. Houser, C.E. Wade, et al. 2003. Hormonal changes associated with the transition between nursing and natural fasting in northern elephant seals (*Mirounga angustirostris*). *General and Comparative Endocrinology* 130(1): 78–83.

Ortiz, R.M., B. Long, D. Casper, et al. 2010. Biochemical and hormonal changes during acute fasting and re-feeding in bottlenose dolphins (*Tursiops truncatus*). *Marine Mammal Science* 26(2): 409–419.

Ortiz, R.M., C.E. Wade, and C.L. Ortiz. 2001. Effects of prolonged fasting on plasma cortisol and TH in postweaned northern elephant seal pups. *American Journal of Physiology-Regulatory, Integrative and Comparative Physiology* 280(3): R790–R795.

Pernia, S.D., A. Hill, and C.L. Ortiz. 1980. Urea turnover during prolonged fasting in the northern elephant seal. *Comparative Biochemistry and Physiology, B* 65(4): 731–734.

Rea, L.D., D.A. Rosen, and A.W. Trites. 2000. Metabolic response to fasting in 6-week-old Steller sea lion pups (*Eumetopias jubatus*). *Canadian Journal of Zoology* 78(5): 890–894.

Rea, L.D., M. Berman-Kowalewski, D.A. Rosen, et al. 2009. Seasonal differences in biochemical adaptation to fasting in juvenile and subadult Steller sea lions (*Eumetopias jubatus*). *Physiological and Biochemical Zoology* 82(3): 236–247.

Rea, L.D., and D.P. Costa. 1992. Changes in standard metabolism during long-term fasting in northern elephant seal pups (*Mirounga angustirostris*). *Physiological Zoology* 65(1): 97–111.

Reilly, J.J. 1991. Adaptations to prolonged fasting in free-living weaned gray seal pups. *American Journal of Physiology-Regulatory, Integrative and Comparative Physiology* 260(2): R267–R272.

Rosen, D.A., and A.W. Trites. 2002. Changes in metabolism in response to fasting and food restriction in the Steller sea lion (*Eumetopias jubatus*). *Comparative Biochemistry and Physiology Part B: Biochemistry and Molecular Biology* 132(2): 389–399.

Rosen, D.A., B.L. Volpov, and A.W. Trites. 2014. Short-term episodes of imposed fasting have a greater effect on young northern fur seals (*Callorhinus ursinus*) in summer than in winter. *Conservation Physiology* 2(1): cou021. doi: 10.1093/conphys/cou021

Rosen, D.A., A.J. Winship, and L.A. Hoopes. 2007. Thermal and digestive constraints to foraging behaviour in marine mammals. *Philosophical Transactions of the Royal Society B: Biological Sciences* 362(1487): 2151–2168.

Sharick, J.T., J.P. Vazquez-Medina, R.M. Ortiz, et al. 2015. Oxidative stress is a potential cost of breeding in male and female northern elephant seals. *Functional Ecology* 29(3): 367–376.

Schmidt-Nielsen, K. 1997. *Animal physiology: Adaptation and environment*, 5th ed., Cambridge: Cambridge University Press.

Verrier, D., R. Groscolas, C. Guinet, et al. 2011. Development of fasting abilities in subantarctic fur seal pups: Balancing the demands of growth under extreme nutritional restrictions. *Functional Ecology* 25(3): 704–717.

Worthy, G.A., and D. Lavigne. 1982. Changes in blood properties of fasting and feeding harp seal pups, *Phoca groenlandica*, after weaning. *Canadian Journal of Zoology* 60(4): 586–592.

Worthy, G.A., and D. Lavigne. 1983. Energetics of fasting and subsequent growth in weaned harp seal pups, *Phoca groenlandica. Canadian Journal of Zoology* 61(2): 447–456.

Worthy, G.A., and D. Lavigne. 1987. Mass loss, metabolic rate, and energy utilization by harp and gray seal pups during the postweaning fast. *Physiological Zoology* 60(3): 352–364.

Daniel E. Crocker
Sonoma State University, Rohnert Park, CA

Birgitte I. McDonald
Moss Landing Marine Labs, San Jose State University, Moss Landing, CA

Daniel E. Crocker

Daniel has been studying the behavior and physiology of marine mammals for more than three decades. He has worked on a wide variety of marine vertebrates with northern elephant seals as his main research species. His work has focused on the metabolic adaptations associated with extended fasting when animals are on shore and movement and foraging ecology when animals are at sea. His recent work focuses on the endocrine system and how stress responses influence health and reproduction.

Birgitte McDonald

Birgitte McDonald is an associate professor at Moss Landing Marine Laboratories, where she has been teaching courses on ecology and physiology of marine vertebrates and scientific skills since 2015. She obtained her Ph.D. from University of California, Santa Cruz and conducted postdoctoral research at Scripps Institution of Oceanography and Aarhus University (Denmark). As a physiological and behavioral ecologist, she investigates adaptations that allow animals to survive in extreme environments. Understanding the mechanisms that allow an organism to interact and survive in its environment is crucial for predicting their response to climate change. Her research has provided opportunities to work with a broad range of species in a diversity of habitats from the Antarctic to the Galapagos.

Driving Question: How do marine mammals provision their young to succeed in the challanging marine environment?

INTRODUCTION

As air-breathing endotherms living in marine habitats, marine mammals have had to overcome the problem of giving birth in an environment that is potentially very challenging to a newborn. As described in previous chapters, the marine environment presents a suite of physiological challenges that are magnified in newborns. For example, newborns have high surface-to-volume ratios and thin blubber layers, resulting in more heat loss compared to adults (**Chapter 5**). Because of these challenges, marine mammals tend to give birth to a single, large, **precocial** young, with many species exhibiting rapid growth and development. Across marine mammal species, offspring birth mass increases

DOI: 10.1201/9781003297468-12

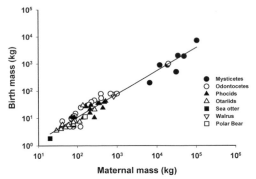

Figure 12.1 The relationship between body mass and offspring birth mass in selected marine mammal species where data or estimates are available. The slope of the line on this log-log plot, or the scaling exponent, is 0.85. Because the slope of this line is less than 1, it shows that smaller marine mammals produce larger offspring relative to their body size. (Figure is modified with additional data from Costa 2009)

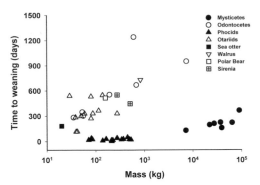

Figure 12.2 The relationship between body mass and time to weaning in selected marine mammal species where data or estimates are available. Mysticete whales and phocid seals have shorter lactation periods for their body size as compared to other marine mammals. (Figure is modified with additional data from Costa 2009)

as a power function of maternal mass, called a **scaling exponent**. This scaling exponent is 0.85 (using data from 58 species), which means that smaller species give birth to relatively large offspring for their body size (**Figure 12.1**).

After birth, marine mammals exhibit high rates of parental investment, delivering large amounts of milk energy to their offspring to help them achieve the large body size required for independent foraging. The transition to nutritional independence is a critical stage in marine mammal development and survival because young animals lack the skills and the physiology (see **Chapter 2**) to efficiently acquire food. In mammals, lactation is the most energetically demanding component of reproduction for the mother, as it simultaneously provides the offspring with energy stores and extends the developmental period.

LACTATION STRATEGIES

Marine mammals exhibit a diverse range of lactation strategies that are influenced by size and environment. Periods of parental investment range widely from four days to several years

(**Figure 12.2**). The most obvious defining characteristic is where species give birth. Cetaceans have behavioral and anatomical adaptations that allow them to give birth, suckle, and nurture their young in water. In contrast, pinnipeds must reproduce and give birth on land or ice.

Underneath this broad difference, the parental strategies and physiological capabilities that shape the movements, breeding systems, foraging strategies, and life history patterns are diverse. In some species (e.g., some mysticetes and many phocid seals), movement to an appropriate breeding environment necessitates a **capital breeding** strategy, where stored body reserves are accumulated between breeding episodes and then used for reproduction. Other species (e.g., odontocetes and otariid seals) use an **income breeding** strategy, where offspring are fed using resources from foraging during the lactation period (**Figure 12.3**).

Cetaceans

The ability to give birth and nurse in the ocean was likely a key feature in allowing the success and wide distribution of cetacean species. Unfortunately, aquatic lactation has also made the postpartum physiology of cetaceans extremely difficult to study, and much of what is known

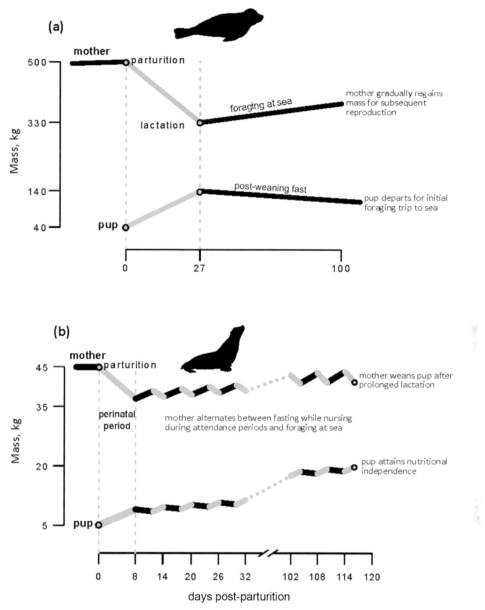

Figure 12.3 Schematic showing the difference between (a) capital breeding strategy (fasting during lactation) used by some phocids and mysticetes, and (b) income breeding strategy (foraging/fasting cycle) used by otariids and some phocids. (Figure is taken from Champagne et al. (2012) and used with permission)

about lactation physiology is based on carcasses from whaling or strandings and from studies of captive animals. For many species, birth has never been observed and knowledge of reproductive and parental behavior is fragmentary or based on comparisons with other species. Despite these limitations, field studies provide some important insights into postpartum behavior and physiology. Cetaceans exhibit low adult mortality and **fecundity** (birth rates), but high rates of infant mortality, and the period of parental care is critical to offspring survival and population health. Many of the maternal traits that support calf survival are behavioral and social, including communication and nursing behavior, but there are also important differences in lactation physiology that influence the life history strategies of cetaceans.

The duration of lactation varies widely both between and within species of cetaceans, with the greatest difference between suborders. For many species, the duration of lactation is estimated from just a few samples, and the methods used to determine lactation duration (i.e., milk in the stomach from harvested individuals or behavioral observations of known individuals) are prone to error. Cetacean offspring can travel with their mothers, suckling frequently with only brief periods of maternal absence. Offspring suckle from two mammary teats located in slits on either side of the genital opening. The calf wraps the tip of its tongue around the nipple from below and pushes against the roof of the mouth. The base of the tongue creates suction by moving up and down. The mother uses muscular contraction of both the **myoepithelial cells** of the mammary gland and surrounding **cutaneous muscles** to cause milk ejection.

Both baleen and toothed whales nurse in the water but they exhibit different lactation strategies. The mysticete approach is a capital breeding strategy resulting in a short lactation period relative to their body size (**Figure 12.1**). Most species lactate for 5–8 months, typically weaning their calves during the summer months when food availability is high. Mysticetes have compartmentalized breeding to a distinct part of the year, requiring high intensity lactation and offspring growth. In the best-studied mysticete whales, individuals forage during summer months in polar waters with high abundance of krill or fish and then move to warmer waters for parturition, nursing, and mating. Feeding may not occur at all at the birthing locations. It has been hypothesized that these breeding areas provide warmer and calmer water for the calves, lowering thermoregulatory costs and increasing growth efficiency. However, more recent studies suggest that baleen whale mothers and their calves are thermal-neutral in cold

water and the migration may be to avoid predation on young calves. An exception to this general pattern is the bowhead whale (*Balaena mysticetus*) that calves at high latitudes in the summer and has an extended lactation period of 12 months (Oftedal 1997). Weaning in mysticetes is abrupt, with little or no overlap with the nursing period.

Odontocetes have an income breeding strategy, with mothers foraging during lactation. Calves are nursed from 8 months to several years, with larger species lactating for the longer periods. Long periods of parental investment may allow training of young in migration, foraging, and social behavior. Weaning in odontocetes is more gradual than in mysticetes, and offspring may stay with the mother in social groups after weaning. In many species, females become pregnant while still lactating and may care for young for a significant portion of gestation. This may allow females to "bet-hedge" against loss of the developing fetus, with the option of continuing investment in the current offspring.

How long lactation lasts in cetaceans is influenced by several factors (Whitehead and Mann 2000). Cetacean young do not fast after weaning. Odontocetes gradually transition from nursing to foraging, while mysticetes have only a brief period where solid food intake overlaps with nursing. The period of maternal investment may reflect: (1) foraging strategies and the time required for the offspring to develop sufficiently to find and catch food; (2) the fitness consequences of continued parental investment; (3) the impact of continued lactation on the next breeding attempt; and (4) occurrence of **weaning conflict** between the mother and calf. Weaning conflict arises from the fact that the mother and offspring may be selected to desire different levels of investment in current offspring.

Pinnipeds

In contrast to the cetaceans, pinnipeds give birth on land or ice, despite spending most of their lives in the water. Most newborn pups require some time on land to develop before being able to cope with life in the sea. This requires an appropriate breeding site, away from potential predators, which can sometimes be distant from their patchy food resources. The strategy for managing the separation of at-sea feeding and terrestrial birth leads to the major differences between phocid and otariid reproductive biology.

Most phocids have a capital breeding strategy where parental investment comes from stored body reserves. This strategy is associated with a short lactation period for their body size (**Figure 12.3**). A few species, such as harbor seal (*Phoca vitulina*), gray seal (*Halichoerus grypus*), bearded seal (*Erignathus barbatus*), harp seal (*Pagophilus groenlandicus*), ringed seal (*Phoca hispida*), and Weddell seals (*Leptonychotes weddellii*), can have a mixed strategy where individuals feed during lactation but this tendency varies with body reserves and the proximity of prey (Lydersen, Hammill, and Kovacs 1995; Lydersen and Kovacs 1996). Some of the smallest phocids, such as harbor seals and ringed seals, feed frequently throughout lactation (Lydersen 1995; Bowen et al. 2001). For those phocids that do forage, the trip is usually just a few hours at a time. This capital breeding strategy in phocids is similar to the short lactation period relative to body size in fasting mysticete whales. Phocids lactate for a mere 4 days in the hooded seal to up to 6–7 weeks in the Weddell seal. Weaning is often abrupt as the mother abandons the pup to return to sea. In species where the mother feeds during lactation, weaning may be less abrupt and pups may spend significant time in the water prior to weaning. The most extreme example of the abbreviated phocid lactation system is the hooded seal (*Cystophora cristata*), where females lactate for only 3–4 days. In this short time, the pup can double its birth mass receiving more than 10 L of milk per day containing an average fat content of 59% (Lydersen, Kovacs, and Hammill 1997). Shortening the lactation period and eliminating foraging allows many phocid seals to use highly dispersed and distant food resources.

Within capital breeding phocids, there is wide variation in the postweaning behavior and physiology of the pups. In some species, the pup remains at the rookery and has an extended fasting after weaning. During this period, the pup begins to develop the physiology needed to dive to appropriate depths for foraging, including development of oxygen stores and control of the cardiovascular system (Burns, Clark, and Richmond 2004). The size of the weaned pup often determines of the duration of the postweaning fast and how soon the pups initiate independent foraging. In many species, pups undergo a long migration to foraging areas after departing the breeding site. In these cases, pup body reserves may still be critical to allowing thermoregulation in cold water and to allow time for development of foraging skills.

Similar to phocids, otariid mothers also fast from food or water while on shore. However, they only remain with their pups for approximately a week, called the **perinatal period**, before returning to sea to forage. The duration of the perinatal period varies with species and environmental conditions, but is typically 5–9 days. The female returns to her pup, which has been fasting during her absence, and suckles it for 1–3 days. This pattern of intermittent suckling and foraging continues for 4 months to around several years. In some species, like subantarctic fur seals (*Arctocephalus tropicalis*), fasting periods for the pup can be greater than 30 days toward the end of lactation (Verrier et al. 2012). The duration of the period of parental care varies with latitude, being shortest in the polar species and longest in the equatorial Galapagos fur seal (*Arctocephalus galapagoensis*) and sea lion (*Zalophus wollebacki*). The need to return frequently to provision their offspring limits otariids to foraging distances within 10s to 100s of kilometers of their breeding rookeries. This constraint places a priority on maximizing the efficiency of foraging in order to minimize the time that the pup is left fasting on shore. For this reason, otariids are often associated with highly productive coastal regions of ocean upwelling. To maximize foraging success during this short period, otariids exhibit high levels of foraging effort and energy expenditure for locomotion while foraging when compared to phocids. The importance of successful foraging is evident in the composition of otariid milk, which increases in fat content (~25% milk in species with average trips of 2 days to ~50% in species with average trips of 6 days) with the duration of the foraging trip (Costa 1991).

Walruses (*Odobenus rosmarus*) have many reproductive features in common with phocids and otariids; however, they exhibit one critical difference. Pups are born on ice or on land, but unlike the other pinnipeds, walrus nurse at sea. The ability to feed at sea allows pups to join their mothers on short foraging trips. The perinatal fasting period is only a few days in walrus and the milk they produce is relatively low in fat (~26%) compared to other pinnipeds. This aquatic lactation strategy has reduced the need to have fasting capabilities and high lactation efficiency. The ability of young pups to move around in the marine environment with their mothers provides them the opportunity to learn about foraging strategies and locations in a way that is not possible in the other pinnipeds. As a result, weaning takes place much more slowly, with the period of maternal investment lasting as long as 2 years.

Sea otters

Although less specialized for the marine environment when compared to other marine mammals, sea otters (*Enhydra lutris*) spend most of their time at sea, including the period of maternal investment. Sea otters give birth, lactate, and rear their young in shallow coastal waters. Females give birth to a single pup and lactate for an average of 6 months. Along the income/capital breeding continuum, sea otters represent the extreme case of an income strategy found among marine mammals. High surface-to-volume ratios cause elevated heat loss and contribute to the highest metabolic rates compared to body size among marine mammals. Because of this, sea otters must consume 20–25% of their body mass in food each day (Costa and Kooyman 1982) and spend as much as 50% of their time foraging, depending on reproductive status and prey availability (Tinker, Bentall, and Estes 2008). Sea otters lack the typical blubber layer of marine mammals and are dependent on increased food intake to supply the nutrients for lactation (see also **Chapter 5**). The increased energy cost of lactation imposes a significant energetic burden on sea otter mothers, with energy demands nearly 100% higher than pre-pregnancy levels by the time pups are weaned (Thometz et al. 2014). Despite increased feeding and low fat reserves, females often lose body mass during lactation (Monson et al. 2000).

Female sea otters breed annually, regardless of body condition or environmental factors. This "bet-hedging" strategy allows an otter to abandon pups early in development if she is not able to replenish the energy cost of lactation through increased foraging (Monson et al. 2000). Females also make decisions about when to wean their pup based on resource availability. Female sea otters must decide whether to continue to invest in their current pup to increase its likelihood of survival or wean them early to save energy for their own self-maintenance and health. In areas of high otter population density, reductions in prey resource availability may make it difficult for mothers to complete lactation. These energetic constraints have been implicated in poorer body condition and higher mortality rates among sea otters nearing the end of lactation on the central California Coast, termed as "end-lactation syndrome" (Thometz et al. 2014).

Polar bears

Polar bears (*Ursus maritimus*) are considered marine mammals because they move over a wide range of sea ice and enter the water to move between ice floes and land. This less marine-adapted lifestyle is reflected in their reproductive and lactation strategy. The postpartum physiology of polar bears has many similarities to that of brown bears. Polar bear offspring are unique among marine mammals in that they are **altricial**, initially dependent on their mothers for thermoregulation.

Pregnant females enter dens on land or, more rarely, on stable ice, and fast through pregnancy and the first several months of lactation. After a short 4-month gestation period, polar bears give birth to a small (~0.5 kg) cub while in their dens. The combined fasting period can be up to 8 months and females lose over 40% of the body mass during denning. Like other fasting adapted marine mammals in general, the bulk of energy reserves used are stored fat (Atkinson and Ramsay 1995). Fatter females produce larger cubs at the time of emergence from the den. Like many pinnipeds, fatter females also have increased ability to spare protein resources in vital body organs while fasting (Atkinson, Nelson, and Ramsay 1996, see also **Chapter 11**). Despite these similarities, there are several important differences between polar bear blubber and that of other marine mammals. Polar bear adipose tissue does not undergo changes in fatty acid composition across lactation (Thiemann, Iverson, and Stirling 2006). Instead, polar bear blubber undergoes dramatic changes in total lipid content to support the production of milk. For example, an elephant seal loses a lot of blubber mass during lactation but the remaining adipose tissue still has a lipid content of ~90%. Polar bear females lose not only blubber mass, but the lipid content of the remaining tissue also drops from 78 to 62%.

ENERGETIC INVESTMENT IN OFFSPRING

The strategies described above result in marked differences in milk composition and energetic investment in young (**Figure 12.4**). In general, capital breeders tend to have more energy-dense milk compared to income breeders. Phocids with the shortest lactation periods have the most energy-dense milk and rapid pup growth.

Cetaceans

Though data from individual species is sparse, the milk of cetaceans has the highest fat and energy content of any mammalian taxa except

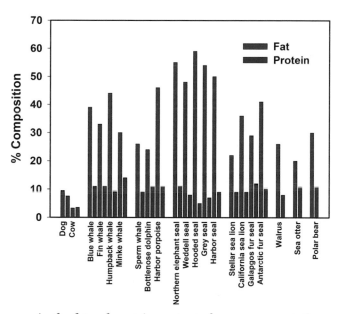

Figure 12.4 Differences in the fat and protein content of some representative species of marine and terrestrial mammal species. In general, marine mammals have higher fat and protein contents of milk than terrestrial species. Capital breeding phocids and mysticetes tend to have the highest milk nutrient contents.

Table 12.1 **Lactation duration and milk composition for selected cetaceans. Milk composition is from about mid-lactation**

SPECIES	LACTATION DURATION MONTHS	MILK COMPOSITION		
		% FAT	% PROTEIN	% SUGAR
Blue whale	6–7	39	11	1.3
Bryde's whale	6–7	30	15	
Fin whale	6–7	33	11	2.4
Humpback whale	10–11	44	9	0.7
Gray whale	7–8	53	6	
Minke whale	5–6	30	14	1.4
Bottlenose dolphin	19	24	11	1.1
Common dolphin	6	30	10	
Harbor porpoise	8	46	11	
Sperm whale	25	26	9	0.1
Spinner dolphin	15–19	26	7	

Source: Original sources for data are in Oftedal (1997).

for the pinnipeds (Oftedal and Iverson 1995; *Table 12.1*). Fat content varies across lactation but is inversely related to duration of parental investment, being ~30–50% at mid-lactation phase in mysticetes and ~10–30% at mid-lactation phase in most odontocetes (Oftedal 1997). While data is sparse, studies suggest that milk fat content peaks near mid-lactation. The high energy cost of lactation is reflected in thinner blubber layers in lactating females. The high milk energy outputs of mysticete whales represent the most extreme examples of parental investment in nature. A pregnant blue whale (*Balaenoptera musculus*) may weigh 119,000 kg, of which as much as 45,000 kg is blubber. During lactation, the mother converts most of this blubber to milk, allowing the calf to gain 17,000 kg by weaning (Lockyer 1981). Most odontocetes have smaller body reserves than mysticetes and their food resources are more varied in time and space. This may limit their ability to maintain the high milk energy outputs of large baleen whales. The duration of lactation in cetaceans may ultimately reflect the spatial and temporal characteristics of the food resources, and whether sufficient body reserves are accrued to permit rapid energy transfer to the offspring in milk.

Pinnipeds

A variety of studies have shown increased maternal investment from larger and fatter pinnipeds (e.g., Crocker et al. 2001). This means that foraging success impacts maternal investment and pupping rates in many species. Females that cannot recoup energy expended through foraging may not have a successful breeding season in the next year.

Phocids and otariids differ in their energetic investment in offspring. Phocids are larger, have more body fat, can fast for longer periods of time, and produce large volumes of milk. Most phocid species that fast during lactation lose 30–40% of body mass during breeding. Because lactation is brief, pup maintenance costs are low and the bulk of milk energy can be used for growth (>80% in elephant seals and gray seals). The short lactation period also results in a lower maternal **metabolic overhead**. Maternal metabolic overhead is the proportion of a female's energy expenditure that goes toward meeting her own metabolic needs. These costs can include maintenance metabolism and the energy costs of milk synthesis and maternal behavior. By compressing the period of parental investment, phocid seals increase the efficiency of lactation, increasing the proportion of energy expenditure given to their pups. Rapid energy delivery means that the pups can put most of that toward body reserves. In contrast, only 20–50% of ingested milk energy is available for growth in otariids as a result of the intermittent feeding strategy exhibited by otariids (McDonald et al. 2012). Otariid mothers expend more energy during the longer lactation period the maintenance needs of herself and her pup. Otariid pups may be smaller than phocid pups at weaning but they actually receive greater total

amounts of maternal energetic investment. For this reason, offspring behavior can also have important impacts on how maternal energy is allocated in otariids (McDonald et al. 2012).

Pinnipeds have the most energy-dense milk of all mammals, allowing for rapid transfer of energy to the pups (*Table 12.2*). Their milk is high in fat and protein content (as high as 61% fat in hooded seals and 14% protein in northern fur seals (*Callorhinus ursinus*)), but very low in carbohydrate content (<1%). By comparison, typical whole milk from a dairy cow is ~4% fat, ~3% protein, and ~5% sugars. Species with shorter lactation durations produce milk with higher fat and energy contents. In many species, milk fat content can increase with stage of lactation, amount of time on shore, and duration of foraging trips (Costa 1991).

Table 12.2 **Lactation variables for selected pinnipeds**

SPECIES	LACTATION DURATION DAYS	MATERNAL FORAGING % TIME AT SEA	MILK COMPOSITION		PUP GROWTH KG/DAY
			% FAT	% PROTEIN	
Bearded seal	24	84	47	10	3.3
Crabeater seal	17		51	11	4.2
Hooded seal	4	0	59	5	7.0
Harbor seal	25	55	50	9	0.6
Harp seal	12	71	50	8	2.2
Gray seal	16	0	54	7	2.5
N. elephant seal	26	0	55	11	4.0
Ringed seal	39	69	38	10	0.4
S. elephant seal	23	0	40	10	3.4
Weddell seal	50	40	48	8	2.0
Australian sea lion	532	60	31	10	0.11
Ca. sea lion	332	73	36	9	0.12
S.A. sea lion	548	59	32	10	0.21
Stellar sea lion	332	76	22	9	0.35
Antarctic fur seal	117	71	41	10	0.08
Galapagos fur seal	540	50	29	12	0.07
Subantarctic fur seal	300	80	43	12	0.04
N. fur seal	118	73	42	14	0.07
Walrus	730	NA	26	8	0.45

Source: Original sources for data are in Schulz and Bowen (2004).

Otters

Only one study of sea otter milk is available for comparison. The milk was ~20–26% fat, ~9–12% protein, and ~1% sugars (Jenness et al. 1981), and milk fat followed the general trend of being higher during late lacation. It has been estimated that in early lacation a female must ingest ~19.5 MJ day^{-1} which would increase to ~26.5 MJ day^{-1} by late in lacation to cover higher costs of milk production. Energy transfer to pup is 3.5-fold greater in late lactation (Cortez et al. 2016).

Polar bears

Denning is an excellent strategy for maximizing lactation efficiency. During denning, polar bear females have low metabolic rates that are less than resting metabolism, despite the costs of milk synthesis. After leaving the den, mother and cub may stay in the area for a while before they make the trek to the sea ice to hunt seals. Polar bears have an extremely long lactation period ranging from 1.5 to 2.5 years, depending on ice conditions, access to food resources, and the ability of the cub to hunt independently. Peak lactation is thought to occur during the denning period or soon after bears emerge from the den. Polar bear milk composition changes throughout denning and feeding, declining in milk fat content from a high of 36% fat and increasing in protein content after females begin to feed (Derocher, Andriashek, and Arnould 1993; *Table 12.1*). Unlike other marine mammals, polar bear milk has substantial carbohydrate content (as high as 5%) that declines across the lactation period (Arnould and Ramsay 1994). As the cubs grow, milk output decreases and females with older cubs may stop lactation during seasonal periods of fasting. Unlike many pinnipeds, the fatty acid composition of polar bear blubber stays consistent across lactation.

NUTRIENT MOBILIZATION

Capital breeding marine mammals need to rapidly mobilize body reserves for milk production. Income breeding pinnipeds must also use stored body reserves during their on-shore periods. Most of what is known about nutrient mobilization for lactation comes from studies on pinnipeds because of their on-land accessibility. In order to produce such energy-dense milk while fasting, pinnipeds must rapidly mobilize lipids stored body fat and protein derived from muscle and vital body organs and deliver it to the mammary gland, while preventing the mobilized protein from being burned as fuel for maintenance metabolism or being used to make sugars. Pinniped milk also contains significant amounts of water. Although the high energy content reduces the water needed for milk production, the bulk of this water must be provided through **metabolic water production (MWP)** from maternal metabolism. This water is produced largely from the oxidation of fats for energy (see in detail **Chapter 11**). Here we consider the role that nutrient mobilization and fasting plays in determining the ability of pinnipeds to produce such energy-dense milk.

The primary determinant of nutrient delivery to the mammary gland is mobilization of fatty acids from stored triglycerides in adipose tissue or **lipolysis**. Fasting and lactating pinnipeds have some of the highest circulating levels of fatty acids found in nature. By the end of lactation, northern elephant seals have plasma fatty acids levels that average over 3 mM, a concentration that would be considered harmful (dyslipidemia) in humans (Crocker et al. 2014). These high levels of fatty acids not only support maternal maintenance energy needs, but also allow rapid uptake and use by the mammary gland. The mechanisms that allow such rapid mobilization of stored fats in seals appear to vary widely, even between species with similar life history patterns which suggests strong evolutionary selection pressure.

The primary endocrine features associated with lactation in seals are a profound reduction in insulin and elevation in cortisol. While insulin's primary role is the regulation of carbohydrate metabolism, it also exerts strong anti-lipolytic effects. This means that insulin promotes the production of new fats and inhibits the mobilization of stored fats. Low insulin levels appear critical to fat mobilization in seals, resulting in a decline in insulin release over lactation when compared to other life history stages (Fowler ct al. 2008). Correlative analysis suggests that low insulin levels are the primary determinant of fatty acid mobilization during lactation in some species. Cortisol helps mobilize fat and protein in most species of animals but usually also leads to increased mobilization of proteins for energy metabolism and for making sugars – features that might lead to vital organ damage during extended fasting. Several species of phocids and otariids elevate plasma cortisol concentrations over their lactation periods. This elevation is strongly associated with both circulating fatty acid levels and milk fat content in some species. Unlike most terrestrial mammals, seals avoid the protein wasting effects of cortisol. Protein oxidation provides a maximum of 4–8% of energy expenditure (as measured from the production of urea or changes in body composition, Crocker et al. 1998). Elevations in cortisol trigger the return to sea to forage (Guinet et al. 2004). Growth hormone, the primary regulator of lipolysis, milk output, and fat content in dairy cattle, appears to play a more minor role in seals.

Several enzymes are important in the process of lipolysis and release of fatty acids from triglycerides. Direct measurements of rates lipolysis during lactation in elephant seals have shown that they are uncoupled from the levels of circulating fatty acids during lactation (Houser, Champagne, and Crocker 2007). The enzymes that allow adipose tissue to reuptake fatty acids for storage – fatty acid translocase (CD36) and fatty acid transport protein 1 (FATP1) – decrease with fasting in seals (Viscarra and Ortiz 2013). This suggests that reducing the reuptake of fatty acids by adipose tissue is an important feature that keeps fatty acids levels high to supply the mammary gland as fasting seals deplete their fat reserves while lactating. In most mammals, the primary enzyme responsible for lipolysis of fats

is hormone-sensitive lipase (HSL). Suprisingly, HSL levels are very low in the blubber of lactating elephant seals. Instead the enzyme **adipocyte triglyceride lipase (ATGL)** that removes one fatty acid from a triglyceride molecule has an increased importance when compared to other mammals (Fowler et al. 2015). In some species, like gray and harbor seals, the levels of an enzyme bound to mammary gland tissues called **lipoprotein lipase (LPL)** is important to allowing the mammary gland to uptake fatty acids from triglycerides in circulation. In these species, LPL levels increase in parallel with milk fat content (Iverson, Hamosh, and Bowen 1995). In other species, like elephant and hooded seals, this enzyme plays a minor role (McDonald and Crocker 2006). These differences suggest that despite similar reproductive strategies, the various seal lineages may have evolved important metabolic differences in lactation physiology under strong evolutionary pressure for efficient lactation while fasting.

Since blubber is the source for fatty acids for milk synthesis, its composition and the way individual fatty acids are mobilized and used for metabolism can potentially influence the composition of milk. It has been reported in many different species of marine mammals that blubber layers are stratified from inner to outer layers (Strandberg et al. 2008). External layers have a higher proportion of medium-chain (≤18C) monounsaturated fatty acids (MUFA), possibly as a **homeoviscous adaptation** for the purpose of maintaining membrane fluidity at the low temperatures encountered at depth. Interior layers in phocid blubber are highly enriched in saturated fatty acids (SFA) and long-chain (≥20C) MUFA. This inner layer is heavily metabolized during fasting and lactation (Fowler et al. 2014). In phocids, the mobilization of specific fatty acids from blubber, and their incorporation into milk, conforms to biochemical predictions based on the number of carbons and saturated bonds. Long-chain (>20C) MUFA are the least mobilized and polyunsaturated fatty acids (PUFA) and SFA are highly mobilized from the blubber. In the mammary gland, fatty acid synthesis from glucose and ketones results in short- and medium-chain fatty acids containing fewer than 12 carbons. However, these short- and medium-length fatty acids are usually not detected in seal milk, suggesting that there is little synthesis of new lipid in the mammary gland. In other words, plasma fatty acid delivery and uptake by the mammary gland is responsible for milk fat content. PUFA availability to the developing pup's muscle tissue may contribute to the development of oxidative capabilities for diving and provides non-shivering thermoregulatory benefits (Trumble et al. 2010). The majority of long-chain MUFA mobilized is directed to milk synthesis and the mother may preferentially use PUFA and SFA for her own metabolism. The proportion of long-chain MUFA in milk increases later in lactation in several species. The MUFA delivered in milk may help pups establish a thermoregulatory blubber layer with optimal characteristics for energy density and thermoregulation.

MATERNAL METABOLIC ADAPTATIONS TO FASTING WHILE LACTATING

The accessibility of pinnipeds when lactating on-shore has allowed the use of cutting-edge techniques to study the metabolic adaptations associated with lactating while fasting. In some cases, tracer techniques have been used to study metabolism in lactating pinnipeds that have not been used in any other wildlife species. These studies demonstrate a remarkable ability to maintain metabolic homeostasis despite extraordinarily high rates of energy mobilization and loss. For example, the combined energy loss for metabolism and milk production in lactating phocid seals can be more than 7 times the predicted standard metabolic rate for a mammal of similar size (Mellish et al. 2000; Crocker et al. 2001). Many of the metabolic features that enable these feats of high-energy-expenditure fasting are also found in

non-lactating conspecifics, but several studies have revealed important alterations of maternal metabolism and physiology associated with lactation. These alterations include changes in the release of hormones that regulate metabolism and evidence for changes in the tissue responses to regulatory hormones. These studies also provide evidence that metabolic features are strongly impacted by the amount of body fat, or **adiposity**, of lactating females. For example, lactating females release less insulin and exhibit reduced glucose clearance when compared to fasting pups. These changes can vary directly with the depletion of adipose tissue reserves while lactating. Unlike humans and most mammals, where being fat is associated with reduced insulin sensitivity, seals appear to develop more diabetes-like features as they deplete their body reserves while lactating. Similarly, responses to metabolic hormones like glucagon, which typically causes production of sugars from stored body protein and release of fatty acids from blubber, are altered in direct relation to depletion of body fat (Crocker et al. 2014). Responses that would hinder the ability to continue fasting and producing milk, like releasing insulin and making sugars from muscle or vital organs get downregulated, while responses that facilitate milk production and fasting, like enhanced lipolysis, get upregulated. This represents a remarkable example of metabolic adaptation for rapid reorganization of systemic and tissue metabolism to provide appropriate nutrients to the mammary gland, while protecting mothers from harmful effects of lactating while fasting.

MAMMARY GLAND PHYSIOLOGY

Another aspect of pinniped lactation physiology that has received recent attention in pinnipeds is the physiological requirements of the pinniped mammary gland. The mammary gland of all mammals undergoes a complex set of changes during lactation that includes proliferation and differentiation of cells, secretion, and ultimately the death and regression of mammary cells after weaning in a process known as **involution**. The unusual composition of pinniped milk and the ability of the otariid mammary gland to sustain function despite long interruptions in suckling provide an important comparative model with which to better understand the regulation of involution in other species.

The milk sugar, lactose, has been detected in only trace concentrations in phocid milk and is absent from the milk of otariids and odobenids. The low levels of milk lactose may reflect a mammary gland that is highly adapted for pinniped lactation systems. Lactose is an important determinant of milk water content as it is unable to diffuse through the mammary cell membrane and draws water into the mammary alveoli by osmosis. Low lactose levels are associated with the ability to concentrate milk solids, allowing the high lipid and protein content of the milk. Low lactose levels may help promote the high fat and protein contents in pinniped milk, in addition to reducing the use of proteins for carbohydrate synthesis in vital organs.

The low lactose levels may also be a result of a mutation in the alpha-lactalbumin gene (LALBA) that normally encodes a protein that plays an important role in lactose synthesis. Mutations in the LALBA gene may help preserve mammary gland functionality during intermittent suckling. During their long foraging bouts, mammary glands of otariids do not undergo involution but remain active and ready to suckle the pup when the female returns to shore. In other mammals, the termination of suckling is associated with accumulation of milk in the gland. Prolonged exposure to factors in the milk cause downregulation of expression of milk protein genes, followed by involution and mammary cell death. The α-lactalbumin protein encoded by the LALBA gene has been identified as an important factor in this process in mice and humans. Mutations have been found in this gene for several otariid species, which may protect the otariid mammary gland from involution while foraging (**Figure 12.5**).

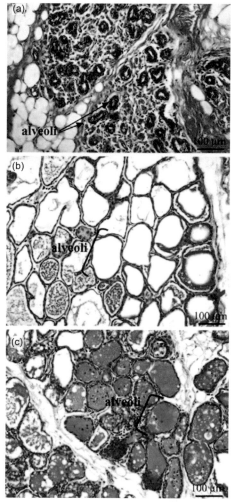

Figure 12.5 Mammary gland morphology of an otariid. Histological sections of a mammary gland from (a) pregnant, (b) lactating on-shore, and (c) lactating while foraging at-sea fur seals. Fat globules within alveoli appear as white bodies. Pink staining within alveoli represent milk components. Residence of milk components in alveoli would lead to involution in most species but otariids have evolved the ability to maintain mammary function in absence of suckling. (Figure is used with permission from Sharp et al. 2007)

In contrast, phocid seals lack this mutation and have shown involution of the mammary gland within 24 hours of weaning (Reich and Arnould 2007). The mutation of the LALBA gene in otariids and walruses may have been an important factor in allowing the divergence of pinniped life history strategies between capital and income breeding strategies.

DIGESTION AND ASSIMILATION BY PUPS

The ability of the mother to rapidly produce such large quantities of high-fat milk must be matched by the digestive and assimilation capabilities of the pup to process and use these nutrients. Maximum energy assimilation rates by the digestive and metabolic systems of animals have been theorized to constrain energy expenditure and growth (Weiner 1992). Seven X basal metabolic rate (BMR) has been hypothesized as an upper limit to assimilation of dietary energy from free-ranging energy budgets (Hammond and Diamond 1997) and a mammalian record of 7.7X BMR was reported for mice under conditions of extreme cold exposure (Johnson and Speakman 2001). An average elephant seal pup assimilates energy at a rate of 8.3X BMR during suckling (Crocker et al. 2001) and a hooded seal pup likely has the highest rate of nutritional assimilation relative to body weight found in nature. These feats require an appropriate lipid digestion and intestinal uptake capacity; mechanisms to transport the digested lipids to adipose tissue for storage; and the ability to withstand extreme blood lipemia or high concentrations of emulsified fat. The digestive adaptations of pinnipeds are not yet studied.

TOOLBOX

The study of lactation and postpartum reproduction in marine mammals has focused largely on pinnipeds because they are accessible on land while nursing their young. These studies have investigated how females invest energy in their offspring, how the offspring allocate the acquired energy toward growth and development, and the metabolic features that enable production of high-energy-density milk while fasting. These studies use a combination of techniques that allow estimation of energy expenditure and energy intake, ranging from determining mass and body condition to using data loggers to measure movement and behavior. As technology advances, we will continue to increase our knowledge about postpartum physiology, particularly in the groups such as cetaceans, where data is lacking.

Doubly labeled water (field metabolic rate and milk intake)

Knowledge on rate of milk production is key to understanding the intra- and interspecific variation in the patterns of energy transfer observed in nature and described in this chapter. To measure milk production or intake, it is essential to have measures of both field metabolic rate and milk compostition. It is possible to measure field metabolic rate over short periods of time (~7–10 days), and this, combined with milk energy output, can be used to investigate lactation and postpartum reproductive effort.

Field metabolic rate can be measured using the **doubly labeled water** (DLW) method (Costa 1987; Speakman 1997). The basis of the DLW method is to follow the decline in enrichment of the isotopes of oxygen (^{18}O) and hydrogen (^{2}H or ^{3}H) in the body water. CO_2 production is measured by injecting known amounts of $^{2}H_2O$ (or $^{3}H_2O$) and $H_2^{18}O$ into the animal (**Figure 12.6**). An initial blood sample is taken after the isotopes equilibrate with the animal's body water (as described above), and it is followed by a final blood sample at the end of the study period, 7–10 days later depending on animal size and metabolic rate. The decline in the hydrogen isotope is a measure of total water influx (TWI), which is composed of MWP and water consumed in the food (i.e., milk or fish). The ^{18}O isotope declines as a function of both water flux and CO_2 production. The difference between the rates of decline of these two isotopes is proportional to the animal's CO_2 production. Energy expenditure can then be calculated from CO_2 production using an

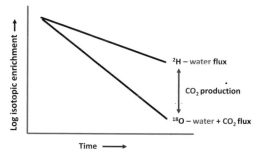

Figure 12.6 Visualization of changes in isotopic enrichment over time for the doubly labeled water (DLW) method of measuring metabolism in free-ranging animals. The hydrogen isotope leaves the animal's body as water. The oxygen isotope leaves the animal's body as both water and CO_2. Turnover of hydrogen isotopes can be used to estimate water influx and milk intake.

appropriate conversion factor depending on the diet of the female or pup (Costa 1987).

Obtaining milk samples from large and/or difficult-to-access species can be challenging, and accurate estimates of milk production may be impractical. Despite this challenge, milk composition has been determined for many marine mammals. From these studies, we know that milk composition can change substantially over the course of lactation and attendance bouts, so it is important to consider the timing of sampling when studying milk composition and intake. In pinnipeds, an intramuscular injection of oxytocin is usually administered to help initiate milk let down before sampling. The lipid, protein, water, and ash components of the milk are measured independently and in duplicate following standard protocols (reviewed in Oftedal and Iverson 1995). Lipids are typically extracted using organic solvents and protein content is usually measured based on the nitrogen content of milk. The water content of milk is easily measured using oven drying. Carbohydrates are often not analyzed because of their established minor contribution to marine mammal milk. If measured accurately, the sum of the individual milk components should total ~100% of the

initial sample mass. The total energy content can be estimated using standard values for the energy density of lipid and protein. Alternatively, if the primary question is the energy content of milk, gross energy content can be determined with high precision using bomb calorimetry of the dry material and water content data.

Milk intake can be measured using labeled water techniques if one assumes that the only influx of water is from MWP and milk intake (Oftedal and Iverson 1987). Because of this assumption, this method is only appropriate when offspring are solely dependent on milk. Milk consumption rates are calculated using the water influx rates (calculated from the decline in labeled H as described above), milk water content, and MWP, ideally derived independently from the metabolic rate of each pup. Milk intake is then converted to energy consumed using the energy content of milk and can be combined with data on maternal energy expenditure and pup energy storage to create a lactation energy budget.

Biologgers

As described in previous chapters, many aspects of biomechanics, diving physiology, and foraging ecology of marine mammals have been uncovered with the use of data loggers. Given the enormous developments in these tags in recent years, we are now able to measure fine scale movements, along with physiological parameters, of behaviors and/or in species that are difficult to observe. Most of what we know about lactation and reproductive effort in marine mammals is based on pinniped studies, but as technology advances we can address more of these questions in cetaceans with the creative use of data loggers.

Stomach temperature data loggers have primarily been used in adult animals not only to investigate at-sea feeding behavior, but have also been used to observe suckling behavior in harbor seals (Hedd, Gales, and Renouf 1995). This approach found that harbor seals commonly suckled in water (which would often be missed with behavioral observations), and that pups started ingesting prey while they were still nursing, suggesting that weaning may be a gradual process in this species (Schreer, Lapierre, and Hammill 2010). This method can be used to study suckling behavior and the transition to nutritional independent in other species.

Time depth recorders (TDR) revolutionized our ability for studying marine mammals at sea, and now many modern tags include accelerometers and magnetometers that measure fine-scale movement that can be used to estimate energy expenditure through measures such as overall dynamic body acceleration and stroke/fluke rate. The use of these tags on lactating females provides information on time activity budgets (both at sea and on land) and improves our understanding of energy allocation during lactation. In addition, accelerometers have also been used to pick up behavioral signals such as feeding events. With visual calibration of animals instrumented with accelerometers, it is possible to identify other behaviors, such as suckling, resulting in a better understanding of reproductive effort in animals that are difficult to observe (**Figure 12.7**; Videsen et al. 2017; Tackaberry et al 2020). If mothers/calf pairs are instrumented at different stages of lactation, it will be possible to estimate how much time the calf spends suckling and how this changes as lactation progresses.

Metabolism

The full scope of biomedical tools for the study of metabolism are available for the study of lactation in marine mammals; although their applicability is currently mostly limited to pinnipeds that are able to be chemically immobilized without substantial handling artifacts (Champagne et al. 2012). The tools include the use of antibody-based tools to measure protein expression in tissue and circulating levels of chemical messengers including hormones and adipokines (hormones made by blubber). Western blots,

enzyme immunoassays (EIA), enzyme-linked immunosorbent assays (ELISA), and radioimmunoassays are used to quantify the concentrations of proteins and chemical messengers. The use of omics tools including genomics, transcriptomics, and proteomics can be used to understand the changes in gene expression in specific tissues of fasting and lactating pinnipeds, usually blubber and muscle. The systemic impacts of these changes in gene expression can be studied using metabolomics, which attempts to measure as many of the metabolites in circulation or tissue as possible, often several hundreds in number (Champagne et al. 2013). Finally, metabolic tracers can be used to quantify metabolite flux during lactation. A variety of techniques have been developed that use turnover of radioisotope or stable-isotope-labeled metabolites to investigate whole-body metabolism. Several of these tracer methodologies that were developed for biomedical studies have been used for the first time in wildlife systems in studies on lactating pinnipeds (Champagne, Houser, and Crocker 2006). These tools reveal the unique features that allow some species of marine mammals to lactate while fasting.

CONCLUSIONS

The need for offspring survival in the physiologically challenging marine environment has led to the evolution of specialized lactation systems in marine mammals. Marine mammals tend to give birth to a single, large, precocial young, with many species exhibiting rapid growth and development and unusually high rates of milk energy delivery. In many species, terrestrial breeding or breeding far from foraging areas has led to capital breeding strategies, where lactation is supported from stored body reserves while fasting and the duration of lactation is relatively short. The need for high rates of milk energy delivery have given rise to some of the highest fat and protein contents for milk found in nature, in association with modified mammary gland function and physiology.

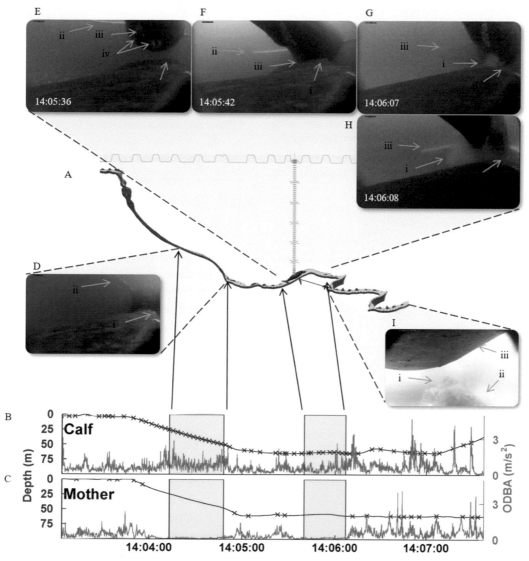

Figure 12.7 Use of video and data loggers to profile nursing behavior in wild cetaceans (From Tackaberry et al 2020, doi:10.7717). (A) Trackplot of the mother. Dorsal side of track (blue and light grey), ventral side of track (dark and light grey), fluke upstroke (red triangles) and fluke downstroke (blue triangles) shown. Dive profile of the calf (B) and its mother (C) with overall dynamic body acceleration (ODBA, green) and fluke strokes (red). Nursing events (yellow) with the first event categorized as a descending phase and the second event as a horizontal phase. Note the calf's increased fluke strokes and ODBA compared to the mother's during nursing events. (D) First nursing event with an image from the calf's tag showing calf nares (i) and mother's left flipper (ii). (E–H) Second nursing event. (E) Calf approaching to nurse with nares (i) and mother's left flipper (ii), hemispherical lobe (iii), and mammary slits (iv) visible. (F) Nursing calf. Calf nares (i), mother's left flipper (ii) and hemispherical lobe (iii) referenced. (G) End of nursing with a visible cloud of milk (i). Calf's nares (ii) and mother's left flipper (iii) are referenced. (H) Calf moving away with expanding milk cloud (i). Calf nares (ii) and mother's left flipper (iii) are referenced. (I) Mother initiates bottom-side rolls, indicating feeding.

REVIEW QUESTIONS

1   What is the difference between capital and income breeding strategies?
2   What is metabolic overhead and how does it differ between capital and income breeders?
3   Why do marine mammals have such lipid-rich milk?
4   Why is marine mammal milk typically low in carbohydrate?
5   Why is lactation particularly energetically difficult for sea otters?
6   What adaptation may help prevent mammary gland involution when sea lions are away from their pups foraging?
7   What is the likely reason that marine mammal blubber is stratified in its composition?
8   The scaling exponent for marine mammal birth mass compared to maternal birth mass is less than one. What does this mean for the relative birth sizes of small and large marine mammals?
9   Why is important for a mysticete whale or phocid seal to be fat before breeding?
10 How do polar bear offspring differ from other marine mammals?

CRITICAL THINKING

1   Is there an optimal lactation strategy?
2   Why might longer lactation periods be beneficial for odontocete cetaceans?
3   How might capital breeding in phocids and income breeding in otariids affect the locations of their breeding colonies?

FURTHER READING

Costa, D.P., and J.L. Maresh. 2022. Reproductive energetics of phocids. In *Ethology and Behavioral Ecology of Phocids*, 281–309. Cham: Springer.
Champagne, C.D., D.E. Crocker, M.A. Fowler, et al. 2012. Fasting physiology of the pinnipeds: The challenges of fasting while maintaining high energy expenditure and nutrient delivery for lactation. In *The Comparative Physiology of Starvation, Fasting and Food Limitation*, ed. M.D. McCue, 309–336. Berlin: Springer-Verlag.
Oftedal, O.T. 1997. Lactation in whales and dolphins: Evidence of divergence between baleen-and toothed-species. *Journal of Mammary Gland Biology and Neoplasia*, 2(3): 205–230.

GLOSSARY

Adipocyte triglyceride lipase (ATGL): An enzyme that removes the first fatty acid from a triacylglycerol molecule. This enzyme has been shown to be the predominant lipolytic enzyme in some marine mammal's blubber.

Adiposity: The degree of fatness of an animal. High adiposity is a feature of most marine mammals.

Altricial: A young animal born in an undeveloped state with limited mobility and requiring a high level of care and feeding by the parents compared to precocial young.

Capital breeding: A breeding strategy in which all or most of the energy used for breeding comes from stored body reserves. Capital and income breeding are two extremes on a continuous scale.

Cutaneous muscles: Muscles that are found in the subcutaneous tissue (tissue under the skin) and attach to the skin.

Doubly labeled water: Isotopically labeled water in which both the oxygen and hydrogen molecules have been labeled with isotopes. The doubly labeled water method uses turnover of the tracer to measure CO_2 production and is an important technique for measuring the field metabolic rates of free-ranging animals.

Fecundity: The ability to produce offspring. Fecundity is usually defined as the rate of births for an individual or species.

Homeoviscous adaptation: Remodeling of the cell membrane structure (lipid composition) that maintains adequate membrane fluidity across the temperature range.

Income breeding: A breeding strategy in which animals forage during the period of reproduction. Capital and income breeding are two extremes on a continuous scale.

Involution: The shrinkage of an organ to its former size when inactive. Involution is usually used in reference to reproductive organs or mammary glands after reproduction.

Lipoprotein lipase: An enzyme that allows tissue uptake of fatty acids from triglycerides in blood circulation.

Lipolysis: The breakdown of fats and other lipids by hydrolysis to release fatty acids. Lipolysis is the key process in mobilizing energy for fasting marine mammals.

Metabolic overhead: The proportion of a lactating female's energy expenditure that goes to her own maintenance metabolism instead of milk energy delivery to the pup or calf.

Metabolic water production: The water that is produced from cellular respiration. For many marine mammals this is the only source of water for milk production and survival during fasting.

Myoepithelial cells: Modified epithelial cells (thin protective layer of cells that line outer surface of organs and blood vessels, as well as inner surfaces of cavities) that are found in glands. The cells have contractile properties that allow ejection of fluids such as sweat and milk.

Perinatal period: The period immediately after birth. In otariids, this period refers to the initial period of lactation before the mother returns to sea to forage.

Precocial: A young animal born in a more developed state with increased mobility and reduced need for parental care compared to altricial young.

Scaling exponent: The exponent of a power function that describes the relationship between two log transformed variables. Many physiological and behavioral variables in biology exhibit power functions with body size across species.

Weaning conflict: The period where the offspring and parent are in conflict about the level of continued parental investment. Offspring may be selected to seek more resources from parents than that which maximizes their ability to invest in future reproduction.

REFERENCES

Arnould, J.P.Y., and M.A. Ramsay. 1994. Milk production and milk consumption in polar bears during the ice-free period in western Hudson Bay. *Canadian Journal of Zoology* 72: 1365–1370.

Atkinson, S.N., and M.A. Ramsay. 1995. The effects of prolonged fasting of the body composition and reproductive success of female polar bears (*Ursus maritimus*). *Functional Ecology* 9(4): 559–567.

Atkinson, S.N., R.A. Nelson, and M.A. Ramsay. 1996. Changes in the body composition of fasting polar bears (*Ursus maritimus*): The effect of relative fatness on protein conservation. *Physiological Zoology* 69(2): 304–316.

Bowen, W.D., S.L. Ellis, S.J. Iverson, et al. 2001. Maternal effects on offspring growth rate and weaning mass in harbour seals. *Canadian Journal of Zoology* 79(6): 1088–1101.

Burns, J.M, C.A. Clark, and J.P. Richmond. 2004. The impact of lactation strategy on physiological development of juvenile marine mammals: Implications for the transition to independent foraging. *International Congress Series* 1275: 341–350.

Champagne, C.D., D.S. Houser, and D.E. Crocker. 2006. Glucose metabolism during lactation in a fasting animal, the northern elephant seal. *American Journal of Physiology-Regulatory, Integrative and Comparative Physiology* 291(4): R1129–R1137.

Champagne, C.D., S.M. Boaz, M.A. Fowler, et al. 2013. A profile of carbohydrate metabolites in the fasting northern elephant seal. *Comparative*

Biochemistry and Physiology Part D: Genomics and Proteomics 8(2): 141–151.

Champagne, C.D., D.E. Crocker, M.A. Fowler, et al. 2012. Fasting physiology of the pinnipeds: The challenges of fasting while maintaining high energy expenditure and nutrient delivery for lactation. In *Comparative Physiology of Fasting, Starvation, and Food Limitation*, ed. M.D. McCue. Berlin Heidelberg: Springer.

Cortez, M., C.E. Goertz, V.A. Gill, et al. 2016. Development of an altricial mammal at sea: II. Energy budgets of female sea otters and their pups in Simpson Bay, Alaska. *Journal of Experimental Marine Biology and Ecology* 481: 81–91.

Costa, D.P. 1987. Isotopic methods for quantifying material and energy intake of free-ranging marine mammals. In *Approaches to Marine Mammal Energetics*, ed. A.C. Huntley, D.P. Costa, G.A.J. Worthy, and M.A. Castellini. Society for Marine Mammalogy. Allen Press Lawrence, KS.

Costa, D.P. 2009. Energetics. 2009. In *Encyclopedia of Marine Mammals*, ed. W.F. Perrin, B. Wursig, and J.G.M. Thewissen. Academic Press, San Diego, CA.

Costa, D.P., and G.L. Kooyman. 1982. Oxygen consumption, thermoregulation, and the effect of fur oiling and washing on the sea otter, *Enhydra lutris*. *Canadian Journal of Zoology* 60(11): 2761–2767.

Costa, D.P. 1991. Reproductive and foraging energetics of pinnipeds: Implications for life history patterns. In *The Behavior of Pinnipeds*, ed. D. Renouf. London: Chapman and Hall.

Crocker, D.E., P.M. Webb, D.P. Costa, et al. 1998. Protein catabolism and renal function in lactating northern elephant seals. *Physiological Zoology* 71(5): 485–491.

Crocker, D.E., J.D. Williams, D.P. Costa, et al. 2001. Maternal traits and reproductive effort in northern elephant seals. *Ecology* 82(12): 3541–3555.

Crocker, D.E., C.D. Champagne, M.A. Fowler, et al. 2014. Adiposity and fat metabolism in lactating and fasting northern elephant seals. *Advances in Nutrition* 5(1): 57–64.

Crocker, D.E., M.A. Fowler, C.D. Champagne, et al. 2014. Metabolic response to a glucagon challenge varies with adiposity and life-history stage in fasting northern elephant seals. *General and Comparative Endocrinology* 195: 99–106.

Derocher, A.E., D. Andriashek, and J.P.Y. Arnould. 1993. Aspects of milk composition and lactation in polar bears. *Canadian Journal of Zoology* 71: 561–567.

Fowler, M.A., C.D. Champagne, D.S Houser, et al. 2008. Hormonal regulation of glucose clearance in lactating northern elephant seals (*Mirounga angustirostris*). *Journal of Experimental Biology* 211(18): 2943–2949.

Fowler, M.A., C. Debier, E. Mignolet, et al. 2014. Fatty acid mobilization and comparison to milk fatty acid content in northern elephant seals. *Journal of Comparative Physiology B* 184(1): 125–135.

Fowler, M.A., D.P. Costa, D.E. Crocker, et al. 2015. Adipose triglyceride lipase, not hormone sensitive lipase, is the primary lipolytic enzyme in fasting elephant seals (*Mirounga angustirostris*). *Physiological and Biochemical Zoology* 88(3): 284–294.

Guinet, C., N. Servera, S. Mangin, et al. 2004. Change in plasma cortisol and metabolites during the attendance period ashore in fasting lactating subantarctic fur seals. *Comparative Biochemistry and Physiology Part A: Molecular & Integrative Physiology* 137(3): 523–531.

Hammond, K.A., and J. Diamond. 1997. Maximal sustained energy budgets in humans and animals. *Nature* 386(6624): 457–462.

Hedd, A., R. Gales, and D. Renouf. 1995. Use of temperature telemetry to monitor ingestion by a harbour seal mother and her pup throughout lactation. *Polar biology* 15(3): 155–160.

Houser, D.S., C.D. Champagne, and D.E. Crocker. 2007. Lipolysis and glycerol gluconeogenesis in simultaneously fasting and lactating northern elephant seals. *American Journal of Physiology - Regulatory, Integrative and Comparative Physiology* 293(6): R2376–R2381.

Iverson, S.J., M. Hamosh, and W.D. Bowen. 1995. Lipoprotein lipase activity and its relationship to high milk fat transfer during lactation in grey seals. *Journal of Comparative Physiology B* 165: 384–395.

Jenness, R., T.D. Williams and R.J. Mullin. 1981. Composition of milk of the sea otter (*Enhydra lutris*). *Comparative Biochemistry and Physiology Part A: Physiology* 70(3): 375–379.

Johnson, M.S., and J.R. Speakman. 2001. Limits to sustained energy intake V. Effect of cold-exposure during lactation in *Mus musculus*. *Journal of Experimental Biology* 204(11): 1967–1977.

Lockyer, C. 1981. Growth and energy budgets of large baleen whales from the southern hemisphere. *Mammals in the Seas* 3: 379–487.

Lydersen, C. 1995. Energetics of pregnancy, lactation and neonatal development in ringed seals (*Phoca hispida*). In *Whales, Seals, Fish and Man*, ed. A.S. Blix, L. Walloe, and O. Ulltang. Amsterdam: Elsevier Science B.V.

Lydersen, C., M.O. Hammill, and K.M. Kovacs. 1995. Milk intake, growth and energy consumption in pups of ice-breeding grey seals (*Halichoerus grypus*) from the Gulf of St. Lawrence, Canada. *Journal of Comparative Physiology B* 164: 585–592.

Lydersen, C., and K.M. Kovacs. 1996. Energetics of lactation in harp seals (*Phoca groenlandica*) from the Gulf of St. Lawrence, Canada. *Journal of Comparative Physiology B* 166: 295–304.

Lydersen, C., K.M. Kovacs, and M.O. Hammill. 1997. Energetics during nursing and early postweaning fasting in hooded seal (*Cystophora cristata*) pups from the Gulf of St. Lawrence, Canada. *Journal of Comparative Physiology B* 167: 81–88.

McDonald, B.I., and D.E. Crocker. 2006. Physiology and behavior influence lactation efficiency in northern elephant seals (*Mirounga angustirostris*). *Physiological and Biochemical Zoology* 79(3): 484–496.

McDonald, B.I., M.E. Goebel, D.E. Crocker, et al. 2012. Biological and environmental drivers of energy allocation in a dependent mammal, the Antarctic fur seal pup. *Physiological and Biochemical Zoology* 85(2): 134–147

Mellish, J.E., S.J. Iverson, and W.D. Bowen. 2000. Metabolic compensation during high energy output in fasting, lactating grey seals (*Halichoerus grypus*): metabolic ceilings revisited. *Proceedings of the Royal Society of London B* 267: 1245–1351.

Monson, D.H., J.A. Estes, J.L. Bodkin, et al. 2000. Life history plasticity and population regulation in sea otters. *Oikos* 90(3): 457–468.

Oftedal, O.T. 1997. Lactation in whales and dolphins: Evidence of divergence between baleen- and toothed-species. *Journal of Mammary Gland Biology and Neoplasia* 2(3): 205–230.

Oftedal, O.T., and S.J. Iverson. 1987. Hydrogen isotope methodology for measurement of milk intake and energetics of growth in suckling young. In *Approaches to Marine Mammal Energetics*, ed. A.C. Huntley, D.P. Costa, G.A.J. Worthy, and M.A. Castellini. Lawrence: Allen Press.

Oftedal, O.T., and S. J. Iverson. 1995. Comparative analysis of nonhuman milks. A. Phylogenetic variation in the gross composition of milks. In *Handbook of Milk Composition*, ed. R. Jensen, 749–789. Academic Press.

Reich, C.M., and J.P.Y. Arnould. 2007. Evolution of Pinnipedia lactation strategies: A potential role for α-lactalbumin?. *Biology letters* 3(5): 546–549.

Schreer, J.F., J.L. Lapierre, and M.O. Hammill. 2010. Stomach temperature telemetry reveals that harbor seal (*Phoca vitulina*) pups primarily nurse in the water. *Aquatic Mammals* 36(3): 270.

Sharp, J.A., C. Lefevre, A.J. Brennan, et al. 2007. The fur seal—a model lactation phenotype to explore molecular factors involved in the initiation of apoptosis at involution. *Journal of Mammary Gland Biology and Neoplasia* 12(1): 47–58.

Speakman, J. 1997. *The Doubly Labeled Water Method*. New York: Chapman and Hall.

Strandberg, U., A. Käkelä, C. Lydersen, et al. 2008. Stratification, composition, and function of marine mammal blubber: The ecology of fatty acids in marine mammals. *Physiological and Biochemical Zoology* 81(4): 473–485.

Tackaberry, J.E., D.E. Cade, J.A. Goldbogen, et al. 2020. From a calf's perspective: Humpback whale nursing behavior on two US feeding grounds. *PeerJ* 8: e8538.

Thiemann, G.W., S.J. Iverson, and I. Stirling. 2006. Seasonal, sexual and anatomical variability in the adipose tissue of polar bears (*Ursus maritimus*). *Journal of Zoology* 269(1): 65–76.

Thometz, N.M., M.T. Tinker, M.M. Staedler, et al. 2014. Energetic demands of immature sea otters from birth to weaning: Implications for maternal costs, reproductive behavior and population-level trends. *The Journal of Experimental Biology* 217(12): 2053–2061.

Tinker, T.M., G. Bentall, and J.A. Estes. 2008. Food limitation leads to behavioral diversification and dietary specialization in sea otters. *Proceedings of the National Academy of Sciences* 105(2): 560–565.

Trumble, S.J., S.R. Noren, L.A. Cornick, et al. 2010. Age-related differences in skeletal muscle lipid profiles of Weddell seals: Clues to developmental changes. *The Journal of Experimental Biology* 213(10): 1676–1684.

Verrier, D., S. Atkinson, C. Guinet, et al. 2012. Hormonal responses to extreme fasting in subantarctic fur seal (*Arctocephalus tropicalis*) pups. *American Journal of Physiology-Regulatory, Integrative and Comparative Physiology* 302(8): R929–R940.

Videsen, S.K.A., L. Bejder, M. Johnson, et al. 2017. High suckling rates and acoustic crypsis of humpback whale neonates maximise potential for mother–calf energy transfer. *Functional Ecology* 31(8): 1561–1573

Viscarra, J.A., and R.M. Ortiz. 2013. Cellular mechanisms regulating fuel metabolism in mammals: Role of adipose tissue and lipids during prolonged food deprivation. *Metabolism* 62(7): 889–897.

Weiner, J. 1992. Physiological limits to sustainable energy budgets in birds and mammals: Ecological implications. *Trends in Ecology & Evolution* 7(11): 384–388.

Whitehead, H., and J. Mann. 2000. Female reproductive strategies of cetaceans. In *Cetacean Societies: Field Studies of Dolphins and Whales*, ed. J. Mann, H. Whitehead, R.C. Connor, et al., 219–246. Chicago: University of Chicago Press.

Claire A. Simeone and Shawn P. Johnson

Sea Change Health, Sunnyvale, CA

Claire A. Simeone

Claire has spent more than a decade in the veterinary field. In that time, she honed her skills as an expert in clinical marine mammal medicine. She has created novel therapies, like a treatment for sea lion eye trauma, and has researched diseases, treatments, and understood health trends for marine mammals across North America. She has worked both in the field and in hospitals around the world to improve animal health and welfare, trained the next generation of health leaders, and advocated for conservation of wildlife. Claire is the Founder and CEO of Sea Change Health, an organization dedicated to safeguarding ocean health and all who rely on it. Prior to starting Sea Change Health, Claire was the Director of the Hawaiian Monk Seal Conservation Program at The Marine Mammal Center, world's largest marine mammal hospital. Claire was the first veterinarian selected as a TED Fellow in 2018. Her TED talk shared the concept of "zoognosis," the knowledge spread between humans and animals.

Shawn P. Johnson

Dr. Shawn Johnson is the Chief Operating Officer and Director of Innovative Medicine at Sea Change Health. An expert in marine mammal medicine and anesthesia with more than two decades of experience, Shawn's passion is to develop innovative treatments and techniques for marine mammal patients. Shawn most recently led the largest marine mammal teaching hospital in the world at The Marine Mammal Center, and he has worked for the U.S. Navy Marine Mammal Program, the National Marine Mammal Foundation, Alaska SeaLife Center, and the Oiled Wildlife Care Network, UC Davis. He has worked with wild marine mammals in the field around the globe and is the author of more than three dozen scientific papers.

Driving Question: Why do some animals get sick while others stay healthy?

INTRODUCTION

We often think of our immune system as the first line of defense in protecting us from getting sick. The specialized cells inside our bodies fight off invaders like bacteria and viruses, and protect us from experiencing illness, injury, and even death. The immune system plays a key role in humans, marine mammals, and other animals. But it is not the only thing that influences our health, and it cannot protect us all by itself. What other things protect marine mammals from disease? And how is an animal's health influenced by the environment around them and vice versa?

DOI: 10.1201/9781003297468-13

BODY SYSTEM ADAPTATIONS

The answer is a series of **adaptations** that allow marine mammals to inhabit their aquatic environment. Manatees (*Trichechus* sp.), for example, have a specialized digestive tract to break down the seagrass they feed on in both fresh and saltwater. Sometimes, these adaptations provide actual protection against things in the environment that could otherwise harm them. For example, cetacean skin wounds heal in a specialized way that keeps water out.

Sometimes, these special adaptations end up predisposing an animal to harm. Polar bears (*Ursus maritimus*), for example, feed almost exclusively on seals. Their fat-rich blubber provides a healthy layer of insulation, but it also stores man-made chemicals that can harm their health in the long term.

And sometimes, it is the marine mammals that influence the health of the environment. Marine mammal species help to keep prey populations balanced and keep the ecosystem healthy. Certain marine mammals play a critical role far beyond what would be expected for their population size. Named a **keystone species**, sea otters (*Enhydra lutris*) are a well-known example. Their high metabolism requires them to eat many invertebrates each day, including mussels, clams, and urchins. Their presence keeps the sea urchin population in check. In places where sea otters are no longer a part of the community, sea urchins quickly overpopulate and destroy entire kelp forests by eating the base of kelp fronds. The entire ecosystem of the kelp forest, and all the organisms that are a part of it, rely on the sea otters to stay healthy and keep this balance in place.

Multiple factors contribute to whether an animal – and the environment it lives in – is healthy. In this chapter, we will explore how the physiologic adaptations that these species have developed can either protect or predispose them to disease.

CETACEANS

The animals in the order *Cetacea* are one of the most radically adapted groups to the aquatic environment. They are fully aquatic species and cannot survive outside of the water because their bodies have undergone extreme changes in anatomy and physiology. Two of the body systems that are the most highly specialized are their **respiratory** and **dermal** systems. These adaptations, which are necessary for a fully aquatic life, also play a significant role in maintaining health in these species.

Physiologic adaptation: respiratory system

The respiratory system is the collection of organs and tissues responsible for breathing. As fully aquatic mammals that live in water but breathe air, the respiratory system of cetaceans is highly adapted for swimming and diving. Their **nares**, or nostrils, are positioned on the top of their head to allow them to breathe at the water surface. Their lungs are highly **compliant**, or able to greatly stretch and expand. The ability to collapse down under immense pressure and spring back open rapidly helps protect them from decompression sickness during repeated deep diving. You can learn more about diving adaptations in **Chapter 4**.

The microscopic features of human lungs that keep small particles out are largely absent in cetacean lungs. They have very little **lymphoid** tissue within the lung, which would normally house the immune cells whose job it is to evict pathogens. The lining of the lower airways contains almost no mucus-producing cells, which typically play a role in trapping foreign particles (Goudappel and Slijper 1958). Even further down the airways, the smallest **alveoli** contain no epithelial lining at all, which usually provides a last layer of protection against invaders. This directly exposes the capillary network to respiratory gases, which allows for a rapid exchange of oxygen and carbon dioxide (Piscitelli et al. 2013). It is suggested that these protective tissues are unnecessary to cetaceans that live in a mostly dust-free environment. However, this leaves the cetacean respiratory system susceptible to injury from a variety of factors and prone to developing pneumonia.

Respiratory threat: oil spills

Oil spills are a threat to marine mammal health because petroleum is a chemical that can cause long-term health effects. Cetaceans appear to be able to detect oil slicks on water, but they do not necessarily avoid them (Gubbay and Earll 2000). Dolphins swimming through an oiled area may be exposed through direct contact with the oil in the water column; through ingestion of contaminated water and prey; and through inhalation and **aspiration** of oil.

When the respiratory tract comes into contact with oil it can cause significant disease. Immediately following an oil spill, **volatile** hydrocarbon vapors are released from the oil. Bottlenose dolphins (*Tursiops truncatus*) exposed to volatile hydrocarbons following the *Deepwater Horizon* oil spill in the Gulf of Mexico in 2010 were five times more likely to have moderate to severe lung disease than counterparts at a reference site in Florida where oil was not observed (Schwacke et al. 2014). Inhaling petroleum vapors has been linked to decreased lung function, chronic **bronchitis**, and airway inflammation in humans (Sekkal et al. 2012). In addition to lung disease, affected dolphins also showed evidence of an impaired stress response, liver damage, and tooth loss (Schwacke et al. 2014).

Respiratory threat: wildfires

Smoke pollution is another threat to respiratory health, and as climate change intensifies the frequency and severity of wildfires, animals and humans are increasingly at risk of illness and death from extreme air pollution episodes. Researchers found that blood carbon dioxide levels were elevated in bottlenose dolphins in the month following a wildfire event, which may have signaled that their lung function was compromised. Dolphins were also three times more likely to have bacterial pneumonia at time of death after exposure to this wildfire smoke when compared to years before the wildfires (Venn-Watson et al. 2013). The lack of mechanisms to protect their respiratory tracts from exposure to particulates can predispose cetaceans to developing disease when these catastrophic events happen.

Physiologic adaptation: skin

Cetacean skin has undergone significant changes to maintain **homeostasis** in the marine environment. The skin is hairless, smooth, and lacks any appendages or glands. If you've seen your fingers get wrinkled after spending too much time in the bath, you can imagine that spending weeks in the water could cause some serious damage to our skin. To protect against the osmotic challenges of living in the sea, cetacean **epidermis** provides a protective layer that is impermeable to water.

To maintain this protection, skin cells turn over and are shed at a rate 8.5 times higher than human skin (Hicks et al. 1985). Cetaceans have also developed an alternative method of repairing skin damage because they are unable to form a scab in the water (Bruce-Allen and Geraci 1985). A breach in the skin allows seawater to contact epidermal cells, causing deterioration and swelling of these cells to provide a protective barrier to the deeper regenerating cells. As the skin wound heals from below, the degraded cells are sloughed off as the wound closes.

In bottlenose dolphins, the skin heals at about the same rate as humans. In other cetaceans, skin wound healing is longer. The time required for a skin wound to heal can be affected by how thick the skin is, as well as environmental conditions such as water temperature and salinity. In the beluga whale (*Delphinapterus leucas*), skin wounds heal in

30–40 days, five times longer than in bottle-nose dolphins. This is primarily because skin of beluga whale is five times thicker than dolphin skin (Geraci and Bruce-Allen 1987). Beluga likely have thicker skin because they live in substantially colder water. When they move to river estuaries during the summer months, they are exposed to water with lower salinity and higher temperature. This leads to molt, when skin sloughing is increased and skin cells grow and move more rapidly. When salinity gets too low for too long, these natural physiologic processes are overwhelmed, leading to the effects discussed in the next subsections.

Skin threat: freshwater disease

The skin is a primary protective barrier against environmental threats in cetaceans, but it is highly susceptible to water quality abnormalities like changes in salinity. When the salinity in seawater drops suddenly and cetaceans live for an extended period in low-salinity water, the skin can break down. Indo-Pacific bottlenose dolphins (*Tursiops aduncus*) in Australia developed skin ulcers covering up to 70% of their body after a large influx of freshwater from rainstorms dropped salinity dramatically. Bacteria, fungi, and algae colonized the damaged skin, and the associated stress on the body eventually led to death (Duignan et al. 2020).

Dolphins along the coastline of the Gulf of Mexico have been observed with similar freshwater skin lesions. The geography of the coast, with its many small bays, its location at the mouth of the Mississippi River, and large hurricanes with inundating rainfall all lead to waterways where salinity ranges widely. Although dolphins have been thought to avoid areas with very low salinity, researchers found no evidence that dolphins moved as salinity changed in this area. And dolphins in low-salinity areas were more likely to have skin lesions and changes in their blood and urine than dolphins in higher-salinity areas (Takeshita et al. 2021). You can learn more about salinity and water balance in **Chapter 8**.

Skin threat: fungal disease

Sometimes skin infections can be a secondary effect of a larger stressor on the whole body. For example, 30% of dolphins living in the southern Indian River Lagoon in Florida developed lacaziosis (lobomycosis), a fungal disease of the skin caused by an yeast-like organism known as *Lacazia loboi* (Reif et al. 2006). This is a substantially higher proportion of dolphins affected with the disease than in other regions of Florida, where the number of cases are closer to 2% (Hart et al. 2011). Water quality of the Indian River Lagoon has dramatically decreased over the past 50 years due to changes in watershed drainage and land development, leaving the local dolphin population susceptible to this naturally occurring fungus. There is also evidence that the population of dolphins affected by lacaziosis may have a suppressed immune system. They show lower production of white blood cells, called lymphocytes, which hinders the body's ability to respond to **pathogens** like fungi.

Dolphins that are infected with lacaziosis may have extensive areas of affected skin that cover large areas of the body. Skin lesions can be found at sites of previous trauma such as shark bite scars, where the skin's normal protective mechanisms may be reduced. Raised, bumpy areas of skin may turn white or discolored and are often found on the dorsal fin, flukes, or head. As lacaziosis is a **zoonotic disease**, humans are also susceptible to infection by *L. loboi* and care should be taken when handling animals with visible skin lesions. Although transmission between animal and human is rare, the fungus can spread through breaks or cuts in the skin.

PINNIPEDS

Seals, sea lions, and walruses each have unique habitats and express unique behaviors, but one behavior that they share is a tendency to haul out or congregate in large groups on land or ice. While they do come on land to rest or molt, the main driver to haul out in large groups is reproduction.

Physiologic adaptation: reproduction and haul-out behavior

Pinnipeds haul out in predictable cycles driven by their **reproductive system**. Nearly all California sea lions (*Zalophus californianus*), for example, return to islands off the coast of southern California and Mexico each year to give birth, raise their pups, and mate again. Animals are packed along the beaches and rocks, and these dense groups are ideal for disease transfer. Diseases spread easily when animals directly touch each other, roll around in the excretions of others, and breathe or cough on each other.

Haul-out threat: leptospirosis

Leptospirosis is a disease caused by pathogenic species within the genus *Leptospira* and can cause a range of clinical manifestations including acute kidney failure. Leptospira is **endemic** in the California sea lion population, and the bacteria is excreted in the urine of chronically infected individual sea lions. It only takes a few sick or chronically shedding animals to infect animals nearby (Prager et al. 2013; **Figure 13.1**).

With endemic pathogens like leptospirosis, cyclical disease outbreaks continue because each season of births brings a new group of uninfected, susceptible animals into the population. Environmental conditions can also have dramatic effects on how this disease spreads, as was witnessed during the unusually warm water in the northeastern Pacific between 2013 and 2016. Changing oceanographic conditions likely disrupted the food sources for the sea lions which led to increased pup mortality and changes in movement and migration patterns. No leptospirosis was detected in the California sea lion population during this time, likely due to lower numbers of susceptible pups around, and changes to how the animals mixed during the summer breeding season (Prager et al. 2018).

Figure 13.1 California sea lions laying on each other on floating docks at Pier 39 in San Francisco, CA. Migratory patterns and close contact among sea lions impact outbreaks of leptospirosis which is excreted in the urine. (Credit: Katherine Prager, NOAA NMFS No. 21422)

Haul-out threat: hookworms

Parasites that depend on soil to find a host are challenged by the marine lifestyle of pinnipeds. Hookworms (*Uncinaria* sp.) must rely on the haul-out behavior and reproductive cycles of the pinnipeds to complete their life cycle. Fur seals (*Callorhinus* sp. and *Arctocephalus* sp.) become infected with hookworms when the worm buries through the skin and migrates into the adult females' mammary glands. Larvae are transferred to pups through the milk, where they mature into adult worms in the pup's intestines (**Figure 13.2**). As the hookworms bury into the intestinal lining to feed on the pup's blood, high hookworm burdens can cause anemia through blood loss. Hookworms can even perforate the intestine, causing gut contents to spill into the abdomen and the bloodstream, ending in death (**Figure 13.3**). Nearly 100% of pups can be infested in a single year, and more than 70% may die (Lyons et al. 2005). Pups that are infected with numerous hookworms are more likely to die, but those that have fewer parasites may survive and clear the infection. Before the pups recover, the hookworm eggs that they shed in their feces into the **rookery** soil hatch as free-living larvae. The larvae cannot survive the extreme weather conditions where the seals reproduce so they must quickly penetrate the skin of the seals and live in their

Figure 13.3 Hookworm hemorrhagic enteritis in a South American fur seal (*Arctocephalus australis*) pup. There are numerous 1.5 to 2.0 cm hookworms (*Uncinaria* sp.) in the distal jejunum. Note the hemorrhagic content and small, pinpoint hemorrhagic spots on the mucosa (hookworm feeding sites). (Credit: Mauricio Seguel)

tissues as parasitic larvae. The life cycle is completed when larvae are reactivated within lactating seals. The larvae have a remarkably short period of time to infect all of the pups born that year, as births happen over a couple of weeks. The hookworms have likely developed a sensitivity to the fur seal hormones, which allows them to overcome the harsh environment by synchronizing with the seals' reproductive cycles (Seguel et al. 2018).

SEA OTTERS

Sea otters are a member of the weasel family, and they are one of the mammals that most recently reentered the marine environment. Because of this recent shift, some of their physiological adaptations are different from cetaceans and pinnipeds. Instead of relying on a thick, internal blubber layer for insulation, sea otters prevent heat loss through a layer of trapped air that is held against the skin by an exceptionally dense fur (Williams et al. 1992). In fact, their fur is the densest of any animal, with close to one million hairs per square inch! This luxurious fur keeps them well insulated on the water surface, and they supplement this haircoat with another internal adaptation: a very high metabolism. You can learn more how marine mammals maintain body temperature in **Chapter 5**.

Figure 13.2 A nursing South American fur seal pup. Hookworms are transmitted only through colostrum during the first days of life of pups. (Credit: Mauricio Seguel)

Physiologic adaptation: metabolism

As an otter dives the air layer is compressed, which reduces the insulating quality of fur at depth. This elevated thermal energetic cost, along with a resting metabolic rate three times the rates observed for terrestrial mammals of a similar size, results in large food requirements of roughly 20–25% of body mass in prey items per day (Costa and Kooyman 1982; Yeates et al. 2007). Simply put, they must eat an enormous amount of food each day. This allows them to not only thrive in the cold marine environment, but it also presents risks, both in the food they eat and when they cannot find enough. You can learn more about food and energy in the **Chapter 9**.

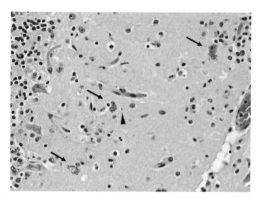

Figure 13.4 *Sarcocystis neurona* **protozoal schizonts (arrows), free merozoites (arrowhead), and associated inflammation in the cerebellum of a Southern sea otter. (Credit: Melissa Miller, California Department of Fish and Wildlife)**

Metabolic threat: protozoal diseases

Sea otters consume a variety of marine invertebrates, including clams, mussels, and snails. As these prey items filter seawater or scrape biofilm off kelp, they may collect and concentrate the infectious **oocysts** of parasites like protozoa (Lindsay et al. 2001). Ingestion of infectious *Sarcocystis neurona* or *Toxoplasma gondii* oocysts can cause severe neurologic symptoms, such as seizures, coma, and death, in otters that consume the contaminated invertebrates (Thomas et al. 2007; Miller et al. 2010). While it only takes a few oocysts to cause disease, a sea otter's ravenous appetite may expose them to lots of parasites (**Figure 13.4**).

Both protozoal diseases are spread by land mammals. *Sarcocystis neurona* is spread by the opossum, while *T. gondii* is transmitted to sea otters mainly by cats (Conrad et al. 2005). Healthy wetlands typically filter out a large amount of these pathogens and prevent them from entering the sea, but in areas where wetlands have become degraded, high levels of protozoal cysts can enter the water. Because sea otters have a relatively restricted territory and live in kelp forests close to the coast, these environmental changes – along with the otters' unique physiology and increased metabolism – have made them susceptible to protozoal diseases as they flow from land to sea. Infectious diseases are the most common cause of death for southern sea otters. A study looking at 15 years of data found that infectious diseases killed nearly two-thirds of otters and protozoal diseases like *Sarcocystis* and *Toxoplasma* played a major role in 20% of those cases (Miller et al. 2020).

Metabolic threat: end-lactation syndrome

Creating and caring for offspring takes a lot of work. Over the course of a year, sea otter females are doing one of three things: mating, producing a pup which then needs to be nursed, and weaning the pup. (Riedman and Estes 1990). Sea otters have a naturally high metabolism anyway, but it takes an immense number of extra calories – nearly double their normal amount – to successfully feed a pup. When the pups are very young, moms forage less often and do not dive as deep. They prioritize parental care, but that puts them in a negative energy balance (Thometz et al. 2016). As pups grow older, moms increase their diving and foraging to try to make up for their weight loss, but as they get ready to wean, they're often emaciated and exhausted. They then only have a short time before they are pregnant again and the cycle starts over.

This balancing act happens on a razor's edge. Biologists discovered that if the cycle happens over and over, or if a female does not regain her body condition before becoming pregnant again, that she can literally starve to death. This massive energy depletion is called *end-lactation syndrome*, and affected otters have severe muscle atrophy and almost no fat left. End-lactation syndrome is a major cause of death for southern sea otters, affecting more than half of adult females (Chinn et al. 2016). Older females and those with many pregnancies are at a higher risk, and the risk is higher still if a female lives in a crowded area with limited food. This is a drastic example of how a specialization like a high metabolism can both allow a sea otter to thrive in the ocean and also lead to its death if the balance is not right.

SIRENEANS

Sirenians including manatees (*Trichechus* sp.) and the dugong (*Dugong dugon*) feed almost exclusively on seagrasses and other water plants, which makes them the only **herbivorous** marine mammals. To effectively use the nutrients inside plants, the digestive tract of herbivores needs the ability to break down plant material. Different herbivores have developed different adaptations to solve this problem. Some terrestrial herbivores, like cows, use bacteria in a specialized part of their stomach called the **rumen** to digest plants. Others, like horses, have adapted a part of their anatomy that is further down the gastrointestinal tract, relying on bacteria in the large intestine to digest their plant diet.

Physiologic adaptation: gastrointestinal tract

Like horses, sirenians have a gastrointestinal tract that is specially formatted to break down plant material inside their elongated large intestine. Manatees also have several specialized glands in their stomach that are suspected to play a role in regulating water–salt balance as they forage for plants in both salt and fresh water (Reynolds and Rommel 1996). Because they have a long digestive tract, eat foods low in fiber, and eat relatively little for their body size, it takes food much longer to pass through the digestive tract of a sirenian. It can take a dugong (*Dugong dugon*) one week to digest a meal (Lanyon and Marsh 1995). This slow transit includes extended time in the large intestine, giving the bacteria inside a chance to completely digest the plants they eat.

Gastrointestinal threat: biotoxins

Because of their specialized diet, manatees are susceptible to diseases that may be associated with the plants they eat. Biotoxins are toxic compounds that are produced by living organisms, and marine algae can produce a variety of biotoxins. In Florida, where one species of manatee lives, more than 70 potentially harmful algal species have been identified, producing toxins such as brevetoxin, domoic acid, okadaic acid, and saxitoxin (Abbott et al. 2009).

As an example, brevetoxin (produced by the diatom *Karenia brevis*) is **lipophilic**, and it will concentrate in seagrass blades and rhizomes and in the **epiphytes** living on the surface of the grass. As manatees ingest the seagrass, they receive a high dose of the toxin which causes listlessness and neurological symptoms such as incoordination or tremors. If the toxin is inhaled, pneumonia and bronchitis may develop (Bossart et al. 1998). While some brevetoxin-exposed manatees have been successfully rehabilitated and released into the wild, large toxic blooms have caused fatalities of more than 100 individuals in a single event. Harmful algal blooms can also lead to large die-offs of the seagrass itself (Morris et al. 2022). Manatees that escape the toxin's direct effects may struggle to find enough food after their main food source disappears.

Biotoxins pose a large risk to marine mammal health. Animals that eat a specific and narrow diet, or live within a relatively small geographic range, can be particularly susceptible to harmful algal blooms and the toxins they produce. You can learn more about biotoxins in **Chapter 14**.

POLAR BEARS

Polar bears (*U. maritimus*) are the largest bear in the world and an **apex predator** in the Arctic. They are a symbol of strength and power, with no natural predators except for humans and other polar bears. Yet recently they've become a symbol of a changing planet. The Arctic is warming even more rapidly than the rest of the planet and loss of sea ice has led to iconic images of once majestic polar bears nearly wasting away.

One of the major challenges to adjusting to these changes in their environment is their highly specialized diet and hunting method. Polar bears are primarily "sit and wait" hunters, adapted to catching seals by waiting for them to surface at breathing holes (Pagano et al. 2018). They are designed to use as little energy as possible, and this strategy works when food rewards are rich in calories (see also **Chapter 2**).

Physiologic adaptation: blubber

Polar bears feed mainly on marine mammals such as ringed seals (*Phoca hispida*), bearded seals (*Erignathus barbatus*), and hooded seals (*Cystophora cristata*). All the seal species that polar bears hunt have a thick blubber layer and polar bears rely on their own blubber too. Blubber is an oily layer of fat that stores energy, insulates heat, and – particularly in the water – protects from heat loss and increases their buoyancy. You can learn more about the thermoregulatory role of blubber in **Chapter 5**.

Blubber threat: contaminants

The polar bear's seal-based diet is high in fat. Lipophilic contaminants, such as PCBs, DDT, and a variety of flame retardants are stored in fat over the lifetime of an animal. These contaminants also transfer from one animal to another when prey is eaten by predator. Thus, the polar bear diet is also high in contaminants. Many contaminants can have effects on reproduction by disrupting hormone pathways, and can cause **immunosuppression**, leaving animals with an increased susceptibility to infectious diseases

or certain types of cancers. Studies over three decades tracked levels of lipophilic contaminants in polar bear fat. Nearly all the legacy organochlorine contaminants (which have been banned or highly regulated around the world) showed a significant decline, suggesting that banning these compounds will lead to them slowly disappearing from the environment (Dietz et al. 2013a). For now, organochlorine contaminants persist in polar bear tissues at levels that continue to have the potential to affect health.

Unfortunately, when studying contaminants for which no international bans currently exist, the trends are reversed. Brominated flame retardant (BFR) concentrations more than doubled on average across the same study period (Dietz et al. 2013b). In addition to the effects described above, BFRs can also cause changes in both the structure and function of the brain, and may affect the liver and thyroid.

Prey is not the only factor to consider when evaluating contaminants. Where a polar bear lives may be just as important. Mercury is a naturally occurring chemical that is toxic to the nervous system and many other organs. Researchers have shown that while polar bears living in certain regions of the Arctic may be at little to no risk of mercury exposure, other bears live in mercury hot spots where they may be at severe risk of accumulating mercury in their bodies (Dietz et al. 2022). You can learn more about contaminants **Chapter 14**.

MARINE MAMMAL POPULATIONS

Individual marine mammals are susceptible to diseases due to a variety of physiological and anatomical adaptations that allow them to inhabit an aquatic environment. Diseases can also have a dramatic negative effect on marine mammal health at the population level and are a major risk to small, isolated populations. Many threats from noninfectious and infectious diseases to marine mammal populations are a direct consequence of degradation of the environment they live in (Burge et al. 2014; VanWormer et al. 2019)

Population threat: cancer

Cancer is rare in marine mammals, except for two distinct populations: beluga whales living in St. Lawrence Estuary in Canada and California sea lions along the California coast. Belugas living in the highly polluted St. Lawrence Estuary had a high prevalence of cancer, as well as high levels of PCB contaminants in their tissues. Tumors were found in 27% of the dead belugas studied from this estuary between 1983 and 1999 (Martineau et al. 2002).

In California sea lions, 23% of stranded adults examined postmortem along the central California coast were diagnosed with urogenital carcinoma, the highest among a pinniped species (Deming et al. 2018). Two factors are known to play a role in the development of cancer in this species: (1) presence of otarine herpesvirus-1; and (2) presence of immuno-suppressive contaminants such as PCBs, which are frequently elevated in tissues from animals with urogenital carcinoma (Gulland et al. 2020; Deming et al. 2021).

Population threat: viral infections

Morbillivirus is a genus that includes measles, canine distemper virus, rinderpest virus, and marine morbilliviruses. Marine morbilliviruses, including phocine distemper virus (PDV), and the cetacean morbilliviruses (CeMV) cause explosive outbreaks of disease (Kennedy 1990). Morbilliviruses cause immunosuppression and after they debilitate an animal, secondary infections such as fungal or bacterial disease are often seen. In addition, high contaminant levels have been associated with animals infected with morbillivirus, suggesting that the immunosuppression caused by a high contaminant load could make animals even more susceptible to infection (Hall et al. 1992). Animals that survive the outbreak do have immunity for the virus, but it does not last for the lifetime of the animal. As immunity wanes, and younger animals that are unexposed enter the population, a threshold is reached that allows the population to be susceptible to a large outbreak once again.

Outbreaks of marine morbilliviruses killed an estimated 18,000 harbor seals in Europe in 1988, several thousand striped dolphins (*Stenella coeruleoalba*) in the Mediterranean from 1990 to 1992, and two outbreaks on the U.S. Atlantic coast in 1987–1988 and 2013–2014 have killed more than 2,500 bottlenose dolphins combined (Kennedy 1990; Aguilar and Raga 1993; Lipscomb et al. 1994; Morris et al. 2015). Such large numbers have the potential to severely affect a population, especially if that population is small and relatively isolated. The spread of PDV into the Pacific Ocean has been linked with the decline of Arctic Sea ice and increased movement of circumpolar seal species from the North Atlantic to the Pacific. While it hasn't caused mass mortality in the Pacific, the disease does appear to have impacted other marine mammal species like sea otters and is a potential risk to sensitive species such as the Hawaiian monk seal (*Neomonachus schauinslandi*) (VanWormer et al. 2019).

Population threat: climate change

As the environment changes from climate change and human pressures, we have seen marine mammals adapt in a variety of ways. Sea level rise has decreased the area that can be used for rookeries by pinnipeds. Northern elephant seals (*Mirounga angustirostris*) are expanding their range as their population grows, but their traditional rookery beaches are shrinking. There are reports of them resting on beaches where they have never been documented before. The endangered Hawaiian monk seal reproduces largely on sandy atolls, which are disappearing with sea level rise, forcing them to choose new sites. In Florida, manatees are dying of starvation after seagrass die-offs related to warming and polluted coastal water. The manatees diet shifted to algae when seagrass became less available which is nutritionally insufficient (Allen et al. 2022).

It is unknown exactly what diseases will be favored as marine mammals change their behavior. Increased concentrations of animals may allow an introduced disease to spread through a naïve population quickly. Animals already stressed with fewer food resources may be susceptible to common pathogens. Scientists must respond to a changing environment by altering the questions they ask, to understand how diseases will affect marine mammals in this new world.

TOOLBOX

Diagnostic tests

Veterinarians use many of the same techniques to diagnose disease in marine mammals as they would in dogs or cats, or as a physician would diagnose disease in a human. Sterile swabs are used to collected samples for bacterial culture; blood tubes are used to collect and store blood samples; tests such as polymerase chain reaction (PCR) and enzyme-linked immunosorbent assays (ELISAs) are used to screen for diseases.

These tests have caveats when used on marine mammal patients. Many marine organisms are difficult to recognize, especially in a laboratory that is used to working primarily with terrestrial pathogens. *Listeria* sp., a bacterium that causes food poisoning in humans, was originally reported in abscesses in pinnipeds, based on the biochemical reactions of the bacteria. However, genetic tests revealed the organism to be *Arcanobacterium*, a marine bacterium that does not cause disease in humans (Johnson et al. 2003).

Tests such as PCR and ELISA are often developed for terrestrial diseases. They may cross-react with marine organisms but may make it difficult to tell from where the disease originated. For instance, seals are susceptible to PDV, a virus that causes pneumonia and encephalitis, and is closely related to CDV. Initial PCR testing showed these seals to be positive to CDV. While there is evidence that some outbreaks of seal distemper may have come from sled dogs with canine distemper, genetic testing revealed that PDV is genetically distinct from CDV (Mahy et al. 1988).

Tuberculosis has a similar story. The causative agent, *Mycobacterium* sp., can cause abscesses in the lungs, lymph nodes, and throughout the body, and it is a zoonotic disease. Original tests of tuberculosis in sea lions described it as *Mycobacterium bovis*, which often infects cows. It wasn't until genetic testing described a new strain, *Mycobacterium pinnipedii*, that it was understood that the tuberculosis strain was unique to pinnipeds – although it could still spread to other species from them (Cousins et al. 2003).

We must take marine mammals' unique physiology into account when analyzing changes to bloodwork. Deep-diving cetaceans or pinnipeds may have a high percentage of red blood cells (described as their hematocrit) in their blood to carry extra oxygen as they dive. If we compared their hematocrit to the normal values for a dog or a human, we would think a healthy marine mammal might be dehydrated or abnormal. It is important to know what normal blood values are for marine mammals, so that we can correctly assess changes in animals with disease. (Learn more about hemodynamics in the *Oxygen* chapter).

Remote sampling

Conservation and management of marine mammals often relies on health monitoring of individuals and populations. Samples provide key information about an animal's health, but if a marine mammal isn't hauled out on land, or stranded on a beach, it can be challenging to collect samples to diagnose disease. Remote sampling may involve using a dart gun to perform a remote biopsy, which collects a small tissue core sample of skin and blubber that can be analyzed. Skin can give us information about the genetics of that animal, and what population it belongs to, and blubber can be analyzed for the presence of lipophilic contaminants.

Newer, noninvasive methods of sampling are being developed using drones to study marine mammal ecology, behavior, health, and movement patterns (Raoult et al. 2020). Capturing the exhaled breath (blow) of a

Figure 13.5 SnotBot® hexacopter hovering over whale to collect exhaled "snot" on petri dishes that are used to evaluate the whale's health and ecology. (Credit: Ocean Alliance)

humpback whale using an unmanned hexacopter drone allowed for the comparison of the microbiome between two humpback whales populations, providing a tool for remote health monitoring at a population scale (Apprill et al. 2017; **Figures 13.5** and **13.6**).

Vaccination and treatments

Many of the medications used to treat marine mammal diseases are the same drugs as those used in humans or other animals. Antibiotics are used to fight bacterial infections, anti-inflammatories control inflammation and associated pain, and anesthetics provide sedation for surgical procedures. However, unique adaptations to marine mammal physiology may change the way drugs are metabolized in the body, translating to altered doses required for treatment. For instance, meloxicam, a nonsteroidal anti-inflammatory drug (NSAID), is typically excreted from the body in roughly 24 hours in dogs and humans. A study

in dolphins showed elevated levels of meloxicam in the blood for more than seven days, meaning that a once-weekly dose likely provides the same level of pain relief in dolphins as a daily dose in humans (Simeone et al. 2014).

Very small populations of marine mammals can be severely impacted by infectious outbreaks. For endangered Hawaiian monk seals, morbillivirus is a pathogen of great concern as it has been associated with marine mammal mass mortality events and was a potential cause of a major die-off of the closely related Mediterranean monk seal (*Monachus monachus*). The goal of any vaccination program is to achieve herd immunity by vaccinating enough animals to prevent an outbreak. Vaccinating free-ranging wildlife can be extremely difficult, but the use of behavioral modeling can be utilized to identify ideal strategies (Robinson et al. 2018). Network modeling has led to an efficient and successful vaccination campaign for much of endangered Hawaiian monk seal population.

Figure 13.6 Aerial of picnic table. Field processing of blue whale breath microbiome sample. The table is set up in the hold of a chartered Chilean fish carrier boat. The hexacopter at right carries a central sterile petri dish. The moisture-laden exhalate is drawn onto the dish by the rotor downdraft. The dish is then sampled once back on the boat for laboratory analysis. (Credit: Daniel Casado)

CONCLUSIONS

The health of marine mammals is driven by a complex combination of physiological adaptation, behavior, and environment. Some animals get sick because physiological adaptations to life in the water also predispose them to disease. Pinnipeds haul out on land in big groups to reproduce, which creates the perfect environment for diseases such as leptospirosis or hookworm parasites to spread. A manatee's stomach and intestines are specialized to digest plants, but the plants themselves can concentrate large amounts of toxins. A changing environment can also play a role in whether an animal gets sick. Warm water anomalies have led to starvation for many marine mammals, and sea level rise is impacting where they can come ashore to rest. When studying the health of an individual animal or a marine mammal population, it is important to consider all the potential relationships that might be at play.

REVIEW QUESTIONS

1 What protective features of the lungs are present in humans but absent in cetaceans, and may predispose them to developing disease like pneumonia?

2 In what ways has petroleum from oil spills been shown to affect the health of dolphins?

3 How is cetacean skin adapted to protect them from the effects of being constantly submerged in water?

4 How do hookworms spread to fur seal pups? Is it different than how it spreads to adult fur seals?

5 What specializations do manatees have in their gastrointestinal tract to help them eat plants in the water?

6   How do sea otters insulate themselves from the cold ocean differently than a seal or a dolphin?

7   How does an apex predator like a polar bear accumulate high levels of man-made contaminants?

8   If you were a veterinarian treating a manatee for brevetoxin poisoning, what kinds of symptoms would you look for?

9   Why are sea otters at particular risk for becoming infected by protozoa?

10 Is cancer common in marine mammals? What factor(s) do the two species with the highest rates of cancer share?

CRITICAL THINKING

1   How can skin disease be a sign of deeper health problems?

2   Is there an ideal, healthy way for a female sea otter to balance eating, being pregnant, and raising a pup?

3   What might be some reasons for the differences seen between marine morbilliviruses in the Atlantic where large, fatal outbreaks are seen, and the Pacific where no outbreaks have been reported, but individual cases are seen?

FURTHER READING

(Cetacean): Sanderfoot, O. V., S. B. Bassing, J. L. Brusa, et al. 2021. "A Review of the Effects of Wildfire Smoke on the Health and Behavior of Wildlife." *Environmental Research Letters* 16 (12).

(Climate Change): Gulland, F. M. D., J. Baker, M. Howe, et al. 2022. "A Review of Climate Change Effects on Marine Mammals in United States Waters: Past Predictions, Observed Impacts, Current Research and Conservation Imperatives." *Climate Change Ecology*: 100054.

(Polar bears): Patyk, K. A., C. Duncan, P. Nol, et al. 2015. "Establishing a Definition of Polar Bear (*Ursus maritimus*) Health: A Guide to Research and Management Activities." *Science of the Total Environment* 514: 371–78.

(Sea otters): Shanebeck, K. M., and C. Lagrue. 2020. "Acanthocephalan Parasites in Sea Otters: Why We Need to Look Beyond Associated Mortality…." *Marine Mammal Science* 36 (2): 676–89.

(Topics): Michael, S. A., D. T. S. Hayman, R. Gray, et al. 2021. "Risk Factors for New Zealand Sea Lion (*Phocarctos hookeri*) Pup Mortality: Ivermectin Improves Survival for Conservation Management." *Frontiers in Marine Science* 8: 881.

GLOSSARY

Adaptation: A change that allows an organism to be better suited to its environment.

Alveoli: Tiny air sacs in the lungs where gas exchange occurs.

Apex predator: A predator at the top of the food chain that isn't preyed upon by any other animals.

Aspiration: Inhaling or sucking a foreign object into the airway.

Bronchitis: Inflammation of the bronchioles or small airways.

Compliant: Able to expand; stretchy.

Dermal: Relating to the skin.

Endemic disease: A disease that is constantly present in a particular place or group.

Endocrine system: The organs that produce hormones in the body.

Epidermis: The top layer of the skin.

Epiphytes: A plant that grows on another plant but is not a parasite.

Herbivore: An animal that eats plants.

Homeostasis: The state of balance that an organism needs to survive.

Immunosuppression: A decrease in the immune system's ability to fight off infections.

Keystone species: A species that holds an ecosystem together. They have an outsized role, and without them the ecosystem would change dramatically.

Lipophilic: "Fat-loving." Something that tends to combine with lipids.

Lymphoid: Relating to the tissue that produces immune cells and antibodies, like lymph nodes or tonsils.

Nares: Nostrils.

Oocyst: Similar to an egg, this is the walled sac that certain parasites use to spread to other hosts.

Pathogen: A microorganism (like a bacterium or virus) that can cause disease.

Respiratory: Relating to the organs involved in breathing.

Rookery: A breeding colony of animals.

Volatile: Easily evaporated.

Zoonotic disease: A disease that can be spread between humans and animals.

REFERENCES

Abbott, G. M., J. H. Landsberg, A. R. Reich, et al. 2009. "Resource Guide for Public Health Response to Harmful Algal Blooms in Florida: Based on Recommendations of the Florida Harmful Algal Bloom Task Force Public Health Technical Panel." *Fish and Wildlife Research Institute Technical Reports* 14.

Aguilar, A., and J. A. Raga. 1993. "The Striped Dolphin Epizootic in the Mediterranean Sea." *Ambio* 22 (8): 524–28.

Allen, A. C., C. A. Beck, D. C. Sattelberger, et al. 2022. "Evidence of a Dietary Shift by the Florida Manatee (*Trichechus manatus latirostris*) in the Indian River Lagoon Inferred from Stomach Content Analyses." *Estuarine, Coastal and Shelf Science* 268 (May): 107788.

Apprill, A., C. A. Miller, M. J. Moore, et al. 2017. "Extensive Core Microbiome in Drone-Captured Whale Blow Supports a Framework for Health Monitoring." *MSystems* 2 (5). https://doi.org/10.1128/msystems.00119-17.

Bossart, G. D., D. G. Baden, R. Y. Ewing, et al. 1998. "Brevetoxicosis in Manatees (*Trichechus manatus latirostris*) from the 1996 Epizootic: Gross, Histologic, and Immunohistochemical Features." *Toxicologic Pathology* 26. https://doi.org/10.1177/019262339802600214.

Bruce-Allen, L. J., and J. R. Geraci. 1985. "Wound Healing in the Bottlenose Dolphin (*Tursiops truncatus*)." *Canadian Journal of Fisheries and Aquatic Sciences* 42 (2). https://doi.org/10.1139/f85-029.

Burge, C. A., C. Mark Eakin, C. S. Friedman, et al. 2014. "Climate Change Influences on Marine Infectious Diseases: Implications for Management and Society." *Annual Review of Marine Science* 6: 249–77.

Chinn, S. M., M. A. Miller, M. Tim Tinker, et al. 2016. "The High Cost of Motherhood: End-Lactation Syndrome in Southern Sea Otters (*Enhydra lutris nereis*) on the Central California Coast, USA." *Journal of Wildlife Diseases* 52 (2): 307–18.

Conrad, P. A., M. A. Miller, C. Kreuder, et al. 2005. "Transmission of Toxoplasma: Clues from the Study of Sea Otters as Sentinels of *Toxoplasma gondii* Flow into the Marine Environment." *International Journal for Parasitology* 35. https://doi.org/10.1016/j.ijpara.2005.07.002.

Costa, D. P., and G. L. Kooyman. 1982. "Oxygen Consumption, Thermoregulation, and the Effect of Fur Oiling and Washing on the Sea Otter, *Enhydra lutris*." *Canadian Journal of Zoology* 60 (11). https://doi.org/10.1139/z82-354.

Cousins, D. V., R. Bastida, A. Cataldi, et al. 2003. "Tuberculosis in Seals Caused by a Novel Member of the Mycobacterium Tuberculosis Complex: Mycobacterium Pinnipedii Sp. Nov." *International Journal of Systematic and Evolutionary Microbiology* 53 (5). https://doi.org/10.1099/ijs.0.02401-0.

Deming, A. C., K. M. Colegrove, P. J. Duignan, et al. 2018. "Prevalence of Urogenital Carcinoma in Stranded California Sea Lions (*Zalophus californianus*) from 2005–15." *Journal of Wildlife Diseases* 54 (3): 581–86.

Deming, A. C., J. F. X. Wellehan, K. M. Colegrove, et al. 2021. "Unlocking the Role of a Genital Herpesvirus, Otarine Herpesvirus 1, in California Sea Lion Cervical Cancer." *Animals* 11 (2): 1–17.

Dietz, R., R. J. Letcher, J. Aars, et al. 2022. "A Risk Assessment Review of Mercury Exposure in Arctic Marine and Terrestrial Mammals." *Science of the Total Environment* 829. https://doi.org/10.1016/j.scitotenv.2022.154445.

Dietz, R., F. F. Rigét, C. Sonne, et al. 2013a. "Three Decades (1983–2010) of Contaminant Trends in East Greenland Polar Bears (*Ursus maritimus*).

Part 1: Legacy Organochlorine Contaminants." *Environment International* 59. https://doi.org/10.1016/j.envint.2012.09.004.

Dietz, R., F. F. Rigét, C. Sonne, et al. 2013b. "Three Decades (1983–2010) of Contaminant Trends in East Greenland Polar Bears (*Ursus maritimus*). Part 2: Brominated Flame Retardants." *Environment International* 59. https://doi.org/10.1016/j.envint.2012.09.008.

Duignan, P. J., N. S. Stephens, and K. Robb. 2020. "Fresh Water Skin Disease in Dolphins: A Case Definition Based on Pathology and Environmental Factors in Australia." *Scientific Reports* 10 (1). https://doi.org/10.1038/s41598-020-78858-2.

Geraci, J. R., and L. J. Bruce-Allen. 1987. "Slow Process of Wound Repair in Beluga Whales, *Delphinapterus leucas*." *Canadian Journal of Fisheries and Aquatic Sciences* 44 (9). https://doi.org/10.1139/f87-203.

Goudappel, J. R., and E. J. Slijper. 1958. "Microscopic Structure of the Lungs of the Bottlenose Whale." *Nature* 182 (4633). https://doi.org/10.1038/182479a0.

Gubbay, S., and R. Earll. 2000. "Review of Literature on the Effects of Oil Spills on Cetaceans." United Kingdom.

Gulland, F. M. D., A. J. Hall, G. M. Ylitalo, et al. 2020. "Persistent Contaminants and Herpesvirus OtHV1 Are Positively Associated with Cancer in Wild California Sea Lions (*Zalophus californianus*)." *Frontiers in Marine Science* 7 (December). https://doi.org/10.3389/fmars.2020.602565.

Hall, A. J., R. J. Law, D. E. Wells, et al. 1992. "Organochlorine Levels in Common Seals (*Phoca vitulina*) Which were Victims and Survivors of the 1988 Phocine Distemper Epizootic." *Science of the Total Environment* 115 (1–2). https://doi.org/10.1016/0048-9697(92)90039-U.

Hart, L. B., D. S. Rotstein, R. S. Wells, et al. 2011. "Lacaziosis and Lacaziosis-like Prevalence Among Wild, Common Bottlenose Dolphins *Tursiops truncatus* from the West Coast of Florida, USA." *Diseases of Aquatic Organisms* 95 (1). https://doi.org/10.3354/dao02345.

Hicks, B. D., D. J. St. Aubin, J. R. Geraci, et al. 1985. "Epidermal Growth in the Bottlenose Dolphin, *Tursiops truncatus*." *Journal of Investigative Dermatology* 85 (1). https://doi.org/10.1111/1523-1747.ep12275348.

Johnson, S. P., S. Jang, F. M. D. Gulland, et al. 2003. "Characterization and Clinical Manifestations of *Arcanobacterium Phocae* Infections in Marine Mammals Stranded Along the Central California Coast." *Journal of Wildlife Diseases* 39 (1). https://doi.org/10.7589/0090-3558-39.1.136.

Kennedy, S. 1990. "A Review of the 1988 European Seal Morbillivirus Epizootic." *The Veterinary Record* 127 (23): 563–67.

Lanyon, J. M., and H. Marsh. 1995. "Digesta Passage Times in the Dugong." *Australian Journal of Zoology* 43 (2). https://doi.org/10.1071/ZO9950119.

Lindsay, D. S., K. K. Phelps, S. A. Smith, et al. 2001. "Removal of *Toxoplasma gondii* Oocysts from Sea Water by Eastern Oysters (*Crassostrea virginica*)." *Journal of Eukaryotic Microbiology* 48. https://doi.org/10.1111/j.1550-7408.2001.tb00517.x.

Lipscomb, T. P., F. Y. Schulman, D. Moffett, et al. 1994. "Morbilliviral Disease in Atlantic Bottlenose Dolphins (*Tursiops truncatus*) from the 1987–1988 Epizootic." *Journal of Wildlife Diseases* 30 (4). https://doi.org/10.7589/0090-3558-30.4.567.

Lloyd-Smith, J. O., D. J. Greig, S. Hietala, et al. 2007. "Cyclical Changes in Seroprevalence of Leptospirosis in California Sea Lions: Endemic and Epidemic Disease in One Host Species?" *BMC Infectious Diseases* 7. https://doi.org/10.1186/1471-2334-7-125.

Lyons, E. T., R. L. DeLong, T. R. Spraker, et al. 2005. "Seasonal Prevalence and Intensity of Hookworms (*Uncinaria* spp.) in California Sea Lion (*Zalophus californianus*) Pups Born in 2002 on San Miguel Island, California." *Parasitology Research* 96 (2). https://doi.org/10.1007/s00436-005-1335-5.

Mahy, B. W. J., T. Barrett, S. Evans, et al. 1988. "Characterization of a Seal Morbillivirus." *Nature* 336. https://doi.org/10.1038/336115a0.

Martineau, D., K. Lemberger, A. Dallaire, et al. 2002. "Cancer in Wildlife, a Case Study: Beluga from the St. Lawrence Estuary, Québec, Canada." *Environmental Health Perspectives* 110 (3). https://doi.org/10.1289/ehp.02110285.

Miller, M. A., P. A. Conrad, M. Harris, et al. 2010. "A Protozoal-Associated Epizootic Impacting Marine Wildlife: Mass-Mortality of Southern Sea Otters (*Enhydra lutris nereis*) Due to Sarcocystis Neurona Infection." *Veterinary Parasitology* 172 (3–4). https://doi.org/10.1016/j.vetpar.2010.05.019.

Miller, M. A., M. E. Moriarty, L. Henkel, et al. 2020. "Predators, Disease, and Environmental Change in the Nearshore Ecosystem: Mortality in Southern Sea Otters (*Enhydra lutris nereis*) from 1998–2012." *Frontiers in Marine Science* 7 (November). https://doi.org/10.3389/fmars.2020.00582.

Morris, L. J., L. M. Hall, C. A. Jacoby, et al. 2022. "Seagrass in a Changing Estuary, the Indian River Lagoon, Florida, United States." *Frontiers in Marine Science* 8. https://doi.org/10.3389/fmars.2021.789818.

Morris, S. E., J. L. Zelner, D. A. Fauquier, et al. 2015. "Partially Observed Epidemics in Wildlife Hosts: Modelling an Outbreak of Dolphin Morbillivirus in the Northwestern Atlantic, June 2013–2014." *Journal of the Royal Society Interface* 12 (112). https://doi.org/10.1098/rsif.2015.0676.

Pagano, A. M., G. M. Durner, K. D. Rode, et al. 2018. "High-Energy, High-Fat Lifestyle Challenges an Arctic Apex Predator, the Polar Bear." *Science* 359 (6375). https://doi.org/10.1126/science.aan8677.

Piscitelli, M. A., S. A. Raverty, M. A. Lillie, et al. 2013. "A Review of Cetacean Lung Morphology and Mechanics." *Journal of Morphology* 274 (12). https://doi.org/10.1002/jmor.20192.

Prager, K. C., D. J. Greig, D. P. Alt, et al. 2013. "Asymptomatic and Chronic Carriage of Leptospira Interrogans Serovar Pomona in California Sea Lions (*Zalophus californianus*)." *Veterinary Microbiology* 164 (1–2): 177–83.

Prager, K., B. Borremans, M. Buhnerkempe, et al. 2018. "Fade-out of an Endemic Pathogen in a Wildlife Host: Did Environmental Stress Cause Leptospira to Disappear from the California Sea Lion Population (*Zalophus californianus*)." In *International Association of Aquatic Animal Medicine*. Long Beach, CA.

Raoult, V., A. P. Colefax, B. M. Allan, et al. 2020. "Operational Protocols for the Use of Drones in Marine Animal Research." *Drones* 4 (4). https://doi.org/10.3390/drones4040064.

Reif, J. S., M. S. Mazzoil, S. D. Mcculloch, et al. 2006. "Lobomycosis in Atlantic Bottlenose Dolphins from the Indian River Lagoon, Florida." *Journal of the American Veterinary Medical Association* 228 (1): 104–8.

Reynolds, J. E., and S. A. Rommel. 1996. "Structure and Function of the Gastrointestinal Tract of the Florida Manatee, *Trichechus Manatus Latirostris*." *Anatomical Record* 245 (3). https://doi.org/10.1002/(SICI)1097-0185(199607)245:3<539::AID-AR11>3.0.CO;2-Q.

Riedman, M. L., and J. A. Estes. 1990. "The Sea Otter (*Enhydra lutris*): Behavior, Ecology, and Natural History." *Biological Report: US Fish & Wildlife Service* 90 (14).

Robinson, S. J., M. M. Barbieri, S. Murphy, et al. 2018. "Model Recommendations Meet Management Reality: Implementation and Evaluation of a Network-Informed Vaccination Effort for Endangered Hawaiian Monk Seals." *Proceedings of the Royal Society B: Biological Sciences* 285 (1870). https://doi.org/10.1098/rspb.2017.1899.

Schwacke, L. H., C. R. Smith, F. I. Townsend, et al. 2014. "Health of Common Bottlenose Dolphins (*Tursiops truncatus*) in Barataria Bay, Louisiana, Following the Deepwater Horizon Oil Spill." *Environmental Science and Technology* 48 (1). https://doi.org/10.1021/es403610f.

Seguel, M., F. Muñoz, D. Perez-Venegas, et al. 2018. "The Life History Strategy of a Fur Seal Hookworm in Relation to Pathogenicity and Host Health Status." *International Journal for Parasitology: Parasites and Wildlife* 7 (3): 251–60. https://doi.org/10.1016/j.ijppaw.2018.07.003.

Sekkal, S., N. Haddam, H. Scheers, et al. 2012. "Occupational Exposure to Petroleum Products and Respiratory Health: A Cross-Sectional Study from Algeria." *Journal of Occupational and Environmental Medicine* 54 (11). https://doi.org/10.1097/JOM.0b013e31825fa6c9.

Simeone, C. A., H. H. Nollens, J. M. Meegan, et al. 2014. "Pharmacokinetics of Single Dose Oral Meloxicam in Bottlenose Dolphins (*Tursiops truncatus*)." *Journal of Zoo and Wildlife Medicine* 45 (3). https://doi.org/10.1638/2013-0281R1.1.

Takeshita, R., B. C. Balmer, F. Messina, et al. 2021. "High Site-Fidelity in Common Bottlenose Dolphins Despite Low Salinity Exposure and Associated Indicators of Compromised Health." *PLoS One* 16 (9 September). https://doi.org/10.1371/journal.pone.0258031.

Thomas, N. J., J. P. Dubey, D. S. Lindsay, et al. 2007. "Protozoal Meningoencephalitis in Sea Otters (*Enhydra lutris*): A Histopathological and Immunohistochemical Study of Naturally

Occurring Cases." *Journal of Comparative Pathology* 137 (2–3). https://doi.org/10.1016/j.jcpa.2007.05.001.

Thometz, N. M., M. M. Staedler, J. A. Tomoleoni, et al. 2016. "Trade-Offs Between Energy Maximization and Parental Care in a Central Place Forager, the Sea Otter." *Behavioral Ecology* 27 (5). https://doi.org/10.1093/beheco/arw089.

VanWormer, E., J. A. K. Mazet, A. Hall, et al. 2019. "Viral Emergence in Marine Mammals in the North Pacific May Be Linked to Arctic Sea Ice Reduction." *Scientific Reports* 9 (1). https://doi.org/10.1038/s41598-019-51699-4.

Venn-Watson, S., C. R. Smith, E. D. Jensen, et al. 2013. "Assessing the Potential Health Impacts of the 2003 and 2007 Firestorms on Bottlenose Dolphins (*Tursiops trucatus*) in San Diego Bay." *Inhalation Toxicology* 25 (9): 481–91.

Williams, T. D., D. Diane Allen, J. M. Groff, et al. 1992. "An Analysis of California Sea Otter Enhydra Lutris Pelage and Integument." *Marine Mammal Science* 8 (1). https://doi.org/10.1111/j.1748-7692.1992.tb00120.x.

Yeates, L. C., T. M. Williams, and T. L. Fink. 2007. "Diving and Foraging Energetics of the Smallest Marine Mammal, the Sea Otter (*Enhydra lutris*)." *Journal of Experimental Biology* 210 (11). https://doi.org/10.1242/jeb.02767.

Zuerner, R. L., C. E. Cameron, S. Raverty, et al. 2009. "Geographical Dissemination of Leptospira Interrogans Serovar Pomona During Seasonal Migration of California Sea Lions." *Veterinary Microbiology* 137 (1–2): 105–10.

John Harley
University of Alaska Southeast, Juneau, AK

Judith M. Castellini
University of Alaska, Fairbanks, AK

Todd O'Hara
Texas A&M University-College Station, TX

John Harley

Dr. John Harley completed his B.S. in marine biology at the University of California, Santa Cruz working with elephant seals at Año Nuevo. He left California for Alaska in 2012 to pursue a Ph.D. in biochemistry at the University of Alaska, Fairbanks where he studied seals, sea lions, and otters. He is now an assistant research professor at the University of Alaska Southeast where his research interests include wildlife toxicology and One Health, harmful algal blooms, and climate adaptation. Dr. Harley has a passion for the outdoors and human-powered adventure, and enjoys being lost in the Alaskan wilderness.

Judith M. Castellini

Maggie (J. Margaret) earned her M.Sc. in zoology from the University of British Columbia in 1984. She recently retired from research and student mentoring at the University of Alaska, Fairbanks. Beginning with the influence of her thesis advisor, Dr. Peter Hochachka, she has had a career-long interest in how animals adapt to different environments. Her more recent work in the UAF Wildlife Toxicology Lab and beyond introduced her to the concept of One Health. Continuing the theme of adaptation, Maggie loves to adapt analytical techniques to unusual conditions; she has set up biochemistry labs on seal haul outs in the Antarctic and adapted methods of mercury analysis to a wide range of animal tissues. She has a not-so-secret love for northern elephant seals, having spent several seasons working on rookeries with the scars to prove it.

Todd O'Hara

Dr. Todd O'Hara's academic training includes a D.V.M. from the University of Wisconsin and a Ph.D. in pharmacology/toxicology from the Medical College of Virginia. Todd is a veterinary toxicologist, was in Alaska from 1995 to 2019 and is now at Texas A&M University. He and his wife went to Alaska for the "opportunity of a lifetime," which included close

DOI: 10.1201/9781003297468-14

encounters with arctic seals, whales, and polar bears with the North Slope Borough Department of Wildlife Management. At the University of Alaska Fairbanks, Todd established the Wildlife Toxicology Laboratory and developed countless collaborations. His major interests (aside from fishing) are environmental/wildlife toxicology, zoonotic diseases, food security, and wildlife conservation and medicine in a One Health context. This work continues with efforts including locations and colleagues in Mexico and Central California.

Driving Question: How are marine mammals exposed to and affected by *toxicants*?

INTRODUCTION

Marine mammals must deal with many classes of toxicants, both natural and ***anthropogenic*** (caused or influenced by humans). There are many potential routes of exposure (**Figure 14.1**),

commonly more than one route at a time. In general, a toxicant is any toxic (harmful) substance that may be naturally occurring or man-made, while ***biotoxins*** are produced naturally by organisms. For the purpose of this chapter the term "toxicant" includes biotoxins. There are appreciable amounts of toxicants present in most ocean environments originating from marine (e.g., methylmercury and biotoxins like domoic acid), atmospheric (e.g., inorganic mercury and volatile organic compounds like diesel fuel), and terrestrial (e.g., DDT, dichloro-diphenyl-trichlorethane) sources. Marine mammals must also cope with additional stressors that result from their unique and often highly specialized physiology, behavior, and diet. This chapter discusses several driving forces of physiological adaptation of marine mammals to toxicants, with examples of different types of toxicants, key ways marine mammals may be exposed, examples of toxicant effects, and systems marine mammals may use for detoxification and/or

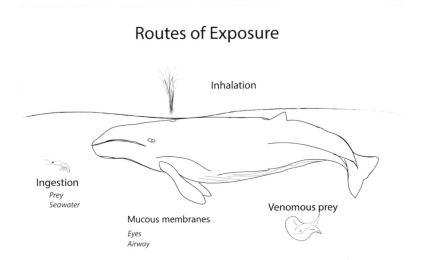

Routes of Exposure

Inhalation

Ingestion
Prey
Seawater

Mucous membranes
Eyes
Airway

Venomous prey

Figure 14.1 A conceptual figure displaying the potential routes of exposure of marine mammals to toxicants. Some of these pathways are often overlooked, including the exposure of mucosal membranes to poisons such as brevetoxins or volatile organic compounds. While venom exposure may be rare, venom may be a potential cause of mortality for some marine mammals that are known to feed on venomous animals (e.g., killer whales and stingrays). (From Harley and O'Hara 2016)

elimination. Physical forms of contamination, such as debris, external oiling, and sound pollution, will be only briefly discussed and these are covered in detail in other chapters (see **Chapter 6** and **Chapter 15**).

TOXICANT GROUPS

Abiotic toxicants

An *abiotic* toxicant is one that is produced via non-biological processes. In some cases, these processes are natural, such as geologic erosion, and in other cases human-derived, such as pesticides or many industrial-based compounds. When investigating these chemicals as toxicants two major approaches for marine mammals have predominated: (1) unexplained adverse effects (e.g., population decline and unusual mortality events, *UME*) with no obvious cause, implicating toxicants; and (2) suspect or relatively high concentrations of a contaminant(s) assessed for potential adverse effects. As the study of marine mammal toxicology evolved, numerous ways to classify these abiotic chemical agents were defined, because they are investigated in very different contexts (e.g., environment, food web, tissue concentrations, potential links with adverse effects). Some of these classifications include grouping by chemical structure (e.g., polyaromatic hydrocarbons, PAHs), pharmacologic or toxicologic mechanism of action or consequence (e.g., carcinogens), environmental or biological persistence or half-life (e.g., persistent organic pollutants, POPs), chemical behavior (e.g., oxidants), and by industrial and clinical use (e.g., antibiotics, flame retardants).

Sometimes classifications reflect a combination of categories such as organochlorine pesticides, brominated (containing bromine) flame retardants, and petroleum hydrocarbons. Some of these terms are not very specific. Crude oils, for example, are complex mixtures of thousands of organic and inorganic compounds affecting many potential target organs, and these vary widely in composition by the location of the oil rig or the type of substrate from which they are extracted.

Marine mammals are exposed to oil pollution in the form of crude, processed, and weathered oils that can directly contact the skin, hair/fur, eyes, mouth (including baleen), and *nares* (nostrils or nasal passages). Mammals can inhale volatile petroleum fractions at the water surface, ingest oil directly, and consume petroleum components in food, representing many routes of exposure for a single animal or population.

Some abiotic toxicants are grouped according to nonessential elements that are part of their molecular structure. Many toxic nonessential elements [sometimes also referred to as "heavy" metals, e.g., mercury (Hg), cadmium (Cd), lead (Pb)] interfere with the biologic action of essential elements such as calcium (Ca) and sulfur (S). Essential elements can be a part of chemical compounds that are required by an organism for healthy growth and proper maintenance of physiologic functions. Even so, high concentrations of some essential elements (e.g., selenium, Se) may be toxic and may produce outright *toxicosis*. Selenium toxicosis has been well documented in domestic mammals grazing in the western parts of North America. We do not promote use of "toxic" elements as a classification and, as physiologists and health professionals, use the terms essential and nonessential elements with the understanding that "essentiality" for marine mammals is not well understood and likely varies between taxonomic groups.

Mercury has both natural and anthropogenic sources, although temporal trends of Hg based on arctic ice-sheet cores indicate that concentrations increased markedly following industrialization of the early 20th century due especially to burning of fossil fuels and increased industrial mineral extraction activities. Information about temporal trends of Hg concentrations in animals is limited and studies are difficult to compare; analyses are complicated by well-described spatial trends and difficulties in comparing different age, sex, and tissues. Some species/locations show increasing concentrations of Hg across multiple decades for species where preindustrial archived tissues exist (for more detail related to temporal trends for Hg in the Arctic see Further Reading AMAP 2011).

When we discuss chemical elements, we include elements incorporated into more complex molecules, some of which are organic (molecule that contains carbon–carbon or carbon–hydrogen bonds). Some **organotin** compounds (organic compounds containing tin, Sn), for example, are used as marine antifoulants on boats, nets, and other industrial equipment. One group of organotins used as antifoulants, the butyltins, concentrate in tissues with high protein-binding capacity, including liver, kidney and brain. Another toxicant of concern in marine environments is monomethyl mercury ($MeHg^+$, commonly referred to as methylmercury), an organic molecule that includes the toxic nonessential element Hg. Methylation of Hg (addition of an organic methyl ($-CH_3$) group) occurs in aquatic environments via bacteria. The $MeHg^+$ produced is neurotoxic (among other effects) and moves readily through the food web.

Halogenated organic molecules, known as **organohalides** (aka organohalogens), contain fluorine, chlorine, bromine, or iodine and include organochlorine pesticides such as polychlorinated biphenyls (PCBs) and DDTs. These are terms used for groups within a general class formerly known as persistent organic pollutants (POPs). Halogens are attached to an organic (carbon) molecule. Many are manufactured, some occur naturally, and their persistence which was once a desired quality as an industrial chemical now presents an ecotoxicology dilemma. They persist in abiotic (nonliving) and **biotic** (living) components of nearly all ecosystems. There are many other examples of chlorinated organic compounds with some as by-products of combustion, PCB manufacture, and other industrial processes. See *Table 14.1* for a few examples.

Many industrial-sourced compounds are biologically novel and relatively new, but there are a number of elements (Hg, Cd, etc.) and marine biotoxins that have likely been present throughout the evolution of marine mammal species.

Table 14.1 **Examples of chlorinated organic contaminants**

COMMON ABBREVIATION	CHEMICAL NAME	TYPICAL SOURCE
PCB	Polychlorinated biphenyls	Organochlorine pesticides
DDT	Dichloro-diphenyl-trichloroethanes	Organochlorine pesticides
PCDF	Polychlorinated dibenzofurans	By-product of combustion (e.g., PCBs)
PCDD	Polychlorinated dibenzodioxins	By-product of combustion, manufacturing
PCQ	Polychloroquater-phenyls	Derived from PCBs (e.g., high heat)
PCN	Polychloro napthalenes	Insulating coating (e.g., for electrical wire), etc.

Called "natural" as a chemical class, it should be emphasized human activities have altered their presence and concentrations in the **biosphere**. As examples, industrial mining, land use changes, physical alterations to streams and rivers, have resulted in changing concentrations and distributions of natural toxicants in the environment.

Biotoxins and harmful algal blooms

Biotoxicology is the study of poisonous organisms and toxins produced by these organisms are biotoxins (derived from biota). Phytoplankton (planktonic plants) form the foundation of the marine food web and like terrestrial plants, they depend on sunlight for photosynthesis. The limited availability of nutrients and solar radiation drives competition between and within phytoplankton species. And just like many terrestrial plants, some phytoplankton have evolved to produce secondary metabolites – organic compounds which are not directly involved in growth or reproduction – in order to deter grazing by other organisms such as zooplankton (e.g., krill) or crustaceans.

Many secondary metabolites produced by plants are intended to deter immediate predation but have little or no harmful effects on nontarget species. For example, caffeine, produced in a number of plants, is a potent insecticide but does not pose a significant risk to humans. However, there are a number of secondary metabolites produced by terrestrial and aquatic plants that are harmful to broad ranges of organisms including both target and nontarget species.

Some phytoplankton produce toxins that are harmful or lethal to marine mammals (*Table 14.2*). Seasonal or atypical proliferations of these species are known as harmful algal blooms (HABs) and have been associated with large-scale die-offs of marine mammals, seabirds, and fish (see **Chapter 13** and **Chapter 15**). Blooms of toxic phytoplankton have likely occurred throughout the history of marine mammal evolution and archeological evidence from marine mammal fossils has suggested HABs were a driver of mass mortalities even millions of years ago (Pycnson et al. 2014). However there is strong evidence to suggest that with warming ocean temperatures due to anthropogenic climate change, the geographic and temporal range of HABs is expanding (Gobler 2020). Increases in the frequency and range of marine mammal stranding and deaths associated with algal toxins have been observed in many regions. In this chapter, we discuss some of the most common HABs and their impacts on marine mammals. For a detailed review of biotoxins that affect marine mammals, see Further Reading, Broadwater, Van Dolah, and Fire 2018, and **Chapter 13**.

Domoic acid (also known as amnesic shellfish poisoning, ASP) is a neurotoxin produced by several phytoplankton species, most notably diatoms of the genus *Pseudo-nitzschia*. Paralytic shellfish toxins (PST; or paralytic shellfish poisons, PSP) are a suite of chemicals produced by numerous phytoplankton species, particularly dinoflagellates of the genera *Alexandrium* and *Pyrodinium*. The most commonly described component of PSTs is saxitoxin, one of the most potent naturally occurring toxins and a Schedule I chemical warfare agent. There are more than 50 chemical forms of saxitoxin which have similar mechanisms of action. Brevetoxins (also known as neurotoxic shellfish poisoning, NSP) are produced by several phytoplankton species,

Table 14.2 **Examples of biotoxins produced by marine and freshwater phytoplankton that may affect marine mammals. The biotoxins presented for each row are generally the most common toxicant associated with each classification, and they all have other structural and/or mechanistic analogs (similar chemical forms)**

BIOTOXIN	BIOTOXIN CLASSIFICATION	PHYTOPLANKTON TYPE	PHYTOPLANKTON GENUS SPP.
Domoic acid	Amnesic shellfish poisoning (ASP)	Diatoms	*Pseudo-nitzschia* spp.
Saxitoxins	Paralytic shellfish toxins (PST) *also known as paralytic shellfish poisoning (PSP)*	Dinoflagellates (marine) Cyanobacteria (freshwater)	*Alexandrium* spp. *Gymnodinium* spp. *Pyrodinium* spp. *Aphanizomenon flosaquae* *Dolichospermum circinalis* *Lyngbya wollei* *Planktothrix* spp. *Raphidiopsis raciborskii*
Brevetoxins	Neurotoxic shellfish poisoning (NSP)	Dinoflagellates	*Karenia brevis* (Florida red tides)
Ciguatoxins	Ciguatera fish poisoning (CFP)	Dinoflagellates	*Gambierdiscus* spp.
Okadaic Acid	Diarrhetic shellfish poisoning (DSP)	Dinoflagellates	*Dinophysis* spp. *Prorocentrum* spp.

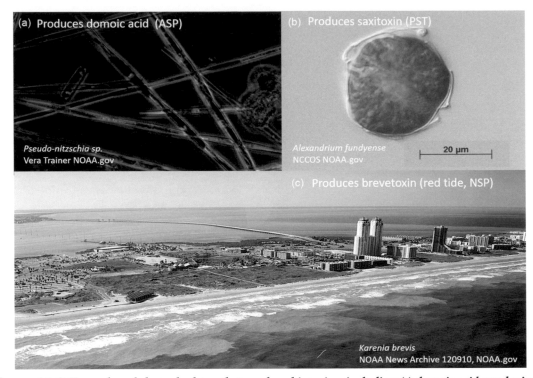

Figure 14.2 Examples of phytoplankton that produce biotoxins, including (a) domoic acid-producing *Pseudo-nitzschia* sp. (amnesic shellfish poison; ASP), (b) saxitoxin-producing *Alexandrium fundyense* (paralytic shellfish poison, PSP), and (c) brevetoxin-producing *Karenia brevis* (neurotoxic shellfish poison, NSP). The biotoxins presented are generally the most common toxicant associated with each disease (e.g., ASP), although they all have other structural and/or mechanistic analogs (chemicals that have similar toxic effects). (Photos from NOAA.gov, accessed August 2022. *Pseudo-nitzschia*, https://www. climate.gov/news-features/event-tracker/record-setting-bloom-toxic-algae-north-pacific; *Alexandrium fundyense*, https://coastalscience.noaa.gov/news/nccos-nos-coast-survey-development-lab-scientists-moving-toward-operational-forecast-system-harmful-algal-blooms-alexandrium-fundyense-gulf-maine/; *Karenia brevis* bloom, https://photolib.noaa.gov/Collections/Fisheries/Other/emodule/1054/eitem/61631)

but are most commonly associated with the dinoflagellate *Karenia brevis*, which causes noxious red tides in coastal regions of the eastern United States and Gulf of Mexico. Brevetoxins, PSTs, and domoic acid (**Figure 14.2**) are the most commonly reported toxins that affect marine mammal health; however, these are far from the only algal toxins marine species are exposed to. Ciguatoxin and okadaic acid (produced by dinoflagellates) have both been detected in marine mammal species although the extent of their effects is largely unknown (see Further Reading, Broadwater, Van Dolah, and Fire 2018). Recently, algal toxins typically associated with

freshwater environments have received increasing attention. Cyanobacteria or blue-green algae occur in freshwater lakes and rivers, and several species produce toxins that pose a risk to human and animal health. Typically these toxins have been associated with cyanobacterial HABs in inland lakes and reservoirs. However, there is growing evidence to suggest that these toxins flow through freshwater systems into estuaries and the marine environment. Toxins such as microcystins, anatoxins, and cylindrospermopsins, produced by a number of freshwater cyanobacteria, have been shown to transfer to marine mammals and important prey species in the marine environment (e.g., Preece, Moore, and Hardy 2015; Further Reading, Broadwater, Van Dolah, and Fire 2018). There are other HABs that can affect marine mammal populations through physical irritation or drastic reduction in dissolved oxygen (dead zones) but are beyond the scope of this chapter.

By-products of diving physiology and biochemistry

Finally, we must consider chemicals specifically created by the physiology and biochemistry of diving. During dives, marine mammals deplete their on-board oxygen supplies (hypoxia) which promotes the formation of acutely toxic by-products such as lactic acid, hydrogen ions, and *reactive oxygen species (ROS)*. Details on diving physiology are presented elsewhere (see **Chapter 3**). In terms of toxicants, it is important to consider that some diving marine mammals must tolerate *anaerobic* cellular conditions as a result of extended periods of apnea and use a battery of biochemically based self-defense mechanisms against potential toxic by-products of diving. These processes may alter the response of diving mammals to other toxicants which induce oxidative stress with similar actions as ROS.

ROUTES OF EXPOSURE

As marine mammals integrate multiple components of an ecosystem, many may be considered *sentinel species* (provides a warning) for environmental toxicants. The main routes of exposure to toxicants for most marine mammals are through diet (foraging, **Figure 14.3**) or maternal transfer (fetus and pup/calf, **Figure 14.4**).

Food web transfer, bioaccumulation, and biomagnification

Unfortunately, many of the characteristics that lead marine mammals to be effective environmental sentinels also result in higher toxicant concentrations than relevant prey species. Many marine mammals are long-lived and occupy high trophic levels (strict predators), especially marine mammal species such as killer whales (*Orcinus orca*), polar bears (*Ursinus maritimus*), leopard seals (*Hydrurga leptonyx*), Steller sea lions (*Eumetopias jubatus*), and walruses (*Odobenus rosmarus*) that prey upon other marine mammals. Numerous studies have reported concentrations of some toxicants, including MeHg+ and organohalides, positively correlated with age (*bioaccumulation*). Tissue bioaccumulation occurs when contaminants accumulate over the life of the animal. Bioaccumulation of certain toxicants may be more pronounced in specific tissues such as liver (e.g., Hg), kidney (e.g., Cd), and blubber (e.g., PCBs). *Biomagnification* is a process where contaminant concentrations increase with increasing trophic level. Bioaccumulation and biomagnification do not occur independently. For example, increased size (age) may also correspond to increased trophic level with the ability to take larger prey. These processes are illustrated in **Figure 14.5**.

Bioaccumulation and biomagnification processes may be complicated by growth/dilution dynamics and reproductive status. For instance, Ross et al. (2004) found increasing concentrations of organochlorines (i.e., PCBs, PCDEs – polychlorinated diphenyl ethers) in older male harbor seals (*Phoca vitulina*), but found decreasing concentrations in females associated with *purity* (number of completed pregnancies) and lactation (i.e., chemical transfer to pup). The effects of reproduction on toxicant accumulation and offloading in marine mammals is an important driver of observed concentrations in adult females and represents a significant exposure route *in utero* and to nursing pups or calves. It is important to note, the exposure to females can still be quite high, like males, but females have additional mechanisms to eliminate these compounds in a manner that is not strategic for the fetus and neonate, as discussed in the section "Maternal Transfer". This represents a limitation of simply assessing tissue concentrations to assess exposure and potential for adverse effects.

Figure 14.3 Many marine mammals are strict predators, consuming invertebrates (e.g., sea otter eating mussels), fish (e.g., sea lion eating fish), and even other marine mammals (e.g., killer whale eating seals). Feeding ecology affects both the types of toxicants to which they may be exposed and the degree of exposure. (Photos from NOAA.gov, accessed August 2022. https://www.photolib.noaa.gov/Collections/NOAAs-Ark)

Trophic transfer of toxicants is often complicated by varying abilities of taxonomic groups, or even individual species to sequester or ***biotransform*** certain chemicals (transform by biochemical means). In some instances, biotransformation can greatly alter the potency and/or mechanism of action of a particular toxicant. For instance, ciguatoxin, which is a potent neurological biotoxin associated with the dinoflagellate *Gambierdiscus toxicus*, enters lower trophic positions as gambiertoxins. As the gambiertoxins are biotransformed by herbivorous

and predatory fish they are oxidized into more potent ciguatoxins (Lewis and Holmes 1993).

Biotoxins are often encountered by marine mammals through diet. However, saxitoxin produced by the dinoflagellate *Gonyaulax catenella* (responsible for some "red tides" along with *Karenia brevis*) can be aerosolized by cell lysis via wave/wind action and has been known to cause respiratory problems in humans and other mammals (Kirkpatrick et al. 2004). Since most marine mammal species do not directly prey upon the toxin-producing phytoplankton, it is

Figure 14.4 Maternal transfer of toxicants to marine mammal pups/calves may occur *in utero* (transplacental via blood) or via nursing (lactational via milk), depending on the chemical nature of the toxicant and the ability for it to cross the placental barrier or be present in milk. This may provide a significant excretory route for reproductive females, but results in exposure to the fetus and pup/calf at particularly vulnerable life stages. (Photos from NOAA.gov accessed August 2022. https://www. photolib.noaa.gov/Collections/NOAAs-Ark)

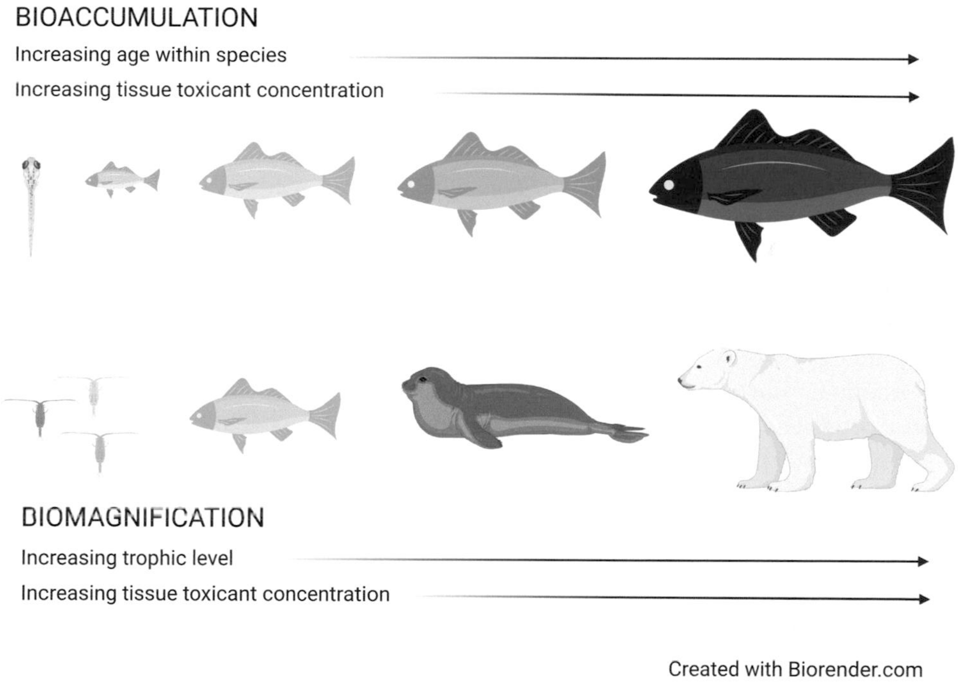

BIOACCUMULATION

Increasing age within species

Increasing tissue toxicant concentration

BIOMAGNIFICATION

Increasing trophic level

Increasing tissue toxicant concentration

Created with Biorender.com

Figure 14.5 A conceptual figure showing bioaccumulation (accumulation of toxicants in tissues with increasing age within a species) and biomagnification (accumulation of toxicants in tissues with increasing trophic level). These processes are not independent. For example, as an animal gets older, it might feed at a higher trophic level. Tissue concentrations may also be affected by physiological processes like fasting and reproduction that can change tissue distribution of toxicants, and growth dilution when rapid growth can lower tissue concentrations of toxicants.

important to consider the trophic transfer and biotransformation of biotoxins through the food web. For instance, Bricelj et al. (2005) found that genetic mutations in wild softshell clams (*Mya arenaria*) meant the clams were able to tolerate much higher concentrations of STX than their "sensitive" wild-type counterparts, which is presumably favorable for the clams but rather unfortunate for the species consuming them.

Marine mammals are exposed to toxins present in HABs in areas where they overlap in time and space. Evidence of PSTs and domoic acid in fecal samples of North Atlantic right whales (*Eubalaena glacialis*) suggest that individuals are exposed to these algal toxins on a regular basis (70–80% contained traces of PST, 25–30% contained traces of domoic acid, and 22% of the fecal samples collected showed exposure to both domoic acid and PST; Doucette et al. 2012). While PST is known to accumulate in some shellfish, the fact that it can accumulate in the tissues of highly motile fish such as cod species raises some interesting ecological considerations. Movement of prey species or a lag following prey consumption created by the compounding effects of bioaccumulation in a prey species and biomagnification across trophic levels may result in observations of toxicoses due to HABs in regions or during times that do not have observable HAB events (see Further Reading, Broadwater, Van Dolah, and Fire 2018).

As an example, during the fall of 1987, 14 humpback whales (*Megaptera novaengliae*) died near Cape Cod from apparent saxitoxin poisoning (Geraci et al. 1989). While plankton and shellfish in the immediate vicinity did not contain traceable concentrations of saxitoxin, stomach content analysis of the whales found a high prevalence of mackerel, which spawn in the waters of the Gulf of St. Lawrence. Concentrations of STX in mackerel liver were 154 μg/100 g tissue, which is nearly double the US Food and Drug Administration threshold for human consumption of mussels of 80 μg/100 g (Geraci et al. 1989). It is tempting to view HABs as limited by the spatial and temporal lifespan of the toxin-producing algae; however, when considering the life span and geographic movement of fish consumers able to accumulate the toxin, the overall scope of the effects of HABs increases dramatically.

Saxitoxin has also been implicated in the death of a number of Hawaiian monk seals (*Monachus schauinslandi*) as it was present in seal tissues and prey species, as well as lung-related trauma consistent with drowning due to paralysis (see Further Reading, Broadwater, Van Dolah, and Fire 2018). However, as with most cases of algal toxin poisoning, the authors note that without reference concentrations it is difficult to definitively establish cause of mortality (diagnosis). There is some speculation that ciguatoxin, a toxin present in reef fish, has contributed to the decline of Hawaiian monk seal populations (see Further Reading Broadwater, Van Dolah, and Fire 2018), although this remains rather speculative. Recent work has shown detectable levels of ciguatoxin activity in blood samples from Hawaiian monk seals, which supports the theory that individuals were indeed exposed to the toxin and transfer to marine mammals is possible.

Mammalian exposure to domoic acid is due primarily to consumption of contaminated prey species (see Further Reading, Broadwater, Van Dolah, and Fire 2018), which in the case of California sea lions (*Zalophus californianus*) appears to be largely attributed to the filter feeding of anchovy. There are several interesting observations that arise from studying stranding data from California sea lions from past 15 years. Strandings associated with acute and chronic domoic acid toxicity were reported every year from 1998 to 2006 except 1999 (Bejarano et al. 2008). This may be explained by an increased stranding response effort or improved diagnostics and surveillance, or both. However, there is also evidence that blooms of *Pseudo-nitzschia* have co-occurred during periods of increased domoic acid poisoning frequency (Bejarano et al. 2008, Further Reading, Broadwater, Van Dolah, and Fire 2018). The majority of strandings associated with domoic acid appear to be of adult female sea lions, likely related to foraging ecology. The breeding season for California sea lions occurs during the summer months in California (May–August), which generally coincides with blooms of *Pseudo-nitzschia*. Adult females spend a large portion of the breeding season foraging at sea, while the adult males are generally hauled out defending territory. Since the majority of toxicant exposure for marine mammals is often through diet, it is intuitive that differences in feeding ecology would drive different exposure scenarios, but it is worth emphasizing here that due to the transient nature of HABs and HAB toxins, even feeding on the same prey species in the same area may result in different HAB toxin exposures due to differences in seasonal feeding patterns.

Maternal transfer to fetus and pup/calf

For many species, especially placental mammals, one of the most sensitive life stages to poisoning is the fetus and neonate. Marine mammal fetuses are exposed to maternal toxicants both *in utero* (transplacental) as well as during nursing as neonates (transmammary; lactational). The placenta is designed to protect the fetus by acting as a semipermeable barrier preventing the transfer of some toxicants and pathogens, but is not fool proof. For instance, $MeHg^+$ is transferred to the developing fetus through the placenta, and while studies of embryonic development in marine mammals are rare, $MeHg^+$ has been shown to impair development and cause adverse health outcomes in fish-eating humans and wildlife (e.g., Johansson et al. 2007; Tonk et al. 2010). Adverse health outcomes in fish-eating humans and wildlife have been used to extrapolate potential adverse effects in marine mammals and to design studies to directly assess marine mammal neonates.

Interestingly, one of the main excretory routes for $MeHg^+$, besides excretion in urine and bile, is through hair growth. Mammalian hair is composed of large amounts of the protein keratin which contains a large number of cysteine amino acid which form the disulfide (S–S) bridges necessary for hair rigidity. Methylmercury binds tightly to the sulfur in cysteine, and thus hair concentrations of $MeHg^+$ are often several-fold higher than blood or other tissues. Both pinniped and cetacean fetuses develop hair *in utero*, which in pinnipeds can persist for several weeks postpartum (*lanugo*). See **Chapter 5** for more discussion of lanugo. Most mammals, including humans, display signs of lanugo development *in utero*, which is likely a conserved trait from common ancestors, and therefore cannot be easily implicated as a direct adaptation of the developing marine mammal fetus to excreting toxicants. However, it is interesting to speculate shedding of contaminants *in utero* via lanugo continues to be observed in marine mammals due to its ability to protect the developing fetus from neurotoxicosis in situations where mercury exposure may be high.

Steller sea lion pups are born with lanugo that is later replaced with new fur grown while pups are suckling. Lanugo collected from young Alaska Steller sea lion pups (<3 months old) had high concentrations of mercury relative to fur from older suckling pups (Castellini et al. 2012), indicating a greater transfer of $MeHg^+$ *in utero* across the placenta, than later from milk. This is different from more **lipophilic** (soluble in lipid; "love fat") toxicants, like PCBs and other organohalides, which are readily transferred during lactation in the lipid-rich milk typical of marine mammals.

The term "blubber" is often misrepresented in both popular culture and occasionally scientific literature, so we will use the definition provided by Reeb, Best, and Kidson (2007) where blubber is integument layer below the epidermis (dermis and hypodermis). As large **endotherms** that make their living in the marine environment, many marine mammals utilize extensive lipid-rich blubber tissue for thermal insulation and other needs (storing nutrients, enhancing hydrodynamics, protection from predators, etc.). Many chapters in this book explore these topics in detail. In migratory species, such as gray whales (*Eschrichtius robustus*), not only are individual animals often subject to large changes in ambient water temperature, but migratory periods are often marked by little or no foraging (i.e., fasting). The laying down and mobilization of energy stored in fatty tissue is truly remarkable in marine mammals. While duration of nursing is often radically different between marine mammal species (ranging from a few days to several years), in some species with relatively short nursing periods, such as the northern elephant seal (*Mirounga angustirostris*), females can lose up to 42% of their initial body mass during nursing (58% of fat stores) (Costa et al. 1986). The energetics of nursing are discussed in **Chapter 9** and **Chapter 12**, however the transfer of toxicants during lactation is interesting to consider from the perspective of both the lactating female and the nursing pup. While nursing young are exposed to a complex mixture of potential toxicants during a sensitive developmental stage, it does represent a significant excretory route for reproductive females (e.g., Frouin et al. 2012). A decrease in maternal contaminant concentrations following pup production has been described in sea otters (*Enhydra lutris*) (Jessup et al. 2010). In particular, females giving birth to their first offspring often transfer a majority of their blubber-based contaminant load to their firstborn. Indeed, Beckmen et al. (2003) found increased blood concentrations of PCB *congeners* (related chemical substances) as well as congeners of DDT in pups born to young northern fur seal (*Callorhinus ursinus*) females (first-born pups) compared to pups born to older females that had presumably produced pups over several years. Similarly, Ylitalo et al. (2001) found higher concentrations of organochlorines in first-born breeding resident male Alaskan killer whales as compared to breeding males from subsequent births. Maternal transfer is, in fact, a leading cause of concern when evaluating population level concentrations of contaminants.

EXAMPLES OF TOXICANT EFFECTS AND DETOXIFICATION MECHANISMS

It is critical to not broadly extrapolate a chemical's adverse effect on "marine mammals" as this is a group with a large diversity of lineages and primarily grouped together based on generalized habitat relationships (i.e., lives in or feeds in the ocean). In the same way, we should not generalize regarding the physiology or life history characteristics of marine mammals; although many marine mammal lineages share traits either conserved from previous common ancestors or a result of *convergent evolution* that allow us to make comparisons. Mammals have evolved complex and often elegant detoxification and sequestering mechanisms for toxicants, and adaptive machinery that is present in marine mammals is often found across mammalian lineages. While particular biotransformation pathways may not be unique to marine mammals, it is important to discuss them in the context of adaptation and acclimation since marine mammals are exposed to different classes of toxicants and at varying concentrations than their terrestrial counterparts.

Toxic oxygen: adaptations to anaerobic diving

While cellular and physiologic adaptations to diving are discussed elsewhere in this book (See especially **Chapter 3**), it is worth noting that some of the most potent toxicants that marine mammals are exposed to come not from environmental exposure but from their own physiology and biochemical by-products. As discussed above (section "By-products of Diving Physiology and Biochemistry"), marine mammals have developed a unique diving physiology that allows them to cope with extended periods of hypoxic and anaerobic activity.

While molecular oxygen (O_2) is necessary for cellular respiration, the lone O^{2-} atom is quite damaging in many forms that are produced during normal physiologic processes and/or as a consequence of some toxicants. These O^{2-}

by-products are known to damage macromolecules (like proteins and lipids), in some cases irreversibly. In particular, the breath-holding behavior (apnea) of many marine mammals (during diving, sleeping) and associated changes in tissue blood flow creates risk of a number of oxygen-related injuries such as *ischemia-reperfusion* injury and the generation of ROS. Voluntary breath holds in northern elephant seal pups have been shown to increase the activity of hypoxanthine (which generates ROS upon reperfusion) in plasma, but without observable effects to biomarkers of oxidative stress (4-hydroxynonenal and 8-iso prostaglandin F2α) (Vázquez-Medina, Zenteno-Savín, and Elsner 2006). Others have found that baseline concentrations of carbon monoxide (CO) and carboxyhemoglobin (COHb) in Weddell seals (*Leptonychotes weddellii*) and northern elephant seals are several times higher than human smokers (Pugh 1959; Tift, Ponganis, and Crocker 2014).

Both Tift, Ponganis, and Crocker (2014) and Vázquez Medina, Zenteno-Savín, and Elsner (2006) suggest that production of CO and hypoxanthine provides protective effects against oxidative damage caused by apnea. Indeed, recent work has shown that arterial endothelial cells in elephant seals are more resistant to oxidative stress than human cells (Allen et al. 2021). Carbon monoxide is a toxic gas that can cause hypoxia and death due to its affinity for heme protein (containing iron; present in hemoglobin and myoglobin. See **Chapter 3** for more information about hemoglobin and myoglobin chemistry. Carbon monoxide lowers blood O_2 saturation levels by preventing oxygen from binding appropriately to hemoglobin. However, CO may be produced within an animal, largely through the breakdown of heme, which also produces biliverdin (bile pigment produced during heme breakdown). While biliverdin is water soluble and rapidly excreted, biliverdin reductase (BVRA) reduces biliverdin to bilirubin which is not water soluble and a well-known neurotoxicant. Many scientists have noted the paradox of the evolution of an energy-requiring process that converts a nontoxic product into a toxic compound. However, CO, as well as other products of the breakdown of heme protein (such as bilirubin and biliverdin) are known to have *antioxidant* properties that can protect against the damaging effects of oxidation, including ischemia-reperfusion injury.

There is some evidence to suggest that marine mammals produce antioxidants (including glutathione, superoxide dismutase and glutathione peroxidase) at higher rates than terrestrial mammals, presumably as an adaptation or acclimation to dealing with periods of apnea (Wilhelm Filho et al. 2002). Increased antioxidants may assist marine mammal responses to environmental toxicants unrelated to molecular oxygen products and by-products of cellular respiration (e.g. Hg, Cd; see section "Essential and Nonessential Elements").

Essential and nonessential elements

The more infamous nonessential elements of concern in the marine environment (Cd, Hg; also sometimes referred to as "heavy" metals, although this is not our preferred term) do not exist in isolation – marine mammals are exposed to these metals through a variety of routes, but generally through their diet. For high-trophic-level marine mammals such as most pinnipeds and odontocetes, this means a diet of fish, invertebrates, and possibly other marine mammals (for instance killer whales and walruses consume seals and sea lions). Other nutritional components of a fish-based diet, such as selenium (Se), may counter some of the potential negative effects of intoxication due to some nonessential elements. Marine mammals that eat fish and other marine mammals often have high concentrations of some elements (e.g., Hg, Se) compared to many terrestrial mammals, while cadmium (Cd) is often associated with a diet dominated by marine invertebrates (Dehn et al. 2005), as illustrated in **Figure 14.6**. Despite reports of concentrations above levels of concern for other mammalian species (Dietz, Nørgaard, and Hansen 1998; Rea et al. 2013; see

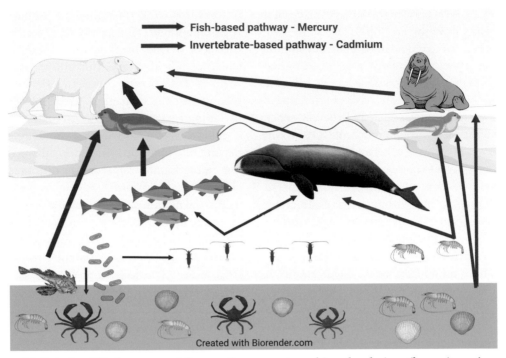

Figure 14.6 A simplified conceptual figure of mercury (purple) and cadmium (brown) moving through an arctic marine food web. Feeding ecology affects toxicant exposure. Movement of mercury through food webs is primarily fish-based while cadmium is primarily invertebrate-based. Food web transfer is more complex – we have not indicated all possibilities in this figure. For example, while the diets of some seals and walrus are primarily invertebrate-based, they may also include fish and/ or seals resulting in associated exposure to mercury. As trophic level increases the arrows get thicker, indicating potential biomagnification.

Further Reading, Dietz et al. 2013), there have been few reports of marine mammal poisoning events (signs of acute toxicosis) associated with these elemets of concern, although this may be due to the difficulties associated with performing detailed clinical assessments and necropsies on marine mammals of suitable condition.

Because methylmercury ($MeHg^+$) readily crosses the placenta and brain barriers it presents risks related to neurodevelopment; breeding females, pups/calves and fetuses are of particular concern. Target organs may also include brain, kidney, liver, and heart in addition to potential effects on immune and reproductive function. Detoxification processes and sequestration of $MeHg^+$ to an insoluble crystal complex with Se (tiemannite; HgSe) occurs in the livers of many marine mammals. The relationship between Hg and Se in mammals has been known for several decades although the mechanisms of demethylation and subsequent detoxification of mercury are only recently beginning to be described (see Further Reading O'Hara and Hart 2018). Tiemannite crystals have been reported in the livers of many cetaceans. Nakazawa et al. (2011) found evidence for tiemannite molecules in liver, kidney, lung, spleen, pancreas, muscle, and brain of striped dolphin (*Stenella coeruleoalba*). It is unclear whether the tiemannite crystals found in lung and other tissues were present due to localized demethylation and sequestration, sequestration from other tissues, or, in the case of lungs, inhalation of the particles directly. As with cetaceans, there is evidence of co-sequestration of mercury and selenium in pinnipeds, particularly in liver (e.g. harbor seal, hooded seal

Cystophora cristata, harp seal *Pagophilus groenlandicus*, Steller sea lions, etc.).

Many toxicants, including MeHg+ (and Cd), are known to generate oxidative stress. In addition to detoxification and sequestration functions, Se is also an antioxidant and a key component of several antioxidants including the glutathione-peroxidase family of enzymes. The enzyme glutathione peroxidase is an antioxidant that catalyzes hydrogen peroxide into water and prevents lipid oxidation (which may cause cell damage). As discussed in the previous section (Toxic Oxygen: Adaptations to Diving), some diving mammals have higher activities of tissue glutathione peroxidase activity compared to terrestrial mammals as an adaptation to diving physiology; it is reasonable to suspect that higher activities of these antioxidant enzymes could also provide at least partial protection against the effects of oxidative stress induced by some elements (e.g., Hg, Cd).

While these processes are also known to occur in some terrestrial animals, the greater exposure to MeHg+ in fish- and mammal-consuming mammals in the marine environment leads to questions related to detoxification thresholds, efficiencies, and potential effects thresholds in marine mammal species (**Figure 14.7**). As described above, some of these physiologic and cellular defenses may utilize one or more

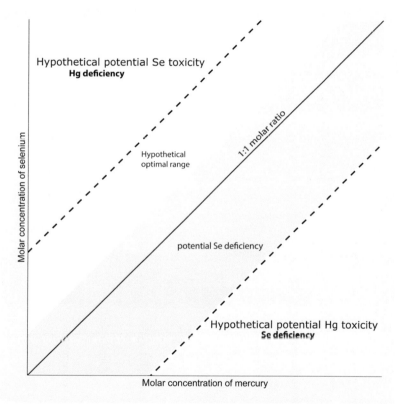

Figure 14.7 Mercury and selenium display a multifaceted relationship in marine mammals, including the formation of Hg–Se (tiemannite, inert) occasionally resulting in a near 1:1 molar correlation. As noted by Khan and Wang (2009), the 1:1 molar ratio can potentially be viewed as Se deficient if Hg–Se is not bioavailable and Se is less available to participate in Se-dependent processes (i.e., glutathione peroxidase, selenoproteins). We emphasize that somewhat counterintuitively, with high concentrations of Se seen in marine mammals that potential Se toxicosis could occur in marine mammals as a result of Hg deficiency. (From Harley and O'Hara 2016; adapted from Khan and Wang 2009)

micronutrients or co-factors and we have stressed the importance of Se. Selenium is present in relatively higher concentrations in marine prey species compared to similar terrestrial trophic levels (**Figure 14.8**). In fact, Se in aquatic organisms often exists in concentrations at or above thresholds for chronic toxicity for some domesticated terrestrial species (Puls 1994; Dietz, Riget, and Johansen 1996), leading to the paradoxical inference that Hg can reduce Se toxicity (Khan and Wang 2009; see **Figure 14.7**). For instance, Dietz, Riget, and Johansen (1996) found concentrations of Se up to 7.69, 4.99,

9.09 μg/g in liver of ringed seals (*Pusa hispida*), beluga whales (*Delphinapterus leucas*), and polar bears respectively; which compare closely to the lower thresholds of chronic toxicity reported in liver for cattle (1.25–7.0 μg/g), pig (4.0–20.0 μ/g), and rabbit (7.04 μg/g) (Puls 1994). Research into the diversity of proteins that incorporate selenium (***selenoproteins***) found that terrestrial animals have a smaller and less diverse assortment of selenoproteins than aquatic species (Lobanov et al. 2007). The authors of the study attribute the diversity in aquatic species at least in part, to the abundance of bioavailable Se in aquatic

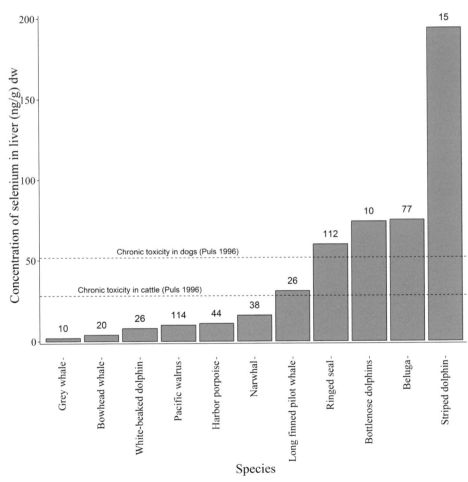

Figure 14.8 Marine mammals display a large variation in selenium (Se) concentrations in liver (original sources of data in Harley and O'Hara 2016). The number of individuals for which the mean concentration is displayed is reported above each bar. Mean concentrations and thresholds of toxicity are presented in dry weight (dw) assuming mean water weight of 75% as described in Harley and O'Hara (2016). (Adapted from Das, Debacker, and Bouquegneau 2003)

(in this case, marine) ecosystems. We emphasize the role of Se in both element detoxification as well as defense against oxidative stress through (1) direct interactions between Se and other elements, and (2) indirect actions of selenoproteins (enzymes and other proteins that include Se active sites).

As noted earlier in the section "Food Web Transfer, Bioaccumulation and Biomagnification", Hg exposure in pinnipeds begins *in utero* and is linked to maternal diet. Growth of hair *in utero* (lanugo) provides an excretory pathway for MeHg⁺ transferred from the dam to her fetus. This is important as the fetus is likely the most vulnerable to adverse health effects and may have limited detoxification and elimination processes available (see Further Reading O'Hara and Hart 2018).

There has been speculation that arctic marine mammals have adapted to high levels of cadmium (Dietz, Nørgaard, and Hansen 1998). Temporal trends assessments for Cd in the Arctic in the last few decades are not clear (Riget and Dietz 2000), and it is conceivable that Cd has been at appreciable levels at least since the 15th century (Hansen, Toribara, and Muhs 1989). The toxic effects of Cd have been known for several decades. One of the most infamous Cd mass poisoning events in humans, "itai-itai" (ouch-ouch) disease, occurred in Japan's Toyama Prefecture in the early 1900s, although the link to Cd from industrial waste was not reported until 1968. Cadmium accumulates in the kidney and liver and is a well-known nephrotoxin, causing damage within the kidney. Chronic Cd poisoning is associated with a number of bone disorders, including osteoporosis and increased occurrence of fractures, both of which appear to be due to decreased rates of bone mineralization. This is likely related to kidney dysfunction (Berglund et al. 2000), although the exact mechanisms are unknown.

Concentrations of Cd in livers of bowhead whales (*Balaena mysticetus*) have been found that are associated with toxic thresholds in domestic animals (Woshner et al. 2001). Rosa et al. (2008) found moderate to severe thickening of the Bowman's capsule in kidneys of bowhead whales associated with age and Cd concentrations, however there was no evidence of renal dysfunction. This supports the conclusion of Dietz, Nørgaard, and Hansen (1998) that given historically high Cd concentrations in the marine environment (as compared to terrestrial, especially arctic regions) marine mammals may have evolved to tolerate or detoxify relatively high concentrations of Cd. Some pinnipeds also have high concentrations of Cd (Dietz, Riget, and Johansen 1996; Dietz, Nørgaard, and Hansen 1998). Dietz, Nørgaard, and Hansen (1998) reported cadmium levels in the kidneys of Greenland ringed seals exceeded thresholds associated with renal damage in other mammals (200 μg/g ww, as determined by WHO 1992). In seal kidneys with very high concentrations of cadmium (up to 726 μg/g Cd in kidney cortex based on wet tissue weight) Dietz, Nørgaard, and Hansen (1998) found no significant differences in kidney structure or necrosis compared to kidneys with low (1.63 – 5.19 μg/g Cd) and intermediate (86.5 – 91.3 μg/g Cd) concentrations.

Metallothioneins (MTs) are a family of proteins, that can bind with many elements and function in uptake, transfer, and excretion of both essential (e.g. Cu and Zn) and nonessential (e.g. Ag, Hg, Cd) elements. It has been suggested that MT may be involved in defending organisms against elements such as Hg and Cd, which bound to MT are less toxic than free.

The relationship between MTs and Hg is not clear – binding of Hg to MTs in kidney and liver of rats has been observed for several decades, and chemical forms of Hg have been clearly demonstrated to induce MT detoxification function for mercury (Roesijadi 1992). However, studies in marine mammals have indicated that a relatively small percentage of mercury is bound to MTs in liver and kidneys. It is possible that MTs represent an acute response (transient) to intracellular MeHg⁺ or inorganic Hg (Hg²⁺), and that temporary binding to MTs assists transport to other organs or eventual biotransformation into inactive forms (e.g. tiemannite).

As with Hg, concentrations of Cd bound to MT vary among marine mammals. Cadmium was found at relatively high concentrations in livers of long-finned pilot whales (*Globicephala melas*), and MT-like proteins were correlated with Cd concentrations in adults (Amiard-Triquet and Caurant 1997). There was seasonal variability, with the percentage bound much higher in summer (51%) than winter (6%). This is in contrast to Wagemann, Hunt, and Klaverkamp (1984), who found 77% of cyto-solic Cd bound to MTs in the livers of nar-whals (*Monodon monoceros*). Das, Debacker, and Bouquegneau (2000) suggested that this might be evidence of an adaptation of arctic narwhals to high concentrations of Cd in prey, since envi-ronmental concentrations of Cd appear not to have fluctuated strongly in northern latitudes (this is in contrast to Hg which has shown an increase since post-industrialization). How-ever, given the large Cd-MT variation seen in pilot whales, it is likely that there are other physiologic and molecular modulators of Cd bound to MT, both as total amount and propor-tion of total MT. These include the presence of other elements in their diet that could compete for MT binding, the induction of hormone sys-tems that are known to impact MT synthesis, or seasonal variation in physiology which could drive transport or sequestration of elements (i.e. nursing, lipid mobilization, etc.).

Induction of MT by various compounds has been shown to protect against Cd induced tox-icity. Interestingly, it has been shown in rats that immature individuals have higher circulating MT concentrations compared to adults (Goer-ing and Klaassen 1984), and similar trends have been described in Franciscana dolphins (*Pon-toporia blainvillei*) where concentrations of MT appeared to be higher in juveniles than adults (Polizzi et al. 2014). The relationship of higher MT concentrations in immature compared to mature animals has also been described in gray seals (*Halichoerus grypus*) (Teigen et al. 1999).

It has been suggested that Cd in females may be transferred via milk, potentially providing a significant source of dietary Cd exposure to the pup during nursing. Whether the elevated levels of MT in immature mammals is caused by maternal contaminant transfer or differences in developmental physiology is unknown (per-haps a higher demand for Cu or Zn), however high concentrations of MT in fetal tissues must certainly confer protections against Cd toxicity.

Organohalides and lipophilic toxicants

Typically, anthropogenic organohalides enter the marine environment through terrestrial run-off or atmospheric transport on particulates fol-lowing human use (e.g., pesticides such as DDT, oil in electric transformers). Many, including some organochlorine pesticides and PCBs are soluble in fat (lipophilic, "love fat") and easily biomagnify in food webs (follow the fat). They tend to concentrate in fatty tissues (brain, liver), including blubber in marine mammals. Not all biomagnifying and bioaccumulating toxicants are lipophilic, as noted in earlier sections. This can often be confusing and is good to appreciate at the outset as there are many physiologic pro-cesses besides association with lipids that can allow some chemicals to accumulate/magnify in tissues (e.g., binding proteins, interactions with elements, non-specific interactions with macromolecules). Many lipophilic agents can have specific receptor-based interactions as well that relate to their biotransformation and/or toxicity; they associate with the lipids and then interact with other macromolecules. Solubility in lipid is very important for toxicant movement within the food web, distribution to tissues, and for crossing tissue barriers such as the blood-brain barrier or the placenta. As with MeHg$^+$, discussed in the previous section (Essential and non-essential elements), the ability to cross such barriers exposes sensitive life stages, the fetus and neonate. Biotransformation of some of the lipophilic chemicals can reduce the fat loving characteristics and/or alter target organs to some transformations can make some parent compounds (ingested form of the chemical) more toxic or impact a different organ system. Increasing elimination potential may actually increase toxicity in some cases. These physical and biochemical processes for lipophilic agents are complex and still require further study in marine mammals.

There are numerous types of organohalides, dozens of organochlorine compounds, and hundreds of congeners (related chemical substances) of some compounds. While we use the PCB class as an example, discussing the organohalides in detail is well beyond the scope of this chapter (see Further Reading O'Hara and Hart 2018). Even focusing on the PCB class, there are 209 congeners of PCBs that are often grouped based on chemical structure (chlorine substitutions). To complicate matters further, individual congeners within a class or group can have vastly different physical properties (different degrees of lipophilicity) and toxicities, as well as various mechanisms of toxicity. The PCBs are known hormone disrupters (receptor-based interactions), and some marine mammals and their associated prey have relatively high concentrations of PCBs as a result of bioaccumulation and biomagnification, some resulting from placental and mammary transfer (diet of dam driving fetal and neonatal exposures) as previously discussed. Lipophilic compounds such as organhalides can occur in particularly high concentrations in the fat-rich milk of a nursing female. Relatively high concentrations of PCB congeners and other organohalides of concern (e.g. DDT) have been associated with adverse reproductive outcomes in marine mammals. While biotransformation of some organohalides is observed in mammals, it has been suggested that small cetaceans have a limited capacity to biotransform certain organohalides compared to terrestrial mammals. It is conceivable that, due to their ancient divergence from terrestrial herbivorous ancestors, some cetaceans have lost certain enzymes as compared to terrestrial mammals. Very small differences in toxicant chemical structure even within a certain class relates to their potential for biotransformation and their persistence and potential for toxicosis, just like any other group (class) of toxicants.

Petroleum hydrocarbons and oil spills

Following catastrophic anthropogenic release of crude oil, such as the Exxon Valdez oil spill (1989) or the Deepwater Horizon (2010), there is a large amount of public and scientific concern devoted to the health and status of local marine mammal populations (See **Chapter 15**). Species that depend on fur or feathers for insulation are at risk of external contamination and subsequent dysfunction, which can result in both poor temperature regulation, transfer to young (nursing), and oral exposure to crude hydrocarbons following grooming/preening. However, cetaceans generally have very thick skin (odontocetes epidermal layer is 10–20 times thicker than terrestrial mammals) and deep blubber tissue (hypodermal, below the skin). The main threat to cetaceans and marine mammals that do not suffer direct contact following oil spills may in fact be inhalation of toxic fumes or irritation of mucous membranes (eyes, nose, blowhole). The vapors of volatile hydrocarbons contain a number of harmful toxicants which can cause localized inflammation of lung tissue, and some compounds can accumulate in blood and filtering organs causing hepatoxicity (related to liver) and neurotoxicity (Geraci and St. Aubin 1982). Polycyclic aromatic hydrocarbons (PAHs) are chemical constituents of crude and refined oil and have been found in a number of marine mammal tissues (e.g., Kannan and Perrotta 2008) and are known to be toxic through the production of reactive chemical intermediates during biotransformation. However, there are limited studies examining the effects of these compounds in marine mammals. These compounds are known to have endocrine disrupting effects, although these effects are generally seen in acute exposure scenarios, and there is little information concerning chronic low-level exposure, as one might expect in marine mammals.

There is some evidence to suggest that bottlenose dolphins (*Tursiops truncatus*) are able to detect oil slicks, however reports of avoidance behavior are mixed and may depend on environmental conditions and the nature of the oil product involved. In an experimental oil spill, three female dolphins appeared to avoid surfacing in oil slicks following initial contact, however in observational studies such as those following the Mega Borg oil spill in the Gulf of Mexico in 1990, 9 groups of bottlenose dolphins did not show consistent avoidance of most types of oil slicks (e.g., Smultea and Würsig 1995). The mechanism of chemodetection of oil slicks is unknown, and may be related to visual signaling, echolocation, tactile or olfactory sensing or combinations of these that may be very dependent on the characteristics of the product spilled or released.

Biotoxins

Attributing mass-strandings and die-offs to marine biotoxins is often complicated by other factors such as viruses, bacteria, and other pathogens as well as nutrition and immunological status of individuals and populations. Rather than being pesky confounding variables, however, these factors likely contribute to the overall health status of the animal, and the balance of two or more factors may well tip the scales from health to outright disease. With respect to some mortality events documented in the USA, there is very strong diagnostic evidence that some HAB-related mortality events are now common within certain regions and species, e.g., California sea lions and domoic acid, manatees (*Trichechus manatus latirostris*) and brevetoxin (See **Chapter 13**).

Toxins that are produced by prey species either as defensive mechanisms or as secondary chemical metabolites play a role in the predator–prey relationship. While examples exist of specific adaptations to poisons in predator–prey relationships, it is often difficult to assign a direct physiologic or behavioral adaptation of a species or taxonomic group as a result of exposure to a poisonous prey.

Given that marine mammals have been exposed to biotoxins for millions of years, it is surprising that cetaceans and pinnipeds do not display more resistance to marine biotoxins. Historically, diagnostic methods have limited the ability to identify or detectably measure algal biotoxins in animals, making it difficult to determine the cause of mass strandings and die-offs. However, with increasingly sensitive and precise assays, there is now good evidence to suggest that algal biotoxins are at least partially responsible for previously unexplained unusual mortality events (*UMEs*) (See **Chapter 13** and **Chapter 15**). The unique nature of HABs, including non-species-specific mortality, has led some to speculate about evidence for mass strandings in fossilized remains from the late Miocene epoch (approximately 7 Mya) (Pyenson et al. 2014). At a site in the Atacama Desert of Chile, Pyenson et al. (2014) found four layers of multi-species assemblages including species of Balaenoptera, Phocidae, and an aquatic sloth (*Thalassocnus natans*). The unique body position and proximity of the remains led the authors to conclude that the individuals died rapidly at sea and were promptly washed ashore by the structure and currents of the supratidal flat (Pyenson et al. 2014). While determining cause of death in six- to nine-million-year-old fossils is undoubtedly uncertain, it is worth noting that HABs have likely contributed to the evolution of marine mammals, and may have even played a role in mass extinctions within the current geologic era (541 million years to the present) (Castle and Rodgers 2009).

Domoic acid has caused or contributed to numerous mortality events (see Further Reading, Broadwater, Van Dolah, and Fire 2018) and, as mentioned earlier in the chapter, is often associated with strandings of California sea lions. In 2015, an unprecedented HAB of *Pseudo-nitzschia* occurred on the US West Coast in conjunction with the Northeast Pacific Marine Heatwave known as "the Blob". Extremely high concentrations of domoic acid entered the food web and resulted in its detection in tissues of whales, dolphins, fur seals, and sea lions; 229 cases of domoic acid poisoning were documented at The Marine Mammal Center in Sausalito, California in 2015 (McCabe et al. 2016). While there are several related forms of domoic acid, it appears that adverse effects are caused by a single toxic species – the same found accumulating to high concentrations in shellfish (Iverson et al. 1989). There is no evidence of biotransformation as may occur with other biotoxins. Domoic acid can be detected in urine and feces (i.e. excreted) and has been associated with acute signs of neurotoxicity, including seizures, erratic head movements, and uncoordinated muscular movements in California sea lions. Recent evidence has suggested a host of adverse effects on a variety of tissues, including heart lesions in California sea lions, may explain rapid death following acute exposure (Gulland et al. 2002).

Paralytic shellfish toxins are particularly difficult to measure quickly and precisely as opposed to other marine biotoxins that are a single compound. One of the main PST compounds, saxitoxin, is produced by both phytoplankton as well as several genera of cyanobacteria (see Further Reading, Broadwater, Van Dolah, and Fire 2018). PSTs have been detected in cetaceans and pinnipeds from many ocean basins from the subtropics to the Arctic Ocean. The toxicokinetics and adverse effects of PSTs in marine mammals are poorly understood. However, as noted previously in this chapter (Routes of exposure) PSTs have been implicated in a number of mass strandings and die offs including Mediterranean monk seals (*Monachus monachus*) and southern right whales (*Eubalaena australis*) (for more details, see Further Reading, Broadwater, Van Dolah, and Fire 2018). An unexplained die off of Alaskan sea otters at Kodiak Islands, Alaska in 1987 may have been associated with saxitoxin, after investigators reported finding concentrations of PSTs in blue mussels (*Mytilus edulis*) more than 50 times above the upper threshold for human consumption in Alaska (80 μg/100 g; DeGange and Vacca 1989).

Brevetoxins have been observed in a number of marine mammal species particularly dolphins including bottlenose dolphins, rough-toothed dolphins (*Steno bredanensis*), and Florida manatees (Fernández et al. 2022; see also Further Reading, Broadwater, Van Dolah, and Fire 2018). Observations of brevetoxicosis and other suspected algal toxin-based events in marine mammals demonstrate the potential for temporal and spatial lags in stranding and mortality events (see also, section "Routes of Exposure"). Due to compounding bioaccumulation in prey species and biomagnification across trophic levels, impacts of HAB toxins on marine mammals and other top predators may lag behind observations of harmful algae in the water. Due to the movement of prey species or a lag between prey ingestion, observations of toxicosis and mortality due to HAB toxins may occur in regions that do not have observable HAB events (see Further Reading, Broadwater, Van Dolah, and Fire 2018).

CONCLUSION

Concern for the well-being of marine mammals leads many students, researchers, and public groups to ask, "how are marine mammals exposed to and affected by toxicants?" By virtue of their marine habitat, their place within the food web, and their physiological and biochemical adaptations to life in the sea, marine mammals face some unique challenges.

There are many toxicants that are unique, or more prevalent, in aquatic or marine environments. Some notable examples include products of marine accidents like oil spills, MeHg⁺ which enters the food web via aquatic bacteria that methylate inorganic mercury, and biotoxins that are a product of HABs. At the same time, marine ecosystems are often contaminated by atmospheric and terrestrial toxicants that are deposited or washed into the ocean, some undergoing long-range transport. Some of these are human-derived such as pesticides, others are naturally occurring such as elements.

Human activity and climate change can increase biologically available concentrations of some naturally occurring toxicants. Naturally occurring toxicants like Cd have likely been present for a very long time, allowing for biota to adapt to their presence. Others, like industrial chemicals produced by humans, are more novel and many persist in the environment for a long time.

Most marine toxicants move freely through the food web and marine mammals are most commonly exposed via diet (foraging) and maternal transfer (fetus, pup/calf). Marine mammals are typically long-lived predators with diets that include invertebrates, fish, and in some cases even other marine mammals. As such, they often have higher tissue toxicant concentrations than their prey, through the combined processes of bioaccumulation and biomagnification. These processes may be complicated by a variety of physiological process such as growth dilution (e.g., rapid pup growth) and accumulation and mobilization of blubber stores. As an example, blubber may serve as a reservoir for many fat-soluble toxicants and when it is used as a fuel source during fasting and reproduction, harmful toxicants may be released and even transferred to the fetus – a particularly sensitive life stage.

The effects of toxicants are varied, dependent on specific chemical structure and properties, and the biotransformation, detoxification, and sequestration/elimination capabilities of individual species. Some key targets for adverse effects include the neurological system, kidney, liver, heart, reproductive, and immunological systems. Many of the biochemical tools available to marine mammals are common to mammals in general (i.e., also to terrestrial mammals). But physiological and biochemical adaptations to diving and ocean living may provide marine mammals with scientifically interesting ways to respond to toxicant challenges. Co-evolution with some marine toxicants allows adaptive strategies to arise that may vary by lineages of marine mammals. The reproductive, fasting, and thermoregulatory strategies of different marine mammals can affect accumulation, elimination, and maternal transfer of contaminants as mentioned above. In most marine mammal species, liver detoxification and sequestration of $MeHg^+$ via interaction with Se, abundant in marine prey, reduces circulating $MeHg^+$ concentrations providing some protection to other target systems. The co-abundance of Hg and Se in the marine ecosystem is key to this process. *In utero* growth of lanugo provides an excretory pathway for marine mammal fetuses exposed via the placenta to $MeHg^+$. Increased concentrations of antioxidants, an adaptation to diving and breath-holding, likely protects marine mammals not only from the toxic by-products of diving, but also from other toxicants that cause oxidative damage that may be abundant in the diet.

Marine mammals face many toxicant challenges that are a function of their environment and biology, and they may at times be negatively affected by toxicant exposure. But their physiological and biochemical adaptations to ocean life also provide mechanisms to help them manage or moderate many of these challenges.

REVIEW QUESTIONS

1 What is a toxicant?
2 What is a biotoxin? Give an example.
3 Give an example of a toxicant that may occur naturally and also be produced by human activity (anthropogenic).
4 Give an example of a toxicant that only (or mostly) occurs in a marine or aquatic environment. Give an example of a marine toxicant that has a terrestrial source.
5 Why do long-lived, high trophic level carnivores often have higher concentrations of toxicants than their prey?
6 Give an example of a toxicant that is transferred from mother to pup/calf and explain briefly how the transfer occurs.
7 What are some ways that selenium (Se) can help marine mammals reduce the negative effects of mercury exposure?

8   Why are there sometimes cases where marine mammals show evidence of biotoxin exposure even when there has been no recent local HAB event?

9   Why do PCBs and some other related chemicals accumulate in blubber?

10 How does adaptation to diving and breath-holding help some marine mammals handle the effects of some toxicants?

CRITICAL THINKING

1   How can we reduce the impact of man-made chemicals on marine ecosystems? How can we persuade others to adopt those changes?

2   How do you think climate change might impact marine mammals exposure to marine toxicants?

3   What do you think would be a good way to find out how a toxicant is affecting marine mammal health?

FURTHER READING

AMAP. 2011. *AMAP Assessment 2011: Mercury in the Arctic*. Oslo, Norway: Arctic Monitoring Assessment Programme.

Broadwater, M.H., F.M. Van Dolah, and S.E. Fire. 2018. Vulnerabilities of marine mammals to harmful algal blooms. In *Harmful Algal Blooms*, ed. S.E. Shumway, J.M. Burkholder, and S.L. Morton, 191–222. Hoboken: Wiley-Blackwell.

Dietz, R., C. Sonne, N. Basu, et al. 2013. What are the toxicological effects of mercury in Arctic biota? *Science of the Total Environment* 443: 775–790.

McHuron, E.A., S.H. Peterson, and T.M. O'Hara. 2018. Feeding ecology tools to assess contaminant exposure in coastal mammals. In *Marine Mammal Ecotoxicology*, ed. M.C. Fossi and C. Panti, 39–74. London: Academic Press.

O'Hara, T.M., and L. Hart. 2018. Environmental toxicology. In *CRC Handbook of Marine Mammal Medicine*, ed. F.M.D. Gulland, L.A. Dierauf and K.L. Whitman, 297–318. Boca Raton: CRC Press, Taylor and Francis Group.

GLOSSARY

Abiotic: Nonliving or derived from a nonliving source.

Anaerobic: Living, active, and occurring in the absence of oxygen.

Anthropogenic: Caused or influenced by humans.

Antioxidant: A chemical compound or substance that acts to slow or prevent oxidation of another chemical; may protect cells from the damaging effects of oxidation.

Bioaccumulation: The accumulation of a substance, such as a toxic chemical, in various tissues of a living organism.

Biomagnification: The increasing concentration of a substance, such as a toxic chemical, in the tissues of organisms at successively higher levels in a food chain.

Biotic: Produced or caused by living organisms.

Biosphere: The part of the earth and its atmosphere in which living organisms exist or that is capable of supporting life.

Biotoxin: A toxin produced by a living organism.

Biotransform: To transform something by biochemical means, especially by enzymes.

Congeners: Related chemical substances.

Convergent evolution: The process in which organisms that are not closely related independently evolve similar features.

Endotherm: An organism that generates heat to maintain its body temperature, typically above the temperature of its surroundings; warm-blooded.

Ischemia: A decrease in blood supply to an organ or tissue caused by constriction or obstruction of blood vessels.

Lanugo: Hair grown *in utero*.

Lipophylic: Soluble in lipid, "loves fat".

Nares: Nostrils or nasal passages.

Organohalide: An organic compound that contains one or more halogens (fluorine, chlorine, bromine, or iodine), includes organochlorines that contain chlorine as the halogen.

Organotin: Organic compounds containing tin (Sn) used as marine antifoulants on boats and nets, and other industrial purposes.

Parity: The number of completed pregnancies to a viable gestational age.

Reactive oxygen species (ROS): Highly reactive chemicals formed from O_2 (e.g., peroxides, superoxides, hydroxyl radical, etc.)

Reperfusion: Restoration of blood flow to an organ or tissue that has had its blood supply cut off.

Selenoprotein: Protein containing an active selenium group.

Sentinel species: Sentinel species can signal warnings, at different levels, about the potential impacts on a specific ecosystem.

Toxicant: Any toxic substance. Toxicants can be poisonous and they may be man-made or naturally occurring. By contrast, a toxin is a poison produced naturally by an organism. The different types of toxicants can be found in the air, soil, water, or food.

Toxicosis: Illness due to poisoning.

UME: Unusual mortality event.

REFERENCES

Allen, K., A. Li, D. Luong, et al. 2021. Elephant seal endothelial cells are resistant to oxidative stress. *The FASEB Journal* 35(S1). https://doi.org/10.1096/fasebj.2021.35.S1.01577

Amiard-Triquet, C., and F. Caurant. 1997. Adaptation of the delphinid *Globicephala melas* (Traill, 1809) to cadmium contamination. *Bulletin de la Société Zoologique de France* 122: 127–136.

Beckmen, K.B., J.E. Blake, G.M. Ylitalo, et al. 2003. Organochlorine contaminant exposure and associations with hematological and humoral immune functional assays with dam age as a factor in free-ranging northern fur seal pups (*Callorhinus ursinus*). *Marine Pollution Bulletin* 46(5): 594–606.

Bejarano, A.C., F.M. Gulland, T. Goldstein, et al. 2008. Demographics and spatio-temporal signature of the biotoxin domoic acid in California sea lion (*Zalophus californianus*) stranding records. *Marine Mammal Science* 24(4): 899–912.

Berglund, M., A. Åkesson, P. Bjellerup, et al. 2000. Metal–bone interactions. *Toxicology Letters* 112–113: 219–225.

Bricelj, V.M., L. Connell, K. Konoki, et al. 2005. Sodium channel mutation leading to saxitoxin resistance in clams increases risk of PSP. *Nature* 434(7034): 763–767.

Castellini, J.M., L.D. Rea, C.L. Lieske, et al. 2012. Mercury concentrations in hair from neonatal and juvenile Steller sea lions (*Eumetopias jubatus*): Implications based on age and region in this northern Pacific marine sentinel piscivore. *EcoHealth* 9(3): 267–277.

Castle, J.W., and J.H. Rodgers. 2009. Hypothesis for the role of toxin-producing algae in Phanerozoic mass extinctions based on evidence from the geologic record and modern environments. *Environmental Geosciences* 16: 1–23.

Costa, D.P., B.J.L. Boeuf, A.C. Huntley, et al. 1986. The energetics of lactation in the northern elephant seal, *Mirounga angustirostris*. *Journal of Zoology* 209(1): 21–33.

Das, K., V. Debacker, and J.M. Bouquegneau. 2000. Metallothioneins in marine mammals: A review. *Cellular and Molecular Biology* 46: 283–294.

Das, K., V. Debacker, and J.M. Bouquegneau. 2003. Heavy metals in marine mammals. In *Toxicology of Marine Mammals*, ed. J.G. Voss, G.D. Bossart, M. Fournier, and T.J. O'Shea. 135–136. Boca Raton: CRC Press.

DeGange, A.R., and M.M. Vacca. 1989. Sea otter mortality at Kodiak Island, Alaska, during summer 1987. *Journal of Mammalogy* 70(4): 836–838.

Dehn, L.-A., G.G. Sheffield, E.H. Follmann, et al. 2005. Trace elements in tissues of phocid seals harvested in the Alaskan and Canadian Arctic: Influence of age and feeding ecology. *Canadian Journal of Zoology* 83(5): 726–746.

Dietz, R., J. Nørgaard, and J.C. Hansen. 1998. Have arctic marine mammals adapted to high cadmium levels? *Marine Pollution Bulletin* 36(6): 490–492.

Dietz, R., F. Riget, and P. Johansen. 1996. Lead, cadmium, mercury and selenium in Greenland marine animals. *Science of the Total Environment* 186(1): 67–93.

Doucette, G.J., C.M. Mikulski, K.L. King, et al. 2012. Endangered North Atlantic right whales

(*Eubalaena glacialis*) experience repeated, concurrent exposure to multiple environmental neurotoxins produced by marine algae. *Environmental Research* 112: 67–76.

Fernández, A., E. Sierra, M. Arbelo, et al. 2022. First case of brevetoxicosis linked to rough-toothed dolphin (*Steno bredanensis*) mass-mortality event in eastern Central Atlantic Ocean: A climate change effect? *Frontiers in Marine Science* 9. https://doi.org/10.3389/fmars.2022.834051

Frouin, H., M. Lebeuf, M. Hammill, et al. 2012. Transfer of PBDEs and chlorinated POPs from mother to pup during lactation in harp seals, *Phoca groenlandica*. *Science of the Total Environment* 417–418: 98–107.

Geraci, J.R., D.M. Anderson, R.J. Timperi, et al. 1989. Humpback whales (*Megaptera novaeangliae*) fatally poisoned by dinoflagellate toxin. *Canadian Journal of Fisheries and Aquatic Sciences* 46(11): 1895–1898.

Geraci, J.R., and D.J. St. Aubin. 1982. *Study of the Effects of Oil on Cetaceans*. Washington, D.C.: Bureau and Land Management.. Prepared for U.S. Department of the Interior.

Gobler, C.J. 2020. Climate change and harmful algal blooms: Insights and perspective. *Harmful Algae* 91: 101731.

Goering, P.L., and C.D. Klaassen. 1984. Resistance to cadmium-induced hepatotoxicity in immature rats. *Toxicology and Applied Pharmacology* 74(3): 321–329.

Gulland, E.M.D., M. Haulena, D. Fauquier, et al. 2002. Domoic acid toxicity in Californian sea lions (*Zalophus californianus*): Clinical signs, treatment and survival. *Veterinary Record* 150(15): 475–480.

Hansen, J.C., T.Y. Toribara, and A.G. Muhs. 1989. Trace metals in human and animal hair from the 15th century graves at Qilakitsoq compared with recent samples. *Meddelelser om Grønland, Man and Society* 12: 161–167.

Harley, J., and T.M. O'Hara. 2016. Toxicology and poisons. In *Marine Mammal Physiology: Requisites for Ocean Living*, ed. M.A. Castellini and J.-A.E. Mellish. 309–336. Boca Raton: CRC Press.

Iverson, F., J. Truelove, E. Nera, et al. 1989. Domoic acid poisoning and mussel-associated intoxication: Preliminary investigations into the response of mice and rats to toxic mussel extract. *Food and Chemical Toxicology* 27(6): 377–384.

Jessup, D.A., C.K. Johnson, J. Estes, et al. 2010. Persistent organic pollutants in the blood of free-ranging sea otters (*Enhydra lutris* ssp.) in Alaska and California. *Journal of Wildlife Diseases* 46(4): 1214–1233.

Johansson, C., A.F. Castoldi, N. Onishchenko, et al. 2007. Neurobehavioural and molecular changes induced by methylmercury exposure during development. *Neurotoxicity Research* 11(3): 241–260.

Kannan, K., and E. Perrotta. 2008. Polycyclic aromatic hydrocarbons (PAHs) in livers of California sea otters. *Chemosphere* 71(4): 649–655.

Khan, M.A.K., and F. Wang. 2009. Mercury-selenium compounds and their toxicological significance: Toward a molecular understanding of the mercury-selenium antagonism. *Environmental Toxicology and Chemistry* 28(8): 1567–1577.

Kirkpatrick, B., L.E. Fleming, D. Squicciarini, et al. 2004. Literature review of Florida red tide: Implications for human health effects. *Harmful Algae* 3(2): 99–115.

Lewis, R.J., and M.J. Holmes. 1993. Origin and transfer of toxins involved in ciguatera. *Comparative Biochemistry and Physiology Part C: Pharmacology, Toxicology and Endocrinology* 106(3): 615–628.

Lobanov, A.V., D.E. Fomenko, Y. Zhang, et al. 2007. Evolutionary dynamics of eukaryotic selenoproteomes: Large selenoproteomes may associate with aquatic life and small with terrestrial life. *Genome Biology* 8(9): R198.

McCabe, R.M., B.M. Hickey, R.M. Kudela, et al. 2016. An unprecedented coastwide toxic algal bloom linked to anomalous ocean conditions. *Geophysical Research Letters* 43(19): 10,366–10,376.

Nakazawa, E., T. Ikemoto, A. Hokura, et al. 2011. The presence of mercury selenide in various tissues of the striped dolphin: Evidence from µ-XRF-XRD and XAFS analyses. *Metallomics* 3: 719–725.

Polizzi, P.S., M.B. Romero, L.N. Chiodi Boudet, et al. 2014. Metallothioneins pattern during ontogeny of coastal dolphin, *Pontoporia blainvillei*, from Argentina. *Marine Pollution Bulletin* 80(1): 275–281.

Preece, E.P., B.C. Moore, and F.J. Hardy. 2015. Transfer of microcystin from freshwater lakes to Puget Sound, WA and toxin accumulation in marine mussels (*Mytilus trossulus*). *Ecotoxicology and Environmental Safety* 122: 98–105.

Pugh, L.G. 1959. Carbon monoxide content of the blood and other observations on Weddell seals. *Nature* 183: 74.

Puls, R. 1994. *Mineral Levels in Animal Health.* 2nd ed. Clearbrook: Sherpa International.

Pyenson, N.D., C.S. Gutstein, J.F. Parham, et al. 2014. Repeated mass strandings of Miocene marine mammals from Atacama region of Chile point to sudden death at sea. *Proceedings of the Royal Society B: Biological Sciences* 281(1781): 20133316.

Rea, L.D., J.M. Castellini, L. Correa, et al.. 2013. Maternal Steller sea lion diets elevate fetal mercury concentrations in an area of population decline. *Science of the Total Environment* 454–455(0): 277–282.

Reeb, D., P.B. Best, and S.H. Kidson. 2007. Structure of the integument of southern right whales, *Eubalaena australis. The Anatomical Record* 290(6): 596–613.

Riget, F., and R. Dietz. 2000. Temporal trends of cadmium and mercury in Greenland marine biota. *Science of the Total Environment* 245(1): 49–60.

Roesijadi, G. 1992. Metallothioneins in metal regulation and toxicity in aquatic animals. *Aquatic Toxicology* 22(2): 81–113.

Rosa, C., J.E. Blake, G.R. Bratton, et al. 2008. Heavy metal and mineral concentrations and their relationship to histopathological findings in the bowhead whale (*Balaena mysticetus*). *Science of the Total Environment* 399(1): 165–178.

Ross, P.S., S.J. Jeffries, M.B. Yunker, et al. 2004. Harbor seals (*Phoca vitulina*) in British Columbia, Canada, and Washington State, USA, reveal a combination of local and global polychlorinated biphenyl, dioxin, and furan signals. *Environmental Toxicology and Chemistry* 23(1): 157–165.

Smultea, M.A., and B. Würsig. 1995. Behavioral reactions of bottlenose dolphins to the Mega Borg oil spill, Gulf of Mexico 1990. *Aquatic Mammals* 21(3): 171–181.

Teigen, Sverre W., R.A. Andersen, H.L. Daae, et al., 1999. Heavy metal content in liver and kidneys of grey seals (*Halichoerus grypus*) in various life stages correlated with metallothionein levels: Some metal-binding characteristics of this protein. *Environmental Toxicology and Chemistry* 18(10): 2364–2369.

Tift, M.S., P.J. Ponganis, and D.E. Crocker. 2014. Elevated carboxyhemoglobin in a marine mammal, the northern elephant seal. *Journal of Experimental Biology* 217(10): 1752–1757.

Tonk, E.C.M., D.M.G. de Groot, A.H. Penninks, et al. 2010. Developmental immunotoxicity of methylmercury: The relative sensitivity of developmental and immune parameters. *Toxicological Sciences* 117(2): 325–335.

Vázquez-Medina, J.P., T. Zenteno-Savín, and R. Elsner. 2006. Antioxidant enzymes in ringed seal tissues: Potential protection against dive-associated ischemia/reperfusion. *Comparative Biochemistry and Physiology Part C: Toxicology & Pharmacology* 142(3): 198–204.

Wagemann, R., R. Hunt, and J.R. Klaverkamp. 1984. Subcellular distribution of heavy metals in liver and kidney of a narwhal whale (*Monodon monoceros*): An evaluation for the presence of metallothionein. *Comparative Biochemistry and Physiology C* 78: 301–307.

WHO. 1992. Cadmium. *Environmental Health Criteria* 134.

Wilhelm Filho, D., F. Sell, L. Ribeiro, et al. 2002. Comparison between the antioxidant status of terrestrial and diving mammals. *Comparative Biochemistry and Physiology Part A: Molecular & Integrative Physiology* 133(3): 885–892.

Woshner, V., T. O'Hara, G. Bratton, et al. 2001. Concentrations and interactions of selected essential and non-essential elements in bowhead and beluga whales of arctic Alaska. *Journal of Wildlife Diseases* 37(4): 693–710.

Ylitalo, G.M., C.O. Matkin, J. Buzitis, et al. 2001. Influence of life-history parameters on organochlorine concentrations in free-ranging killer whales (*Orcinus orca*) from Prince William Sound, AK. *Science of the Total Environment* 281(1): 183–203.

Michael Castellini

University of Alaska, Fairbanks, AK

Michael Castellini

Mike earned his Ph.D. in 1981 in marine biology on the diving and medical physiology of marine mammals. Recently retired from the University of Alaska, Fairbanks, his graduate students worked worldwide on marine mammal biology. He keeps in touch with students across North America and around the world about the interactions of humans and marine mammals. From fishery issues to climate change, and policy to public awareness, he has heard thousands of questions about these issues. Sometimes, he even has answers.

Driving Question: Do human activities influence marine mammal physiology?

INTRODUCTION

From climate change concerns to pollutants, disease, noise and more, most chapters of this book identify some way in which human activities influence marine mammal physiology either directly or through indirect impacts on their ocean environment. Some species are endangered from direct fishery conflicts (vaquita; *Phocoena sinus*) and ship strikes (right whales sp.), while others face habitat loss (manatees; *Trichechus* sp.), and some suffer acutely from oil spills (sea otters; *Enhydra lutris*), or noise contamination (beaked whales; Family ziphidae). The scale ranges from global concerns of climate change to very specific regional situations such as busy shipping ports. Human interactions have also changed over time, as with the reduction in the hunting of large whales in the Southern Ocean, whereas new concerns have arisen such as plastics in the ocean. The focus of this book is marine mammal physiology and it does not deal specifically with political and economic issues such as government regulatory policies or shipping practices. However, many human activities affect marine mammals at the physiological and biochemical levels. For example, ship strikes can injure a marine mammal and result in severe medical, physiological, and metabolic issues, or can cause immediate death. Contaminants may cause a range of biochemical, immune, or endocrine reactions. Loud underwater noises may result in deafness or impaired acoustical sensory responses. This chapter focuses on some of the main known interactions and what we can do to try to reduce them.

When talking about this subject, the most common question from students is *What is the biggest threat to marine mammals?* This is inevitably followed by the hopeful *What can I do to help?*

DIRECT INTERACTIONS

Pollution

The impacts of man-made pollution on the environment are global and not limited to the oceans, though the oceans represent 95% of the **biosphere** on earth. Broadly speaking,

DOI: 10.1201/9781003297468-15

pollutants can affect marine mammals through direct injury, disruptions in physiological processes or food web interactions. **Chapters 13 and 14** cover these topics in detail.

Ocean trash

Direct injuries caused by ocean or beach trash are the clearest to observe and document. For example, marine mammals may eat plastic bags or become entangled in plastic rings, discarded trash, and old fishing gear (**Figure 15.1**). Social campaigns describing the impact of pollution on marine mammals (and birds and turtles) often use these disturbing and graphic images of discarded human-created products that directly kill or injure marine species. Efforts to clean up beaches, reduce trash, and prohibit ocean dumping are popular worldwide. Are these methods effective? To some degree, yes, especially for focused regional problem areas. Tons upon tons of marine debris have been removed

from beaches and there are new and innovative methods such as open ocean autonomous devices to sweep up floating trash. However, the sheer volume of the problem is the challenge. Humans are simply overwhelming our ability to deal with the vast amounts of trash we create. This is one area where individual actions make a tangible difference. Reduce, Reuse, and Recycle your trash. Cut up plastic soda can rings, pick up trash on your local beach, and participate in organized river, beach, and estuary cleanups. Support ordinances to reduce single-use plastic containers. Most of all, be aware of the problem and talk to others about it.

Micro- and nanoplastics

A new and evolving threat to the planet is the problem of **micro- and nanoplastics**. The process of creating plastic products and their eventual breakdown results in extremely small particles. Microplastics are less than 5 mm in

Figure 15.1 A Hawaiian monk seal resting in a large derelict fishing net at Southeast Island, Manawai (Pearl and Hermes), Hawaii. (Photo credit: NOAA Fisheries/Richard Chen (Permit #22677) NOAA Fisheries. Accessed August, 2022, https://www.fisheries.noaa.gov/feature-story/success-2021-mission-clean-marine-debris)

diameter and nanoplastics are less than 100 nm in diameter (Wang et al. 2021). Many of these end up in both fresh and salt water and their impacts on biochemistry and physiology of all animals, including marine mammals, are not yet known. At the macro level, plastics are inert and are used in many food and health products from packaging to medical devices such as syringes, catheters, and surgical implants. But what about micro- and nanoplastics? Are they inert at that level, or can they influence chemical reactions and processes? Can they impact enzyme reactions; do they influence the absorption of food items across the gut, or in filtering blood through the liver? The issue of microplastics will be with us until we reduce our reliance on plastic products. As with larger ocean trash issues, reading, learning, and becoming aware is a critical path forward (for more information see Mofijur et al. (2021)).

Oil spills

A great deal of the world's production of oil is moved by oceanic tanker vessels. In addition to the problem of increased ship noise, transport of oil by sea also carries a real risk of a spill. Fortunately, significant vessel oil spills are rare, and modern improvements in hull design, navigation, safety, personnel training, and oil spill response have been important for this field. When large spills do occur, they can be devastating to many marine species, including marine mammals. The Exxon Valdez Oil Spill (EVOS) released 11 million gallons of oil into Prince William Sound, Alaska in 1989. The spill directly killed countless animals and had long lasting impacts on multiple species. Local populations of sea otters were heavily damaged from the spill because the oil compromised the thermal insulation of their fur (see **Chapter 5**). What was not anticipated was the damage created by breathing heavy oil fumes at the ocean surface. Breathing these fumes killed many otters and may have been the driving force behind the disappearance of many resident orca. We learned a great deal about cleaning sea otters during the EVOS recovery to restore the thermal insulation of their fur coats (Williams and Davis 1990). Standing sea otter rescue teams are now trained in current methods for any future spills.

As shipping routes in the Arctic Ocean open due to climate change, there will be an increase in large vessel traffic. Tour boats are already traversing the Northwest Passage and Russian commercial ships heavily traffic the Northern Sea route. In this region of the world, the weather is extreme, access by clean-up crews is almost impossible because there are so few ports, there are few if any airports, and the interaction of oil in ice-laden waters is almost unknown. The largest accidental oil spill in history was from the 2010 DeepWater Horizon well in the Gulf of Mexico. This event leaked at least 134 million gallons of oil and killed up to a million shorebirds, up to 20% of the local sea turtle population, and up to 50% of the local bottlenose dolphin population. These impacts occurred despite the spill occurring in an accessible, warm water environment. If a spill occurs in the far north in potentially ice-laden seas, it will be almost impossible to clean up or help impacted marine mammals. Efforts to improve engineering, navigation, ship design, international agreements, and safety are essential for oil spill mitigation and prevention.

Chemical pollutants

Except for the cases of acute chemical impact (oil spills, for example), determining how marine pollutants influence marine mammal health is difficult. This work usually involves complex chemistry, medical and veterinary training. **Chapter 14** discusses how dissolved metals, persistent organic pollutants, and other chemicals in the ocean can negatively impact the health of all marine species. Not only are the analyses of these chemicals in the environment and animals challenging and expensive, the interpretation of the impact on biochemistry, reproduction, or survival is not always clear (**Chapter 13**).

Second-level interactions from pollutants are also important. For example, river run-off of fertilizers can alter microorganism populations in bays and estuaries and induce toxic algal blooms, which in turn can harm local marine mammal populations. As the ocean warms through climate change, the number of toxic algae blooms seem to be increasing, and these have caused significant marine mammal mortalities (Elorriaga-Verplancken et al. 2022). Furthermore, paralytic shellfish poisoning (PSP) is a critical health issue for humans in the United States Pacific Northwest. Shellfish may accumulate a range of microorganism toxins under certain environmental conditions and can be lethal if eaten by humans. Some of these poisons are also dangerous or lethal to marine mammals, such as domoic acid (**Chapter 13 and Chapter 14**). There are many cases of California sea lions (*Zalophus californianus*) suffering or dying from algal poisons (Cook et al. 2016) and sea otters may have chronic levels of these toxins given their shellfish diet (Burek Huntington et al. 2021). As with most forensic studies of poisoning, the further away the chemical is from a direct impact, the more difficult it becomes to trace the pathway of disease or injury.

Noise

Chapter 6 discusses how the use and production of sound is essential for many marine mammals. There is a growing concern that constant background noise can impact marine mammals at many levels of behavior and physiology (Marotte et al. 2022).

At the most basic level, sounds made by marine mammals can be masked by the volume of the background noise. This problem exists in both small seas and bays and at an ocean basin scale. How do we test and quantify this problem? How do we know that animals are avoiding noisy areas, or altering their behavior in response to noise? Through wide-ranging field studies, research teams have been measuring responses of marine mammal populations as they move through various sound fields. This is not easily done and requires past knowledge of marine mammal use of the area, and responsible tagging of animals to determine diving and distribution patterns and quantification of the noise field. Studies that produce artificial sounds through underwater speaker arrays are not the same as ship sound coming from all directions.

The world's global economy is extremely dependent on ship traffic and it is unlikely that shipping across the ocean basins will decrease with time. The goal should be to reduce the amount of noise per vessel by improving hull, engine, and propulsion designs and by altering the routes of vessels in and out of ports to avoid known marine mammal habitat. This will require innovative work by engineers, sound technicians, and material specialists.

There is also direct damage to hearing by extremely loud short-duration underwater noise. Examples include underwater blasting, pile driving, or the use of high-power SONAR by navy vessels. In these situations, there can be direct medical, physical, and physiological damage from human-produced sounds. There have been cases where beaked whales have mass stranded soon after naval exercises that included SONAR deployments (Simonis et al. 2020). Medical examinations (necropsy) of the beached whales suggested damage in their acoustic systems was caused by the SONAR. How do you prove that SONAR causes tissue damage in the beaked whales? Direct exposure of marine mammals to SONAR is not ethical and such experiments would never be conducted. The anatomical findings of tissue damage are consistent with known loud noise damage on other mammal species and the evidence is supportive of the possibility. The US Navy recognizes the issue and has protocols to avoid marine mammals when using high power SONAR in peacetime scenarios. There are also data that suggest beaked whales alter their diving patterns when near SONAR, which may negatively affect their physiological dive responses (**Chapter 3**). See Jacobson et al. (2022) for more background on SONAR and beaked whales.

Prey competition with humans

Humans consume a vast amount of protein from the ocean. This can create direct competition for food resources with marine mammals in certain areas. Harp seals (*Pagophilus groenlandicus*) around Newfoundland and Labrador in Canada consume more fish than the entire regional fishery. Harbor seals (*Phoca vitulina*) and California sea lions consume salmon near the mouths of major rivers on the west coast of the Americas. The marine mammals are protected and fishing boats are allowed only minimal deterrence methods (e.g., underwater firecrackers). Sperm whales (*Physeter macrocephalus*) in Alaska take black cod directly off fishing lines. Sea otters take massive amounts of shellfish and invertebrates from regional bays. These close and direct interactions between humans and marine mammals can result in **by-catch**, entanglement and ship-strike issues. In situations like the Steller sea lions (*Eumetopias jubatus*) that feed heavily on Alaska walleye pollock (*Gadus chalcogrammus*), the fishery is heavily managed and often subject to political and economic pressure.

By-catch

Capture or entanglement in fishing gear (by-catch) kills hundreds of thousands of marine mammal worldwide every year. This is direct mortality rather than an indirect impact to physiology or biochemistry. In the 1970s, there was a major focus on dolphin by-catch in the open ocean tuna net fishery including boycotts and a great number of legislative initiatives in the United States. In 1993, it was estimated that some populations of dolphins had been reduced by more than 80% in the Eastern Tropical Pacific specifically due to by-catch in the tuna fleet. The number of dolphins taken in the US fishery has been reduced by almost 99% thanks to improvements in fishing methods and management. However, there continues to be dolphin by-catch from foreign vessels that do not use modern fishery methods and the US bans tuna imports from these vessels. The tuna fishery provides a great deal of food for humans, and hopefully even newer and improved methods will reduce the remaining dolphin by-catch. See Ballance et al. (2021) for a history of these issues.

There are other types of fisheries where marine mammals and fleets interact, and in some cases, the marine mammals **depredate** (take fish) directly from the catch. This leads to direct conflict between the fishing boats and the marine mammals. Possible solutions to reduce these interactions can involve behavioral and physiological options. For example, sperm whales have learned how boat engines make certain noises during active fishing. As discussed in Chapters **6** and **7**, the acoustic sensitivity and accuracy of many cetaceans is highly developed and sperm whales can easily hear these boats and move toward them to take parts of the catch. In Southeast Alaska, research teams and boat captains have come together to develop tracking systems where scientists localize the whales and let the boat captains know where they can fish undisturbed (Straley et al. 2015). Trying to confuse sperm whales with gear modifications (O'Connell et al. 2015) and annoying decoy sounds (Wild et al. 2017) have not been as successful. The whales are difficult to deter because the draw of free food is strong. Similar acoustic decoys have been tested to keep sea otters out of certain coastal areas. Sea otters were not responsive to loud or obnoxious noises, but seemed repelled by the recorded sounds of orca (Davis, Awbrey, and Williams 1987). Trying to fool marine mammals with synthetic or recorded natural sounds to keep them away from fishing fleets is probably not a long-term solution. The reduction of by-catch needs neuroscientists, animal behavior teams, economists, fishing experts, marine technicians and public awareness.

Ship strikes and entanglements

In areas where large ships and marine mammals are in close proximity, direct ship strikes are of great concern (**Figure 15.2**). Well-known examples are off the northeast coast of the United States where this is an issue with North Atlantic

Figure 15.2 A humpback whale calf in Hawaii with an injury from a ship strike. (Photo credit: Courtesy of Ed Lyman, NOAA, under Marine Mammal Health and Stranding Response Program (MMHSRP). Permit #923–1489)

right whales (*Eubalaena glacialis*). From 2017 to 2021, at least 34 right whales have been killed and 15 seriously injured by either ship strikes or entanglement (NMFS 2022). Less than 400 right whales remain, and the loss of every single animal is of consequence. Efforts to minimize the strikes include changing the ship routes and vessel speeds, setting up interactive data with the ships about whale locations in critical areas and even a Whale Alert smartphone application. Decoys or obnoxious sounds to redirect the whales are not long-term solutions. The whales are not likely to move to different areas and the shipping fleet is not going to stop. In the long term, the impacts will likely only be achieved by modernizing the routes, altering ship schedules, and updating avoidance processes by the large ships. Even in more remote waters ship strikes still occur, such as Icy Strait, Alaska, which has many cruise ships, fishing boats, pleasure crafts, and many marine mammals (**Figure 15.3**). Identifying animals,

following their behaviors, and understanding how they sense an oncoming ship is critical to solving this problem, and is a rich field for future students.

When there is fishing gear in the water near populations of marine mammals, there is also the possible entanglement of the animals in that gear. By working with fishing groups and engineers off the coast of New England, **NOAA-NMFS** has provided potential methods for protecting whales including weak links in net lines, releasable crab pot lines, reducing vertical lines, modifying fishing areas, and identification marks on nets to understand where the animals might be picking up the nets/line (NMFS 2022). Public and private stakeholders, fishery experts, boat captains, engineers, and whale behavior experts are working together to minimize the risk to whales.

Because ocean fishing will remain a vital food security and economic driver around the world, the ultimate solutions to conflicts with marine mammals will involve a balance of conservation,

Figure 15.3 This seven-year-old humpback whale was first documented as a calf in Glacier Bay National Park in 2011, and first documented with these propeller injuries at age two. (Photo credit: Courtesy of Christine Gabriele, National Park Service NMFS ESA/MMPA Permit No. 21059. Whale SE Alaska Catalog #2474 Location: Ice Strait, Alaska)

economic, and political approaches (Moore 2021). To resolve human **food security**, it is reasonable to predict that we will increase the amount of food taken from the seas and direct interactions with marine mammals will rise. This topic may be regional and species specific, but can be highly charged and contentious when human food, jobs, and economics are threatened by fishery closures because of potential or actual marine mammal interactions. In the United States, the National Marine Fisheries Service is charged with both enhancing fisheries and protecting marine mammal populations. These missions can be in conflict with one another and the legal components of conducting both can be difficult. Advances in this field will require the efforts of not only physiologists, marine mammal behavior and diet specialists, but also policy analysts, economists, fishery experts, and human behavior professionals.

INDIRECT INFLUENCES

Climate change

Human-driven climate change is a critical threat to the planet that is altering weather and storm patterns, oceanic water temperature, sea level rise, droughts, fire risk, and more. If we focus on the oceanic habitat of marine mammals, the impact however does not come directly from an increase in water temperature. **Chapter 5** explained how marine mammals could adjust their physiology to a few degrees of water temperature differences, on both hourly and seasonal time scales. By contrast, most of their prey are marine **ectotherms** (cold-blooded) animals and a few degrees of water temperature change can be fatal to these prey species. For these reasons, the most significant threat of climate change for marine mammals comes from shifts in prey distribution and changes in habitat.

Shifts in prey distribution

Many of the chapters in this book discuss the feeding habits, swimming patterns, and metabolic needs of marine mammals as they hunt for prey. Because almost all of the food of marine mammals are ectothermic (e.g., squid, fish, krill and more), the distribution and health of the prey themselves are vital to the health and survival of marine mammals. For example, we know that as the North Pacific Ocean warms, the fish and plankton that need cooler water are either moving deeper in the water column or extending their range further north (Rooper et al. 2021). In both cases, the energy expended by a marine mammal to catch each prey item increases. Eventually, it will become too expensive for the seal or whale to catch that prey, and they will either **prey switch** or abandon that food source. In the extreme cases of food shortages, such as El Niño events off California, the number of beached and starving sea lions increases significantly because their prey moved or disappeared (Lawler et al. 2021).

Sometimes marine mammals do not strand on beaches, but instead simply disappear. For example, the cascading pattern over the past 30 years of disappearing populations of otters, harbor seals, and Steller sea lions in the North Pacific has been the focus of massive field and laboratory studies trying to determine feeding patterns, prey distribution, medical health, pollutant levels, and other possible factors for their sequential declines. Major research directions asking "Is it Food?" have looked at animal health, fasting and starving biochemistry, thermoregulation abilities, changes in hunting by orca and sharks, changes in the abundance or quality of prey, and a range of other topics aimed at determining if the problem is related to food. As with most complex food web scenarios, there is no single suspect, the answers are neither simple nor straightforward and relating these to climate change is even more difficult (Fritz et al. 2019).

Change in habitat

Climate change can alter the physical habitat of marine mammals. The clearest examples are Polar animals that rely on annual sea ice for breeding, raising young, hunting, or taking refuge. Sea ice cover and thickness in the Arctic Ocean have been declining for at least 40 years (**Figure 15.4**). As discussed in several earlier chapters, the minimum sea ice level which occurs in October every year is now half of what it was in 1979. Polar bears (*Ursus maritimus*) depend on this ice surface for hunting and raising their cubs. Ringed seals (*Pusa hispida*) dig caves in the snow and ice for pup rearing. Harp and gray seals depend on sea ice for nursing their pups. Walruses (*Odobenus rosmarus*) float and hunt along with the ice pack to support their calves. In the Antarctic, Weddell (*Leptonychotes weddellii*), crabeater (*Lobodon carcinophaga*), and Ross seals (*Ommatophoca rossii*) depend on the sea ice to breed and raise pups. For whales that move through the broken pack ice, or feed along the open ice leads, its presence or absence determines their migration patterns and food resources. Leopard seals (*Hydrurga leptonyx*) hunt penguins that are standing on the ice edges, and orca move through ice leads searching for prey. In **Chapter 5**, we describe these animals as **pagophilic**. The term is derived from the Ancient Greek *pagos* meaning "sea ice," and philos meaning "loving."

If the sea ice continues to disappear, what will happen to these pagophilic marine mammals? At what point is there not enough ice to support the polar bear population? Models suggest bears may move to the few **sea ice refugia** sites in the various archipelagos of Canada and Greenland in the next few decades. These refugia may not have enough space or food to support the relocated population. What happens if the walruses move to the same sea ice refugia? Walruses and polar bears being confined to same geographic area is a problem because polar bears take walruses as prey. Polar bears have been seen moving onto land to find remaining food sources and are a serious threat to land-breeding sea bird populations. Unlike their omnivorous relatives the brown bears, polar bears are not adapted to eat vegetation (**Chapter 10**) and eat a high-fat diet consisting mostly of ice seals.

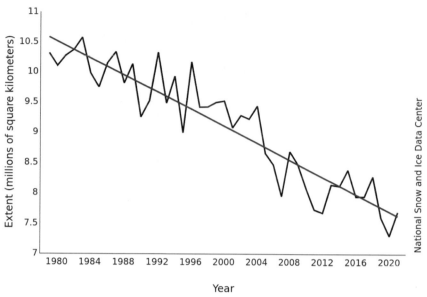

**Average Monthly Arctic Sea Ice Extent
July 1979 - 2021**

National Snow and Ice Data Center

Figure 15.4 Arctic Ocean minimum sea ice cover in September–October since 1979. This is the time of year when the sea ice is at its lowest after the summer. Note the continued decline over time. Current ice cover values in 2021–2022 are almost half of the level seen in 1979. (Image from the National Snow Ice and Data Center)

Several Arctic pagophilic marine mammal species (polar bears, bearded; *Erignathus barbatus*, and Arctic ringed seals) have been declared Threatened under the US Endangered Species Act because their ice habitat is disappearing (NMFS 2022). Solutions to mitigate the effects of climate change will need to come from a range of fields including, but not limited to science, because this is a worldwide geopolitical problem. For further reading on marine mammals and climate change, see Gulland et al. (2022).

CONCLUSION

Humans and marine mammals live in a shared space with increasing overlap. There are numerous specialties and opportunities for traditional biological and veterinary research identified in the chapters of this book. Beyond work in sciences, make yourself aware of local, national, and global needs, policies, and economics. Advocacy for protected spaces takes many forms and can start with personal education and communicating with peers.

At an individual level, there are tangible steps that can be taken, such as:

1 Reduce, reuse, and recycle your trash.
2 Educate yourself on local, national, and global scale.
3 Communicate with your peers.
4 Use your training to further advocate for cleaner habits; limit the effects of climate change, and understanding of our interactions and their consequences with marine mammals.

REVIEW QUESTIONS

1 How does climate change affect a sea lion that lives off the western coast of Mexico?
2 How does an oil spill impact a sea otter, and how is that different from its impact on a dolphin?

3   What is the cause of most background noise in the ocean?

4   How can cleaning up a beach help a marine mammal?

5   What are microplastics and where do they come from?

6   What is a marine protected area and name one that is of interest to you. Why?

7   Why is an oil spill in the Arctic Ocean such a danger to marine life?

8   Where is the Northwest Passage and why is it important?

9   Why are some marine turtles in danger and how does that relate to marine mammals?

10 Why does decreasing sea-ice create a problem for pagophilic species of marine mammals?

CRITICAL THINKING

1   What is the most critical human-marine mammal issue that you have heard about or seen in the last year? Why do you consider it critical?

2   If climate change is so dangerous, why do you think it is so hard to find solutions?

3   I have been asked why we should care about polar bears. "They are only one species and species go extinct all the time." "Survival of the fittest", etc. Given what you have learned about food webs, the complex interactions between species, their environment and human impacts, what is your answer about why we should care?

FURTHER READING

Dedden, A.V., and T.L. Rogers. 2022. Stable isotope oscillations in whale baleen are linked to climate cycles, which may reflect changes in feeding for humpback and southern right whales in the Southern Hemisphere. *Frontiers in Marine Science* 9. doi.org/10.3389/fmars.2022.832075

Hidalgo, M., V. Bartolino, M. Coll, et al. 2022. 'Adaptation science' is needed to inform the sustainable management of the world's oceans in the face of climate change. *ICES Journal of Marine Science* 79(2): 457–462.

Lincoln, S., B. Andrews, S.N. Birchenough, et al. 2022. Marine litter and climate change: Inextricably connected threats to the world's oceans. *Science of the Total Environment* 837: 155709.

Orgeret, F., A. Thiebault, K.M. Kovacs, et al. 2022. Climate change impacts on seabirds and marine mammals: The importance of study duration, thermal tolerance and generation time. *Ecology Letters* 25(1): 218–239.

Roberts, J.O., and H.R. Hendriks. 2022. Potential climate change effects on New Zealand marine mammals: A review. Department of Conservation. DOC Research and Development Series, 366: 47 pp. Wellington, New Zealand.

GLOSSARY

Biosphere: The "livable" space on earth, including land, lakes, air, and ocean. The oceans make up 95% of the earth's biosphere.

By-catch: Accidental catch of species not targeted in fisheries. For example, the accidental catch of dolphins in tuna nets.

Depredation: The process of marine mammals taking caught fish on lines or in nets.

Ectotherms: "Cold-blooded" animals like fish and squid that do not control their body temperature. Their body temperature is that of the surrounding water.

Food security: Reliable access to a sufficient quantity of affordable, nutritious food.

Microplastics: Small plastic particles from the production or breakdown of larger plastics. Usually less than 5 mm in diameter.

NOAA-NMFS: The US National Oceanic and Atmospheric Administration and one of its divisions, the National Marine Fisheries Service.

Nanoplastics: Same as microplastics above, but much smaller. Less than 100 nm in diameter.

Pagophilic: "Ice loving" animals that depend on sea-ice for survival.

Prey switching: Changing to a different prey when the preferred prey is no longer available.

UME: Under the Marine Mammal Protection Act, an unusual mortality event (UME) is defined as "a stranding that is unexpected; involves a significant die-off of any marine mammal population; and demands immediate response".

REFERENCES

Ballance, L.T., T. Gerrodette, C. Lennert-Cody, et al.. 2021. A history of the tuna-dolphin problem: Successes, failures, and lessons learned. *Frontiers in Marine Science* 8. doi.org/10.3389/fmars.2021.754755

Burek Huntington, K.A., V.A. Gill, A.M. Berrian, et al. 2021. Causes of mortality of northern sea otters (*Enhydra lutris kenyoni*) in Alaska from 2002 to 2012. *Frontiers in Marine Science* 8: 105.

Cook, P.F., C. Reichmuth, A. Rouse, et al. 2016. Natural exposure to domoic acid causes behavioral perseveration in wild sea lions: Neural underpinnings and diagnostic application. *Neurotoxicology and Teratology* 57: 95–105.

Davis, R., F. Awbrey, and T. Williams. 1987. Using sounds to control the movements of sea otters. *The Journal of the Acoustical Society of America* 82(S1): S99.

Elorriaga-Verplancken, F.R., C.J. Hernandez-Camacho, L. Alvarez-Santamaria, et al. 2022. Largest mortality event to date of California sea lions in Mexico might be linked to a harmful algal bloom. *Aquatic Mammals* 48(1): 59–67.

Fritz, L., B. Brost, E. Laman, et al. 2019. A re-examination of the relationship between Steller sea lion (*Eumetopias jubatus*) diet and population trend using data from the Aleutian Islands. *Canadian Journal of Zoology* 97(12): 1137–1155.

Gulland, F.M., J. Baker, M. Howe, et al. 2022. A review of climate change effects on marine mammals in United States waters: Past predictions, observed impacts, current research and conservation imperatives. *Climate Change Ecology* 3: 100054.

Jacobson, E.K., E.E. Henderson, D.L. Miller, et al. 2022. Quantifying the response of Blainville's beaked whales to US naval sonar exercises in Hawaii. *Marine Mammal Science* 38(4): 1549–1565.

Lawler, D.F., B. Tangredi, K. Matassa, et al. 2021. Proximate aspects of starvation-related morbidity and mortality among young California sea lions (*Zalophus californianus*). *Journal of Veterinary Anatomy* 14(2): 39–56.

Marotte, E., A.J. Wright, H. Breeze, et al. 2022. Recommended metrics for quantifying underwater noise impacts on North Atlantic right whales. *Marine Pollution Bulletin* 175: 113361.

Mofijur, M., S. Ahmed, S.A. Rahman, et al. 2021. Source, distribution and emerging threat of micro- and nanoplastics to marine organism and human health: Socio-economic impact and management strategies. *Environmental Research* 195: 110857.

Moore, M.J. 2021. *We are all whalers: The plight of whales and our responsibility*, Chicago: University of Chicago Press.

NMFS. 2022. *Recovering threatened and endangered species, FY 2019–2020 Report to Congress*. National Marine Fisheries Service (Silver Spring, MD).

O'Connell, V., J. Straley, J. Liddle, et al. 2015. Testing a passive deterrent on longlines to reduce sperm whale depredation in the Gulf of Alaska. *ICES Journal of Marine Science* 72(5): 1667–1672.

Rooper, C.N., I. Ortiz, A.J. Hermann, et al. 2021. Predicted shifts of groundfish distribution in the Eastern Bering Sea under climate change, with implications for fish populations and fisheries management. *ICES Journal of Marine Science* 78(1): 220–234.

Simonis, A.E., R.L. Brownell Jr., B.J. Thayre, et al. 2020. Co-occurrence of beaked whale strandings and naval sonar in the Mariana Islands, Western Pacific. *Proceedings of the Royal Society B* 287(1921): 20200070.

Straley, J., V. O'Connell, J. Liddle, et al. 2015. Southeast Alaska Sperm Whale Avoidance Project (SEASWAP): A successful collaboration among scientists and industry to study depredation in

Alaskan waters. *ICES Journal of Marine Science* 72(5): 1598–1609.

Wang, X., N. Bolan, D.C. Tsang, et al. 2021. A review of microplastics aggregation in aquatic environment: Influence factors, analytical methods, and environmental implications. *Journal of Hazardous Materials* 402: 123496.

Wild, L., A. Thode, J. Straley, et al. 2017. Field trials of an acoustic decoy to attract sperm whales away from commercial longline fishing vessels in western Gulf of Alaska. *Fisheries Research* 196: 141–150.

Williams, T.M., and R.W. Davis. 1990. *Sea otter rehabilitation program: 1989 Exxon valdez oil spill*, San Diego: International Wildlife Research.

Michael Castellini

University of Alaska, Fairbanks, AK

Michael Castellini

Mike earned his Ph.D. in 1981 in marine biology on the diving and medical physiology of marine mammals. He recently retired from the University of Alaska, Fairbanks where his students worked worldwide on marine mammal biology. He has stayed up all night working with sleeping seal pups, collected field samples from hundreds of pinnipeds on beaches from the Arctic to the Antarctic, and endured long expeditions at sea on research vessels. From aged whales to measurements of diving biochemistry, he has learned a lot about marine mammals, but still has many questions about the mysteries of this interesting group of mammals.

Driving Question: What are new and exciting fields of study?

INTRODUCTION

This book investigates what adaptations make up the design of a marine mammal? How does it swim, breathe, drink, find food, dive, or adapt to climate change? The chapters discuss those answers in detail, but there are still many unknowns about the physiology and biochemistry of this large group of mammals. Some of these mysteries include sleep, aging, stress levels in wild populations, and development of pups

and calves. This chapter introduces a broad range of topics where future research may lead.

Stress hormones

Because stress creates a suite of medical, physiological, and biochemical problems in mammals, measuring it is important for determining population health and for diagnoses in veterinary medicine. In addition to classical physiological measurements such as heart or breathing rates, the measurement of stress-related circulatory or tissue hormones such as **cortisol** is an important component of assessing animal health. Cortisol is commonly known as the "fight or flight" hormone and is involved in the chemical systems that respond to stress or danger. However, the very process of collecting samples can be stressful. For example, capturing a sea lion on the beach or a small cetacean in a net or pen can elevate circulating cortisol levels beyond "normal." Then how can baseline levels be obtained? Captive marine mammals that are trained to let their blood samples to be collected can provide baseline levels but it may not represent normal values in wild populations. See Atkinson et al. (2015) in **Further Reading** for a review of stress measurements in marine mammals.

In the wild, elevated levels of stress hormones can be examined by the collection of blood and tissue samples from compromised animals, such as beach stranded or entangled individuals. Normal levels are harder to collect. Current methods to obtain resting or normal values include the

DOI: 10.1201/9781003297468-16

measurement of cortisol in fecal samples, blubber biopsies, blowhole exhalations from cetaceans, and examining the wax in ear plugs of whales. Each of these methods provides a different view of stress levels. For example, collecting multiple fecal samples on a beach may help tell the story of the overall stress level for a group of animals under those conditions, but cannot help with the health or medical status of any particular animal, nor describe hormone levels when the animals were at sea. Blubber biopsies from freely swimming whales are a good comparison to beached or entangled whales but are difficult to obtain in large numbers. Unmanned aerial systems (UAS) or drones can now fly over the exhaled vapor plume of an undisturbed whale (**Chapter 13**). This has great promise for the collection of baseline samples, but still requires ship time, drone pilots, and sophisticated sampling devices. All of the above collection methods can provide a view of stress levels in current time, but they do not look backward at stress history.

Analyzing the earwax found in the earplugs of baleen whales is an innovative method for looking at cortisol levels backward through time in these whales (Trumble et al. 2018). Wax in the earplugs is laid down over the lifetime of the whale in an almost tree ring process allowing contaminants and hormones to be extracted from the time-based wax layers. Using this method, it is possible to measure cortisol concentrations in baleen whales over years of their life history. This tool can be used both on museum specimens for historical information and during current field collections, such as from stranded whales. Correlating wax cortisol levels to circulating concentrations is challenging and this method can only be used on dead animals. There is potential for this to become a powerful tool for looking at relative changes in hormones or contaminants over time.

eDNA

As animals move through their environment, they shed skin, hair, or mucus and drop fecal material. Modern DNA extraction methods allow the identification of species present, sex, and sometimes individuals from environmental samples. Environmental DNA (eDNA) is rapidly gaining traction in several fields, including fisheries management (Kirtane et al. 2021) and marine mammal biology (Baker et al. 2018). For some of the same reasons as stress hormone collection, eDNA samples can be difficult obtain in many field conditions. Beach fecal samples are easy to collect, but hard to attribute to any individual animal. DNA from skin samples of at-sea whales can be obtained from remote biopsy but requires well-trained field teams and complex logistics. Fecal samples from some whales can be readily dip-netted from the water but requires ship time and personnel (**Figure 16.1**).

As with all eDNA work, research is required to determine the viability of the tissue before the DNA degrades, added to the effects of seawater drifting for at-sea samples. That is, skin samples may have been shed at location X, but drifted with ocean currents to location Y. Depending on the question asked by the research, this drifting problem can be either of little consequence (e.g., evidence that sperm whales have been in the general area at any timepoint), or very sensitive (have sea lions been in this particular area within a specific timeframe?).

Pup and calf development

Whale calves and pinniped pups are not good divers nor breath-holders as newborns. Many do not have sufficient blubber to provide thermal protection or the blood and tissue characteristics to hold great amounts of oxygen. Some young animals do not learn to swim until older. Yet, they will develop all of these abilities, plus many more, as they mature into juvenile and adult marine mammals. Many of the chapters in the book refer to calf or pup biology, but there remain fields of study around those developmental patterns. For example, several chapters refer to **echelon swimming (Chapter 1; Chapter 2)**, where a cetacean calf swims close by its mother to gain an energetics advantage or reduce drag. How does the calf learn to do this?

Figure 16.1 NOAA scientists collecting orca fecal samples by dip netting from small boat. (Photo credit: Courtesy of Kim Parsons, NWFSC, NOAA Fisheries, NMFS permit #21348).

At what ages (or size) does the advantage no longer exist? Other chapters refer to the ability of some pinniped pups to rapidly gain mass during lactation, so that they can survive when their mothers wean them and they do not yet know how forage by themselves. During the same time that they are in their postweaning fast, seal pups are also developing the biochemical and physiological abilities to hold their breath for long periods. They are not yet good divers and need time to develop the cardiovascular mechanism that facilitate diving. Both cetacean and pinniped young do not have high levels of myoglobin for holding oxygen in their tissues, and need time to develop those stores.

One of the mysteries in this field is how to coordinate all of these developmental processes so that the young animal, as a juvenile, has a reasonable chance of surviving until 2–3 years old. Juvenile survivorship is very low for some species and even those that do make it through their first year as a pup or calf appear thin and in poor body condition when seen as 1 year olds. Burns et al. have studied the diving biology of young seals and found that diving biochemistry

(e.g., tissue **acid buffering** discussed in later section; Lestyk et al. 2009), myoglobin levels (Shero et al. 2019), red blood cell chemistry, and more follow developmental patterns that eventually allow the seals to become successful divers. Even **sleep apnea** (discussed below) is not well developed in seal pups. A young seal pup on the beach is not really yet a marine mammal that can dive, forage, and navigate in the ocean. See Noren (2020) in **Further Reading** for more information about the physiological development for diving in young marine mammals.

Physiological health (population level)

Assessing the health of a marine mammal in a veterinary or captive situation is a well-developed field and integrates laboratory and medical methods (**Chapter 13**). Many of those methods can also be used in field cases, such as strandings or entanglements, when research teams can work with individual animals over time. Examining the health of entire groups or populations of animals in the field requires different physical and analytical methods.

The case of the decreasing population of Steller sea lions (*Eumetopias jubatus*) across the north Pacific is a good example to consider. Steller sea lions range from northern California up to British Columbia, across the Gulf of Alaska, out the Aleutians and into coastal Siberia. Since about 1979, the population of Steller sea lions has been decreasing across a great deal of their range, especially in the western regions. They were listed under the US Endangered Species Act (ESA) in 1990 with the western population categorized as endangered, and the eastern population listed as threatened. What is the distinction between these two groups that drives that difference? Is the difference in endangered status driven by disease, food, contamination, climate differences or reproductive failure? Furthermore, how can teams assess those possible differences with wild populations that are difficult to reach? See Fritz et al. (2019) in **Further Reading** and **Chapter 15** for more about Steller sea lion populations.

To try to answer these questions, over a thousand Steller sea lion pups, juveniles, and adults (mostly females) have been captured over the past 30 years and held for short periods of time on beaches, boats, or aquaria to take blood samples, measure blubber thickness, fur and milk samples, and more. While some of these studies held animals for a few months to assess health conditions (Mellish et al. 2006; Skinner, Tuomi and Mellish 2015), most have been one-time captures, and the results of those measurements analyzed for normal vs. high/low values. But what does a sick or compromised Steller sea lion look like? Using blood values as an example, large databases show patterns of a **normal distribution**, where about 68–70% of the values fall within ±1 **standard deviation of the mean.** The mean value and its SD are then compared between populations using routine statistical comparative methods. Values outside of the normal curve (**outliers**) can be compared for differences. In this way, whole populations can be assessed for "health." Rea et al. (2013) discuss how this method showed differences in

Steller sea lion contaminant loads and that the western population of sea lions may have higher amounts of mercury in their tissues.

Applied to other marine mammal populations, the same general approach can be used to measure a broad range of physiological or morphometric (weight, length, girth, blubber thickness) values, but requires significant and multiple field deployments, extremely large databases, and statistical analyses of population patterns.

Biochemistry of diving

All marine mammals are divers, though they vary greatly in how deep they dive and what those dives look like. **Chapter 3** considers the distribution and consumption of oxygen during diving and explains the production of **lactic acid** and high levels of carbon dioxide during energy production under low or no oxygen conditions. Lactic acid causes cramps (as many human runners have experienced) and high CO_2 can cause respiratory distress, headaches, and more. How do marine mammals tolerate lactic acid and high CO_2 levels? Are they adapted to better produce energy under low oxygen conditions than terrestrial mammals? There is some evidence that they also may produce high levels of oxidants in their blood after diving and reoxygenation at the surface (see **Chapter 14**).

Marine mammals can tolerate high levels of lactic acid, because they have higher levels of **acid-buffering** compounds in their muscles compared to terrestrial mammals. (Noren 2004; Lestyk et al. 2009). A buffer works by neutralizing acid so that the tissue pH does not decrease significantly. During long diving beyond their aerobic dive limits (**ADL; Chapter 3**) when lactate and CO_2 are building up in the tissues, the muscle is able to buffer the acid and reduce tissue damage from low pH. High tissue levels of buffers are also found in extreme animal athletes like racehorses, cheetahs and others. See Dolan et al. (2019) in **Further Reading** for a review on muscle buffering across species.

Do marine mammals have different biochemical abilities in their tissues to better produce energy during diving conditions than terrestrial mammals? No differences have been found across many marine mammal species for novel or different **metabolic pathways** to produce low oxygen energy. By contrast, many invertebrate species that experience low oxygen have different biochemical systems to produce energy (Somero and Hochachka 2014). Marine mammals use the same biochemical low-oxygen pathways as terrestrial mammals but have fine-tuned them to produce energy more efficiently or at higher rates while diving. Marine mammals may also have more efficient oxidative metabolic pathways to manage low amounts of oxygen before they become oxygen stressed (Chicco et al. 2014). Further studies have shown that marine mammals may produce antioxidants to reduce the oxidative damage that can occur during **reperfusion** of low-oxygen tissues when breathing heavily at the surface after a dive (**Chapter 14**). Finally, marine mammals may have evolved differences in organ cell membranes to essentially shut down some tissues during diving, reduce metabolism, and hold that pattern until they are again at the surface. See Hochachka and Guppy (1987) in **Further Reading** for more about cellular adaptations to hypoxia in animals.

Finally, the impact of extremely high pressures, while diving to depth (up to 200–250 atmospheres, over 3,000 psi), on biochemical reactions, cell membranes, and metabolic pathways in marine mammals have only been minimally addressed, see **Chapter 4**.

The generally high levels of oxygen carrying hemoglobin in their muscles (**Chapter 3**), along with these 3–4 biochemical enhancements for the efficient production of energy during diving, reduced acid damage, and the ability to reduce cell metabolism, all facilitate marine mammal remarkable underwater and breath-hold tolerances compared to terrestrial mammals.

Sleep

Imagine an adult elephant seal during the foraging phase of its annual cycle. Elephant seals dive in a pattern where they are underwater for about 30 minutes, and at the surface for only a few minutes before diving again. This means that they spend about 90% of their time underwater for many months in a row. Beyond the amazing diving and swimming physiology associated with this pattern, when do they sleep? By contrast, many elephant seals appear to be sleeping most of entire time on the beach for their breeding season (**Figure 16.2**)

Now imagine a dolphin that appears to be resting on the surface of the water. The ocean surface is a dangerous place to be asleep due to the possibility of attacks by sharks, humans (ship strikes). How do they remain vigilant if sleeping?

Finally, consider any diving mammal asleep at depth. What keeps them from accidentally taking a breath while sleeping underwater? How are breath-holding, diving, and sleeping tied together?

The partial answer involves the co-evolution of diving, sleeping, and neural physiology. In the example of the elephant seal at sea, our best evidence from diving patterns is that they fall asleep during the descent phase of some dives, drifting downwards. Is there a pressure sensor that lets them know they are now too deep, it is time to wake up, and either return to the surface to breathe or continue their dive? When on land, elephant seals show extreme bouts of **sleep apnea**, and hold their breath while sleeping on the beach. Why? Current theories suggest because these seals sleep underwater at sea, their biochemical, physiological, and neural systems for diving, sleeping, and breathing are co-mingled and have become a set of linked adaptations. For example, it would be dangerous for a sleeping (Lyamin et al. 2008) seal at 400 m deep underwater to take a breath. Similarly, a sleeping seal in apnea on the beach will activate physiological diving pathways and its heart rate and circulatory patterns are as if it were diving underwater. Why? It can take a breath any time it wants on the beach. The answer, once again, is these adaptations for breath-holding, sleeping, and diving have co-evolved to be associated with one another (Castellini 1996).

Figure 16.2 Male northern elephant seals sleeping on the beach near Santa Cruz, California. The males are molting last year's fur/skin and the new dark fur/skin is seen underneath. (Photo M. Castellini. MMPA permit 496)

What about the vigilant dolphins? Dolphins and at least 5–6 other small cetaceans use **unihemispheric** brain sleep, which means they sleep with only half their brain at a time (Lyamin et al. 2008). This allows them to slowly swim and pay attention to surface conditions, while also sleeping with the other parts of the brain. They can sleep and remain vigilant at the same time. This feature is not unique to the dolphins and was first studied in many birds, that can also sleep, yet remain vigilant. There is no evidence, yet, that seals show unihemispheric sleep, though data so support this ability in fur seals (Kendall-Bar et al. 2019).

It has been proposed that sperm whales sleep suspended heads-up, in ocean just below the surface, but no physiological recordings of brain waves have been taken to confirm sleep, nor do we know if they are exhibiting unihemispheric sleep during this behavior. They appear to be sleeping, but we do not know for sure (Miller et al. 2008).

Both Davis (2019) and Lyamin and Siegel (2019) in **Further Reading** provide excellent current reviews of sleep in aquatic mammals.

Aging

Classic aging of toothed marine mammals involves the sectional analysis of layers in the teeth, much like the study of tree rings. However, this method does not work for baleen whales that do not have teeth. It is possible to age the baleen itself, but even in long-lived whales, the baleen is continually replaced and is from only the past 20–25 years. Native indigenous bowhead whale hunters in Alaska have found stone harpoon points in whales, which means that the whale must have been at least 100 years old (George, Moore, and Thewissen 2020). Using the natural rate that some tissue amino acids change over time in eye lenses, it has been found that some bowhead whales in Alaska could be over 200 years old (Wetzel et al. 2017). Molecular methods are now being used to provide yet another estimate of age in marine mammals and can analyze compounds in the skin to age animals (Bors et al. 2021). These molecular methods can be applied to both baleen and toothed whales.

Given that some whales may live to over 200 years of age, the questions that immediately come to mind are "how do they live so long?" and "how to they age in their diving, exercise, reproductive or other physiological/biochemical processes?" The first point of how do some whales live so long remains a mystery. Do they have different methods of DNA/tissue repair? Do they slow down natural degradation rates of neural tissue loss? Do they avoid cancer? All of these questions are open for new investigators.

The second question of what physiological processes might slow down as marine mammals age has been studied in some pinniped species (Allen et al. 2019). In a well-known and tagged population of Weddell seals (*Leptonychotes weddellii*) in Antarctica, studies have shown that muscle aging may impact diving performance, but that behavioral adjustments and age-independent body oxygen content may be able to compensate for these seals up to their oldest known age of about 25 years (Hindle et al. 2009; Hindle, Mellish, and Horning 2011).

TOOLBOX

There are so many different areas covered in this chapter that a list of tools would be almost as long as the chapter itself. Therefore, we are considering only two major areas.

Molecular biology and genetics

There are at least three different tools that are critical for genetics and **eDNA** research on marine mammals. The first area includes the refinement of remote water collection systems that can sample ocean water near or around marine mammals. This is a combination of ocean engineering systems and molecular methods. Many of these systems are already developed and deployed by research programs, but mainly for collecting eDNA that focuses on fish and fisheries management. Building such systems to get near at-sea aggregations of marine mammals will involve specializing their tracking and deployment for fast-moving mammals, with the ability to collect both sloughed skin cells and fecal samples and preserving the specimens. The

second need for tool development is the molecular biology approach to characterize the DNA of hard to sample marine mammal species. From archived samples in museums to strandings and captive animals, many of the genetic sequences of more common species are either in hand or easily obtained as markers. But, for species that are not easily seen, such as many of the beaked whales, collecting the background genetics samples will be a challenge. The third area is the collection, processing, and database development for land-based fecal samples for pinnipeds and water-based collections by hand nets for cetaceans. Fecal samples from these species have been collected for years. But, usually not with the modern methods necessary for eDNA sampling, collection, and database development. As eDNA research and analysis becomes more common, the hope is that the applicable sampling techniques for marine mammal fecal collection will also be standardized.

Diving physiology and biochemistry

As mentioned in several chapters, the development of more sophisticated, smaller and robust dive recorders is critical to this field. For example, **Chapter 3** discusses the challenges of designing and building *in vivo* blood lactate analysis devices. In addition, an important tool is the development of recorder deployment strategies for animals that are rarely seen or handled. Recorder attachment on open ocean cetaceans has been carried out for decades, but the methods for attaching them have improved significantly, including recent work to deploy recorders via hexacopter drones. Suction cups to hold the dive recorders on cetaceans are common, but do not last long. Methods that use barbs to attach to the skin or into the blubber are more difficult to deploy and last much longer, but risk tissue damage.

For pinnipeds captured on the beach, in the water, or on ice floes, there are more options to attach recorders, and these have evolved mainly to the use of waterproof glues that attach the instruments to the pelage or skin. The challenge is the recovery of the instrument if the animal may never be found again. The use of satellite links for data transmission, or devices that release remotely and float to the surface are the most common solutions to that particular problem.

CONCLUSIONS

There are many fields of marine mammal studies that cross over between medicine, veterinary science, physiology, and biochemistry. While this book covers many of those fields, there are always new questions that arise, or old questions that cannot be answered with current technology. Blood lactate levels during diving in field animals remains unanswered. What are the biochemical features that allow some whales to be 200-years old or more? How do divers navigate at great depths in total darkness? Why do some whales get stranded on beaches and others do not? This chapter alone covers only half a dozen of those unanswered or mystery topics. The list of physiological information we do not know about marine mammals is long and provides many avenues of future research for new students and scientists. Even in peripheral fields, new technology or tools can have significant impacts on marine mammal science. For example, the invention of waterproof long-lasting epoxies allowed the attachment of dive recorders to many species of pinnipeds in the place of older harnesses, sutures, etc. The rapid development of GPS navigation devices was quickly taken up in the design of dive recorders. Portable ultrasound machines developed for human medicine are now routinely used in marine mammal studies.

In short, there remain so many questions and unknowns about marine mammals that new students and scientists have vast fields of possible future studies available to them.

STUDY QUESTIONS

1 What is environmental DNA and where can it be found for marine mammals?
2 What is an earplug in a cetacean and why is it useful to study the history of that animal?

3 What is cortisol and why is it important to assess the health of a marine mammal?
4 Why is it important to measure blood lactate in a diving mammal?
5 Why are new seal pups not very good at diving?
6 How do scientists assess population health from blood samples of pinnipeds?
7 What is unihemispheric sleep?
8 Why is the eye lens an important tissue for determining the age of a large whale?
9 How long can some whales live?
10 How are hexacopters important to the study of marine mammals?

CRITICAL THINKING

1 What are some advantages of sleeping at depth for a pinniped and how is that related to their diving physiology? Why would they also hold their breath while sleeping on land?
2 Blubber has been discussed throughout this book from streamlining, to fasting, thermoregulation, and more. How do you suppose evolutionary pressure works on all of those simultaneously? Is one feature more important and another?
3 How is calf/pup development related between thermoregulation, diving, tissue chemistry, and behavior?

FURTHER READING

Atkinson, S., D. Crocker, D. Houser, et al. 2015. Stress physiology in marine mammals: How well do they fit the terrestrial model? *Journal of Comparative Physiology B* 185(5): 463–486.

Davis, R.W. 2019. *Marine mammals: Adaptations for an aquatic life*, Cham: Springer Nature.

Dolan, E., B. Saunders, R.C. Harris, et al. 2019. Comparative physiology investigations support a role for histidine-containing dipeptides in intracellular acid–base regulation of skeletal muscle. *Comparative Biochemistry and Physiology Part A: Molecular & Integrative Physiology* 234: 77–86.

Fritz, L., B. Brost, E. Laman, et al. 2019. A re-examination of the relationship between Steller sea lion (*Eumetopias jubatus*) diet and population trend using data from the Aleutian Islands. *Canadian Journal of Zoology* 97(12): 1137–1155.

Hochachka, P.W., and M. Guppy. 1987. *Metabolic arrest and the control of biological time*, Cambridge: Harvard University Press.

Lyamin, O.I., and J.M. Siegel. 2019. Sleep in Aquatic Mammals. In *Handbook of behavioral neuroscience*, 375–393. London: Elsevier.

Noren, S.R. 2020. Postnatal development of diving physiology: Implications of anthropogenic disturbance for immature marine mammals. *Journal of Experimental Biology* 223(17): jeb227736.

GLOSSARY

Acid buffering: A chemical that resists pH changes upon addition of acid.

ADL: Aerobic Dive Limit is the maximum dive time in a marine mammal below which no blood lactic acid is measured during recovery.

Cortisol: Hormone created in the adrenal glands that regulates many metabolic pathways, and in this case, increases during stress to prepare the body for the "flight or fight" response.

Echelon swimming: Cetacean calf swimming in close proximity to the mother, as if the mother is carrying the calf and reducing the cost of swimming for the calf.

eDNA: Environmental DNA. DNA that found in the environment from sloughed tissues, skin, or fecal matter.

Lactic acid: Small three-carbon acid end product of low oxygen metabolism in most vertebrates.

Metabolic pathways: The paths and directions of molecules as they move through chemical reactions in the body.

Normal distribution: In population statistics, the distribution of values above and below the mean value. In a perfect normal distribution, this is in the shape of a bell curve.

Outliers: In population statistics, the group of values that lay outside of two standard deviations from the mean. The equates roughly to values outside (high or low) of 95% of all values.

Reperfusion: Return of blood flow to tissues after a dive.

Sleep apnea: In marine mammals, the pattern of breath-holding (apnea) that occurs while the animal is sleeping.

Standard deviation: In statistics, the range of values that encompass 68% of the population. Usually expressed as "+/- One SD".

Unihemispheric sleep: Seen in cetaceans and birds, brain activity during sleep where one hemisphere of the brain is active, and the other hemisphere is resting/sleeping.

REFERENCES

Allen, K.N., J.P. Vázquez-Medina, J.M. Lawler, et al. 2019. Muscular apoptosis but not oxidative stress increases with old age in a long-lived diver, the Weddell seal. *Journal of Experimental Biology* 222(12): jeb200246.

Baker, C.S., D. Steel, S. Nieukirk, et al. 2018. Environmental DNA (eDNA) from the wake of the whales: Droplet digital PCR for detection and species identification. *Frontiers in Marine Science* 5: 133.

Bors, E.K., C.S. Baker, P.R. Wade, et al. 2021. An epigenetic clock to estimate the age of living beluga whales. *Evolutionary Applications* 14(5): 1263–1273.

Castellini, M.A. 1996. Dreaming about diving: Sleep apnea in seals. *News in Physiological Sciences* 11: 208–214.

Chicco, A.J., C.H. Le, A. Schlater, et al. 2014. High fatty acid oxidation capacity and phosphorylation control despite elevated leak and reduced respiratory capacity in northern elephant seal muscle mitochondria. *Journal of Experimental Biology* 217(16): 2947–2955.

George, J., S. Moore, and J. Thewissen. 2020. NOAA Arctic Report Card 2020: Bowhead whales: Recent insights into their biology, status and resilience. https://doi.org/10.25923/cppm-n265

Hindle, A.G., M. Horning, J.-A.E. Mellish, et al. 2009. Diving into old age: Muscular senescence in a large-bodied, long-lived mammal, the Weddell seal (*Leptonychotes weddellii*). *Journal of Experimental Biology* 212(6): 790–796.

Hindle, A.G., J.A. Mellish, and M. Horning. 2011. Aerobic dive limit does not decline in an aging pinniped. *Journal of Experimental Zoology Part A: Ecological Genetics and Physiology* 315(9): 544–552.

Kendall-Bar, J.M., A.L. Vyssotski, L.M. Mukhametov, et al. 2019. Eye state asymmetry during aquatic unihemispheric slow wave sleep in northern fur seals (*Callorhinus ursinus*). *PLoS One* 14(5): e0217025.

Kirtane, A., D. Wieczorek, T. Noji, et al. 2021. Quantification of environmental DNA (eDNA) shedding and decay rates for three commercially harvested fish species and comparison between eDNA detection and trawl catches. *Environmental DNA* 3(6): 1142–1155.

Lestyk, K.C., L. Folkow, A. Blix, et al. 2009. Development of myoglobin concentration and acid buffering capacity in harp (*Pagophilus groenlandicus*) and hooded (*Cystophora cristata*) seals from birth to maturity. *Journal of Comparative Physiology B* 179(8): 985–996.

Lyamin, O.I., P.R. Manger, S.H. Ridgway, et al. 2008. Cetacean sleep: An unusual form of mammalian sleep. *Neuroscience & Biobehavioral Reviews* 32(8): 1451–1484.

Mellish, J., D.G. Calkins, D.R. Christen, et al. 2006. Temporary captivity as a research tool: Comprehensive study of wild pinnipeds under controlled conditions. *Aquatic Mammals* 32(1): 58.

Miller, P.J., K. Aoki, L.E. Rendell, et al. 2008. Stereotypical resting behavior of the sperm whale. *Current Biology* 18(1): R21–R23.

Noren, S.R. 2004. Buffering capacity of the locomotor muscle in cetaceans: Correlates with postpartum development, dive duration, and swim performance. *Marine Mammal Science* 20(4): 808–822.

Rea, L.D., J.M. Castellini, L. Correa, et al. 2013. Maternal Steller sea lion diets elevate fetal mercury concentrations in an area of population decline. *Science of the Total Environment* 454: 277–282.

Shero, M.R., P.J. Reiser, L. Simonitis, et al. 2019. Links between muscle phenotype and life history: Differentiation of myosin heavy chain composition and muscle biochemistry in precocial and altricial pinniped pups. *Journal of Comparative Physiology B* 189(6): 717–734.

Skinner, J.P., P.A. Tuomi, and J.-A.E. Mellish. 2015. The influence of time in captivity, food intake and acute trauma on blood analytes of juvenile Steller sea lions, *Eumetopias jubatus. Conservation Physiology* 3(1): 1–12.

Somero, G.N., and P.W. Hochachka. 2014. *Biochemical adaptation*, Princeton: Princeton University Press.

Trumble, S.J., S.A. Norman, D.D. Crain, et al. 2018. Baleen whale cortisol levels reveal a physiological response to 20th century whaling. *Nature Communications* 9(1): 1–8.

Wetzel, D., J. Reynolds III, P. Mercurio, et al. 2017. Age estimation for bowhead whales, *Balaena mysticetus*, using aspartic acid racemization with enhanced hydrolysis and derivatization procedures. *Journal of Cetacean Research and Management* 17: 9–14.

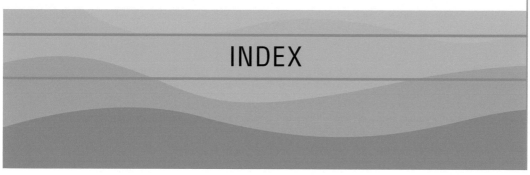

Note: Locators in *italics* represent figures and **bold** indicate tables in the text